Mixture Formation in Spark-Ignition Engines

H. P. Lenz

in collaboration with

W. Böhme, H. Duelli, G. Fraidl,
H. Friedl, B. Geringer, P. Kohoutek,
G. Pachta, E. Pucher, and G. Smetana

Springer-Verlag
Wien GmbH

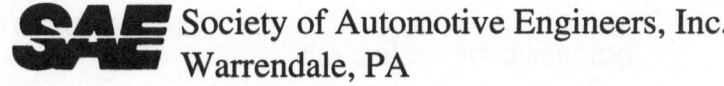
Society of Automotive Engineers, Inc.
Warrendale, PA

Univ.-Prof. Dipl.-Ing. Dr. sc. techn. Hans Peter Lenz
Institute for Internal Combustion Engines and Automotive Engineering
Technical University of Vienna, Austria

© 1992 Springer-Verlag Wien
Originally published by Springer-Verlag/Wien in 1992
Softcover reprint of the hardcover 1st edition 1992

With 352 Figures

ISBN 978-1-56091-188-3   ISBN 978-1-4899-2762-0 (eBook)
DOI 10.1007/978-1-4899-2762-0

Printed on acid-free paper

Preface

Twentyfour years have gone by since the publication of K. Löhner and H. Müller's comprehensive work "Gemischbildung und Verbrennung im Ottomotor" in 1967 [1.1].

Naturally, the field of mixture formation and combustion in the spark-ignition engine has witnessed great technological advances and many new findings in the intervening years, so that the time seemed ripe for presenting a summary of recent research and developments. Therefore, I gladly took up the suggestion of the editors of this series of books, Professor Dr. H. List and Professor Dr. A. Pischinger, to write a book summarizing the present state of the art.

A center of activity of the Institute of Internal-Combustion Engines and Automotive Engineering at the Vienna Technical University, which I am heading, is the field of mixture formation - therefore, many new results that have been achieved in this area in collaboration with the respective industry have been included in this volume.

The basic principles of combustion are discussed only to that extent which seemed necessary for an understanding of the effects of mixture formation.

The focal point of this volume is the mixture formation in spark-ignition engines, covering both the theory and actual design of the mixture formation units and appropriate intake manifolds. Also, the related measurement technology is explained in this work.

The intention of this book is to provide the engineer who is either engaged in scientific research or active in industry, as well as the student who wishes to deepen his knowledge in this field, with a survey of the present state of research and state of the art.

It should be pointed out that, after more than a hundred years' development in the field of engine design, intensive research work is being done worldwide, today more than ever, and great technological progress has been achieved. Therefore, the total of approximately 400 quoted direct sources of literature represent only a portion of the entire literature available.

In view of the complexity of the subject, numerous experts have contributed to this volume, especially former and present staff members. I wish to take this opportunity to express my gratitude for elaboration of individual sections to: Dipl.-Ing. Dr. Walter Böhme, Dipl.-Ing. Dr. Heinz Duelli, Dipl.-Ing. Dr. Günter Fraidl, Dipl.-Ing. Dr. Hubert Friedl, Dipl.-Ing. Dr. Bernhard Geringer, Dipl.-Ing. Dr. Georg Pachta, Dipl.-Ing. Dr. Ernst Pucher, and Dipl.-Ing. Dr. Günter Smetana.
I thank Professor Dr. Hellmuth Schindlbauer for his revision of the chapter "Fuel".

I also express deep appreciation for the great support and courtesy shown by the following publishers and enterprises in permitting me to quote from copyright material including some illustrations:

Society of Automotive Engineers, Inc., Warrendale,
Verlag des Vereins Deutscher Ingenieure, Düsseldorf,
Adam Opel AG, Rüsselsheim,
AUDI AG, Ingolstadt,
BMW AG, Munich,
Dr. Ing. h.c. F. Porsche AG, Stuttgart,
Fritz-Hintermayer GmbH., Nuremberg,
Mercedes-Benz AG, Stuttgart, Pierburg GmbH.,
Neuss, R. Bosch GmbH., Stuttgart,
Volkswagen AG, Wolfsburg

This book was originally published for the German-speaking public. In view of the vivid interest in this volume, I took up the suggestion that it should be translated for the English-speaking world. Most of the symbols and abbreviations used in illustrations, equations, and charts of the German edition have been retained in the English translation. Explanations are given in the corresponding indices.

Grateful acknowledgement is expressed to Mrs. Hedwig Riegler, assisted by Mag. Gertrude Maurer and Mag. Christine Hetzendorfer, for translating the book into English.

A special thanks goes to Mr. Ronald W. Bell, AC Rochester Austria, Vienna, who spared some of his valuable time for revision of the English manuscript.

The author is indebted to Dipl.-Ing. Dr. Ernst Pucher for editing and directing the translation project.

Dipl.-Ing. Peter Kohoutek deserves special mention for managing the layout and graphical presentations.

Vienna, October 1991
o. Univ. Professor Dr. sc. techn. Dipl.-Ing. Hans Peter Lenz

Contents

Symbols, Abbreviations, and Subscripts

Symbols

a,b	[m]	Distance
A	[m²]	Area (various dimensions)
b	[m.s⁻²]	Acceleration
b	[g.kW⁻¹.h⁻¹]	Specific fuel consumption
B	[l.h⁻¹]	Fuel consumption
c	[m.s⁻¹]	Sonic speed
c_v, c_p	[J.kg⁻¹.K⁻¹]	Specific heat capacity with v = const. or p = const.
$d\chi/d\varphi$	[°KW⁻¹]	Energy conversion rate
D,d	[mm]	Diameter
D_{32}	[μm]	Sauter diameter
D_{32}	[μm]	Sauter mean diameter
$E(x)/E_{max}$	[-]	Normalized scattered light energy
E	[J]	Energy
f	[-]	Factor
f	[Hz]	Frequency, number of ignition sparks
f	[mm]	Focal length, lens focal length
f_{SE}	[mm².kW⁻¹]	Specific individual-pipe cross-section factor
F	[%]	Film-like fuel fraction
F	[N]	Force
F_R	[N.kg⁻¹]	Specific friction (related to ρ.A.dx)
g	[m.s⁻²]	Gravitational acceleration
h	[m]	Height
$(h/c)_{Mass}$	[-]	Hydrogen/carbon mass ratio
H	[kJ.m⁻³]	Heating value
HC	[g]	Total emission of unburned hydrocarbons
H_o	[kJ.kg⁻¹]	Specific heating power (or high heating value)
H_G	[kJ.m⁻³]	Mixture heating value
H_u	[kJ.kg⁻¹]	Specific heating value (or low heating value) of liquid fuel
H_u	[kJ.m⁻³]	Specific heating value (or low heating value) of gaseous fuel
i_{HA}	[-]	Transmission of driven axle
$I(x)/I_{max}$	[-]	Normalized scattered light intensity
I_H	[A]	Heating current
k_s/d	[-]	Referenced pipe roughness
K	[-]	Constant
L, l	[m]	Length
m	[kg]	Mass

$\dot{m}$	[kg.s^{-1}]	Mass flow rate, throughput
m	[mm^3.(ms)$^{-1}$]	Steady flow
m	[-]	Vibe factor
m_F	[%]	Fuel film deposit
M	[-]	Mixture ratio
M_d	[N.m]	Engine torque
n	[-]	Number
n	[min^{-1}]	Torque
N	[kW]	Power, engine power
NO_x	[g]	Total emission of nitric oxides
$(o/c)_{Mass}$	[-]	Oxygen/carbon mass ratio
O_h	[-]	Ohnesorge number
p	[N.m^{-2}]	Pressure
p_s	[N.m^{-2}]	Manifold vacuum (various dimensions)
q	[mm^3]	Injected fuel volume per cycle
q_A	[J.kg^{-1}]	Dissipated specific heat
q_B	[J.kg^{-1}]	Added specific heat
Q	[J]	Heat
Q_A	[%]	Squish area portion
r	[m]	Radius
r	[mm]	Measurement radius in focal plane
R	[Ω]	Impedance
R	[J.kg^{-1}.K^{-1}]	Specific gas constant
R_e	[-]	Reynolds number
S	[J.K^{-1}]	Entropy
s	[J.kg^{-1}.K^{-1}]	Specific entropy
s	[m]	Stroke, travel
t	[s]	Time
t	[s]	Pulsation period
T	[$^{\circ}$K, $^{\circ}$C]	Temperature
T_i	[ms]	Pulse length
T_i	[ms]	Pulse timing
T_m	[ms]	Extension of pulse length by correction
T_p	[ms]	Basic injection time
T_t	[ms]	Droplet travel time
T_u	[ms]	Extension of pulse length by voltage compensation
T_v	[ms]	Time delay
U	[V]	Voltage
U_λ	[mV]	Probe voltage
U_v	[V]	Control voltage
U_v	[V]	Valve control voltage
v	[m.s^{-1}]	Velocity
v	[m^3.kg^{-1}]	Specific volume
V	[cm^3]	Volume (various dimensions)

V	[mm^3.ms^{-1}]	Steady flow volume
V̇	[m^3.h^{-1}]	Flow rate, throughput
V$_C$	[m^3]	Compression volume
V$_F$	[m.s^{-1}]	Mean fuel film velocity
V$_H$	[dm^3]	Displacement volume (various dimensions)
V	[%]	Coefficient of variation
w$_e$	[kJ.dm^{-3}]	Specific effective work
W$_e$	[-]	Weber number
x	[m]	Location coordinate
x	[-]	Water or fuel content of the air
x	[-]	Normalized measurement radius
y	[m]	Location coordinate
α	[-]	Flow index
α	[°]	Angle
α$_z$	[°KW]	Ignition angle
δ	[μm]	Wall film thickness
δ	[°]	Angle, bending angle
Δκ	[%]	Deviation of individual-cylinder air-fuel equivalence ratio from the mean equivalence ratio of all cylinders
ΔO$_T$	[m^2]	Surface enlargement due to jet breakup
Δp	[mbar]	Pressure loss (various dimensions)
(Δp/l)$_{R,zp}$	[mbar.m^{-1}]	Referenced frictional pressure loss
Δp$_R$	[N.m^{-2}]	Frictional pressure loss
ε	[-]	Engine compression ratio
ε	[-]	Width/length ratio
Φ	[-]	Pulsation parameter
γ	[kg.m^{-3}]	Density
η	[-]	Efficiency
η$_D$	[-]	Nozzle efficiency
η	[Pa.s]	Dynamic viscosity (various dimensions)
φ	[°CA]	Temporal phase of energy conversion
φ	[-]	Ratio of volume rise
φ	[°CA]	Crank angle
φ$_B$	[°CA]	Burning time, combustion time
φ/φ$_B$	[-]	Normalized crank position (burning time)
κ	[-]	Specific heat ratio
κ	[-]	Air-fuel equivalence ratio
χ	[-]	Fraction of fuel energy released (burning function)
λ	[-]	Relative air-fuel ratio
λ	[mm]	Wavelength, light wavelength
λ$_a$	[-]	Air consumption
λ$_R$	[-]	Pipe friction coefficient
μ	[-]	Constriction index
ν	[m^2.s^{-1}]	Kinematic viscosity

Θ	[°]	Scattering angle
ρ	[kg.m^{-3}]	Density
σ	[N.m^{-2}]	Surface tension
τ	[N.m^{-2}]	Shear stress
ξ	[-]	Oxygen/carbon atomic ratio in the fuel
ψ	[-]	Ratio of pressure rise
ψ	[-]	Flow function
ψ	[-]	Hydrogen/carbon atomic ratio in the fuel
ζ	[-]	Drag coefficient

Abbreviations and Subscripts

0	Standard condition
0	Reference condition (ambient condition)
0	At rest
A	Buoyancy
A	Surge tank
A	Exhaust
A	Exhaust valve
A	Dissipated
A	Intake
abs, Abs	Absolute
atm	Atmospheric
ABS	Anti-blocking system
AEGS	Electronic gear controller
A/F	Air/fuel
AFI	Air-fuel injection system
AS	Exhaust valve closes
ACS	Anti-blocking system with stability check
ASTM	American Standard Test Method
AVL	Ges. für Verbrennungskraft-maschinen u. Meßtechnik m.b.H., Prof. Dr. Dr. h.c. Hans List
*	Critical condition
(x)	Normalized measurement radius

B	Fuel
B	Added
B	Vessel
BA	Acceleration enrichment
BDC	Bottom dead center
C	Compression
CA	Crank angle
CFR	Cooperative Fuel Research Committee, USA
const	Constant
CVCC	Compound Vortex Controlled Combustion
CN	Cetane number
Cyl	Cylinder
D	Vapor
DFR	Dynamic flow range
DKA	Throttle valve switch
DIN	German Institute for Standar-dization
DK	Throttle valve
DME	Digital engine control electronics
DS	Saturated vapor
DSM	Dividing control oscillator
DVG	Deutsche Vergaser-Gesellschaft (German Carburetor Society)

e, eff	Effective		LL	Idle
E	Inlet		LPG	Liquified petroleum gas
E	Inlet valve		LT	Air funnel
ECE	Economic Commission for Europe		M	Mixing chamber
ECU	Electronic control unit		M	Mean value
EMK	Electromotive force		M	Engine
EML	Electronic engine management controller		m	Mass flow
			m	Mean
EÖ	Inlet valve opens		max	Maximum
ES	Inlet valve closes		min	Minimum
ES	Output stage		MON	Motor octane number
ETBE	Ethyl tertiary butyl ether		MSR	Engine overrun control
EV	Injector		MTBE	Methyl tertiary butyl ether
f	Humid		MTL	Tetramethyl lead
Fig	Figure		NA	Post-start enrichment
Fl	Liquid		NTC	Negative temperature coefficient
FON	Front octane number		OH	Hydroxyl group
FTP	Federal Test Procedure		ÖNORM	Austrian Standard
G	Mixture formation system		opt	Optimum
G	Mixture		OT(TDC)	Top dead center
G	Gas		ON	Octane number
ges	Total		PCI	Pre-chamber-injection
GS	Saturated mixture		PLU	Pierburg Aviation Union
H	Hot wire		proz	In percent
HA	Rear axle		PTC	Positive temperature coefficient
HLM	Hot-wire air mass flowmeter		R	Impedance
HSP	Hot spot		R	Friction
i	Indicated		R	Resonance
i	Internal		red	Reduced
i	Pulse		rel	Relative
I	Current		RON	Research octane number
ISA	International Standard Atmosphere		S, s	Intake manifold
inst	Transient		SA	Voltage increase for starting
IS	Heating current		SAS	Overrun fuel cut-off
IVK	Institute for Internal-Combustion Engines and AutomotiveEngineering		SCS	Stratified charge system (Porsche)
			spez	Specific
			SON	Street octane number
IC	Integrated circuit		stat	Steady-state
K	Fuel		stö	Stoichiometric
k	Coolant		SU	Adder stage
KD	Fuel nozzle		T	Droplet
l	Liters		t	Dry
L	Air		Tab	Table
LK	Air compensation		TBI	Throttle body injection

TDC	Top dead center		VL	Full load
th	Thermal		VZ	Preliminary diffuser
TL	Part-load		W	Wall
TOP	Thermodynamically optimized		WA	Warm-up enrichment
	Porsche engine		WD	Water vapor
Ü.OT	Overlap top dead center		ZOT	Ignition top dead center
UT(BDC)	Bottom dead center		zu	Added, supplied
VK	Full-load correction		Z, Zyl	Cylinder
VD	Initial throttle valve		ZZP	Ignition point

1. Basic Principles of Combustion

1.1. General

Combustion engines can be functionally defined as follows: Combustion engines are machines utilizing combustion to convert the chemical energy contained in a fuel into the internal energy of a gaseous working medium, and finally transforming this stored energy into mechanical work output. This action can be viewed as sequenced thermodynamic processes occurring in a cycle. All engines considered here are internal combustion engines as it is assumed that combustion occurs within the working medium.

The first stage of this cycle is a combustion process involving the conversion of chemical energy into the internal energy of the gasous medium. This can be quantified by using calorimeters and the resulting number is the heating value, or enthalpy of formation, of the fuel which is a defined measure of the energy contained in the fuel.

The second stage concerns the conversion of the stored internal energy of the gaseous medium into mechanical work and is accomplished through a variety of simple thermodynamic processes arranged in a working cycle. It is useful to proceed in two steps with this stage in order to determine what type of working cycle would optimally utilize the stored internal energy of the working medium:

- A. Determine the theoretically optimum cycle

- B. Determine a technically feasible cycle which satisfactorily approximates this optimum

Matters relating to the technical feasiblity of specific engine designs are irrelevant for step A, which is only concerned with determining the theoretically optimal cycle amongst all the potential cycles. Theoretical calculations are especially useful in step A as a basis of comparison for a wide variety of potential cycles of operation since the thermodynamic analyses are relatively simple to carry out.

This optimal theoretical cycle now serves as the objective of step B, which seeks a technically feasible cycle as an approximation of this theoretical cycle, thereby resulting in the optimal utilization of the fuel energy. Step B determines the optimum achievable performance of a given engine design in practice and this performance cannot be evaluated by the simple thermodynamic analyses of step A. Many other considerations such as engine material properties, and fuel and lubricant characteristics, also play a decisive role. It should also be noted that other characteristics of combustion have to be taken into detailed consideration in step B. Among these are the effect of temperature on nitrous oxide emissions and the rate of pressure rise on engine noise.

1.2. Determination of the Heating Value

The heating value, or enthalpy of formation, is a measure of the chemical energy contained in the fuel. It is usually determined in a bomb calorimeter, **Fig. 1.1**, by completely burning a given amount of fuel and air and evaluating the energy released by combustion from the change in temperature of the water surrounding the bomb.

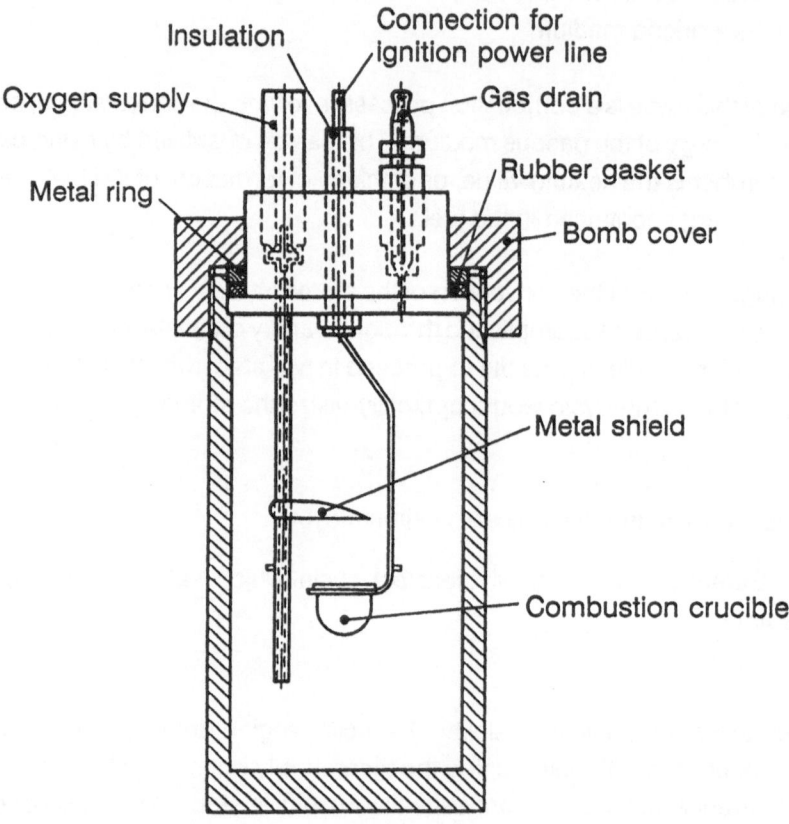

Fig. 1.1. Schematic drawing of a bomb calorimeter [1.33]

1.3. Engine Cycle Fundamentals

1.3.1. Ideal Cycles and Simulation Cycles

A great number of physical and chemical processes are interacting and influencing each other during the working cycle of an engine.

These interrelationships lead to computational difficulties in engine simulation and comparison of engine cycles. It is only possible to clarify this complex situation by an isolated examination of simplified and idealized approximations to the processes within the cycle. The use of such idealized processes has proven to be very helpful for the comparison of engine cycles.

Such simplified processes are used to determine the basic characteristics of the engine cycles that utilize processes of a similar nature in real engine designs. The results of such theoretical simulations will simulate engine performance the better the more the idealized processes approximate the real engine processes. On the other hand, better approximations to the real processes necessarily increase the computational effort and an overview of the cycle becomes more complex and less clear. Therefore, the idealized processes have to be chosen carefully for each engine cycle to be considered. An accurate model considering all the phenomena that actually occur within an engine is not yet possible. It is useful to distinquish between:

A) Ideal cycles based on the following assumptions:

- The working medium is an ideal gas having the properties of standard air. Thus, the specific heats at constant pressure c_p, and at constant volume c_v, and the specific heat ratio are constants;

- The combustion process is replaced by a heat addition which results only in an increase in the internal energy of the working medium;

- The charge exchange processes are replaced by a heat dissipation;

- The compression and expansion processes are adiabatic.

B) Simulation cycles with processes that represent an intermediate stage between the idealized processes and the actual processes. Possible simulation process modifications are:

- Consideration of dissociation;

- Use of real gas or mixture properties (c_p, c_v, $\kappa = f(T)$);

- Use of an arbitrary schedule of heat addition;

- Consideration of incomplete combustion;

- Consideration of incomplete charge exchange;

- Consideration of heat transfer losses.

Examples of ideal cycles and of some simulation process modifications are given in the following subsections.

1.3.2. The Carnot Cycle (Ideal Cycle)

In the Carnot cycle both heat addition and dissipation occur at constant temperature as shown in **Fig. 1.2**. The theoretical thermal efficiency η_{th} is given by:

$$\eta_{th} = \frac{\text{added specific heat } q_B \text{ - dissipated specific heat } q_A}{\text{added specific heat } q_B} \qquad (1.1)$$

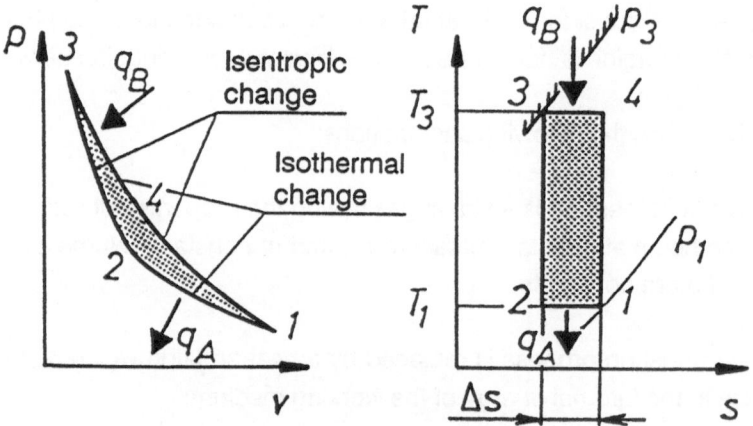

Fig. 1.2. Carnot cycle in the P-V diagram and T-S diagram

$$\eta_{th} = \frac{q_B - q_A}{q_B} = \frac{\Delta s \cdot (T_3 - T_1)}{\Delta s \cdot T_3} = 1 - \frac{T_1}{T_3} \qquad (1.2)$$

The thermal efficiency is dependent only on the temperatures of heat addition and dissipation, and is not dependent upon the properties of the working gas used. This cycle achieves the highest possible value of thermal efficiency. However, the work area of the Carnot cycle is too small to be suitable for practical combustion engine applications. The necessary heat addition (per cycle) q_B derives from such a low fuel rate and such a lean mixture that less work (per cycle) is released in practice than is required to overcome friction. Therefore, the Carnot cycle is not a technically feasible cycle.

1.3.3. The Constant-Volume Cycle (Ideal Cycle)

In the constant-volume cycle heat addition and dissipation both occur at constant volume as shown in **Fig 1.3**.

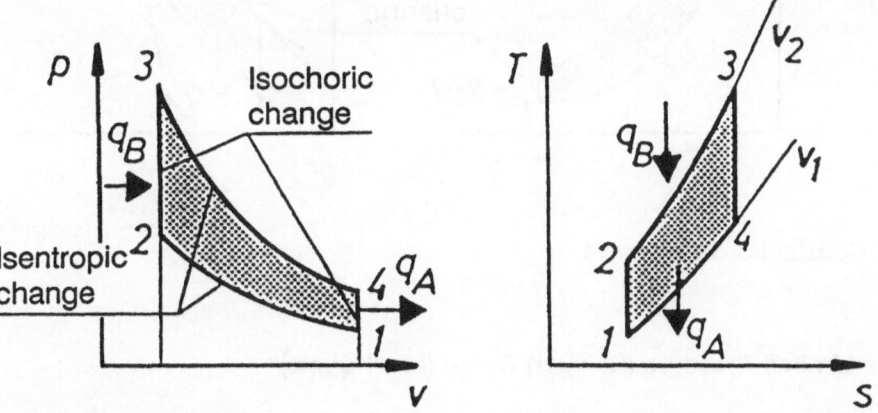

Fig. 1.3. Constant-volume cycle

The thermal efficiency η_{th} of this cycle is evaluated from the same definition as for the Carnot cycle and is given by:

$$\eta_{th} = 1 - \frac{1}{\varepsilon^{\kappa - 1}} \tag{1.3}$$

ε [-]   ..... engine compression ratio, $\varepsilon = v_1/v_2 = v_4/v_3$

κ [-]   ..... specific heat ratio, $\kappa = c_p/c_v$

1.3.4. The Constant-Pressure Cycle (Ideal Cycle)

In the constant-pressure cycle, as shown in **Fig. 1.4**, heat is now added at constant pressure and dissipated at constant volume.

The thermal efficiency η_{th} is evaluated as above and is given by:

$$\eta_{th} = 1 - \frac{\varphi^{\kappa} - 1}{\varepsilon^{\kappa - 1} \cdot \kappa \cdot (\varphi - 1)} \tag{1.4}$$

φ [-]   ..... ratio; $\varphi = v_3/v_2$; Fig. 1.4.

The efficiency now depends not only upon the parameters ε and κ but also depends on engine load via φ.

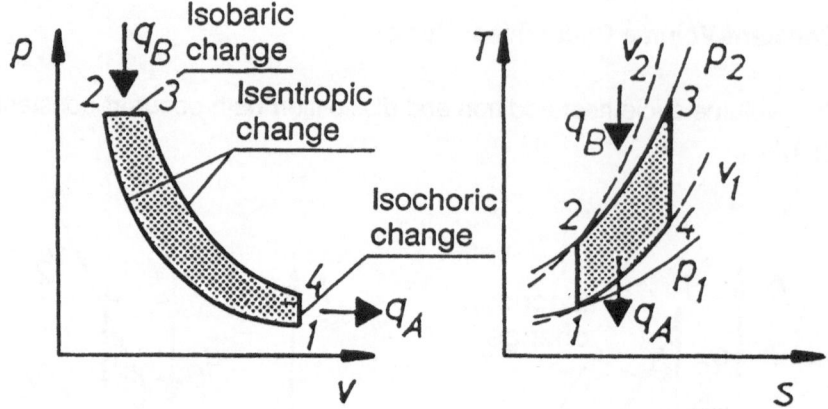

Fig. 1.4. Constant-pressure cycle

1.3.5. The Limited-Pressure (Seiliger) Cycle (Ideal Cycle)

The Seiliger cycle, as shown in **Fig. 1.5**, is an ideal cycle with a limited maximum pressure. The heat addition is controlled such that combustion is isochoric up to a fixed maximum pressure and then becomes isobaric. Such a form of combustion is sought in practice due to material strength limitations. This cycle is more suitable for representing a combustion process in an engine than either the constant-volume or constant-pressure cycle.

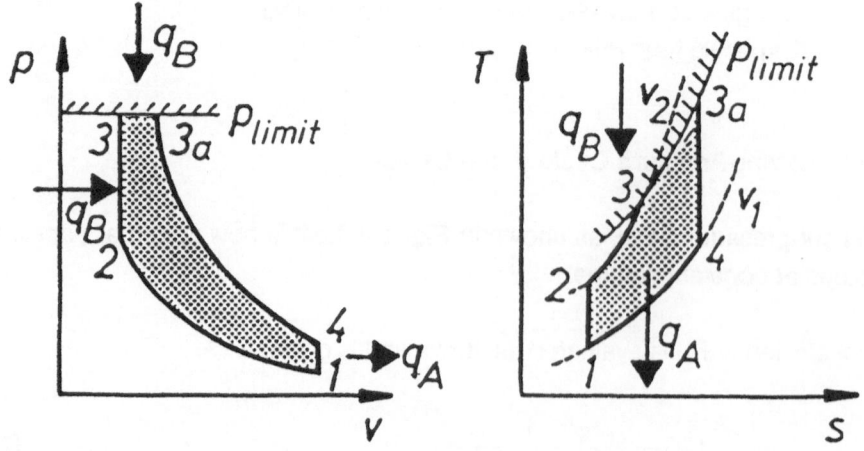

Fig. 1.5. Limited-pressure cycle (Seiliger cycle)

The efficiency of this cycle is given by:

$$\eta_{th} = 1 - \frac{\varphi^{\kappa} \cdot \psi - 1}{\varepsilon^{\kappa - 1} \left[(\psi - 1) + \kappa \cdot \psi \cdot (\varphi - 1) \right]} \qquad (1.5)$$

ψ [-]   ..... ratio; $\psi = p_3/p_2$
φ [-]   ..... ratio; $\varphi = v_{3a}/v_3$

The efficiency depends upon the factors ψ and φ, which depend upon the operating point of the engine.

1.3.6. The Vibe Function

The combustion process in a real engine can be best approximated as a heat addition by a burning function which varies arbitrarily with the crank angle. A commonly used mathematical expression is the Vibe function [1.5, 1.6, 1.7]:

$$\chi = 1 - e^{\left[-f \cdot (\varphi/\varphi_B)^{m+1} \right]} \qquad (1.6)$$

where

χ [-]   ..... is the fraction of fuel energy released
f   [-]   .... is the Vibe coefficient (for spark ignition engines f = 6.908)
φ_B [°KW]   ..... is the combustion duration in °CA
φ [°KW]   .,.. is the independent variable in °CA after start of combustion
m   [-]   ..... is the Vibe exponent

The derivative of the burning function with respect to time yields the rate of energy conversion.

Several such burning curves are plotted in **Fig. 1.6** for typical values of the Vibe exponent m using a normalized independent variable (with respect to φ_B) [1.5, 1.31]. Pressure and temperature curves vs. crank angle can be derived using this burning function. The results may be plotted as a p-v diagram of the simulated cycle as shown in **Fig. 1.7**.

Various assumptions have to be made as to heat dissipation and charge exchange processes (law of heat transfer, valve timings and lifts, flow coefficients, gas dynamic effects in the intake and exhaust manifolds, etc.).

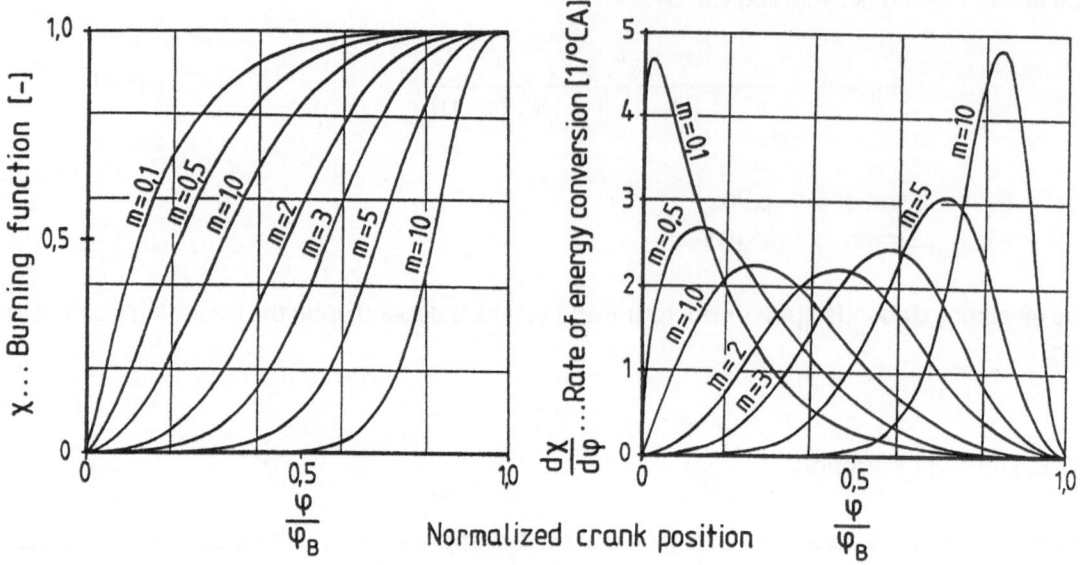

Fig. 1.6. The Vibe function and its derivative in normalized plotting for various Vibe factors m
[1.5 and 1.31]

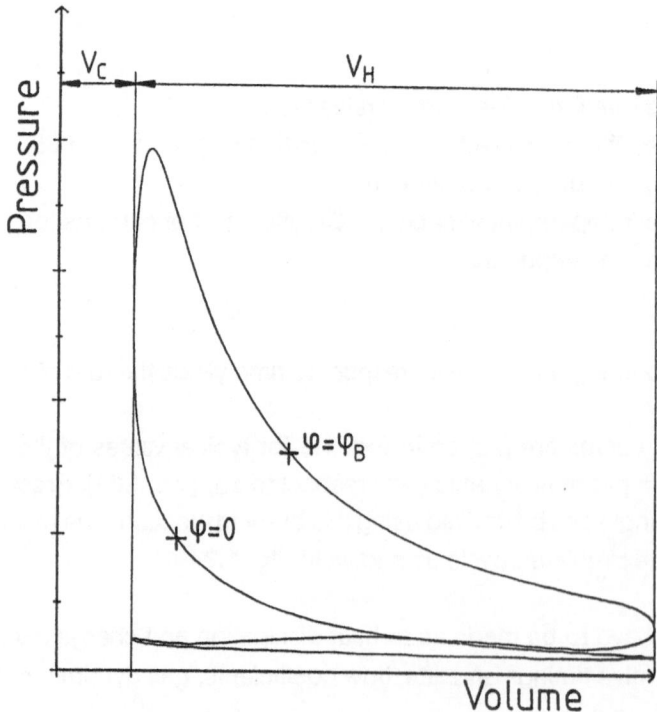

Fig. 1.7. P-V diagram of a comparative process. The Vibe function applies in the range $0 < \varphi < \varphi_B$

1.3.7. Dissociation

The molecules of most polyatomic gases break up into simpler molecules at increased temperatures, and at even higher temperatures break down into atoms and ions.

The following table shows the effect of temperature at atmospheric pressure [1.3]:

Gas	Temperature	Dissociation into	Portion in %
CO_2	2 000 K	CO, O_2	1,5
	3 500 K	CO, O_2	86
O_2	2 000 K	O-Atoms	0,04
	4 000 K	O-Atoms	61

When dissociation is present, the energy released by combustion is only partially converted into internal energy of the working medium since some energy is lost in dissociation. Therefore, the temperature of the medium (as determined by the specific heat) is reduced whenever the medium dissociates.

1.3.8. Process Simulation and Computation

The complex calculations relating to engine combustion processes are carried out today using engine simulations which consider the heat transfers and combustion process in great detail. These simulations also treat the entire charge exchange process in detail and include the events occurring during mixture intake and compression. Despite the importance of this topic, the reader must be referred to other literature for further information so that this book does not become too bulky. Reference is especially made to the book [1.2] preceding this volume. This book refers to various fundamental works in this field, e.g. of Woschni, as well as other works of the current author.

1.4. Details of the Combustion Process in Spark-Ignition Engines

1.4.1. Preflame Reactions

To initiate combustion in a spark-ignition engine, the mixture of fuel and air at a flammable mixture ratio is compressed and ignited by an electric spark. The flame then propagates from the point of ignition throughout the combustion chamber.

In the compression stroke, the pressure and temperature of the mixture consisting of air, residual gases, and fuel vapor is raised to high values. The final compression temperature decreases with an increasing fuel content in the mixture, since the specific heat of the mixture increases with an increasing fuel vapor portion. At the end of compression, the fuel vapor will be superheated, causing evaporation in a warm engine even of the heavier fuel fractions. This results in reactions between the fuel and the air already at this stage.

The temperature rise from induction to the beginning or end of compression, respectively, is approximately the following:

	Beginning of compression	End of compression
Water-cooled cylinder:	40 - 90 °C	460 - 730 °C
Aircooled cylinder:	100 -120 °C	600 - 820 °C

Some hydrocarbon compounds, e.g. benzenes and alcohols, are not changed chemically by the high final compression temperatures during the short time they are exposed to them in the engine. Other fuel constituents, however, start to break down or react with oxygen already at these temperatures, leading, for instance, to changes in the fuel structure by cracking and dehydration, polymerization of products of oxidation, and formation of partially oxidized constituents.

1.4.2. Preignition

Localized hot spots in the combustion chamber, such as spark plug electrodes, exhaust valves, or red-hot deposits, are the causes of preignition. To initiate preignition, the temperature at the point of ignition must be higher than the ignition temperature of the mixture, and the mixture must be exposed to this temperature for a certain amount of time.

The following factors affect preignition:

- Compression ratio: A higher compression ratio accentuates the tendency to preignite due to the higher pressures and temperatures.

- The relative air/fuel ratio λ: The highest tendency of preignition is observed at lambda values of approx. 0.9.

- Greater engine speeds also contribute to preignition, since they lead to increasing combustion chamber wall temperatures. The same applies to higher engine loads.

- The formation of deposits also increases the tendency of preignition.

There are only a few possibilities of avoiding the build-up of deposits in the combustion chamber. The most effective measure is to use fuels with a low boiling range and oil with a low coking tendency. The use of additives may have a favorable effect. As far as the engine is concerned, good cooling of the exhaust valves, rapid warm-up of the combustion chamber after starting, and avoiding excessively hot spots in the combustion chamber, where lead compounds tend to adhere, can be of help.

1.4.3. Ignition

In a spark-ignition engine, ignition normally occurs when a small portion of the mixture is caused to react by the passage of a spark at a temperature between 3000 °C and 6000 °C. The effect of heat is of secondary importance, the decisive factors for initiating ignition are molecular excitation and ionization. If the spark energy is too low, the reaction will not cause any mixture portions beyond those directly passed by the spark to ignite; a certain activation energy will be required.

The flame speed, i.e. the speed of flame propagation, shortly after ignition depends on the properties and composition of the mixture as well as on pressure, temperature, and flow conditions.

Fig. 1.8 plots flame speeds in combustion bombs, i.e. in a quiescent medium.

Fig. 1.9 illustrates the substantially higher flame speeds due to charge motion in an engine. The flame speed decreases in those areas where the lambda value approaches the misfire limits. Flame propagation is possible only within a certain range of fuel concentration in air, which is specific to each fuel.

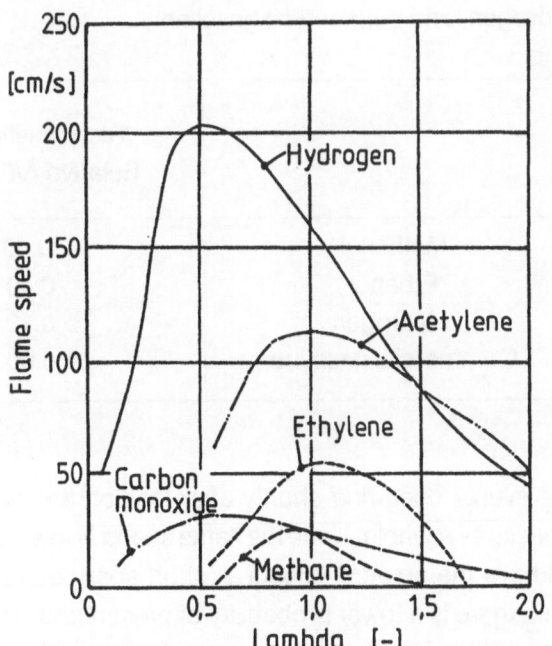

Fig. 1.8. Flame speeds in combustion bombs [1.1]

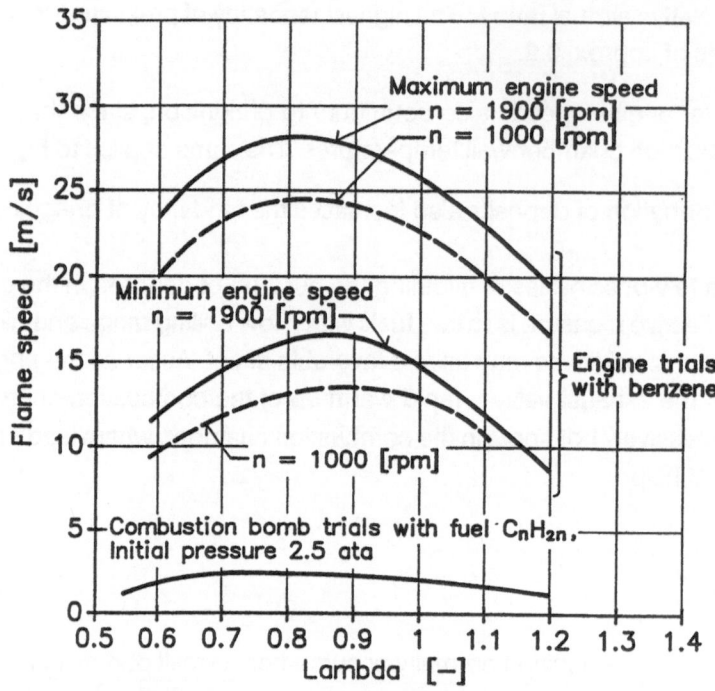

Fig. 1.9. Flame speeds for benzene and gasoline vapor/air mixtures [1.1]

The term "lean" misfire limit is used in the case of fuel deficiency and "rich" misfire limit in the case of overabundance of fuel. The table below gives the two misfire limits for methanol, ethanol, hydrogen, and commercial gasoline:

	"Rich" misfire limit Relative A/F ratio λ	"Lean" misfire limit Relative A/F ratio λ
Methanol	0.40	2.40
Ethanol	0.40	1.90
Hydrogen	0.12	9.00
Commercial gasoline	0.60	1.60

The events occurring shortly after ignition are of paramount importance for the combustion process as a whole, since the flame speed is lowest at the beginning and any delays at this point will have the greatest impact. A short spark duration is unfavorable for igniting lean mixtures, since there is a lower probability of presence of flammable mixture at the point of ignition. Misfiring will lead to higher emissions of unburned hydrocarbons (HC), see **Fig. 1.10**. The ignition delay decreases with a longer spark duration, with otherwise unchanged parameters, see **Fig. 1.11**.

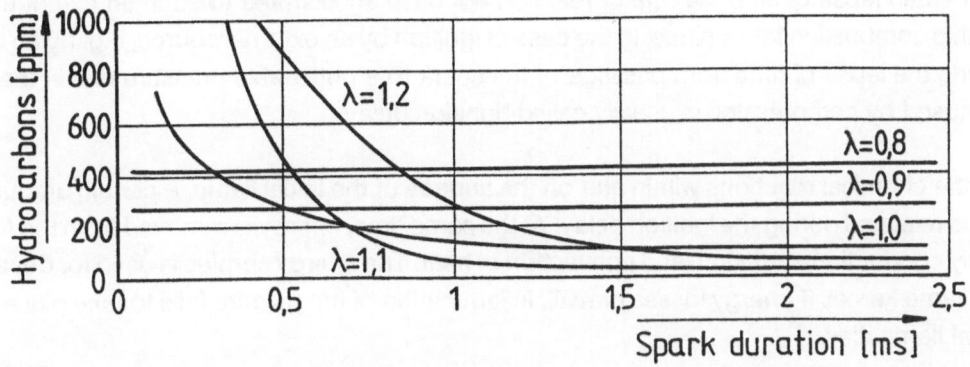

Fig. 1.10. Effect of spark duration on emissions [1.34]

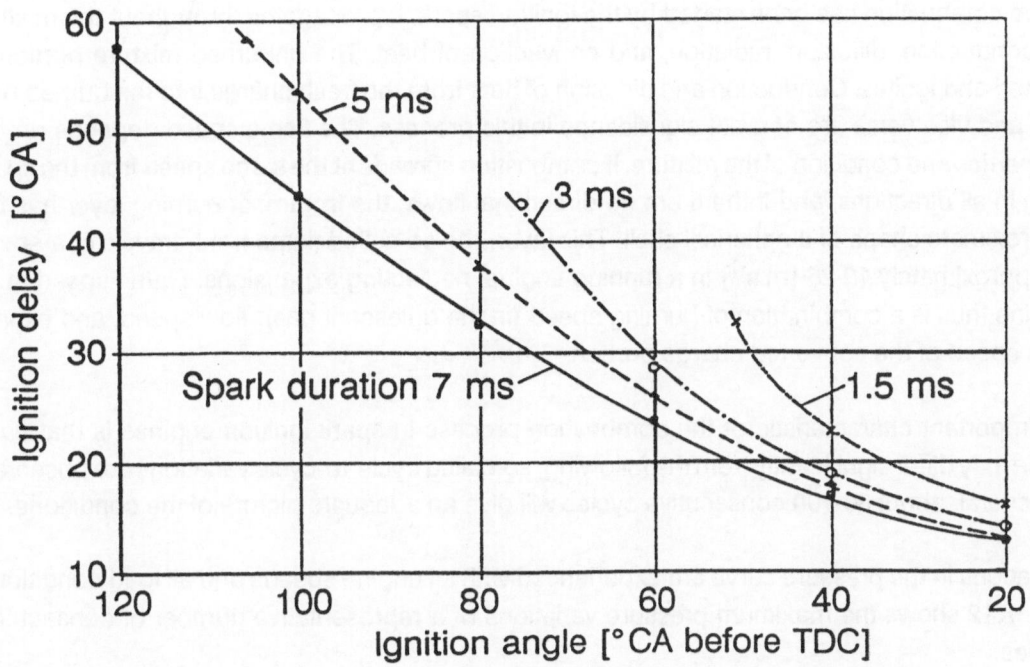

Fig. 1.11. Ignition delay versus crank angle for different spark durations [1.34]

1.4.4. Ignition Delay

When the ignition temperature is attained, the air-vapor mixture ignites by autoignition. In the case of ignition by an external source, inflammation is caused by localized intense heating, for instance by an electric spark. Inflammation of the mixture occurs when a certain amount of time, the so-called ignition delay, has elapsed. Inflammation is considered to be the first appearance of flame or the beginning of a noticeable rise in pressure. The ignition delay is caused by chemical reactions that start to take place when the combustible mixture attains high temperatures.

After a certain lapse of time the rate of reaction will have accelerated to such an extent that a noticeable combustion takes place. In the case of ignition by an external source, e.g. by an electric spark, the lapse of time from passage of the spark to a noticeable pressure rise in the cylinder caused by compression is usually called "ignition delay".

Due to the chemical reactions within and on the surface of the initial flame, a certain amount of energy is released during the ignition delay. At the same time, energy is removed from the flame kernel by conduction, radiation, and convection of heat. The energy surplus is used for developing the flame kernel. If energy losses prevail, inflammation of the mixture fails to take place and the initial flame dies.

1.4.5. Combustion Process and Charge Motion

Once combustion has been started by the ignition spark, it propagates throughout the mixture by conduction, diffusion, radiation, and convection of heat. The unburned mixture portion is heated and ignites. Conduction and diffusion of heat from the fresh charge into the burned portion and vice versa are of great significance in this process. The flame speed depends on the properties and condition of the mixture. If combustion spreads at the same speed from the spark plug in all directions, and if there are no directional flows, the frontmost burning layer has the approximate shape of a spherical shell. This layer, the so-called flame front, travels at a speed of approximately 10-25 $[m.s^{-1}]$ in a running engine, neglecting expansions. Flame speed in an engine thus is a combination of burning speed (in the quiescent gas), flow speed, and expansion speed of the converted charge portion.

An important characteristic of the combustion process in spark-ignition engines is that each cycle may differ significantly from the following, so-called cycle-to-cycle variations are observed. In general, about 50-100 consecutive cycles will give an adequate picture of the conditions.

Variations in the pressure curve are experienced with all engine speeds and all load conditions. **Fig. 1.12** shows the maximum pressure variations of a representative number of consecutive cycles.

Although a change in peak pressure does not necessarily effect a corresponding change in the mean pressure, the peak pressure variation may be considered as an indication of repeatable combustion.

Fig. 1.13 shows peak pressure variations at minor cylinder pressure variations, i.e., in an operating range with stable combustion. **Fig. 1.14**, in comparison, shows the indicated mean pressures of the individual combustion cycles.

Fig. 1.15 and **Fig. 1.16**, on the other hand, show the corresponding plots for engine operating conditions with strong cycle-to-cycle variations of combustion.

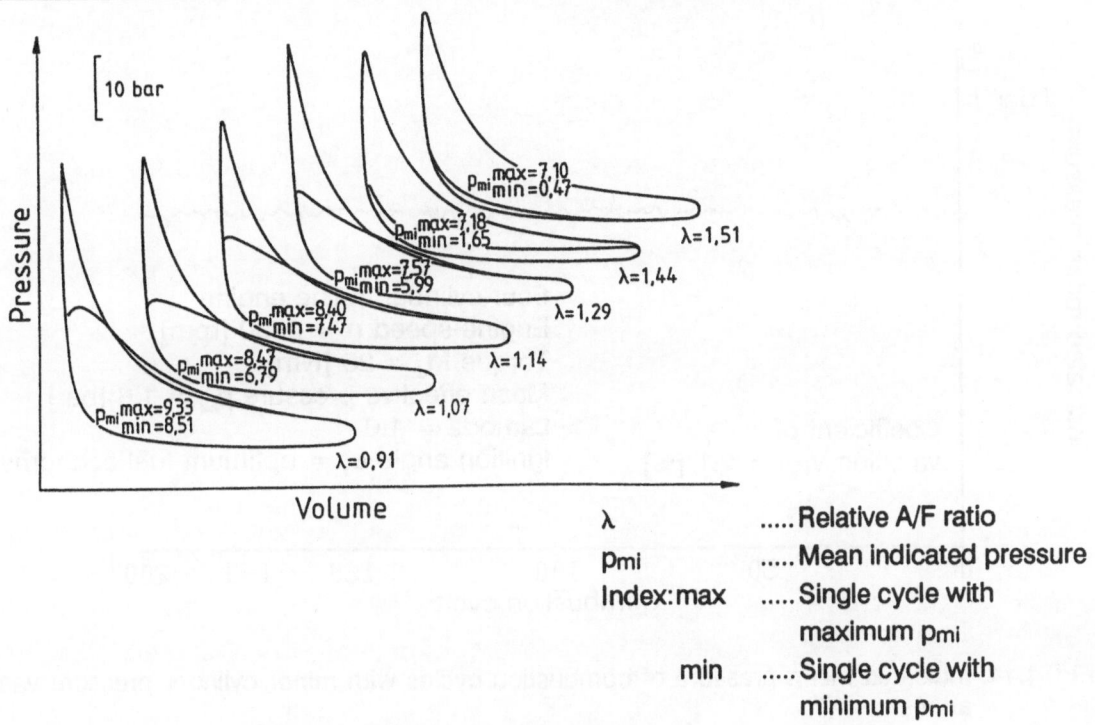

λ .....Relative A/F ratio

p_{mi} .....Mean indicated pressure

Index: max .....Single cycle with maximum p_{mi}

min .....Single cycle with minimum p_{mi}

Fig. 1.12. Maximum pressure variations in the P-V diagram

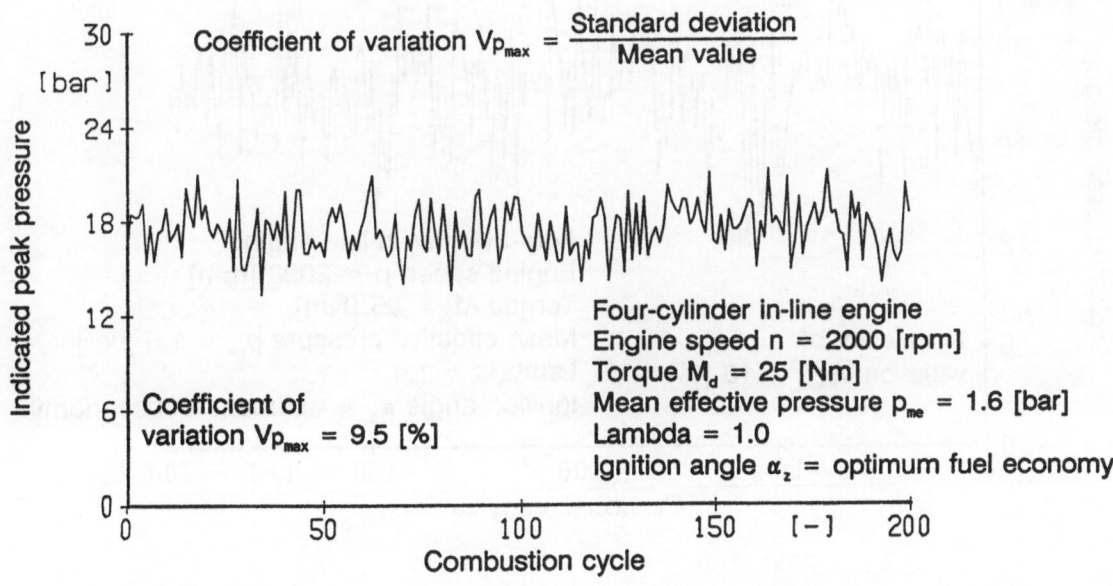

Coefficient of variation $V_{p_{max}} = \dfrac{\text{Standard deviation}}{\text{Mean value}}$

Coefficient of variation $V_{p_{max}}$ = 9.5 [%]

Four-cylinder in-line engine
Engine speed n = 2000 [rpm]
Torque M_d = 25 [Nm]
Mean effective pressure p_{me} = 1.6 [bar]
Lambda = 1.0
Ignition angle α_z = optimum fuel economy

Fig. 1.13. Peak pressure variations in an engine operating range with stable combustion, relative A/F ratio λ = 1.0

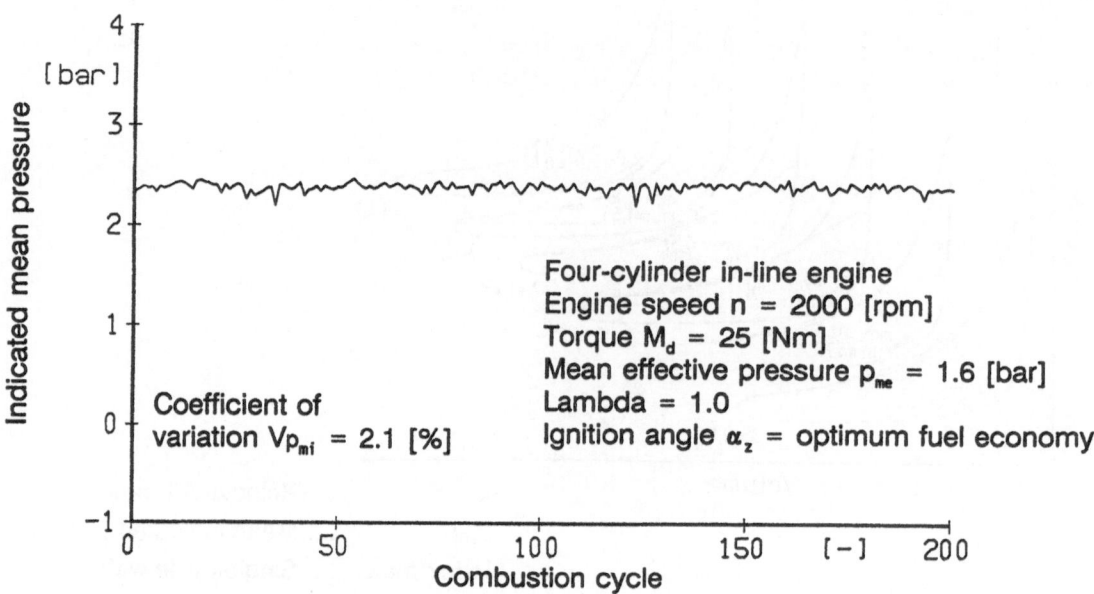

Fig. 1.14. Indicated mean pressure of combustion cycles with minor cylinder pressure variations

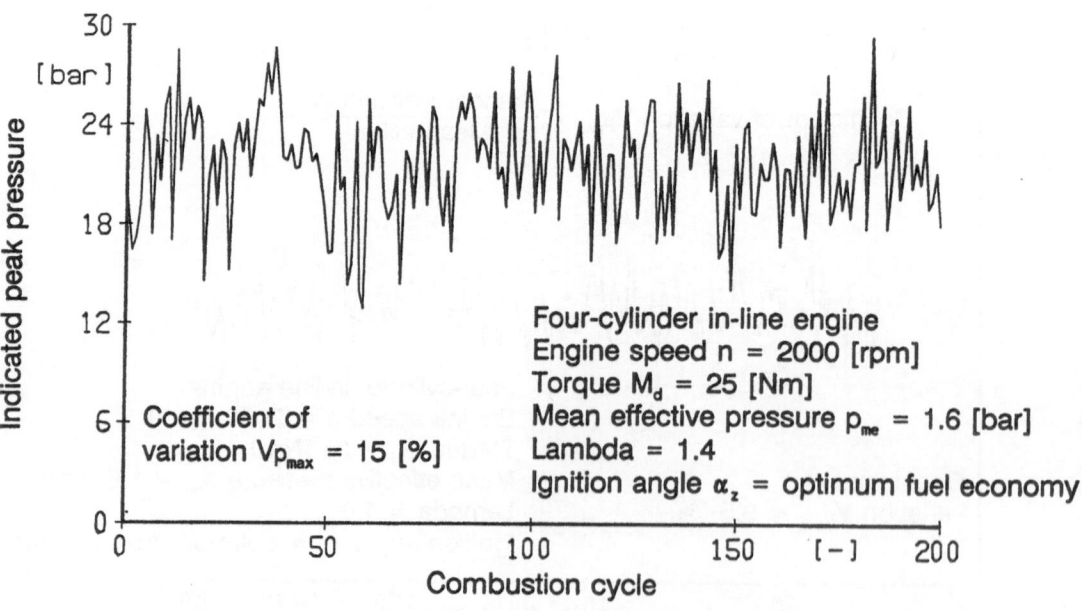

Fig. 1.15. Peak pressure variations in an engine operating range with strong cycle - to - cycle variations of combustion, relative A/F ratio λ = 1.4

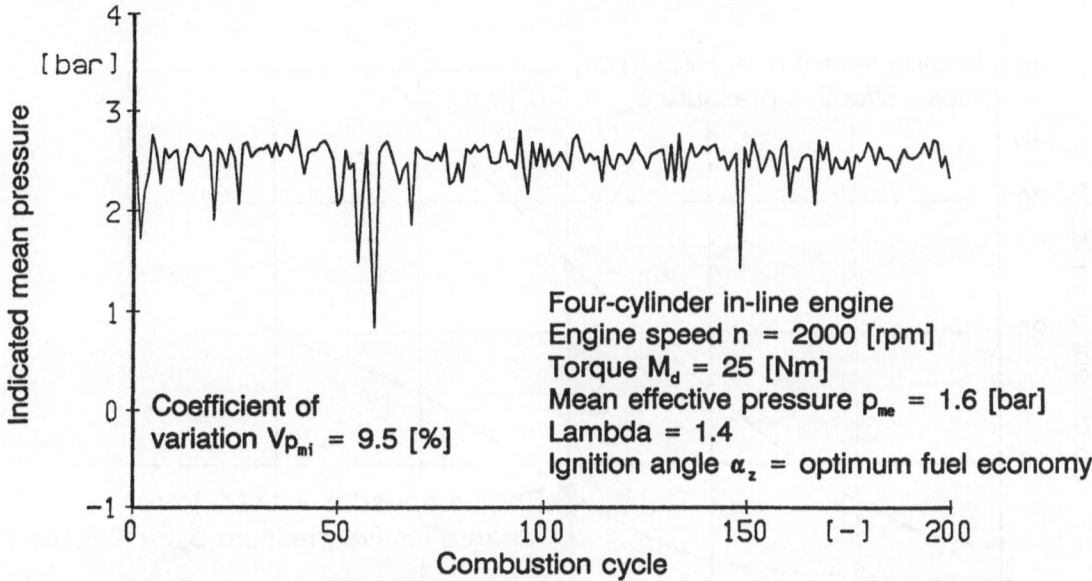

Fig. 1.16. Indicated mean pressure of combustion cycles in accordance with Fig. 1.15, i.e. combustion with strong cycle-to-cycle variations

The effects of relative A/F ratio λ, combustion chamber shape, and rate of exhaust gas recirculation on cycle-to-cycle variations of the combustion process, expressed as coefficient of variation, which is the standard deviation divided by the mean value [1.30], are shown in **Fig. 1.17** and **Fig. 1.18**.

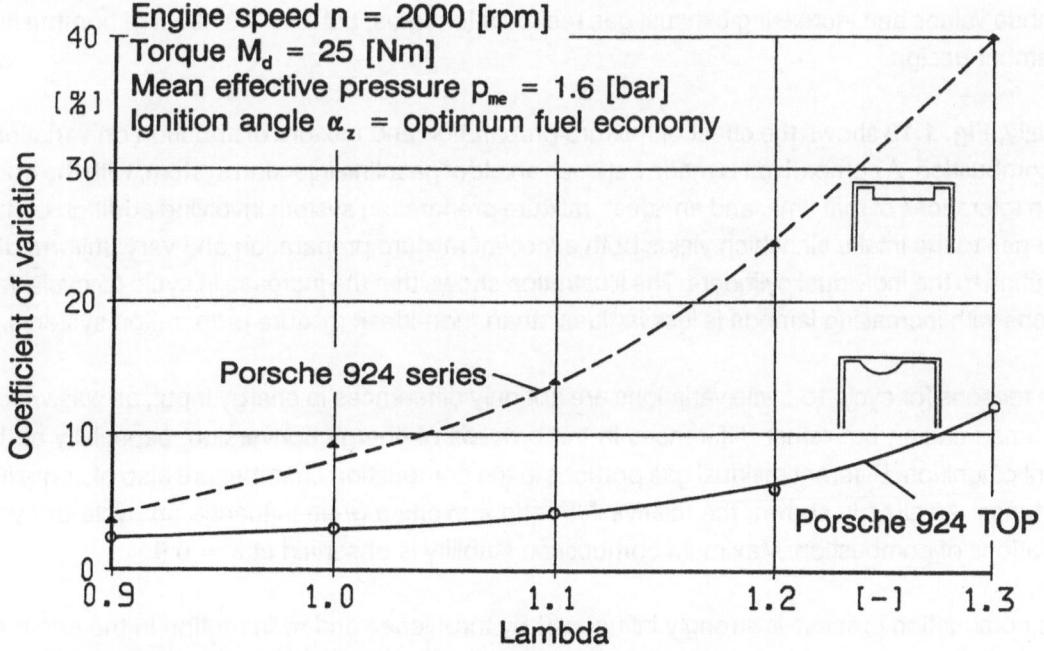

Fig. 1.17. Coefficient of variation of the indicated mean pressure for various combustion chamber designs and lambda values

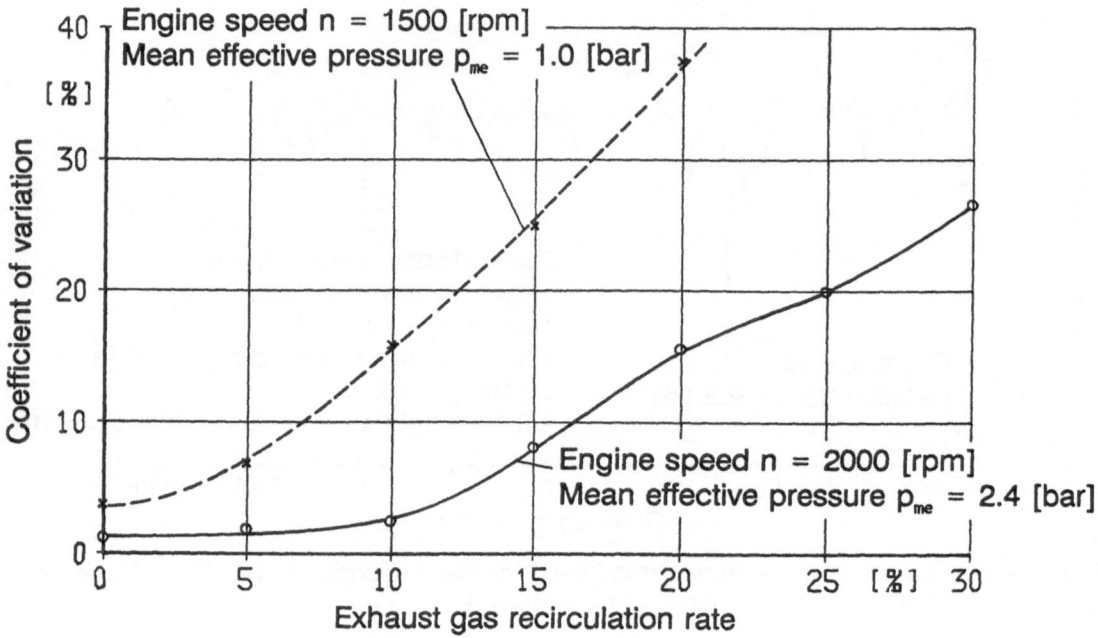

Fig. 1.18. Coefficient of variation of the indicated mean pressure versus rate of exhaust gas recirculation at different operating points [1.9]

The illustration shows not only a strong increase in combustion instability both with increasing lambda values and increasing exhaust gas recirculation rates, but also the effect of combustion chamber design.

Finally, **Fig. 1.19** shows the effect of mixture preparation and mixture distribution on variations of combustion. A comparison is made between an older gasoline injection system, with the common tolerances of that time, and an "ideal" mixture preparation system involving addition of natural gas to the intake air, which yields both excellent mixture preparation and very uniform distribution to the individual cylinders. The illustration shows that the increase in cycle-to-cycle variations with increasing lambda is less in "ideal" than "non-ideal" mixture preparation systems.

The reasons for cycle-to-cycle variations are not only differences in energy input, as was widely assumed earlier, but rather differences in the process of energy conversion, especially at the point of ignition. Different residual gas portions in the combustion chamber are also of some importance. As already shown, the relative A/F ratio λ exerts a great influence on cycle-to-cycle variations of combustion. Maximum combustion stability is observed at $\lambda = 0.9$

The combustion process is strongly influenced by turbulence and swirl motion in the combustion chamber. A higher burning speed is achieved by the increased burning area as a result of the flame front being agitated by turbulence. Air turbulence produced by inlet flow velocities of

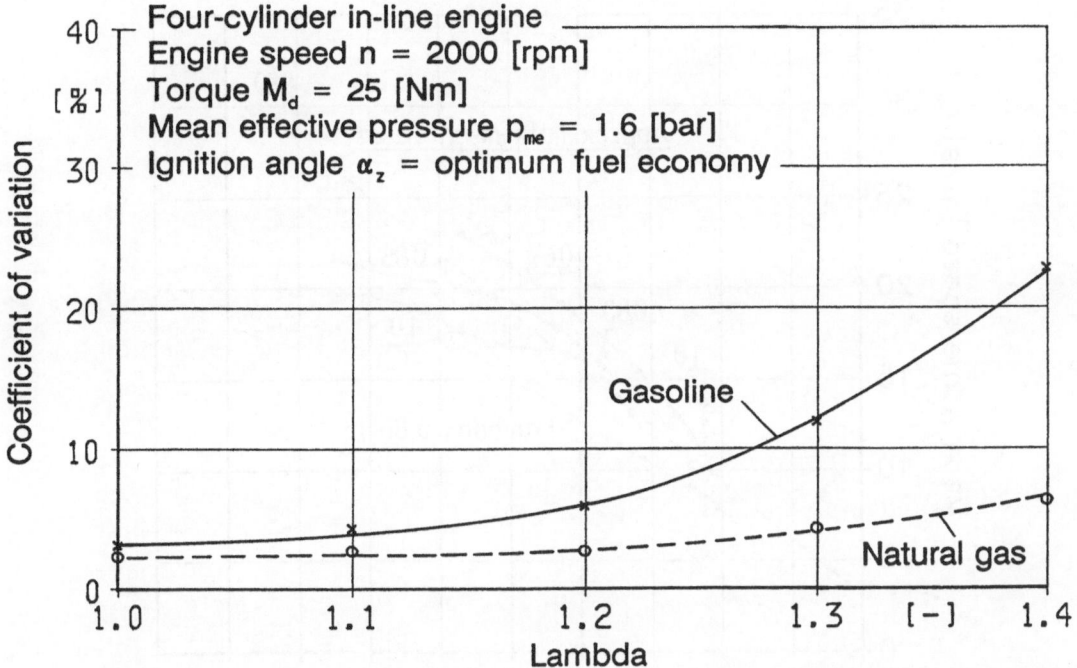

Fig. 1.19. Comparison of the coefficients of variation versus relative A/F ratio λ between gasoline-fueled and natural-gas-fueled engines

up to 100 m.s^{-1} through the valve will still be present at the end of compression. This means that flame speeds in engines are about 10 times higher than in a combustion bomb with a quiescent mixture. The flame speed increases as engine speed increases, since turbulence agitates and consequently enlarges the burning surface, see **Fig. 1.20**.

Charge motion will occur in any engine. It can be intensified by certain measures in the intake system (external turbulence) or in the combustion chamber (internal turbulence).

External charge motion or turbulence can be produced by orifice plates, slides, swirl ports, shrouded valves, and inlet valve throttling. Internal turbulence can be influenced by combustion chamber geometry (piston bowls, squish gaps, divided combustion chamber).

Looking at the effects of externally produced turbulence on the combustion process in greater detail, the following can be established, for instance, when using a throttle to create turbulence immediately before the inlet valve:

The throttle increases the inlet flow between the throttle and inlet valve. Further, the distribution of flow speeds differs in a real- engine inlet tract with a throttle from that of an undisturbed pipe

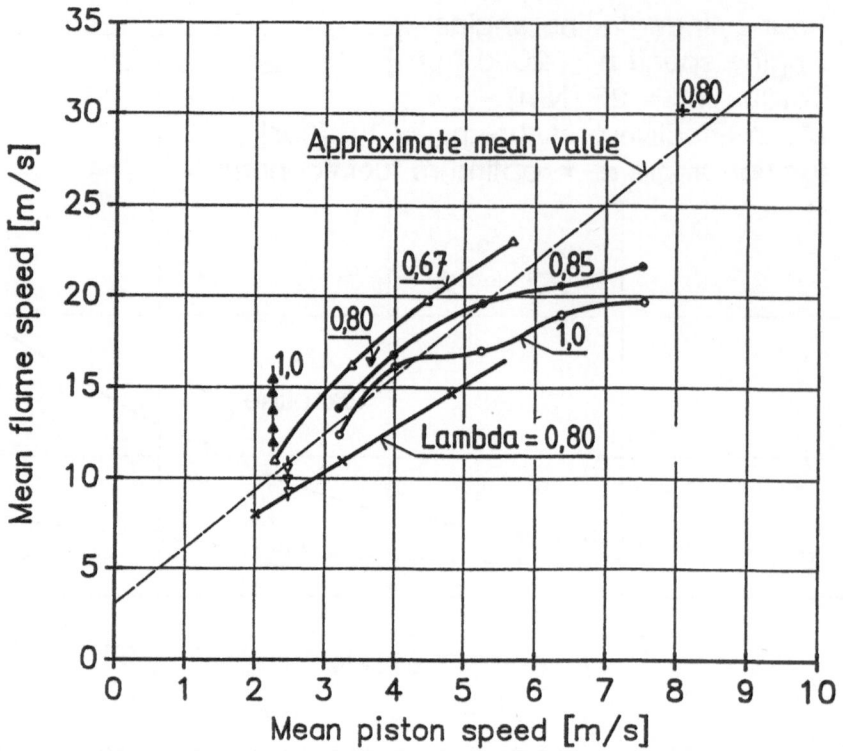

Fig. 1.20. Flame speed as a function of piston speed [1.1]

flow. This may affect mixture preparation, charge exchange, and charge flow conditions in the cylinder.

An improvement in mixture preparation can, of course, be achieved by the above measure only if fuel and combustion air have not yet blended very well. Throttling before the inlet valve may, however, have an adverse effect on charge exchange. This is due to a possibly increased flow resistance or to a pipe resonance calibration - that means choice of an intake manifold design that will have a manifold charging effect at a specific engine speed - may be displaced toward other engine speeds. Another possible effect is a reduced portion of residual gas in the fresh charge when valve overlap is extensive and when the throttle reduces the back-flow of exhaust gas into the inlet duct.

Many works contain research on this subject: Akhlaghi [1.10], for instance, studied the effect of individual ports in the vicinity of inlet valves and found that creating turbulence at the edge of an air stream emitting from an individual port produces the following effects:

- Improved mixture preparation;

- Enhanced exchange processes at the flame front due to small-scale turbulence;

- Enlarged burning surface due to large-scale turbulence;

These effects permit the following:

- Extending the lean limit;

- Retarding the optimum ignition point;

- Shifting the best economy part-load lambda value toward a "leaner" region and

- Thus reducing the specific fuel consumption;

- Improving combustion stability.

Either improvements or a deterioration regarding emissions of nitrous oxides or unburned hydrocarbons may occur, depending on the design and operating load of the engine.

Lawton [1.11] and Lucas [1.12] studied the effect of the so-called "Vortex Generator", i.e. a cone directed upstream with perforations in its envelope, and also observed increased combustion speeds.

Two authors, Goeschel [1.13] and Prescher [1.14], describe the effect of a reduced inlet valve stroke on the combustion process: They observed that intensified inlet turbulence accelerates the combustion process.

Quissek [1.15] also dealt in great detail with the effects of enhanced external turbulence, created by orifice plates in the intake duct near the inlet valve, on the combustion process. **Fig. 1.21** shows that, when using a slit plate to create turbulence, it is possible, despite a later ignition point and delayed start of combustion, to achieve an earlier end of the burning process, that means in total to have faster combustion. The ignition delay is shortened by approximately 45% by displacing the start of combustion nearer to top dead center. Two thirds of this reduction in ignition delay are caused by the increased external turbulence, one third being attributed to a later position in the compression cycle and the resulting higher pressure and higher temperature.

Quissek's [1.15] finding that, with inlet valve strokes greater than 4 mm, the change in flow before the inlet valve will be continued in the cylinder and cause increased turbulence within the cylinder, also appears to be of importance.

In the part-throttle range, the knock limit and optimum ignition point are shifted - both to the same extent - toward top dead center due to the increased external turbulence, whereas under wide-open throttle conditions the knock limit is displaced further than the optimum ignition point, which moves into the range of knocking combustion.

In combustion chambers where the charge is accelerated by squish produced by a specific piston shape, maximum squish intensity is attained shortly before top dead center.

Regarding the effect of intake swirl on the combustion process, Hofbauer [1.16] observed the following:

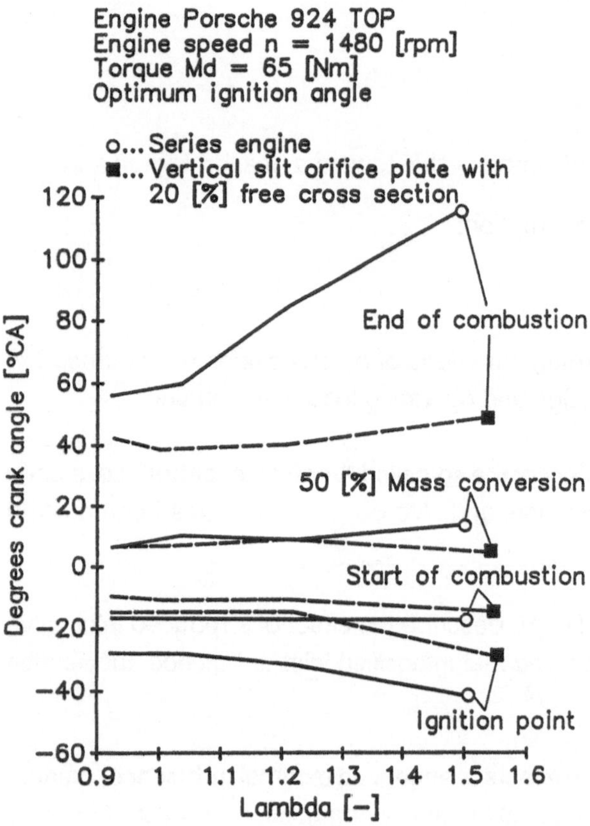
Engine Porsche 924 TOP
Engine speed n = 1480 [rpm]
Torque Md = 65 [Nm]
Optimum ignition angle
o... Series engine
■... Vertical slit orifice plate with 20 [%] free cross section
End of combustion
50 [%] Mass conversion
Start of combustion
Ignition point
Degrees crank angle [°CA]
120
100
80
60
40
20
0
-20
-40
-60
0.9
1
1.1
1.2
1.3
1.4
1.5
1.6
Lambda [-]

Fig. 1.21. Effect of external turbulence on the combustion process [1.15]

Compact combustion chamber designs such as Heron and spherical bowl combustion chamber, in combination with intake swirl, lead to an appreciable increase in combustion speed as compared to a standard combustion chamber design. Higher specific work (also referred to as mean effective pressure), lower specific fuel consumption, and smaller cycle-to-cycle variations at higher peak pressures and higher rates of pressure rise are attained under full-load conditions. In Heron combustion chambers, the combined effect of swirl and squish leads to a reduction in HC emissions to values below those of spherical bowl combustion chambers. In Heron and spherical bowl combustion chambers NO_x emissions are higher, due to swirl, than in series-type combustion chambers. In the part-load range, emission values are similar. Only the NO_x emissions of spherical bowl and Heron combustion chambers with air swirl are increased still higher above the values of the series-type combustion chamber.

Prescher [1.14] uses shrouded valves to produce intake swirl. They cause the mean flow speed in the vicinity of the cylinder walls to be quadrupled at the time of ignition; turbulence is also increased. With constant volumetric efficiency under wide-open throttle conditions, the maximum specific work increases with increasing swirl. Flame propagation speed, combustion heat conversion, and peak pressure are also increased. Cycle-to-cycle variations decrease with increasing swirl, whereas they remain constant when the flame propagation speed is increased on account of a reduced inlet valve stroke.

Similar to Hofbauer [1.16], Nagayama [1.17] studied the combined effect of intake swirl and squish. His findings coincide with the results of Hofbauer [1.16]. He found that the combined effect of increased swirl and intensified squish is always greater than the sum of the individual effects. Swirl improves mixture preparation and reduces both ignition delay and cycle-to-cycle variations. Squish has a major influence on the main combustion process, so that a combination of the two seems useful.

Thring and Overington [1.18, 1.19] show that the optimum degree of charge motion depends on the compression ratio of the engine. Engines with high compression ratios require less turbulence. Also Bloss [1.20] demonstrated the benefits of a combined use of swirl and squish in a spark-ignition engine with a high compression ratio. When increasing the swirl motion beyond a given optimum, heat losses to the wall increase on account of the fast burning process. The knocking tendency decreases with increasing swirl intensity.

Finally, **Fig. 1.22**, according to Carstensen [1.21.], shows the effect of intake swirl on the maximum heat release rate and burning time in the mean part-load range in a single-cylinder experimental engine. The diagram illustrates the decrease in burning time with increasing swirl at different lambda values and the increase in maximum heat release rate.

The effect of different combustion chamber shapes - in terms of gap dimensions - on squish and thus on charge motion is shown in **Fig. 1.23**. It is evident that decreasing gap dimensions and an increasing squish area lead to an appreciable rise in maximum air velocity.

Regarding the effect of the relative A/F ratio λ on the burning speed, the following relation can be established, as shown in previous diagrams: lambda values below the stoichiometric ratio yield the highest burning speeds and the shortest ignition delays. For this reason, such mixtures require the least spark advance.

The effect of ignition timing on the burning speed is complex: Optimum spark advance represents the best compromise between the thermodynamic requirements on combustion and the demands on crankshaft kinematics.

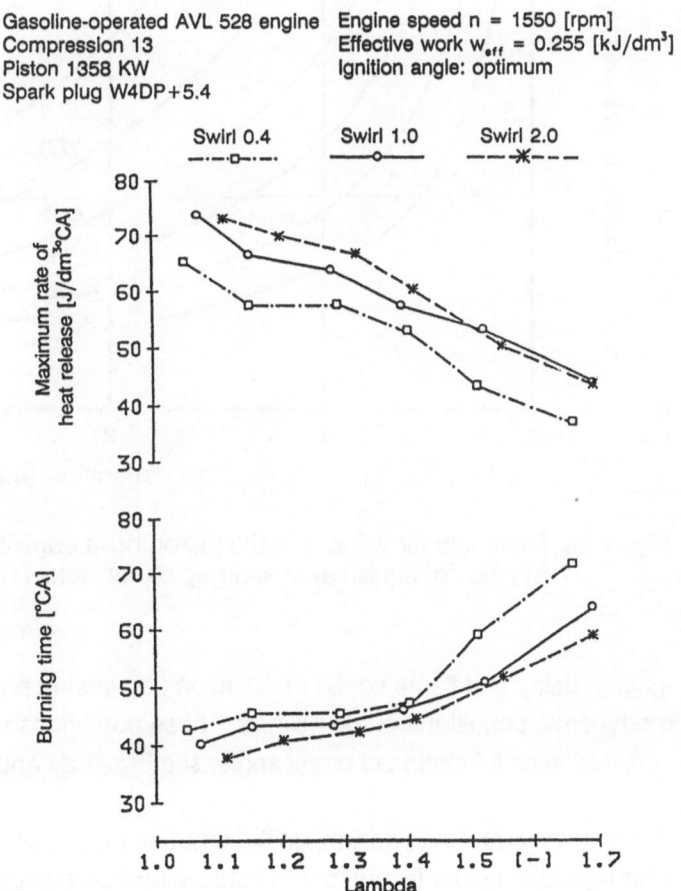

Gasoline-operated AVL 528 engine
Compression 13
Piston 1358 KW
Spark plug W4DP+5.4

Engine speed n = 1550 [rpm]
Effective work w_{eff} = 0.255 [kJ/dm³]
Ignition angle: optimum

Fig. 1.22. Effects of the relative air-fuel ratio lambda on the maximum heat release rate and burning time in the mean part-throttle range; parameter: intake swirl [1.21]

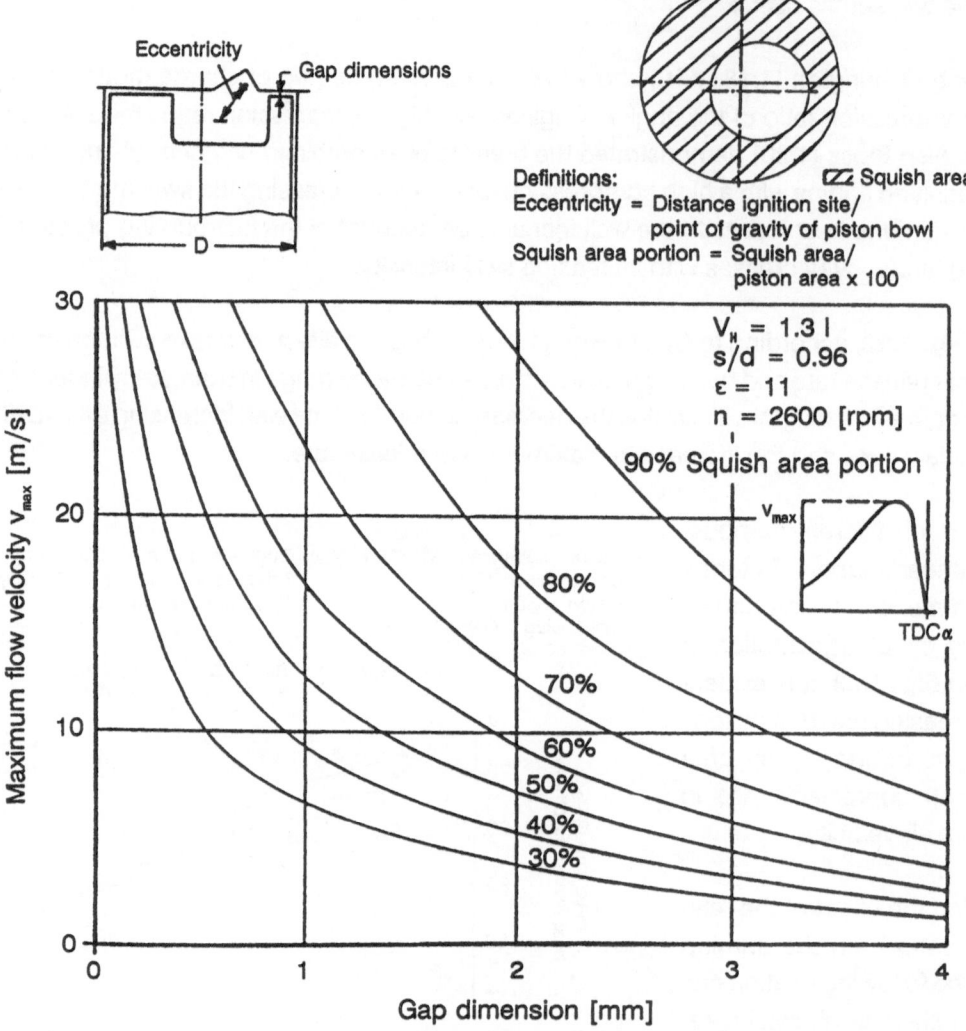

Fig. 1.23. Maximum air velocity at the piston bowl edge of a Heron combustion chamber as a function of squish area and gap dimensions [1.22]

Ignition delay and flame speed of the main conversion process depend on the prevailing thermodynamic conditions of the mixture. These conditions are determined by the position of the conversion as a function of crank angle, see **Fig. 1.24** and **Fig. 1.25**.

With a given crankshaft design, on the other hand, a specific position of the maximum gas force after top dead center is required to achieve best efficiency.

With optimum spark advance, the position of maximum pressure in most engines is 10° to 15° crank angle after top dead center. Effective power and effective efficiency attain their maximum at this point.

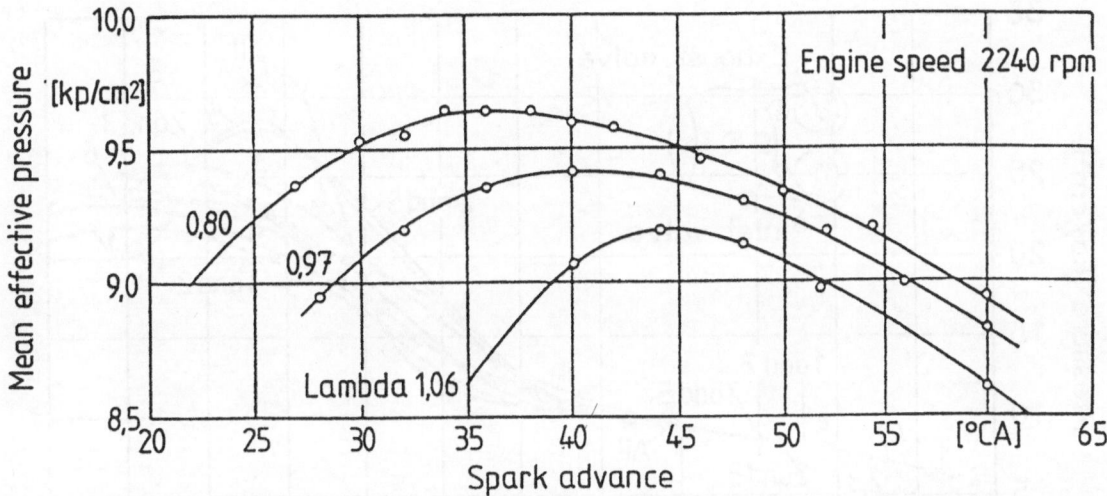

Fig. 1.24. Mean effective pressure as a function of spark advance under wide-open throttle conditions for various relative air-fuel ratios λ [1.1]

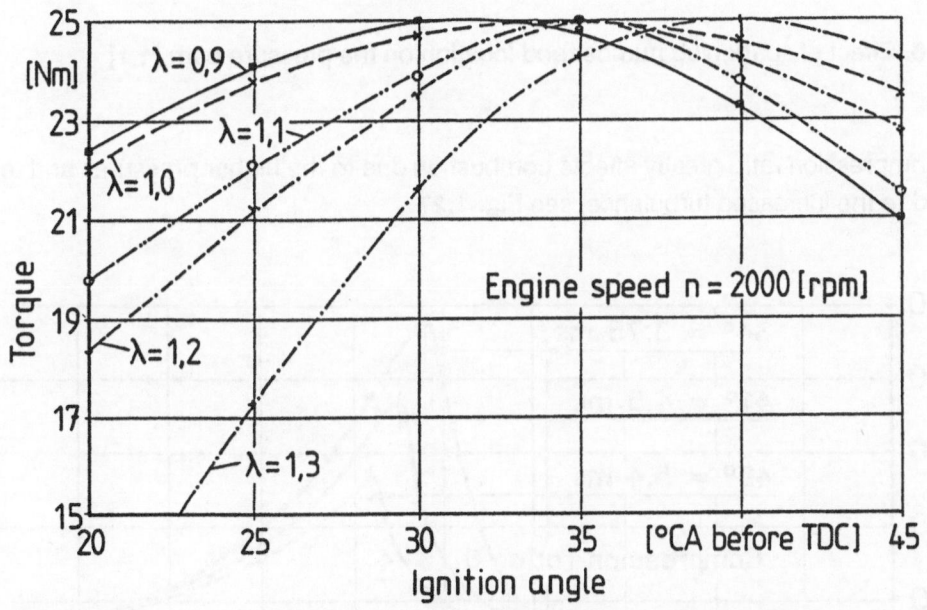

Fig. 1.25. Effective torque versus ignition point under part-throttle conditions of a four-cylinder engine for various relative air-fuel ratios λ

The number and location of spark plugs strongly influence the combustion process. **Fig. 1.26** plots pressure curves versus crank angle for six different spark plug locations. Shorter flame travel, in the case of several points, or a central point of ignition gives the fastest conversion.

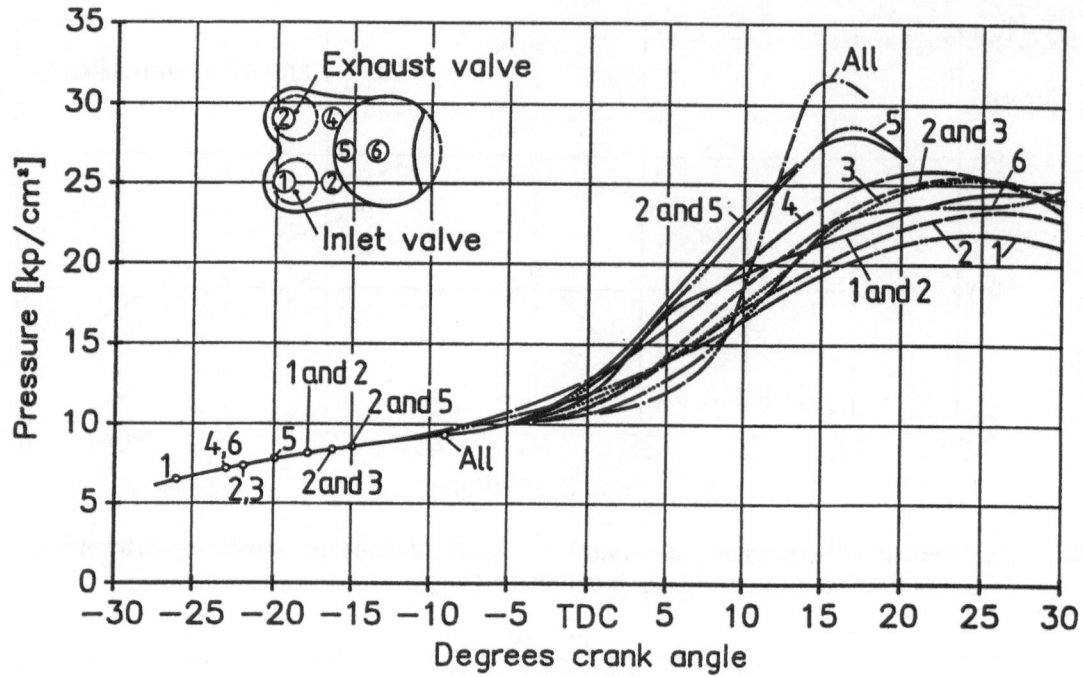

Fig. 1.26. Effect of spark plug number and location on the pressure path [1.1]

A high compression ratio greatly affects combustion due to the higher pressures and temperatures and to the increased turbulence, see **Fig. 1.27**.

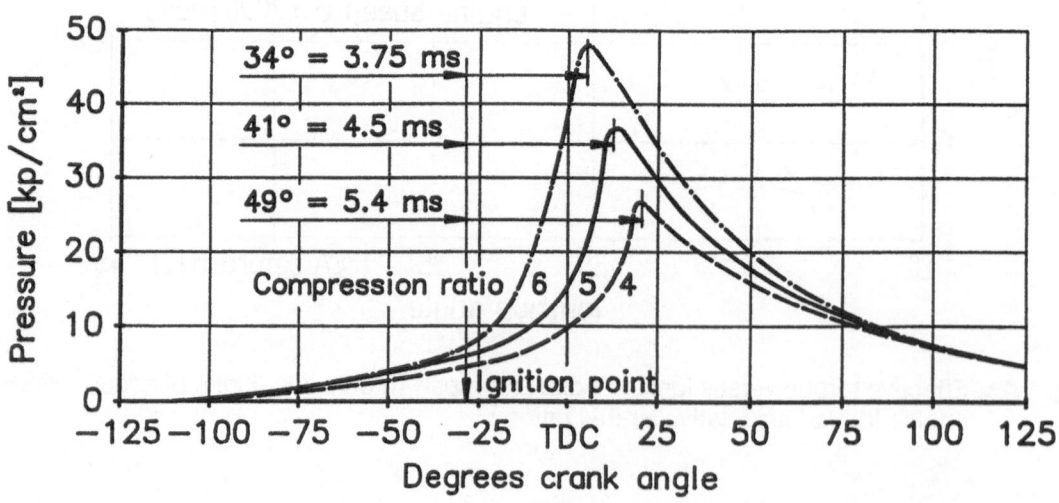

Fig. 1.27. Effect of the compression ratio on combustion at a stoichiometric lambda value [1.1]

1.4.6. Combustion Knock

The terms "knocking" or "pinking" are used to describe a near detonating combustion as compared to a normal combustion process. An extremely steep pressure rise - up to about 8 bar per degree crank angle or up to 50,000 bar per second- is observed during this event.

Combustion knock is caused by spontaneous oxidizing reactions in the hot unburned charge portion. This remaining charge portion is compressed first by the piston movement and then by the moving flame front. Knocking combustion means nearly instantaneous ignition of part of the remaining mixture, see **Fig. 1.28**. Unusually high local pressures accompanied by shock waves and pressure oscillations with an increased heat transfer occur on account of this unduly accelerated combustion process. This may cause engine damage due to high mechanical and/or thermal strain on engine components.

Depending on its intensity, knocking combustion may range from soft pinking to violent thumping. In the beginning this may be without consequence in terms of damage to bearings, pistons, piston pins etc. Knocking that only occurs during short periods of acceleration may be considered as less harmful than a persistently knocking combustion.

While normal combustion speeds are about 10 to 25 m.s^{-1}, knocking combustion is characterized by speeds of 250 to 300 m.s^{-1}. It should be noted that, despite this high increase in speed in comparison to "normal" combustion, the mentioned values are still much lower than the speeds observed during detonation, which may be as much as 2,000 to 3,000 m.s^{-1}.

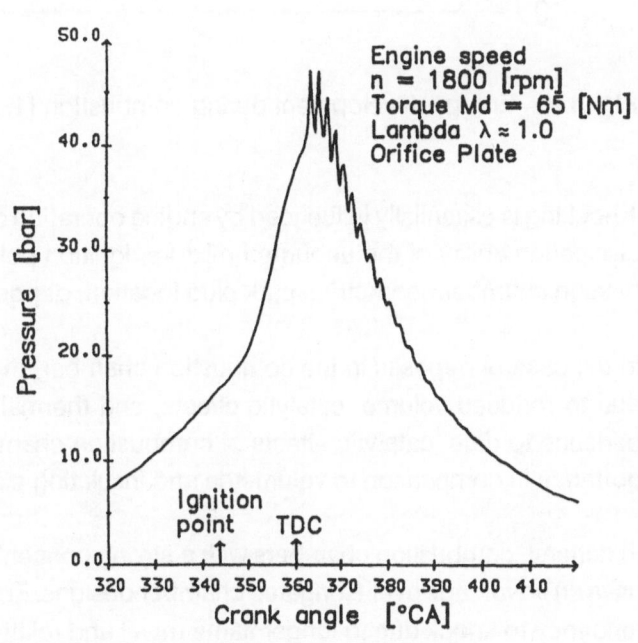

Fig. 1.28. Combustion pressure versus crank angle in a knocking combustion process

In a normal combustion process, inflammation is propagated by means of heat transfer and convection; in the case of detonation, however, this happens on account of a compression shock (adiabatic compression).

Fig. 1.29 shows the energy development in different combustion processes.

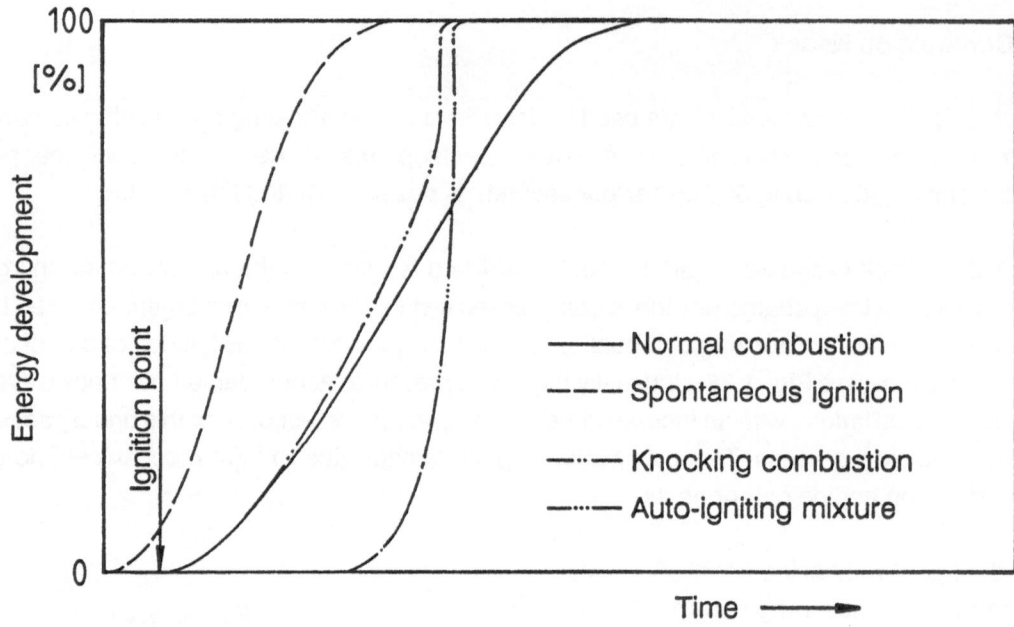

Fig. 1.29. Energy development during combustion [1.1]

Knocking is essentially influenced by engine operating conditions (warm-up, compression, heat dissipation ability of the unburned mixture, ignition point, etc.), by the fuel used, and by combustion chamber geometry (spark plug location, deposits).

In the case of deposits in the combustion chamber, the increased knocking tendency may be due to reduced volume, catalytic effects, and thermal insulation. In accordance with the experience to date, catalytic effects of combustion chamber deposits appear to be of minor importance in comparison to volumetric and insulating effects.

In general, combustion chambers with a strong concentration of the charge near the spark plug have an advantage over elongated chamber designs. Engines with large cylinders have a greater tendency to knock due to longer flame travel and relatively lower dissipation than engines with smaller cylinder dimensions.

In particular, the knocking tendency increases with the following operating or design characteristics:

- Low engine speed, due to a prolonged warm-up period for the remaining unburned mixture;

- Higher loads;

- Higher compression;

- Higher air density;

- Reduced cooling due to increasing temperatures and pressures in the combustion chamber;

- Other than central spark plug location in the combustion chamber;

- Elongated combustion-chamber design, because of longer flame travel and extended warm-up periods for the remaining unburned mixture.

1.4.7. Working Cycle

Combustion in a spark-ignition engine can be greatly influenced by combustion-chamber design. This influence is brought about by the effect that combustion-chamber geometry has on charge motion and, possibly, also on local charge composition (stratified-charge engines).

In the Porsche-TOP engine [1.23], Fig. 1.17, for instance, a piston bowl and squish gaps permit an increase of the compression ratio from $\varepsilon = 9.3$ to $\varepsilon = 12.5$, when using premium-grade gasoline, and an extension of the lean operating range, due to accelerated combustion by high turbulence.

Other common combustion-chamber designs are the result of providing a piston bowl with squish gaps or valve pockets with a flat cylinder head.

Examples of present-day combustion-chamber design are shown in **Fig. 1.30**: The top illustration shows stages of development of the latest BMW double-valve combustion chamber [1.24]. In the optimized design version (extreme right), the largest part of the combustion space (70%) is located in the cylinder head, the rest is in the piston bowl. By suitable shaping of piston and cylinder head, a squish area is created which accounts for about 13% of the total piston area. The spark plug location is near the valve. The bottom illustration of Fig. 1.30 shows the combustion chamber of a four-valve engine [1.25]. The combustion space is contained totally in the cylinder head. A central spark-plug location is made possible by providing four valves per cylinder; this arrangement ensures fast and uniform combustion.

Fig. 1.31 shows the effect of squish area on burning time. By changing from combustion-chamber design A, with a 43% squish area portion, to design B (73%), it has been possible to reduce the burning time from 55° crank angle to 35° crank angle.

In the May-Fireball combustion chamber, **Fig. 1.32**, a specific swirl motion of the charge is generated during the compression stroke. The spark plug is located at the outer circumference of the combustion chamber portion which extends below the exhaust valve.

There is a great number of commonly known configurations of divided combustion chambers for spark-ignition engines. **Fig. 1.33** shows a cross-sectional view of a Volkswagen PCI (Pre-Chamber-Injection) engine. A gasoline nozzle directly injects into the prechamber, where ignition is also initiated.

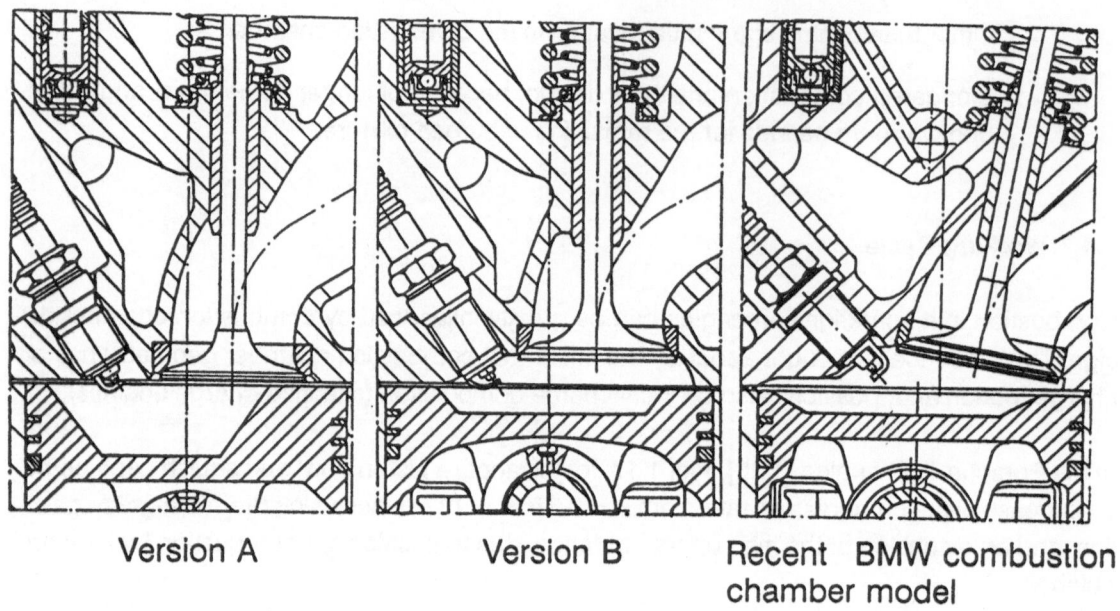

Version A   Version B   Recent   BMW combustion
                                          chamber model

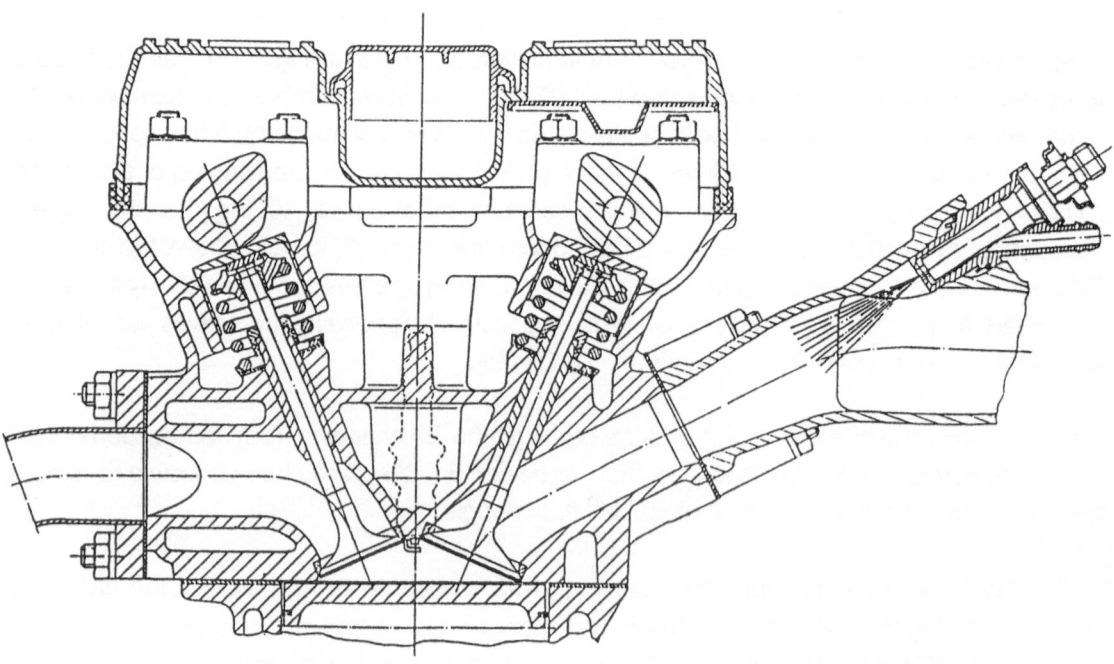

Fig. 1.30. Recent combustion-chamber designs of two- and four- valve engines [1.24, 1.25]

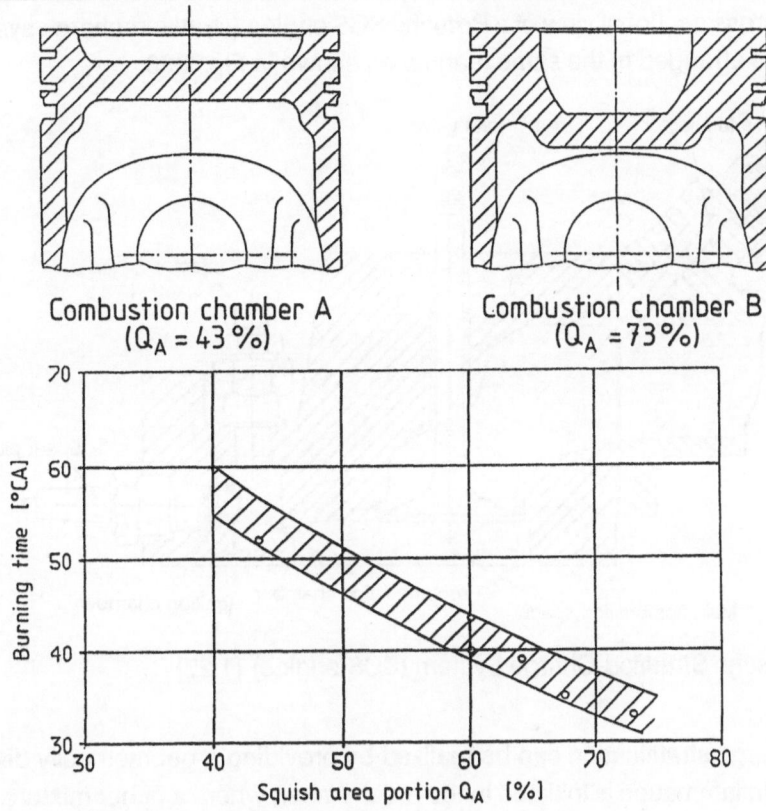

Combustion chamber A
(Q_A = 43 %)

Combustion chamber B
(Q_A = 73 %)

Fig. 1.31. Effect of squish area on burning time

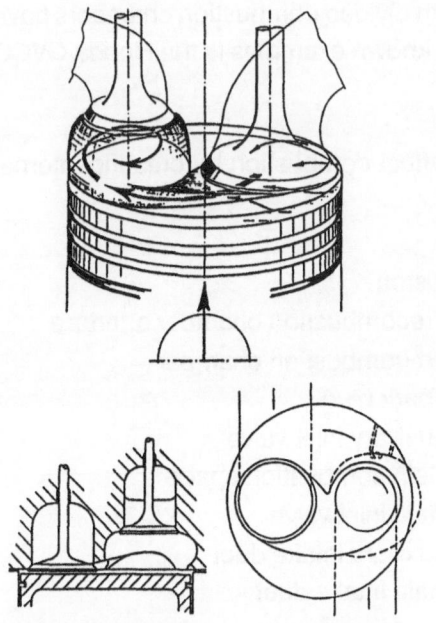

Fig. 1.32. May-Fireball combustion-chamber design [1.28]

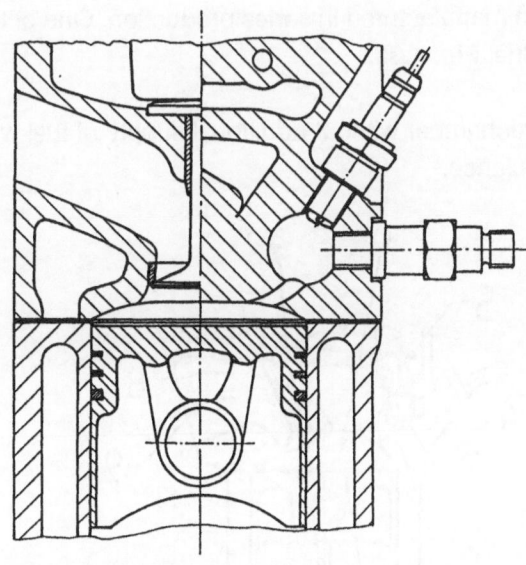

Fig. 1.33. Cross-section of a Volkswagen PCI engine [1.26]

Fig. 1.34 is a cross-sectional view of a Porsche SCS engine (stratified-charge system). Injection and ignition are arranged in the same manner as shown in Fig. 1.33.

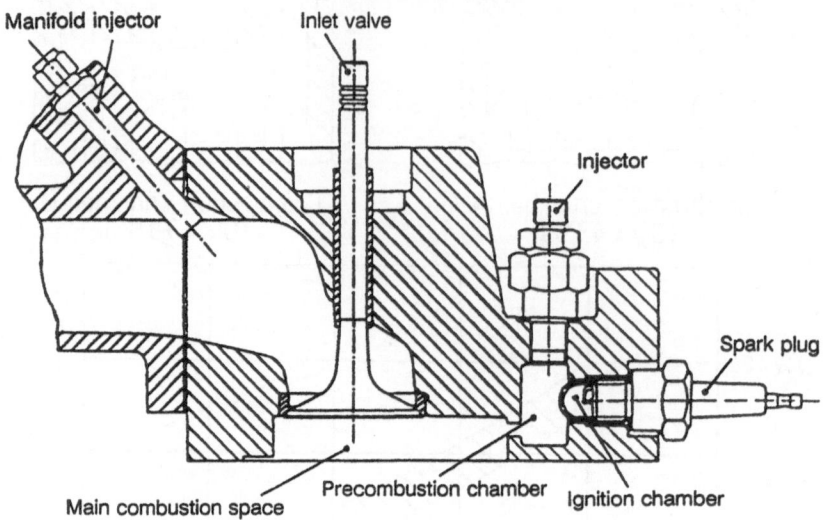

Fig. 1.34. Porsche Stratified-Charge System (SCS engine) [1.27]

The idea of charge stratification can be realized by providing a geometrically divided combustion chamber. Inflammation is initiated in the prechamber, where a richer mixture is present; the main conversion is then introduced into the main combustion space - which contains a lean mixture - by a flame jet. Charge stratification permits a higher lambda value of the total mixture and leads to lower concentrations of the pollutants NO_x and HC.

So far, only a small number of stratified-charge engines with divided combustion chambers have been manufactured in series production. One of the best known examples is the Honda CVCC engine, **Fig. 1.35.**

A prechamber without additional supply of fuel will also affect combustion by creating internal turbulence.

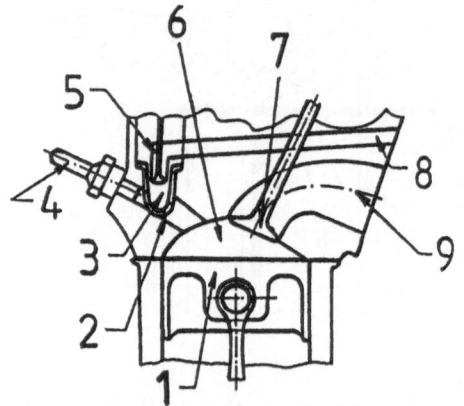

1 Piston
2 Precombustion chamber aperture
3 Precombustion chamber
4 Spark plug
5 Auxiliary inlet valve
6 Main combustion space
7 Main inlet valve
8 Auxiliary intake duct
9 Main intake duct

Fig. 1.35. HONDA CVCC engine [1.29]

1.4.8. Efficiency

Efficiency is a very important factor in judging engine performance. It can be classified as:

- Thermal efficiency η_{th} (based on the idealized process);
- Indicated efficiency η_i (based on the values measured in the cylinder);
- Effective efficiency η_e (based on the values measured at the crankshaft).

If the efficiency is known, specific fuel consumption b of the engine can be computed as follows:

Since the value obtained from multiplication of efficiency η by the lower heating value H_u stands for work per amount of fuel put in, and since specific fuel consumption b is defined as the amount of fuel put in per unit of work, specific fuel consumption b can be expressed as:

$$b = \frac{1}{\eta \cdot H_u} \qquad (1.7)$$

When the efficiencies that have been indicated in sections 1.3.3, 1.3.4, and 1.3.5 for the constant-volume, constant-pressure, and limited-pressure cycles are calculated as a function of the compression ratio ε - which represents a potential means of substantially influencing efficiency - and when presenting the obtained data in a graphical form, **Fig. 1.36** results.

From Fig. 1.36 it is apparent that efficiencies generally increase with increasing compression ratios. Further, starting out from the limited-pressure cycle, efficiency increases toward the constant- volume cycle and decreases toward the constant-pressure cycle.

In order to clearly show the difference between a real gas and an ideal gas in terms of efficiencies achieved, the efficiency values can be calculated according to the constant-volume cycle, for instance, and a mean value for the corresponding temperature range may be chosen for the specific heat ratio. Similarly, real characteristics of a common fuel can be used in the computation, see F.A.F. Schmidt [1.3], for example. **Fig. 1.37** shows computed results of efficiencies according to reference [1.3], for various lambda values versus compression ratio ε. The increase in efficiency with increasing excess air in Fig. 1.37 is caused by changes in specific heats. The decrease in the air deficiency region is caused by the shortage of air. In the range of stoichiometric lambda values, dissociation has the strongest effect.

The effect of the ratios of pressure rise and volume increase on efficiency in a limited-pressure cycle is shown in **Fig. 1.38**, where efficiency has been plotted as a function of the compression ratio for various ratios of volume and pressure rise. It is evident that the ratio of volume rise has a considerable effect on efficiency, whereas the effect of the ratio of pressure rise is less significant.

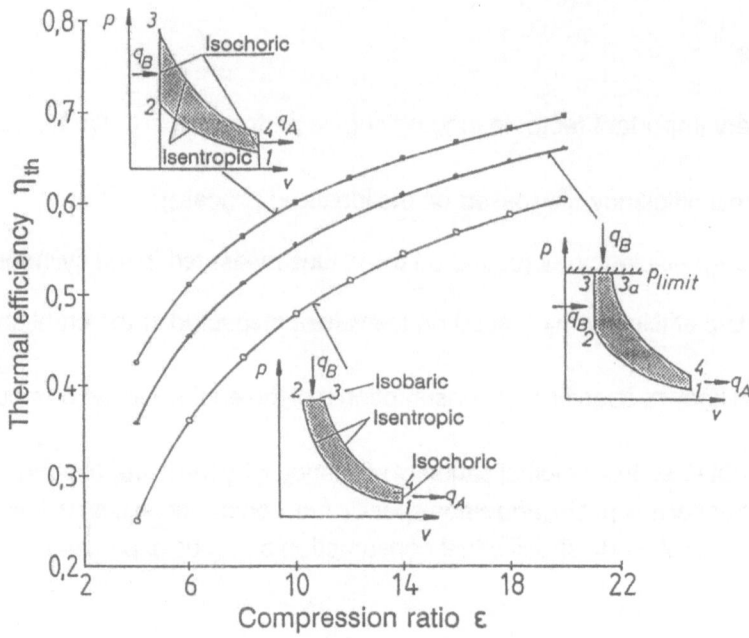

Fig. 1.36. Efficiencies in the constant-volume, constant-pressure, and limited-pressure cycles as a function of compression ratio ε

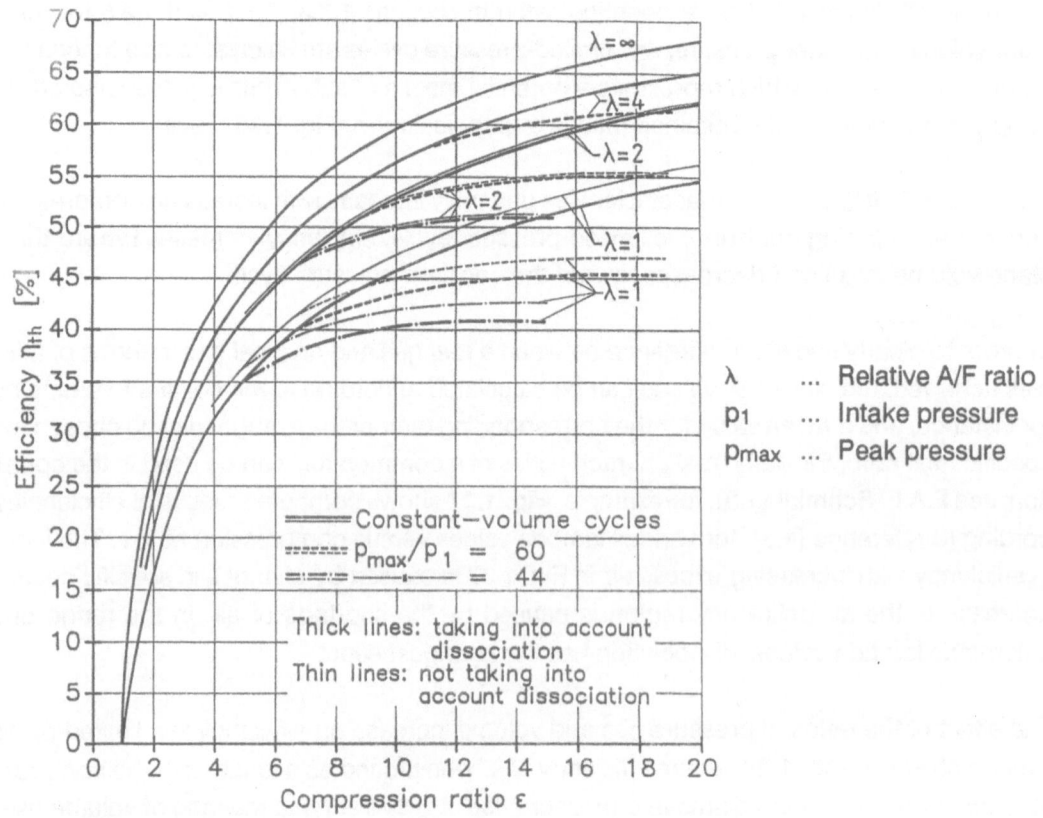

λ	... Relative A/F ratio
p_1	... Intake pressure
p_{max}	... Peak pressure

Fig. 1.37. Efficiencies η_{th} for the constant-volume cycle and limited- pressure cycle including and not including dissociation [1.3]

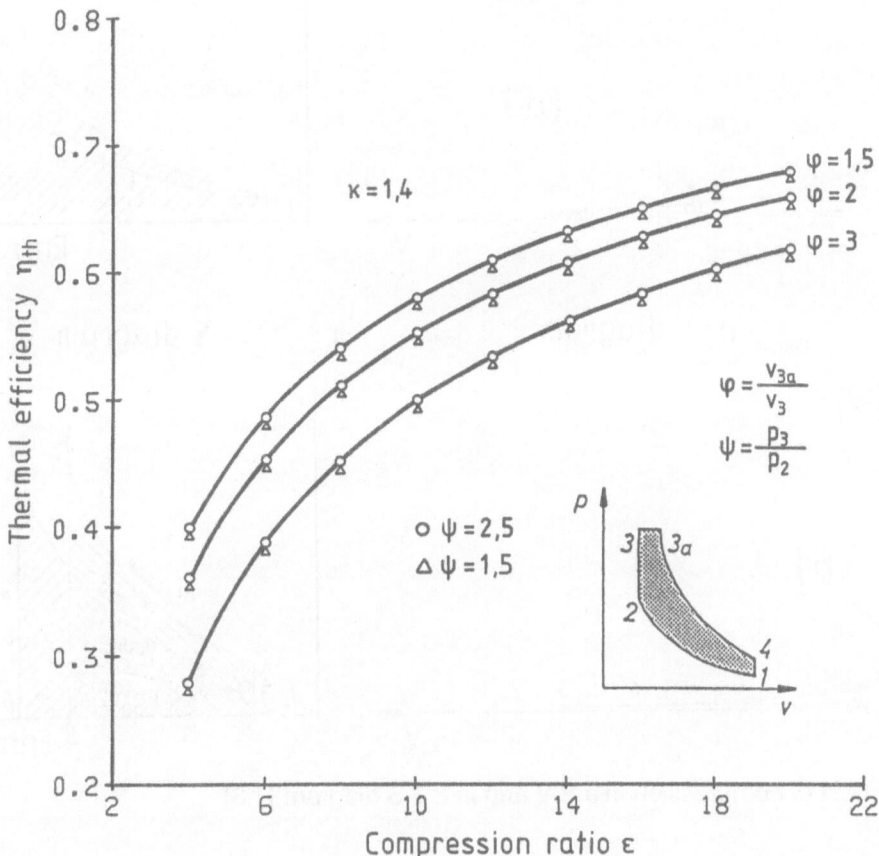

Fig. 1.38. Efficiency of the limited-pressure cycle versus compression ratio for various ratios of pressure rise ψ and volume rise φ

Generally, idealized cycles can be used to illustrate the effects of compression on efficiency and the losses due to incomplete expansion, as shown in the following.

F.A.F. Schmidt [1.3] used the constant-volume cycle as an example to show these effects. **Fig. 1.39** presents the data schematically, **Table 1.1** indicates the amounts of heat added, the amounts of heat dissipated, and the losses due to incomplete expansion.

The heat input has been assumed equal in all cases. A comparison of the areas shows that the amount of heat dissipated is smallest in the high compression cycle.

The loss due to incomplete expansion is caused by the fact that the working gas cannot expand to the point of ambient pressure during the power stroke. In an engine with crankshaft drive, the expansion ratio equals the compression ratio. With increasing compression, expansion also increases, and the loss due to incomplete expansion is reduced.

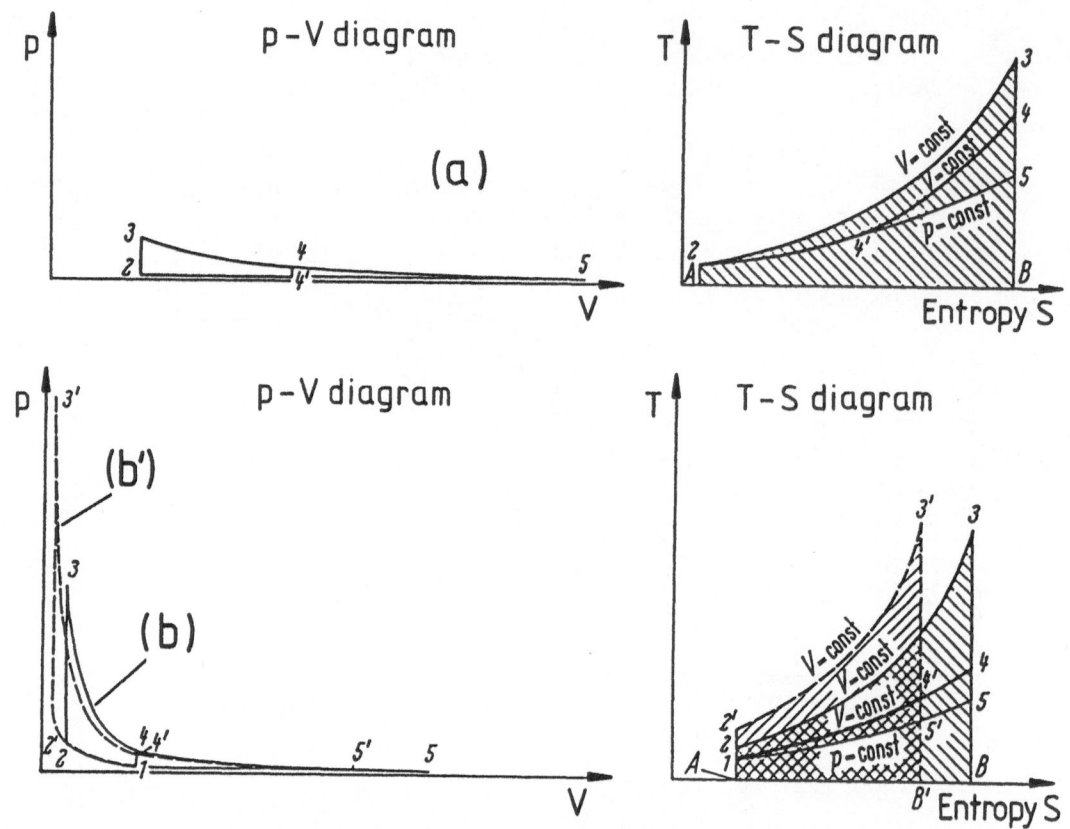

Fig. 1.39. Effect of compression in a P-V and in a T-S diagram [1.3]

Cycle	Heat added	Heat dissipated	Loss by incomplete expansion
no compression (a)	A-2-3-B-A	A-2-4'-4-B-A	4-5-4'-4
low compression (b)	A-2-3-B-A	A-1-4-B-A	4-5-1-4
high compression (b')	A-2'-3'-B'-A	A-1-4'-B'-A	4'-5'-1-4'

Table 1.1. Amounts of heat added and dissipated in accordance with Fig. 1.39 [1.3]

Contrary to the above theoretical considerations, efficiency in real spark-ignition engines increases along with the compression ratio only up to values of approximately 13 to 15.

The quality of the fuel used, combustion chamber geometry, charge motion, charge temperature, cylinder diameter, spark plug position, materials used for cylinder head and block should be included as parameters which affect the optimum compression ratio, since knocking combustion should always be avoided. Excessive compression increases the knocking tendency of the fuel under wide-open throttle conditions - and would require a later point of ignition, which in turn would have an adverse effect on efficiency.

2. Basic Principles of Mixture Formation

2.1. Air

The oxygen required for combustion of fuel in an internal-combustion engine can be supplied to the working cylinder in basically two forms:

- Chemically bound - either as a fuel constituent or as an additional compound capable of releasing oxygen;

- In a free, molecular form (O_2) contained in the intake air.

The advantage of the second alternative - that is, the fact that oxygen need not be carried in the vehicle, thus saving considerable weight -is the reason why the oxygen demand of internal-combustion engines in practice is covered almost exclusively by atmospheric oxygen. The addition of oxygenized constituents, as used today in spark-ignition-engine fuels in portions of up to 10 percent by volume, therefore does not have the purpose of increasing the oxygen content of the combustible mixture, but is used in part on account of their good price per liter (e.g. methanol) and in part on account of their favorable effect on the antiknock quality of fuels for spark-ignition engines (e.g. some ethers). This will be discussed in further detail in chapter "Fuels".

The composition of dry atmospheric air in altitudes up to approx. 15 km is shown in **Table 2.1**.

Gas		Port. by Vol. [%]	Port. by Weight [%]	Mole Mass [g]
Dry air		100	100	28.96
Nitrogen	N_2	78.09	75.52	28.02
Oxygen	O_2	20.95	23.15	32.00
Argon	Ar	0.93	1.28	39.95
Carbon dioxide	CO_2	0.03	0.05	44.01
Neon	Ne	0.0018	0.0012	20.18
Helium	He	0.0005	0.00007	4.003
Krypton	Kr	0.0001	0.0003	83.80
Xenon	Xe	0.000009	0.00004	131.3
Hydrogen	H_2	0.00005	0.000004	2.016

Table 2.1. Composition of dry atmospheric air (mean values taken from references [2.1, 2.2, 2.3, 2.4])

The 20.95% by volume O_2-content shown in the above table starts to decrease at altitudes higher than about 15 km by approx. 0.2 % by volume per km of altitude [2.2]. Up to this altitude, the composition of air indicated in Table 2.1 can be considered as constant.

The density of dry air at a temperature of 0 °C and an air pressure of 760 torr (1013 mbar) is 1.2928 kg.m^{-3}. The air density can be converted to other reference conditions using the equation:

$$\rho_{L,t} = 1{,}2928 \cdot \frac{p_L}{760} \cdot \frac{273}{273 + T_L} \qquad (2.1)$$

$\rho_{L,t}$ [kg.m^{-3}] ..... density of dry air
p_L [Torr]   ..... air pressure
T_L [°C]   ..... air temperature.

Since pressure and temperature vary widely with time and place, national and international reference conditions have been defined.

Air pressure and air temperature at sea level are defined as 760 torr and 15 °C by the "International Standard Atmosphere" (ISA).

Up to an altitude of 11 km, the following relationships apply to pressure and temperature [2.3]:

$$T_L = 15 - 0{,}0065 \cdot h \qquad (2.2)$$

$$p_L = 760 \cdot \left(1 - \frac{0{,}0065}{288} h \right)^{5{,}255} \qquad (2.3)$$

T_L [°C]   ..... air temperature
h [m]   ..... altitude above sea level
p_L [Torr]   ..... air pressure

Table 2.2 shows the variation of air pressure, temperature, and density with altitude based on ISA conditions.

However, the values of the International Standard Atmosphere apply more precisely to the free atmosphere than to earthbound roads. In the latter case, the linear temperature decrease of 6.5 °C per 1000 m altitude resulting from Equation (2.2) and laid down for air at ISA condition appears to be too high.

Therefore, the use of an "isothermal atmosphere" is more adequate for vehicle application. Table 2.3 shows the corresponding values for air pressure and air density as a function of altitude for an isothermal atmosphere of 15 °C.

h [m]	p_L [Torr]	T_L [°C]	ρ_L [kg/m³]
-200	778,2	16,30	1,2492
-100	769,1	15,65	1,2373
0	760,0	15,00	1,2255
100	751,0	14,35	1,2138
200	742,1	13,70	1,2021
500	716,0	11,75	1,1677
1000	674,1	8,50	1,1121
1500	634,2	5,25	1,0583
2000	596,2	2,04	1,0068
3000	525,8	-4,50	0,9094
4000	462,2	-11,00	0,8194
5000	405,1	-17,50	0,7363
6000	353,8	-24,00	0,6598
8000	266,9	-37,00	0,5252
10000	198,2	-50,00	0,4124

Table 2.2. Pressure p_L, temperature T_L, and density ρ_L of air at different altitudes h according to ISA conditions [2.3]

h [m]	p_L [Torr]	ρ_L [kg/m³]	ρ_L [%]
0	760	1,226	100
500	717	1,156	94
1000	675	1,088	89
1500	637	1,026	84
2000	600	0,966	79
2500	566	0,911	74
3000	533	0,859	70
3500	503	0,809	66

Table 2.3. Pressure p_L and density ρ_L of air at different altitudes h at isothermal atmosphere conditions at 15°C.

In addition to pressure and temperature, humidity also slightly changes air density. Based on the lower molecular weight of water vapor (18.02) in comparison to air (28.96), humid air has a lower density than dry air. This has to be taken into account in flow calculations.

$$\rho_{L,f} = \rho_{L,t} \cdot \left(1 - 0{,}378 \cdot \varphi_{rel} \cdot \frac{p_{DS}}{p_L} \right) \tag{2.4}$$

$\rho_{L,f}$ [kg.m^{-3}] ..... density of humid air
$\rho_{L,t}$ [kg.m^{-3}] ..... density of dry air
ρ_{rel} [-] ..... relative humidity
p_{DS} [Pa] ..... saturated vapor pressure of water
p_L [Pa] ..... air pressure

Moreover, the percentage of free oxygen in the air decreases with increasing water vapor pressure.

In order to create a basis of comparison for power evaluations of internal-combustion engines, reference conditions for power computation in different engine categories have been established in different countries.

In the Federal Republic of Germany (DIN 70 020 [2.46]) and in Austria (NORM V 5 003 [2.47]), the reference conditions for air pressure $p_{L,0}$ and air temperature $T_{L,0}$ for the category of motor-vehicle engines have been defined as

$$p_{L,0} = 760 \text{ Torr and } T_{L,0} = 220\,^{\circ}C$$

Engine powers that have been determined under operating conditions that differ from the above reference conditions, are converted, in accordance with the mentioned standards, using the equation

$$N_0 = N \cdot \frac{760}{p_L} \cdot \left(\frac{273 + T_L}{293} \right)^{0,5} \tag{2.5}$$

N_0 [kW] ..... engine power reduced to reference conditions
N [kW] ..... engine power measured at air pressure p_L [torr] and air temperature T_L [C]

As can be seen from the above equation (2.5), humidity has not been taken into account in this conversion.

For stationary engines, marine engines, and engines for rail-borne vehicles, however, different reference conditions are applicable, which in part include humidity. For instance, in the Federal Republic of Germany standard DIN 6270 [2.48] defines the reference conditions for "general-purpose engines" as:

$$p_0 = 736 \text{ Torr}, \ T_0 = 20\,^{\circ}C, \ \varphi_{0,rel} = 60\%$$

$\varphi_{0,rel}$ [%] ..... relative humidity

For comparison of the reference conditions defined for motor-vehicle engines in the Federal Republic of Germany and in Austria with those of other countries, the values of the corresponding reference conditions in the United States of America, stipulated in SAE J 245, are indicated to give just one example:

$$p_0 = 746.2 \text{ Torr (29.38 inch. Hg)}$$
$$T_0 = 29.4 \,^{\circ}\text{C (85 }^{\circ}\text{F)}$$
$$p_D = 9.6 \text{ Torr (0.38 inch. Hg)}$$

p_D [Torr]   ..... partial water vapor pressure

As can be seen from the indicated data, US stipulations provide for an inclusion of humidity when converting engine power to the reference conditions.

Occasionally, there are considerations as to the feasibility of using combustion air with a higher oxygen content than normal air in internal-combustion engines, since a modified composition of the combustion air would influence exhaust emissions and fuel consumption. The development of new techniques such as molecular sieves and membrane technology provides the possibility of producing air with a higher oxygen content than normal air. The effects of oxygen enrichment of the air have been investigated by reference [2.128].

It was found that enriching the air with oxygen leads to an increased formation of nitric oxide, whereas the formation of hydrocarbons is reduced. The causes of these effects are increased temperature, a high temperature gradient, and the higher concentration of oxygen. Owing to the throttle control used in spark-ignition engines, a higher fuel consumption is observed in the part-load range, despite the enrichment with oxygen. Under wide-open throttle conditions, enrichment of the air with oxygen permits a power increase corresponding to the increase in heating value, however in this case problems of knocking, preignition, and backfiring into the intake duct will arise. Therefore, based on the present state of research, an enrichment of the intake air with oxygen in spark-ignition engines in general seems of little value.

2.2. Fuels

2.2.1. Requirements on Fuels

The object of burning fuels (combustibles) in internal-combustion engines is to convert the chemical energy contained in a fuel into output work. This means, foremost, that fuels of high energy content and, in addition, sufficient availability are required.

If internal-combustion engines are used for driving vehicles, the ideal fuel further should meet the requirement of low mass, little volume, and easy, non-hazardous storage. Moreover, it must be capable of reacting with the free molecular oxygen contained in the ambient air, so that carrying an oxidizing agent, in addition to the fuel, can be avoided.

When using fuels in spark-ignition engines, a number of additional fuel properties, which in part even seem contradictory, are required, based on the principle of operation of such engines.

By definition of a spark-ignition engine [2.49], a combustible, largely homogenous mixture must be available before or at the time of ignition by an external source. If the fuel is present in liquid - not gaseous - form, a high readiness to atomize and to evaporate is necessary in order for the fuel to meet the above requirements. Highly volatile fuels seem to be particularly suitable, however they also involve the risk of vapor bubble formation in the mixture-formation mechanism.

Compression of the air-vapor mixture in the cylinder after the intake stroke leads to overheating of the fuel vapors. Since, in a spark- ignition engine, ignition is not supposed to take place before passage of a spark at the spark plug, the auto-ignition temperature of the mixture must be higher than the final compression temperature. Moreover, the auto-ignition temperature should be higher than the temperature attained by the last portion of the mixture to burn (end mixture) when being compressed by the moving flame front, otherwise "knocking" will be the result.

A set of minimum requirements for liquid spark-ignition-engine fuels has been laid down in various countries as standards, in order to facilitate, as far as possible, standardization of the design and worldwide operation of spark-ignition engines. These standards normally comprise maximum and minimum permissible limits, or ranges, for density, knocking resistance, vapor pressure at a given temperature, distillation behavior, distillation and evaporation residues, and contents of sulphur, lead, and other substances hazardous to health (e.g. benzene) as well as for oxygenized constituents (alcohol, ether, etc.).

In the Federal Republic of Germany, standard DIN 51600 [2.50] stipulates minimum requirements for leaded and standard DIN 51607 [2.51] for unleaded spark-ignition-engine fuels. In Austria, the minimum requirements for leaded premium-grade fuels are laid down in NORM C 1103 [2.52], for unleaded premium-grade fuels in NORM EN 228 [2.130]. The reader is referred to the standard specifications for specific values. Especially in view of the increasingly stringent exhaust gas regulations in various countries, efforts are made to specify tolerance ranges in these standards which are as narrow and as uniform as possible throughout the world.

2.2.2. Composition and Structure of Fuels

2.2.2.1. General

The many requirements placed on a spark-ignition-engine fuel greatly limit the wide range of common fuels.

Carbon does have a high energy content (heating value), but it burns very slowly and is not easily prepared.

Hydrogen, on the other hand, has a high heating value and, in addition, burns quite rapidly, but its storage, both in gaseous and liquid form, is not easy - heavy steel gas cylinders, high pressures or low temperatures are required. Research and development on metal-hydride tanks has been in progress for considerable time now, but it is yet to be seen whether this will lead to an improvement of the situation in the future; nevertheless, the stored mass per unit of energy to be transported in the vehicle will still be comparatively high.

Carbon-hydrogen compounds, the so-called hydrocarbons, in comparison, have proved best in meeting the described demands. They are available in a natural form in mineral oil and natural gas, but can also be produced synthetically.

The number of possible hydrocarbon compounds is nearly unlimited, since the tetravalent carbon is able to form straight chains and branched chains as well as a great variety of ring-type structures. The free carbon molecule valences are saturated with monovalent hydrogen atoms.

The lengths and structures of hydrocarbon chains or rings have a decisive influence on the mixture-formation and burning characteristics of fuels. For this reason, efforts are made to engineer HC molecular structures that are suitable for use in spark-ignition engines by subjecting the natural gasoline resulting from crude petroleum distillation to special processing techniques such as cracking, reforming, isomerization, polymerization, hydrogenation, or alkylation. By these procedures, special types of gasoline are obtained which differ from each other in that certain hydrocarbon structures are predominant. In accordance with the technique by which they are produced, these types of gasoline are called, for instance, distilled, cracked, reformed, polymer, isomer, alkylate, pyrolisis, or benzol mixture. A spark-ignition- engine fuel with highly specific properties can be produced by blending different types of gasoline.

If the fuel, after the mentioned processing methods, should still lack the desired properties, or if any further processing should not be justified economically, the gasoline may be "improved" by blending it with nonpetroleum fuels such as alcohols and/or ethers and so-called "fuel additives".

In the following a brief classification and description of pure hydrocarbons and major oxygenized hydrocarbons are given and the most important properties of the different molecular structures are discussed; this is followed by a survey of the most common fuel additives.

2.2.2.2. Pure Hydrocarbons

Depending on their types of bond, hydrocarbons are classified as:

- 1. Chainlike (aliphatic) hydrocarbons;

- 2. Ring-type (cyclic) hydrocarbons.

Each group is subdivided into saturated and unsaturated compounds.

2.2.2.2.1. Chainlike (Aliphatic) Hydrocarbons

They are grouped into:

- Chainlike saturated hydrocarbons: paraffins (alkanes);

- Chainlike unsaturated hydrocarbons:
  Olefins (alkenes);
  Diolefins (dienes);
  Acetylene (alkines).

Paraffins (Alkanes)

These are chainlike hydrocarbons where all free valences of the straight-chain or branched tetra-carbon atoms are saturated with monovalent hydrogen atoms. Since there are no double bonds in the molecular structure, paraffins have the highest possible number of hydrogen atoms.

These compounds therefore have the empirical formula C_nH_{2n+2}, n = 1 to approx. 10

Straight-chain - or better: unbranched - hydrocarbons are called n- paraffins (n stands for normal),

For instance: n-Heptane (C_7H_{16})

Short-chain molecules (i.e. few C atoms) have a strong bond between the atoms, which becomes weaker as the chains grow longer. Therefore, short-chain n-paraffins are characterized by a bad ignitability (high ignition temperature) and a high antiknock quality (e.g. ethane (C_2H_6): RON = 112). These properties are inversed with increasing chain lengths. For instance, the above-shown n-heptane (C_7H_{16}) exhibits a flammability that is already high enough for the substance to be used as the lower reference fuel (RON = MON = 0) in octane ratings of spark-ignition- engine fuels. Hexadecane (cetane) ($C_{16}H_{34}$) is a still better example: due to its high readiness to ignite, it is used as the higher reference fuel (CN = 100) when determining the cetane rating of diesel fuels.

Moreover, density and distillation point increase with larger molecule sizes. At atmospheric temperatures and pressures, the n-paraffins methane (CH_4), ethane (C_2H_6), propane (C_3H_8), and butane (C_4H_{10}) are in a gaseous state and the larger varieties such as pentane (C_5H_{12}), hexane (C_6H_{14}), heptane (C_7H_{16}), octane (C_8H_{18}) etc. in a liquid state. Beyond cetane ($C_{16}H_{34}$) they are solid in their pure form. Normal paraffins are the predominant type of hydrocarbons in mineral oil.

Saturated carbon chains that do not have straight-chain but branched structures are called i-paraffins (iso-paraffins). Iso-paraffins have the same empirical formula as normal paraffins, but they are characterized by a more compact structure. Therefore, iso-paraffins have a lower ignitability or, respectively, a considerably higher knocking resistance.

An example of this is the 2,2,4-trimethylpentane shown below, which, as one of the many possible structures of iso-octane (C_8H_{18}), has a RON of 100 as compared to a RON of -18 obtained by extrapolation for n-octane (C_8H_{18}).

i-Octane (C_8H_{18}) (Trimethylpentane)

Olefins (Alkenes)

These are chainlike unsaturated hydrocarbons (straight-chain or branched) with a double bond between two C atoms. Therefore, they contain two hydrogen atoms less than the respective paraffins (alkanes); based on this fact they are called "unsaturated hydrocarbons".

The empirical formula of olefins is: C_nH_{2n}

E.g. Ethene (Ethylene) (C_2H_4) Butene (Butylene) (C_4H_8)

Olefins have predominantly straight-chain structures; branched structures are not so common. The double bond makes the carbon chain more robust, giving it a better knocking resistance than exhibited by comparable paraffins.

Diolefins (Dienes)

These are chainlike unsaturated hydrocarbons (straight-chain or branched) with two double bonds between C atoms. Therefore, they are still poorer in hydrogen than mono-olefins.

Empirical formula: C_nH_{2n-2}

E.g. Butadiene (C_4H_6)

Acetylene (Alkines)

These are chainlike unsaturated hydrocarbons (straight-chain or branched) with a triple bond between the C atoms.

Empirical formula: C_nH_{2n-2}

E.g.Acetylene C_2H_2

Olefins, diolefins, and acetylene are not contained in most mineral oils but may be formed in the refining processes.

2.2.2.2.2. Ring-Type (Cyclic) Hydrocarbons

They are classified as:

- Ring-type saturated hydrocarbons: Naphthenes (cycloparaffins, cycloalkanes);

- Ring-type unsaturated hydrocarbons: Unsaturated naphthenes (cycloolefins, cyclo-alkenes);

- Aromatics.

Naphthenes (Cycloparaffins)

These are saturated hydrocarbon compounds with the C atoms arranged in a ring. This means they have two hydrogen atoms less than paraffins.

Empirical formula: C_nH_{2n}

E.g.   Cyclohexane (C_6H_{12})

These compounds have a low ignitability and higher knocking resistance than the comparable n-paraffins.

Unsaturated Naphthenes (Cycloolefins, Cycloalkenes)

These are ring-type unsaturated hydrocarbons with one or two double bonds between C atoms.

Empirical formula:

a) C_nH_{2n-2} (1 double bond)

b) C_nH_{2n-4} (2 double bonds)

The natural content of naphthenes in crude oil strongly depends on the origin of the oil.

E.g.   to a) Cyclohexane (C_6H_{10})   to b) Cyclopentadiene (C_5H_6)

Aromatics

Six C atoms are arranged in a ring such that there are three double bonds in addition to three single bonds, which yields the fundamental structure of the aromatic series of hydrocarbons: benzene (C_6H_6).

Benzene (C_6H_6)   Abbreviated formula:

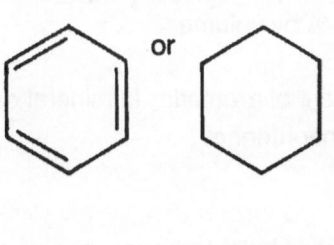

or

By addition (or condensation) of benzene rings, almost any number of aromatic hydrocarbon compounds may be created, however with widely varying properties, depending on the number and arrangement of the rings.

Examples:

Naphthaline ($C_{10}H_8$) Anthrazene ($C_{14}H_{10}$) Phenantrene ($C_{14}H_{10}$) Pyrene ($C_{16}H_{10}$)

Besides added benzene ring assemblies, structures may occur where individual H atoms of the basic substance benzene (C_6H_6) are replaced by alcohol groups, above all by the methyl substituent (CH_3). These "benzene homologs", as with benzene, are expressed by the general empirical formula:

$$C_nH_{2n-6}$$

E.g.   Toluene (C_7H_8)   Xylene (C_8H_{10})

The aromatic series are the hydrocarbons poorest in hydrogen. The compact, symmetrical benzene structure imparts excellent thermal stability or high antiknock characteristics to the aromatic series, which is the reason why aromatics are highly valued as a component in spark-ignition-engine fuels. However, since some aromatic compounds have been proved to be highly toxic or carcinogenic, their use in fuels is limited in some countries. In Austria, for instance, the benzene content in spark-ignition-engine fuels (regular and premium) is limited by legislation to a maximum of 5 % by volume.

The natural content of aromatics in mineral oil also strongly depends on the origin of the oil, as is the case with naphthenes.

2.2.2.3. Oxygenized Hydrocarbons

Oxygenized hydrocarbons are alcohols, phenoles, aldehydes, ketones, and ethers.

For blending with fuels, only alcohols and ethers are of some importance, therefore only these two groups will be discussed in the following section.

2.2.2.3.1. Alcohols (Alkanols)

Alcohols are hydrocarbons which, in addition to carbon and hydrogen atoms, also contain an OH (hydroxyl) group. Depending on the number of OH groups, they are classified as monohydric, dihydric, and trihydric alcohols.

Some of the most common monohydric alcohols are: methanol (CH_3OH), ethanol (C_2H_5OH), propanol (C_3H_7OH), and butanol (C_4H_9OH); an example of the dihydric alcohols is glycol ($C_2H_4(OH)_2$); an important example of the trihydric alcohols is glycerol ($C_3H_5(OH)_3$).

Only the monohydric alcohols - primarily the lower monohydric alcohols such as methanol (CH_3OH) and ethanol (C_2H_5OH) - are used as "alternative fuels" or fuel additives. Higher monohydric alcohols are normally used only as diluents for cost-saving purposes.

Methanol (Methyl Alcohol) (CH_3OH)   Ethanol (Ethyl Alcohol) (C_2H_5OH)

$$H-\underset{\underset{H}{|}}{\overset{\overset{H}{|}}{C}}-OH \qquad H-\underset{\underset{H}{|}}{\overset{\overset{H}{|}}{C}}-\underset{\underset{H}{|}}{\overset{\overset{H}{|}}{C}}-OH$$

Owing to their oxygen content, alcohols have an appreciably lower heating value than pure hydrocarbons - e.g. methanol approximately one-half. However, their air requirement to give a stoichiometric mixture ratio is also lower. Thus, the mixture heating value is similar to that of gasoline. Alcohols readily combine with water and can be mixed with water at any ratio. Their good solubility in water, however, leads to corrosion problems in engines and makes blending with gasoline more problematic. A water content of only less than 1% by volume in alcohol makes stable blends with gasoline impossible, unless a "diluent" (normally a higher alcohol or ether) is used.

Other unfavorable properties of alcohols are their high heats of evaporation at temperatures still in the low range (cold start and warm-up problems) and their tendency - especially of methanol - to corrode plastics (e.g. elastomers). For this reason, Austrian legislation limits the content of methanol in unleaded spark-ignition- engine fuels to a maximum of 3% by volume. Proposed EC regulations envisage the same limit.

The mentioned high heat of evaporation of alcohols and the resultant cooling of the mixture may be of advantage in some cases though, for instance in racing cars.

The most important advantage of alcohols, when used in their pure form as fuels, is their outstanding antiknock quality. For instance, the research octane numbers (RON) of pure methanol, ethanol, propanol, and butanol range from 110 to 118. However, they also exhibit a high "sensitivity" (i.e. difference between RON and MON) [2.5]. The magnitude of this sensitivity strongly depends on the molecular structure of the respective alcohol (position of the OH group). Therefore, before using alcohols in spark-ignition engines, it is an absolute necessity to determine their motor octane numbers (MON) together with the RON.

2.2.2.3.2. Ethers

Ethers are hydrocarbons which, in addition to carbon and hydrogen atoms, also incorporate one oxygen atom.

The merit of ethers for use as spark-ignition-engine fuels - similar to that of alcohols - is their high antiknock quality. The rather expensive production of ethers, however, makes them attractive only for blending with gasoline. Mainly methyl tertiary butyl ether (MTBE), ethyl tertiary butyl ether (ETBE), diisopropyl ether, and - above all in the United States of America - metylisoamyl ether are used for this purpose. (MTBE), for instance, has 115 to 120 RON with approx. 101 MON [2.5].

Methyl tertiary butyl ether (MTBE)

```
              H
              |
          H—C—H
    H         H
    |         |
H—C—O—C—C—H
    |     |   |
    H     H   H
          H—C—H
              |
              H
```

2.2.2.4. Fuel Additives

The purpose of fuel additives is to enhance inherent fuel properties or to provide a given fuel with additional characteristics. The quantities added are less than 1%.

Depending on the specific purpose for which they are used, additives to spark-ignition-engine fuels are categorized as:

- Knock suppressors (antiknock agents);

- Deposit modifiers;

- Antioxidants (oxidation inhibitors);

- Metal deactivators;

- Anticorrosives;

- Deicers;

- Detergents and dispersants.

Knock Suppressors (Antiknock Agents)

Knock suppressors improve the antiknock quality, that means they increase the octane number of a fuel.

There are organometallic and non-metal knock suppressors; the latter are subdivided into organic nitrogen and inorganic oxygen compounds.

In addition, there are so-called lead synergists, which have a synergistic (enhancing) effect on the knock-suppressing characteristic of lead compounds.

In the complete periodic system, only about 15 elements have an antiknock effect similar to that of lead compounds.

Among the organometallic substances, only the lead compounds, e.g. lead alkyls tetraethyl lead (TEL) $Pb(C_2H_5)$ and tetramethyl lead (TML) $Pb(CH_3)$ as well as blends of these two, are used at present. **Fig. 2.1** shows the effect of the lead content in gasoline on its antiknock quality.

So-called "scavengers" are added to the antiknock lead alkyls in order to largely prevent deposits of lead oxides (e.g. PbO, Pb_2O_3 etc.) resulting from the combustion process. The scavengers commonly used are halogens such as ethylene dichloride ($C_2H_4Cl_2$) or ethylene dibromide ($C_2H_4Br_2$), which convert the lead oxides formed during combustion into volatile lead halides or lead oxide halides.

Other metal compounds with antiknock properties are iron and manganese compounds. Iron pentacarbonyl and manganese hexacarbonyl, for example, were used preferably in the past; their use has been abandoned because of their wear-increasing side effects.

Among the non-metal antiknock agents, the mentioned oxygenized hydrocarbons and aromatic amines (e.g. aniline) are suitable to a certain extent.

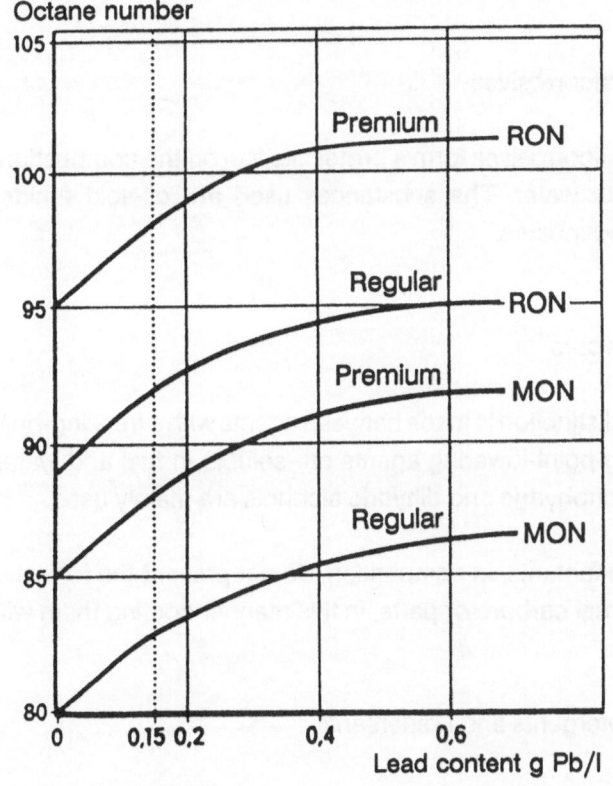

Fig. 2.1. Antiknock quality as a function of lead content [2.6]

Deposit Modifiers

Deposit modifiers are based on phosphorus and boron and are used for modification of combustion chamber deposits, which mainly consist of carbon and additive residues.

Antioxidants (Oxidation Inhibitors)

Antioxidants prevent the formation of resinous, sticky deposits ("gum"), which are formed by olefinic hydrocarbons in the fuel under the influence of atmospheric oxygen. The substances used for this purpose are phenols and aromatic amines.

Metal Deactivators

Metal deactivators neutralize the catalytic effect of metal ions, which are always present in fuels, thus preventing gum formation. Compounds based on salicylaldehyde are suitable substances.

Anticorrosives

Anticorrosives form a protective film on the combustion chamber walls, thus preventing contact with water. The substances used are oil-acid amides, petroleum sulfonates, or petroleum phosphates.

Deicers

A distinction is made between agents with a freezing-point-lowering effect and surfactants. Freezing-point-lowering agents are soluble in fuel and water and lower the freezing point of water. Monohydric and dihydric alcohols are mainly used.

Surfactants, in comparison, do not prevent the formation of ice crystals but they deposit on internal carburetor parts, in this manner coating them with a protective film.

Detergents and Dispersants

Dispersants prevent deposits in the carburetor (e.g. oil vapors vented by the crankcase) by coating the metal surface with a protective layer.

The effect of detergents, on the other hand, is based on peeling-off or soaking existing deposits. In both cases, the mentioned oil-acid amides have proved to be useful agents ("multipurpose additives).

2.2.3. Properties and Technical Characteristics of Fuels

2.2.3.1. Spark-Ignition-Engine Fuels (Gasoline)

Distillation Characteristics, Volatility

As already mentioned in the previous chapter, carburetion fuel is a mixture of a variety of hydrocarbons with widely varying distillation points. Therefore, commercial gasoline does not have a distillation point but, as shown in **Fig. 2.2**, a distillation range, which is approximately 30-200°C.

The distillation curve shown in Fig. 2.2 indicates the volume of fuel evaporated at a given temperature in a standardized distillation vessel. Stipulations regarding the analysis of distillation characteristics are contained in standard DIN 51751 [2.54] for the Federal Republic of Germany and in standard NORM C 1160 [2.55] for Austria.

The trace of the distillation curve is of paramount importance for engine operation. The 10%-point (i.e. 10% of the fuel volume has been evaporated) should be located in the lower temperature range for easy starting of the cold engine. At the same time, it should not be too low, since vigorous vaporization of the fuel contained in the float chamber or fuel pump would result when the hot engine is switched off. When starting the engine again, hot-start problems due to excessive enrichment of the mixture would occur. Moreover, the formation of vapor bubbles in the fuel pump and fuel line may cause a disruption of the fuel supply.

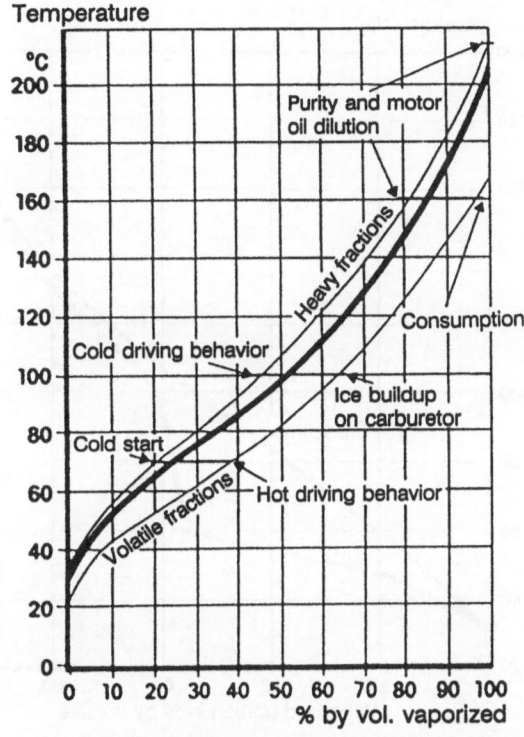

Fig. 2.2. Distillation curve of a commercial fuel and its significance for engine operation [2.6]

The medium range of the distillation curve, especially the 50%-point, is of significance for good transient operation of the cold engine and the tendency of ice formation on the carburetor. It should be located approximately at or slightly above 100 °C for summer fuels and somewhat below 100 °C for winter fuels. In a completely warmed-up engine, the 50%-point is of minor importance.

An excessively high 90%-point should be avoided for the sake of minimizing wall film buildup in the manifold and motor oil dilution.

In accordance with the DIN or, respectively, NORM standards, the final distillation temperature of a standard-conform carburetion fuel (regular and premium) must not exceed 215 °C.

Vapor Pressure

In addition to the distillation curve, the vapor pressure is also a measure of the volatility of a fuel. Since fuels with high vapor pressures at low temperatures have an increased tendency to form vapor bubbles, the permissible vapor pressure at 37.8 °C (Reid vapor pressure) is limited in nearly all countries. In the Federal Republic of Germany and in Austria, the vapor pressures measured by the method proposed by O. Reid and laid down in the US standard ASTM D 323 must not exceed 0.7 bar for summer fuels and 0.9 bar for winter fuels.

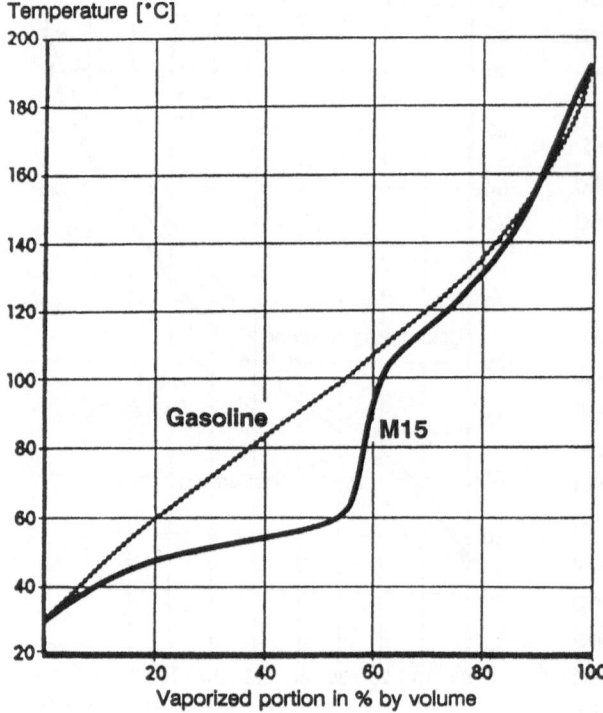

Fig. 2.3. Effect of a 15% admixture of methanol on the distillation characteristics of a spark-ignition-engine fuel [2.6]

When fuels are blended with alcohol, it should be noted that this causes a steep rise in vapor pressure, especially in the lower distillation range, which increases the tendency to form vapor bubbles. **Fig. 2.3** shows the different distillation characteristics of a fuel without any alcohol and after addition of 15% methanol.

Heat of Vaporization

Vaporization of the fuel atomized in the carburetor causes the air-vapor mixture to cool down. Without adding heat, the temperature drop in a stoichiometric mixture is approximately 20 °C for gasoline, 30 °C for benzene, 80 °C for ethanol (ethyl alcohol), and 140 °C for methanol (methyl alcohol). On the one hand, this temperature drop causes higher charging of the cylinder and increased internal cooling of the engine, on the other hand, it may be the reason for ice formation on the carburetor.

The heat of vaporization of commercial gasoline is approx. 380-500 kJ/kg. Ethanol, in comparison, has a value of 904 kJ/kg, methanol even 1110 kJ/kg [2.7].

Density

Owing to the influence of density on volumetric fuel consumption, the fuel standards in most countries specify densities in the range of 0.705 to 0.780 kg/liter [2.134]. Unleaded fuel usually has a higher density than leaded fuel, due to the generally higher portion of aromatics contained in it.

Heating Value

The heating value is a measure of the energy content of a fuel. In practice, a distinction is made between the "high heating value" or specific heating power and the "low heating value" or specific heating value.

The high heating value H_0 is a measure of the heat of combustion, when the water produced in the combustion process condenses and, as a result, heat of condensation is released.

The low heating value H_u expresses the corresponding when the water has vaporized.

Since the water produced by engine combustion always leaves the combustion chamber in a vaporized form, only the low heating value H_u is of importance here. Owing to its somewhat higher density, the specific heating value H_u of premium-grade gasoline is slightly higher than that of regular-grade gasoline. The values usually range from 42,700 to 43,500 kJ/kg. The difference between maximum work to be gained and low heating value amounts to several percent [2.135].

Heating Value of the Mixture

The mixture heating value H_G is the amount of heat released during combustion of 1 normal cubic meter of a fuel-air mixture. The unit of the mixture heating value is $[kJ/m^3]$.

For gasoline and diesel fuel, the fuel volume is negligible compared to the air volume. Thus,

$$H_G = \frac{H_u \cdot \rho_{L,0}}{\lambda \cdot m_{L,stö}} \qquad (2.6)$$

H_G $[kJ.m^{-3}]$ ..... heating value of the mixture
H_u $[kJ.kg^{-1}]$ ..... specific heating value of the liquid fuel
λ $[-]$ ..... relative air-fuel ratio lambda
$\rho_{L,0}$ $[kg.m^{-3}]$ ..... density at standard conditions (0 °C, 760 torr)
$m_{L,stö}$ $[-]$ ..... stoichiometric air requirement (kg air/kg fuel) - refer to page 65

For gaseous fuels (permanent gas, liquefied petroleum gas), the gas volume has to be taken into account. This leads to the expression:

$$H_G = \frac{H_u}{1 + \lambda \cdot m_{L,stö} \cdot (\rho_{K,0}/\rho_{L,0})} \qquad (2.7)$$

H_G $[kJ.m^{-3}]$ ..... heating value of the mixture
H_u $[kJ.m^{-3}]$ ..... specific heating value of the gaseous fuel at standard conditions
$\rho_{K,0}$ $[kg.m^{-3}]$ ..... gaseous fuel at standard conditions

When comparing different fuels using the same engine, the changes in full-load engine power to be expected can be determined from the ratios of the mixture heating values.

Ignition Temperature

The ignition temperature is the temperature at which the fuel auto-ignites when coming into contact with air and continues to burn. It is not a physico-chemical constant of a given fuel, but it depends on the prevailing conditions and, additionally, is strongly influenced by foreign particles. For spark-ignition-engine fuels it lies in the range of 300 to 400 °C [2.7].

Flash Point

The flash point is the lowest fuel temperature at which, in a defined vessel, a flammable mixture of air and fuel vapor is formed at the fuel surface which, when being approached by a source of ignition, flashes with a bright flame that does not continue to burn but is immediately extinguished. The flash point is important, above all, with regard to storage and transport regulations for fuels and determines the hazard class assigned to a fuel.

Spark-ignition-engine fuels have flash points below 21°C, which puts them in hazard class A I.

Diesel fuels have flash points above 55°C, which assigns them to hazard class A III. It should be noted that, due to the general practice of blending diesel fuel with gasoline during winter time, the diesel flash point is considerably lowered. Already a few percent of spark-ignition-engine fuel in diesel fuel will lower the flash point to near room temperature, which then advances the diesel fuel to hazard class A I.

Antiknock Quality

The octane rating expresses the antiknock quality of a spark-ignition- engine fuel, a high octane number representing a high resistance to knocking. Two different procedures are used internationally to determine the octane number of a fuel: the research method according to ASTM D 2699 for determination of the research octane number (RON) and the motor method according to ASTM D 2700 for determination of the motor octane number (MON).

Besides these, there are two other ratings, the street octane number (SON) and the front octane number (FON).

For the RON, MON, and FON ratings, the fuel is tested in a standardized single-cylinder engine, the so-called CFR test engine (CFR standing for Cooperative Fuel Research Committee, USA). The spark-ignition-engine fuel to be tested drives the single-cylinder engine whose compression ratio is varied during engine operation by lifting and lowering the cylinder, until a specific knocking intensity has been adjusted.

The knocking intensity is measured with a Philips detonation meter. Then, a reference fuel consisting of n-heptane (C_7H_{16}) (RON = MON = 0) and iso-octane (C_8H_{18}) (RON = MON = 100) is tuned to the same knocking characteristic; the mixture ratio being required for this (portion of iso-octane in % by volume) represents the octane number of the tested fuel. Octane ratings for numbers above 100 are carried out, in accordance with standard DIN 51 788 (ASTM D 1656), by comparing the test fuel with a mixture of iso-octane and tetraethyl lead, a certain octane number being assigned to the content of tetraethyl lead.

The differences in the above-mentioned test methods originate from the different operating conditions of the test engine.

The motor method differs from the research method in preheating of the mixture, higher engine speed, and variable ignition timing adjustment, thus subjecting the test fuel to higher thermal loads. The MON values therefore are somewhat lower - by approx. 10 octane numbers (see Fig. 2.1) - than the RON values.

The motoring knock behavior is evaluated using the "Modified Union Town Method", the corresponding octane number being called street octane number (SON). Generally, the SON is close to the RON, but it is very dependent on engine design.

The SON is determined in vehicle road testing: The vehicle is run in high gear from a rolling idle, with wide-open throttle acceleration, approximately between 25 and 95 km/hour, with ignition timing being chosen so that knocking becomes just audible. Reference fuels with a known antiknock rating are used for rating the test fuel.

The front octane number (FON) corresponds with the RON of those fuel fractions that can be overdistilled up to 100°C, i.e. the volatile fuel fractions. It is an indication of the fuel knocking characteristics during acceleration, since the liquid, heavier fuel fractions partially fall out on the cold manifold walls (fuel segregation) during mixture delivery to the cylinders.

The numerous possibilities of increasing the antiknock quality of spark-ignition-engine fuels (processing techniques of natural gasoline, addition of ether or knock suppressors) have already been described in Section 2.2.2. Maximum permissible lead contents in fuels are stipulated in nearly all countries, the portions of benzene and oxygenized compounds are limited in some countries. The higher minimum octane ratings required by fuel standards for premium-grade fuels are achieved - if the permissible lead levels have already been attained - by addition of a higher portion of aromatics (e.g. toluene, xylene) in comparison to regular-grade gasoline.

Regarding the minimum octane ratings for spark-ignition-engine fuels stipulated in individual countries, the reader is referred to the respective standard sheets.

If diesel fuel is added to gasoline, as it may happen inadvertently during filling at the gas station, the antiknock quality of the mixture drops drastically, as **Fig. 2.4** illustrates.

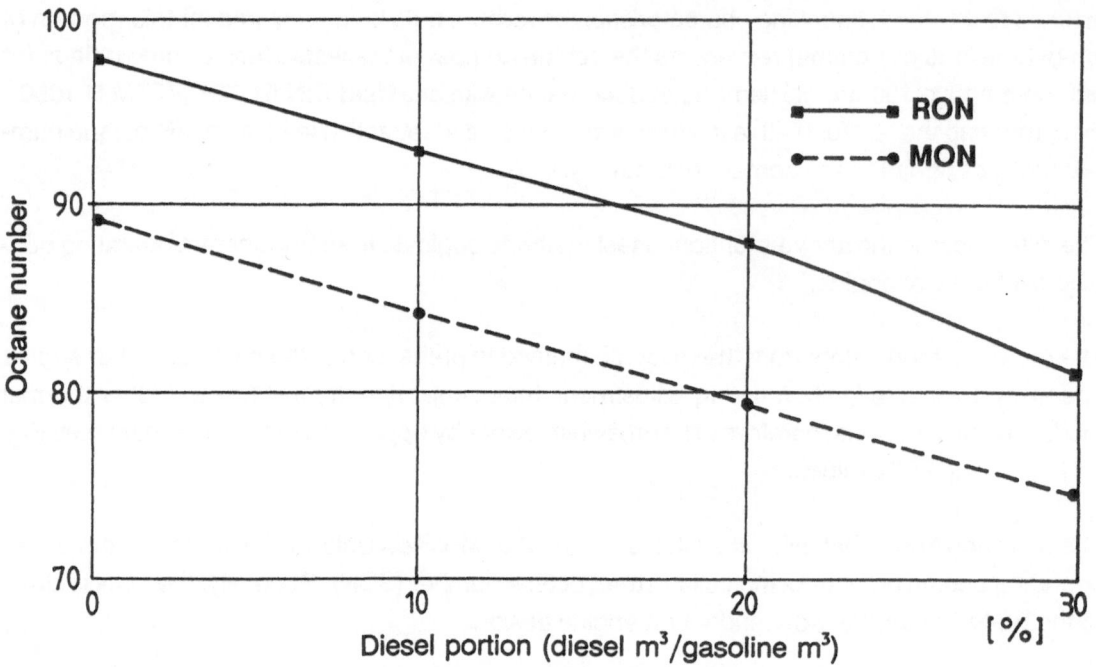

Fig. 2.4. Effect of diesel fuel admixture to premium gasoline on the antiknock quality of the blend

Purity, Combustion Residues

Spark-ignition-engine fuels must be free from solid impurities which may clog, or cause excessive wear of lines, pumps, or nozzles. Moreover, fuels must not produce residues (e.g. resins), which is important in preventing the formation of deposits in the induction system. Evaporation residues are determined according to DIN EN 5 in the Federal Republic of Germany and according to NORM EN 5 in Austria, the maximum permissible value being 5 mg/100 ml of fuel in both standards.

Similarly, the sulphur content in spark-ignition-engine fuels should be as low as possible, in order to minimize the SO_2 content in exhaust emissions, to prevent corrosion damage, and to avoid interference with the action of knock suppressors.

In practice, the sulphur contents usually remain far below the maximum permissible limit of 0.1 % by mass stipulated in the Federal Republic of Germany and Austria; in most cases they are lower than 0.01%.

2.2.3.2. Permanent Gas

Gases that can be liquified only at very low temperatures or high pressures are called permanent gases. The most important permanent fuel gases are methane (CH_4), ethane (C_2H_6), and hydrogen (H_2). In combination with other permanent gases such as nitrogen (N_2), or carbon dioxide

(CO_2), they result in the permanent gas fuels commonly used in combustion engines, e.g. natural gas, town gas (coal gas), or biogas.

Natural gas mainly consists of methane (CH_4) (approx. 80-90%); nitrogen (N_2)and ethane (C_2H_6) account for the remainder [2.9, 2.10]. Owing to its high methane portion, natural gas excels by its high antiknock quality (RON 130) and low pollutant and smoke emissions. The high lean misfire limit of methane (maximum lambda value $\lambda \approx 2.0$) permits a very lean operation, which implies low pollutant emissions and good fuel economy. Moreover, the gaseous state of the fuel ensures good mixture formation at any time, however it makes storage in the vehicle more difficult (normally, steel cylinders with 150-200 bar pressure are used). Owing to the lower mixture heating value when using natural gas, as compared to gasoline, the maximum power rating of a gasoline engine operated with natural gas is about 10-15% less.

Natural gas is available all over the world; the remaining natural deposits are estimated to be about the same order of magnitude as the remaining oil deposits [2.12].

Town gas (coal gas) consists of approximately 55% hydrogen (H_2) and 25% methane (CH_4), the remainder being mainly carbon dioxide (CO_2) and nitrogen (N_2) [2.9, 2.11]. The high H_2 portion in town gas effects an appreciable loss in antiknock quality in comparison to natural gas, however, at the same time, the lean misfire limit is extended to lambda values of about $\lambda = 5$. Also in this case, engine operation is characterized by low pollutant emissions and good fuel economy.

The mixture heating value is approximately 10% less for town gas than for natural gas. Therefore, a loss in maximum power by about the same percentage will have to be accepted when operating the same engine with town gas.

Biogas consists of approximately 60% methane and about 40% carbon dioxide (CO_2) [2.10]. Biogas has properties similar to those of natural gas, but its availability is confined to specific regions. Low combustion temperatures and an extended lean misfire limit ensure low pollutant emissions. The mixture heating value for biogas is also 10% less than that for natural gas.

To gain further knowledge on the internal-combustion process in a biogas-operated pilot injection engine, thermodynamic studies on a single-cylinder experimental engine were carried out in comparison to a natural-gas-operated pilot injection engine and to a normal diesel engine. The high inert gas portion in biogas, mainly carbon dioxide, causes a longer ignition delay in comparison to natural gas and a higher degree of incomplete burning of the bulk gas-air mixture, which implies a higher specific fuel consumption [2.16].

2.2.3.3. Liquified Petroleum Gas

Liquified petroleum gas (LGP) is a mixture of propane (C_3H_8) and butane (C_4H_{10}). It is a by-product of oil or gas mining and refinery processes and therefore has a limited availability.

Liquified petroleum gas, as opposed to permanent gases, can be liquified at ambient temperatures and at pressures as low as about 20 bar, which makes storage in a vehicle less problematic. Optimum compression ratios and optimum ignition points comparable to those of gasoline [2.12] permit an alternate operation of LPG and gasoline, if an air-gas mixer is provided. However, when switching over to gas operation a power loss of up to 15 percent (lower mixture heating value) has to be accepted. Extended misfire limits of LPG ($\lambda = 0.4 - 2.0$) and good mixture formation allow for lean engine operation and thus good fuel economy and low pollutant emissions. The antiknock quality of LGP is somewhat higher than that of premium-grade gasoline, especially in the lower engine-speed range. Depending on the mixture ratio of propane to butane, the research octane number is about 100 or slightly higher, the motor octane number is approx. 95 [2.10, 2.12, 2.13].

It is impossible to give technical data of general validity for liquified petroleum gas, since the composition of LPG greatly differs from country to country and there are no standardized guidelines. To indicate a few examples, properties of various gases are given in **Table 2.4** and compositions of LPG in different countries are listed in **Table 2.5**.

	Propane C_3H_8	Propene C_3H_6	n-Butane C_4H_{10}	i-Butane C_4H_{10}	LPG-standard in discussion	Premium gasoline, winter DIN 51600
Reid vapor pressure at 37,8 °C [bar]	13,0		3,6	5,6		0,6-0,9
Distillation temperature at 1013 mbar [°C]	-42	-48	-0,5	-12		
Vapor pressure at 70 °C [bar]	26	30	8,3	12	max. 31 min. 15	
Liquid density at 15 °C [kg.m^{-3}]	507	523	585	563		730-780
Gas density at 0 °C, 1013 bar [kg.m^{-3}]	2,01	1,91	2,71	2,70		
Heating value H_u [MJ.kg^{-1}]	46,35	46,06	45,72	45,60		~43,5
Relative heating value 15 °C, liquid [%]	72		81			100
Mixture heating value [MJ.m^{-3}]	3,67	3,85	3,71	3,70		~3,7
RON	111	102	94	100		>98
MON	97	85	89	98		>88

Table 2.4. Properties of various liquified petroleum gases [2.129]

Country	Ratio propane/butane	
	Summer	Winter
A	20/80	Up to 80/20
CH	Predominantly propane	
D	Predominantly propane	
DK	50/50	Up to 70/30
F	40/60	
GB	Predominantly propane	
I	20/80	
NL	30/70	Up to 70/30
USA	Propane	

Table 2.5. Compositions of LPG [2.129]

2.3. Stoichiometric Mixture Ratio; Relative Air/Fuel Ratio Lambda

The combustion of fuel in an engine is a process of oxidation during which the combustible fuel constituents, carbon and hydrogen, react with oxygen. During this process, the chemical energy bound in the fuel is converted into heat.

The minimum air mass that is necessary to completely combust 1 kg of a given fuel can be computed on the basis of the chemical composition of the fuel. The mass of oxygen possibly bound in the fuel has to be taken into account in this calculation: it will reduce the minimum air requirement.

If, starting from a spark-ignition-engine fuel of the general formula $C_xH_yO_z$, the fuel is then numerically reduced to C_1, a fuel of the formula $CH_\psi O_\xi$ is obtained.

Thus,

$\psi = y/x$ [-] ..... hydrogen/carbon atom ratio in fuel $C_xH_yO_z$

$\xi = z/x$ [-] ..... oxygen/carbon atom ratio in fuel $C_xH_yO_z$

1 mole (1 mole corresponding to $6.023.10^{23}$ elementary units, which may be molecules, atoms, or ions) of fuel $CH_\psi O_\xi$ requires $(1 + \psi/4 - \xi/2)$ moles of O_2 in order to oxidize completely to carbon dioxide CO_2 and water H_2O.

Thus,

$$1\ CH_\psi O_\xi + (1 + \psi/4 - \xi/2)\ O_2 \dashrightarrow 1\ CO_2 + (\psi/2)\ H_2O$$

Since, according to Table 2.1, dry atmospheric air contains only 20.95% by volume of oxygen, are required to completely combust 1 mole of $CH_\psi O_\xi$.

$$(100/20{,}95).(1 + \psi/4 - \xi/2) = 4{,}773\ (1 + \psi/4 - \xi/2)\ \text{moles of air}$$

When multiplying the above number of moles by the corresponding mole weights of air and fuel,

Weight of 1 mole of air: 28.96

Weight of 1 mole of fuel: $12{,}01 + 1{,}008.\psi + 16.\xi$

and when forming the ratio of the resulting products, the air/fuel mass ratio required for complete combustion is obtained:

$$\left(\frac{m_L}{m_K}\right)_{st\ddot{o}} = \frac{28{,}96 \cdot 4{,}773 \cdot (1 + \psi/4 - \xi/2)}{12{,}01 + 1{,}008 \cdot \psi + 16 \cdot \xi} \qquad (2.8)$$

$(m_L/m_K)_{st\ddot{o}}$ [-] ..... stoichiometric air/fuel mass ratio or stoichiometric mixture ratio

The fuel atom ratios ψ and ξ included in Equation (2.8) are not known in most cases, but can be easily computed on the basis of the known elementary analysis (mass analysis). Thus,

$$\psi = \frac{y}{x} = \frac{12{,}01}{1{,}008} \cdot \left(\frac{h}{c}\right)_{mass} \qquad (2.9)$$

$$\xi = \frac{z}{x} = \frac{12{,}01}{16} \cdot \left(\frac{o}{c}\right)_{mass} \qquad (2.10)$$

$(h/c)_{mass}$ [-] ..... hydrogen/carbon mass ratio

$(o/c)_{mass}$ [-] ..... oxygen/carbon mass ratio.

For a common fuel with a c/h ratio of 86/14, without any oxygen contained in the fuel, this yields a chemically correct mixture ratio of $(m_L/m_K)_{st\ddot{o}} \approx 14.8$

The combustion of carbon can be expressed as:

$$C \quad + \quad O_2 \quad = \quad CO_2 + 395{,}0 \text{ MJ (at 20 }^\circ C)$$
$$12 \text{ kg} + 32 \text{ kg} \quad = \quad 44 \text{ kg}$$
$$1 \text{ kg} \quad + 2{,}667 \text{ kg} = 3{,}667 \text{ kg}$$

The combustion of hydrogen can be expressed as:

$$2H_2 \quad + \quad O_2 \quad = \quad 2H_2O + 484{,}4 \text{ MJ (at 20 }^\circ C)$$
$$4{,}03 \text{ kg} + 32 \text{ kg} \quad = 36{,}03 \text{ kg}$$
$$1 \text{ kg} \quad + 7{,}94 \text{ kg} = 8{,}94 \text{ kg}$$

The so-called relative air-fuel ratio λ is the ratio of the amount of air m_L actually supplied to the engine per amount of fuel m_K to the amount of air m_K theoretically required to completely combust fuel quantity $m_{L,st\ddot{o}}$.

$$\lambda = \frac{(m_L/m_K)}{(m_L/m_K)_{st\ddot{o}}} = \frac{(m_L/m_K)}{(m_{L,st\ddot{o}}/m_K)} = \frac{m_L}{m_{L,st\ddot{o}}} \qquad (2.11)$$

λ [-]   ..... relative air-fuel ratio (also called excess-air factor)

A relative air-fuel ratio $\lambda > 1$ therefore means excess air ("lean" engine operation), a relative air-fuel ratio $\lambda < 1$ means deficiency of air ("rich" engine operation).

2.4. Fuel Management

In a spark-ignition engine, fuel management comprises four major processes:

- Metering (fuel quantity, mixture composition);

- Mixture preparation (atomization, vaporization, blending);

- Mixture delivery;

- Mixture distribution.

To the greater part, metering and preparation take place in the mixture formation unit, whereas the intake manifold - in addition to further preparation (vaporizing, blending) - has the function of transporting and distributing the mixture.

In this section, the fundamental principles of fuel metering will be discussed. Mainly the fuel-metering requirements of spark-ignition engines will be treated here, whereas the possibilities of realization and practical design of metering systems will be discussed in Chapter "Types of Mixture Formation Systems".

Fuel metering mainly comprises two tasks:

- Metering of the fuel quantity in accordance with the required power output;

- Control of the mixture composition in accordance with the relative air/fuel ratio λ desired for a given load point.

2.4.1. Mixture Quantity

The metered quantity depends on the power output desired by the driver.

As is evident from the air-mass-flow-rate map of a series-type 2- liter spark-ignition engine plotted in **Fig. 2.5**, a ratio of approx. 1 : 30 between minimum and maximum flow rate must be covered by the throttle. Similarly, this applies to the maximum ratio of mixture flow rates in carbureted engines.

Most spark-ignition engines use throttle valves for throttling; other throttling devices such as slides are found only in small-displacement motorcycle engines and in some carbureted engines of special design.

The main advantage of the throttle valve is its simplicity with regard to seating, actuation, and sealing, its main disadvantage being its great influence on mixture distribution in induction systems with single-point mixture formation.

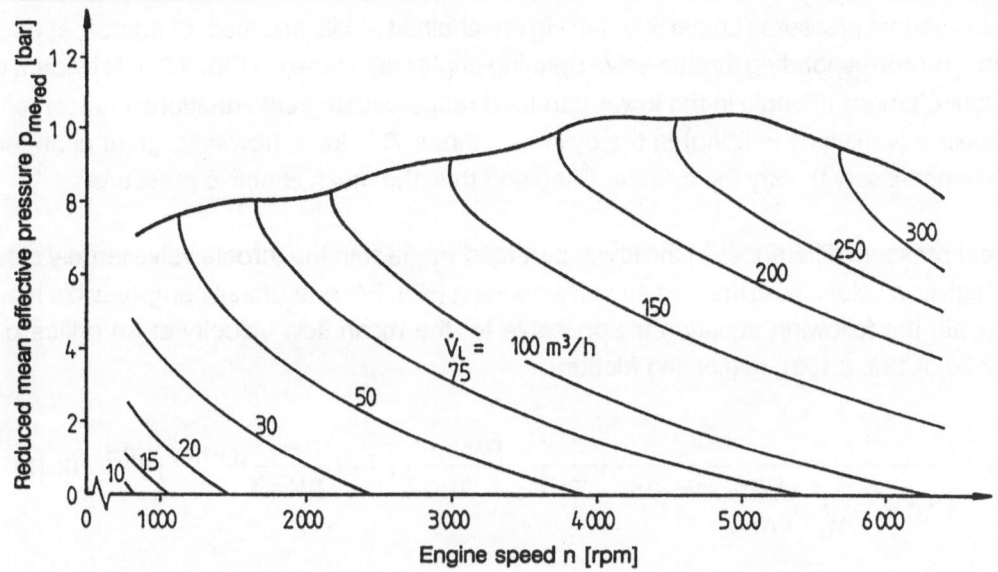

Fig. 2.5. Lines of constant air flow $\dot{V}_L$ in the characteristic map of a 2.0-liter fuel-injected engine.

The function of the throttle valve is to reduce the pressure in the intake manifold under part-load conditions to such an extent that the cylinders, which induct a constant volume of mixture per cycle, receive a charge filling reduced in mass in correspondence with the desired part-load point. As shown in **Fig. 2.6** in a manifold depresssion map of a 2-liter spark-ignition engine,

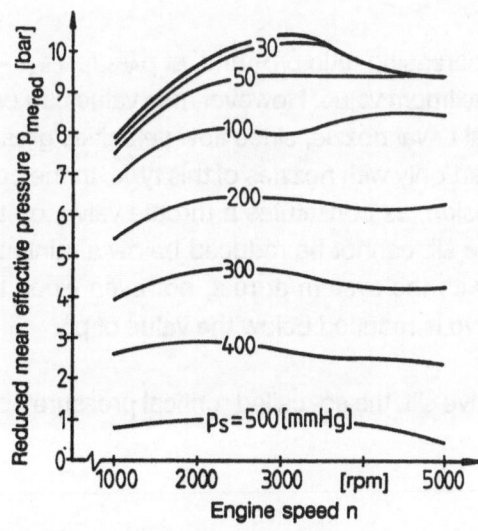

Fig. 2.6. Lines of constant intake manifold depression p_S in the characteristic map of a 2.0-liter carbureted engine [2.20]

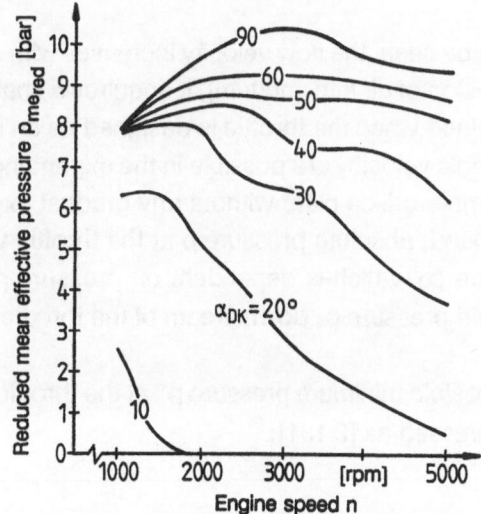

Fig. 2.7. Lines of constant throttle valve opening angles α_{DK} in the characteristic map of a 2.0-liter carbureted engine [2.20]

manifold vacuum pressures above 500 mm Hg are attained at idle and also, of course, at engine overrun. The corresponding throttle-valve opening angles are shown in **Fig. 2.7**. It is evident that only slight changes in angle in the lower part-load range cause great variations in mean effective pressure (variations in filling) in the cylinder. At near full- load, however, great changes in angle are necessary to vary the cylinder filling and thus the mean effective pressure.

The great pressure differences in the lower part-load range with the throttle valve largely closed cause high flow velocities in the throttle-valve opening (slit). For one-phase compressible media, such as air, the following equation is applicable for the mean flow velocity at an orifice plate [2.18, 2.20, 2.132, 2.133], neglecting friction:

$$v_m = \frac{1}{[\,1 - \mu^2(\frac{A}{A_1})^2\,(\frac{p_{Abs.}}{p_{Abs.,1}})^{2/\kappa}\,]^{0,5}} \cdot \{\,\frac{2\cdot\kappa}{\kappa - 1}\cdot\frac{p_{Abs.,1}}{\rho_1}\cdot[\,1 - (\frac{p_{Abs.}}{p_{Abs.,1}})^{(\kappa-1)/\kappa}\,]\,\}^{0,5} \quad (2.12)$$

v_m	..... mean flow velocity in the throttle-valve slit
A	..... free area of the throttle-valve slit
A_1	..... free duct area upstream of the throttle valve
μ	..... constriction index, which includes constriction of the air due to the abrupt contraction of the duct
p_{Abs}	..... absolute pressure in the throttle-valve slit
$p_{Abs.,\,1}$	..... absolute pressure in the duct upstream of the throttle valve
ρ_1	..... air density in the duct upstream of the throttle valve
κ	..... specific heat ratio, for air: $\kappa \approx 1.4$

As can be seen, the flow velocity increases with a decreasing ratio p/p_1 and, at $p_{Abs.}/p_{Abs.,1} = 0$ (expansion of air into vacuum), it acquires a final maximum value. However, this value can only be attained when the throttle is designed as an ideal Laval nozzle, since flow velocities greater than sonic velocitiy are possible in the expanding part only with nozzles of this type. In the case of a simple orifice plate without any gradual expansion, as constitutes a throttle valve, on the other hand, absolute pressure p at the throttle-valve slit cannot be reduced below a minimum pressure p_0, which is dependent on pressure p^* with the system at rest, not even when the manifold pressure p_s downstream of the throttle valve is reduced below the value of p^*.

The possible minimum pressure p^* at the throttle-valve slit, the so- called "critical pressure", can be expressed as [2.131]:

$$p^* = p_0 \cdot (\frac{2}{\kappa + 1})^{\kappa/(\kappa-1)} \quad\quad\quad (2.13)$$

p^*	..... critical pressure
p_0	..... pressure with the system at rest (ambient pressure)

The flow velocity v* that occurs in the slit at this pressure always equals sonic velocity c* [2.131]:

$$v^* = (\frac{2 \cdot \kappa}{\kappa + 1} \cdot \frac{p_0}{\rho_0})^{0,5} = c_0 \cdot (\frac{2}{\kappa + 1})^{0,5} = c^* \qquad (2.14)$$

c*.....sonic velocity in the throttle-valve slit
c₀.....sonic velocity with the system at rest

For air and other two-atom gases κ can be set 1.4, which, after substituting in Equation (2.13), yields a cricital pressure p*

$$p^* = 0,53 \cdot p_0 \qquad\qquad\qquad\qquad\qquad (2.15)$$

Since conventional spark-ignition engines have absolute intake manifold pressures p_s of approx. $0.30 \cdot p_0$ to $0.34 \cdot p_0$ at idle (also see Fig. 2.6) - i.e. pressures far below the critical pressure of $0.53 \cdot p_0$ - sonic velocity actually always occur in the throttle-valve slit at idle and in the lower part-load range. As discussed in detail in Chapter 2.5, this is the reason for excellent fuel atomization in the part-load range in virtually all single-point mixture formation systems.

As mentioned at the beginning of this chapter, metering of the mixture quantity is governed by the power output desired by the driver. Under load operating conditions, metering is controlled by the driver (accelerator), whereas control of the minimum fuel quantity required for maintaining smooth idle operation is handled by the mixture- formation system.

2.4.2. Mixture Composition

For each speed and load point of an engine there is an optimum air-to-fuel ratio of the fresh charge, which depends on the engine operating condition (cold or operating temperature). This ratio depends on the priority requirement in a given operating range.

The priority requirement in case of wide-open throttle operation generally is high power output, whereas good fuel economy and compliance with the pertinent emission standards for the respective vehicle category are the priority requirements in the part-load operating range.

Meeting the requirement of high power output at full-load calls for a mixture composition with a relative air-fuel ratio λ slightly below stoichiometry (λ ≈ 0.9 to 1.0), since such a mixture is characterized by a maximum burn rate and minimum ignition delay. However, as can be seen in **Fig. 2.8**, engine operation in this range implies high specific fuel consumption and high CO and HC emissions.

Meeting the above-mentioned requirements in the part-load range, on the other hand, calls for a considerably leaner mixture. As shown in Fig. 2.8, reducing the richness of the mixture to

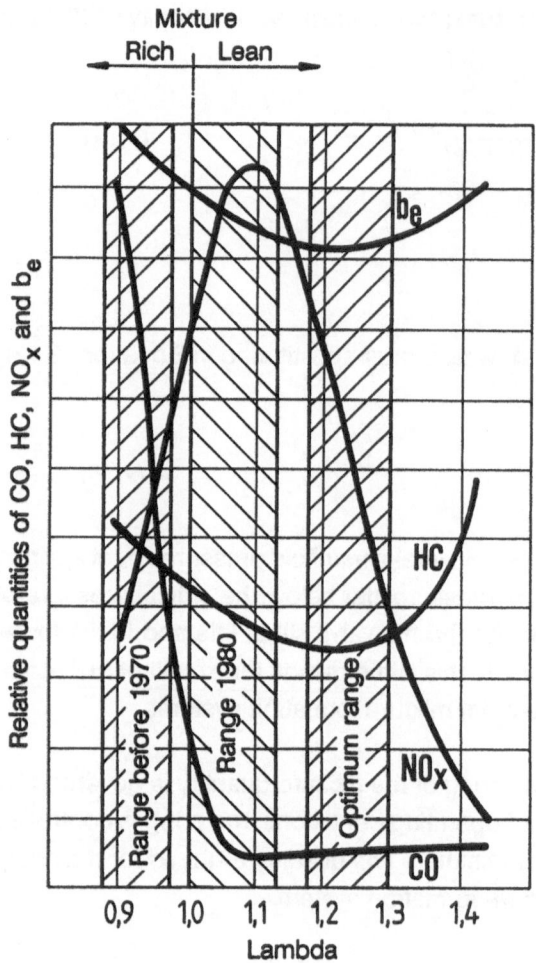

Fig. 2.8. Effect of the relative air-fuel ratio λ on CO, HC, and NOx emissions and on fuel consumption bₑ of a spark-ignition engine [2.14]

lambda values of $\lambda > 1$ leads to an appreciable reduction in fuel consumption and CO as well as HC emissions. The emission standards stipulated in Europe at the beginning of the 70's - which first only limited CO and HC emissions of passenger cars - in combination with the oil shortage experienced for the first time during the same period, lead to a change in part-load mixture tuning to relative air-fuel ratios of $\lambda = 1.0$ to 1.1, as illustrated in Fig. 2.8.

When, additionally, NOx emission limits for passenger cars were introduced at the end of the 1970's, a further mixture lean-out was envisaged besides other NOx-reducing measures, based on the fact that NOx emissions reach peak values in the relative air-fuel ratio range of $\lambda = 1.0$ to 1.1

However, such further, substantial cut in richness is possible only if it is accompanied by comprehensive design measures concerning the combustion chamber, the ignition system, etc., since in conventional engines a dramatic rise in HC emissions due to misfires and delayed combustion would ensue. A major goal in engine development therefore is to design combustion chambers and high-performance ignition systems which facilitate trouble-free engine operation at relative air-fuel ratios of $\lambda > 1.2$. This could lead to a further reduction in fuel consumption as compared to the engine designs used presently for reasons of exhaust-gas treatment based on lambda closed-loop control and three-way catalyst converter - under the condition, of course, that comparable exhaust emission values can be attained.

If the optimum lambda values of a present-design spark-ignition engine, at operating temperature, without lambda closed-loop control ($\lambda = 1.0$) are plotted versus engine speed and load, a three-dimensional area, as shown in **Fig. 2.9**, is obtained. This area in space thus represents the steady-state metering requirements that an engine of this type, at operating temperature, places on the mixture formation unit.

Idle enrichment of the mixture, as shown in Fig. 2.9, can be explained by the low pressure and high portion of residual gas in the cylinder.

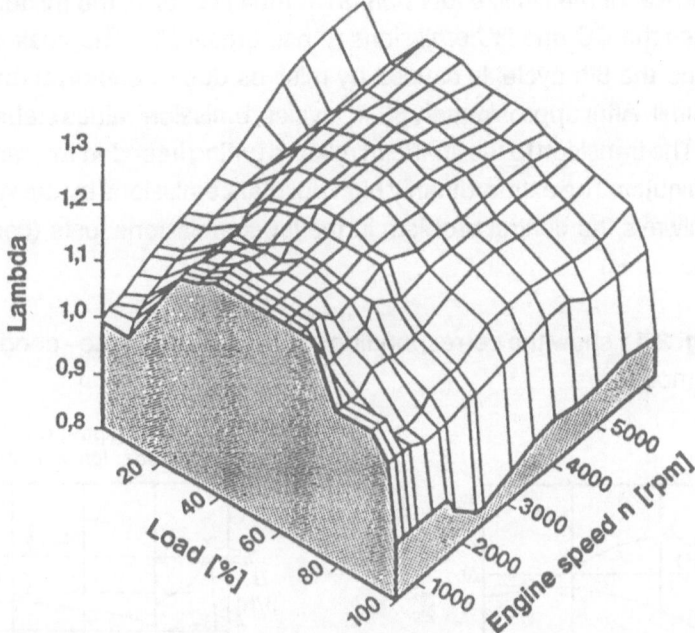

Fig. 2.9. Lambda map of a fuel-injected engine without lambda closed- loop control [2.45]

If, however, the engine has not yet attained its operating temperature, increased quenching of the flame front on the cold combustion chamber walls, less favorable mixture distribution due to higher wall-film deposits, and also reduced vaporization of the fuel have to be counterbalanced by a corresponding increase in fuel quantity.

So, for instance, enrichment factors of 1.5 at +20 °C starting temperature, 2 at 0 °C starting temperature, and 2.7 at -20 °C starting temperature are indicated for the Bosch K-Jetronic injection system. Enrichment decreases with increasing running time of the engine; for instance, at +20 °C it is applied for about 90 seconds, at 0 °C for 120 seconds, at -20 °C for 160 seconds, phasing out after the indicated time. With a Bosch KE-Jetronic, an enrichment factor of 2.5 applied for about 1 second is set when actuating the starter, according to Maisch [2.127]. Warm-up enrichment proper will depend on the cooling water temperature; for -30 °C it amounts to a factor of 1.7, for +20 °C it is 1.3, returning to 1 after 20 seconds.

In engines with single-point mixture formation, the enrichment factors are appreciably higher, since a relatively large amount of fuel will be deposited due to the large intake manifold surfaces.

This causes a shift of the mixture composition toward smaller lambda values, leading to considerably higher CO and HC emissions, as shown in Fig. 2.8.

The illustrations of **Fig. 2.10** show the exhaust emissions of a 2-liter fuel-injected engine measured shortly after cold start (starting temperature 20 °C). After activation of the cold enrichment system and arrival of the filmlike fuel portion at the inlet valve, the mixture is strongly enriched, which causes the CO and HC emissions to rise drastically. The peak in HC emissions, which occurs around the 8th cycle, is caused by misfires due to a short-term drop below the lower (rich) misfire limit. After approximately 30-40 cycles, emission values stabilize at a still comparably high value. The transition to the final lean mixture tuning intended for warm engine operation takes several minutes. The extraordinarily high pollutant emissions by the vehicle during this phase are almost always the central problem in the legal emissions tests (certification testing and periodical inspections)

The diagrams of **Fig. 2.11** show the corresponding - considerably worse - conditions in an older carbureted engine model.

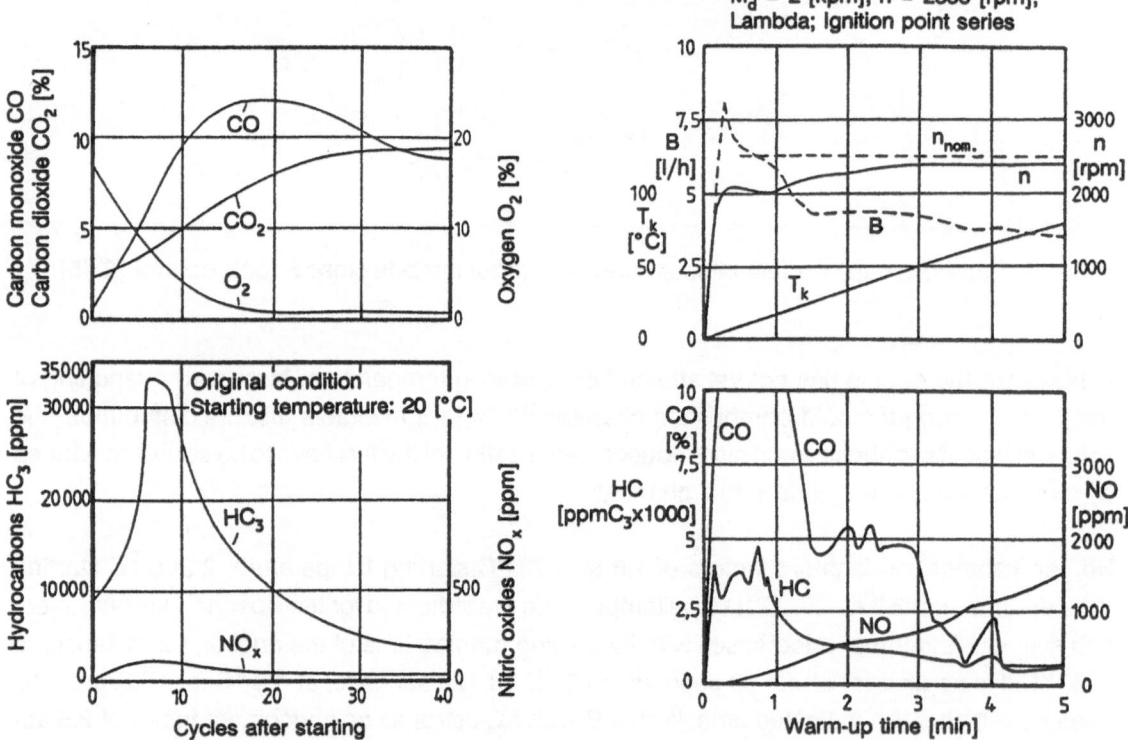

Fig. 2.10. Exhaust-gas emissions of a 2-liter fuel-injected engine shortly after cold start (starting temperature 20 °C) [2.13]

Fig. 2.11. Emission and consumption characteristics of an older model of a 1.6-liter carbureted engine during warmup (starting temperature 0 °C) [2.15]

Spark-ignition engines with three-way catalyst converter systems should, as far as possible for the sake of smooth running, be operated with a stoichiometric mixture at any operating condition, since - in accordance with **Fig. 2.12.** - only relative air-fuel ratios around $\lambda = 1.00$ yield catalyst conversion rates of more than 80% for all three exhaust-gas constituents (CO, HC, and NO_x).

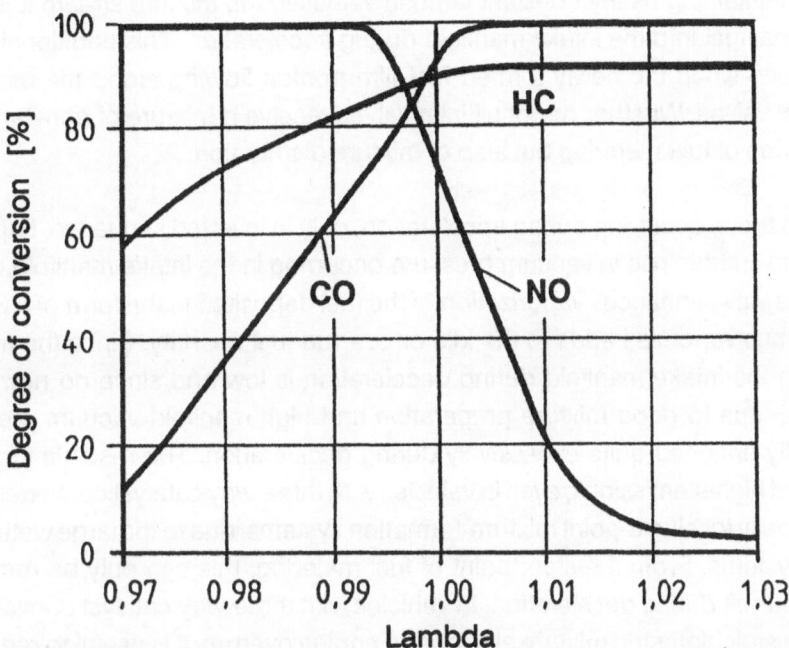

Fig. 2.12. Effect of the relative air-fuel ratio λ on the conversion efficiency of a three-way catalyst converter

The requirements on mixture composition discussed so far were concerned with steady-state operation. In the case of transient engine operation, the problem of optimum metering in the entire characteristic map is much more complicated. For instance, the rapid opening of the throttle valve during acceleration causes a sudden pressure rise in the intake manifold, which leads to a reduced vaporization of the fuel droplets and of the wall film flowing along the bottom of the manifold or, respectively, to an increased tendency of condensation of the vaporized fuel. Further, the fact that secondary fuel atomization at the throttle valve, which occurs in single-point mixture-formation systems, is missing in the multi-point systems discussed here leads to the formation of substantially larger droplets with a higher tendency to impinge when the mixture stream is deflected in the intake manifold.

Either mechanism thus causes wall-film buildup during acceleration accompanied by a corresponding leaning-out of the remaining mixture stream. As a matter of fact, this especially applies to single-point mixture formation systems. However, it means that, in the first seconds of the acceleration process, the cylinders are not sufficiently supplied with fuel, since the wall film has a substantially lower mean flow velocity than the rest of the mixture. This deficiency in fuel supply will be the greater the lower the manifold wall temperature and the greater and steeper the load increment in the engine map. The consequences for the combustion process range from a decreased combustion speed and less favorable position of the point of gravity of combustion to misfires when misfire limits are exceeded: the driver will become aware of this by increased consumption and higher emission values, a less favorable response behavior, a rough engine and ultimately even stalling of the engine.

Therefore, for maintaining nearly constant lambda values in the mixture stream it is necessary to inject additional fuel into the intake manifold during acceleration. This additional fuel supply has to be stopped when the newly formed fuel film portion flowing along the manifold walls reaches the inlet valves. Whether or not all inlet valves receive a mixture of constant lambda is not only a question of fuel metering but also of mixture distribution.

Events similar to those occurring during acceleration, only in inverted sequence, happen during deceleration. The sudden rise in vacuum pressure occurring in the intake manifold due to closing of the throttle valve enhances vaporization of the fuel deposited in the form of a wall film; the amount of fuel thus vaporized adds to the idle or overrun fuel quantity. Since the mass flow of mixture entering the intake manifold during deceleration is low and since no new wall film is being deposited, due to good mixture preparation and high manifold vacuum pressures, the mixture is usually enriched quite excessively during deceleration. The result is increased fuel consumption and higher emissions, even in vehicles with three-way catalyst converter. This again holds true especially for single-point mixture-formation systems, due to the large wetted manifold areas in such systems. From the standpoint of fuel metering, this can only be remedied by a complete fuel shut-off during deceleration. In vehicles with three-way catalyst converters, which should receive a stoichiometric mixture also during engine overrun, it is useful to reduce the fuel supply only to such an extent as is necessary for maintaining a relative air- fuel ratio of $\lambda \approx 1$.

2.5. Mixture Preparation

Combustion is greatly influenced not only by the metering process, but also by mixture preparation. In a spark-ignition engine with direct fuel injection, both metering and mixture preparation are controlled by the mixture formation unit (injection nozzle), whereas in mixture formation processes that take place in the intake manifold these two parameters are additionally influenced by mixture flow. Especially in single-point mixture formation systems there is a very strong interrelation of quantitative and qualitative parameters of mixture formation, based on the fact that mixture flow and distribution are very dependent on the quality of mixture preparation.

Generally, the fuel is inducted into the intake manifold in the form of droplets, however, a partial transition to the other two phases in the manifold (fuel vapor and wall film) already takes place on the way to the inlet valve or manifold wall, refer to **Fig. 2.13**. Depending on the original size spectrum of the fuel droplets, on the location of the mixture formation unit, the manifold geometry, and on the temporal relation between fuel induction and inlet process into the cylinder, a more or less large portion of the fuel droplets is deposited in the form of a fuel film along the manifold walls or on the inlet valve face and vaporizes there due to the high surface temperatures, whereas another portion of the droplets either vaporizes in the air stream or - especially during the inlet valve opening time in multi-point manifold injection systems - enters the combustion chamber still in droplet form. Since bits of fuel film are torn off at the inlet valve orifice, an additional fuel portion enters the combustion chamber in droplet form.

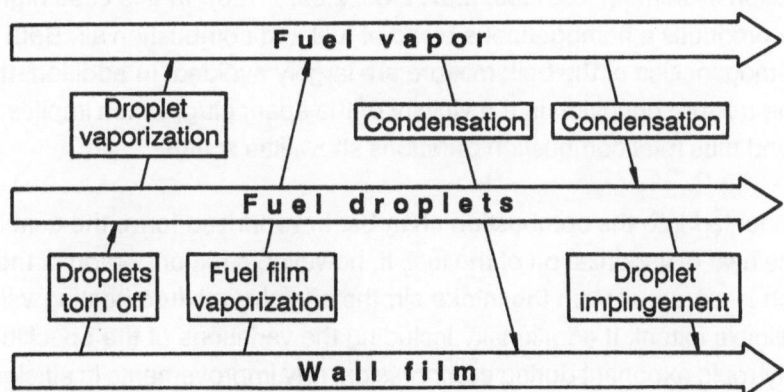

Fig. 2.13. Phase transitions in an intake manifold of a spark-ignition engine

Whereas in single-point mixture formation systems the main factors of influence are atomization at the throttle valve and fuel vaporization on the hot manifold walls or heating elements, in common multi-point manifold injection systems great significance is attributed to mixture formation at the hot inlet valve and the wetted manifold wall areas.

The mixture formation process is very dependent on marginal engine conditions. The degree of atomization provided by the mixture formation unit, positioning of the mixture formation unit in relation to intake manifold geometry, fuel inlet timing in accordance with the cylinder inlet process in intermittent injection systems, conditions of vaporization (distillation behavior of fuel, pressure and temperature in the intake manifold, etc.), and, finally, engine operating conditions affect not only the final condition of the mixture but also the relative portions of the different mixture phases. In many cases there are different effects of the mixture formation process on steady-state and on transient engine operation.

In single-point mixture formation systems, the quality of fuel preparation has a major effect on fuel metering to the individual engine cylinders even under steady-state operation. The extent to which the fuel droplets follow the air stream, thus influencing fuel film buildup on walls and, as a consequence, also mixture distribution, is determined essentially by the droplet size - also refered to in Chapter 2.6.

This means, an ideally prepared mixture is one prepared with a maximum of intensity (high vapor portion and very fine atomization). In multi-point injection systems, on the other hand, the degree of preparation has little influence on fuel flow and, above all, mixture distribution to the individual cylinders, although it does have influence on the combustion process.

When considering only conventional spark-ignition engine combustion processes, which are based on a homogeneous charge, a fuel that has completely vaporized before induction into the combustion chamber will usually be looked at as an ideally prepared one also in multi-point mixture formation systems [2.65, 2.66, 2.67, 2.68, 2.69, 2.126]. In this case high turbulence at the inlet valve produces a homogeneous blend of fuel and combustion air. Both local and temporal non- homogeneities of the bulk mixture are largely avoided. In addition, this means little variation of the mixture condition in the vicinity of the spark plug, which implies that inflammation periods and thus total combustion durations show little scatter.

If the fuel is inducted into the combustion chamber in vaporized form, the usable energy is increased by the heat of vaporization of the fuel. If, however, a major portion of the required heat of vaporization is extracted from the intake air, the effective mixture heating value is changed only to a negligible extent. If additionally including the variations of the specific heat constant and of the polytropic exponent during compression, any improvements in efficiency due to fuel vaporization are caused, above all, by reduced cycle-to-cycle variations of the combustion process.

Improved mixture preparation normally leads to reduced variations, rarely however to a shorter period of inflammation and combustion process [2.70]. If the fuel enters the combustion chamber not as fuel vapor but as droplets, the high lambda gradients which occur during vaporization will not have completely vanished by the time ignition starts, leaving some of the mixture incompletely blended both in the macroscopic and in the microscopic range.

Depending on the spatial and size distribution of the fuel droplets and depending also on the timing of their induction into the combustion chamber and on charge motion, different forms of non-homogeneous mixture distribution may be generated in the combustion chamber, see **Fig. 2.14**.

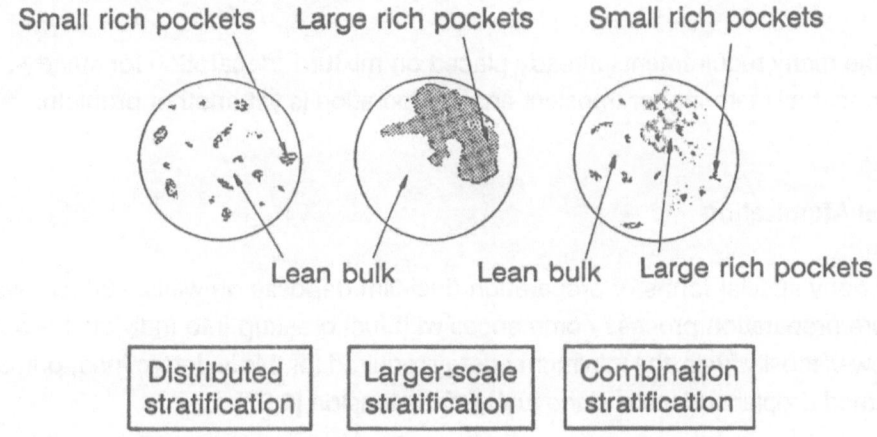

Fig. 2.14. Forms of non-homogeneous mixture distribution in the combustion chamber [2.71]

If single large droplets are inducted into a combustion chamber with little charge motion, their vaporization leads to single, small rich pockets in a very lean bulk mixture, see Fig. 2.14 left. Random charge stratification, as shown in the illustration, represents an unfavorable situation for combustion, as there is less chance of encountering sufficiently rich mixture portions near the spark plug and uniform burning throughout the charge can be problematic.

Large-scale stratification, as shown in Fig. 2.14 middle, may be expected when the fuel is inducted into the combustion chamber at a preferred local and/or temporal position, and when this initial stratification of the charge is maintained or enhanced by charge motion.

Some portions of charge motion (certain forms of squish) in the microscopic range facilitate localized good blending, although, when looking at the overall distribution across the combustion chamber, there are marked differences in concentration. If ignition is initiated in a localized rich mixture portion, the lean remainder of the charge will burn evenly through, as experience has shown in stratified-charge engines. Charge stratification of this type can be used intentionally to extend the engine operating range significantly into the lean range - it represents the principle of operation of stratified-charge engines with undivided combustion chambers. In real engine operation, however, a variety of combinations of different types of non-homogeneities are found. Since the fuel normally enters the combustion chamber both as vapor and as droplets, segregation of these individual phases can be achieved by a directed charge motion, so that the fuel portion that is available already vaporized at the combustion chamber inlet will yield an extensive, coherent stratification, whereas the vaporizing fuel droplets will cause small, localized, excessively rich mixture portions, see Fig. 2.14 right.

Additional problems have to be expected if there are large film-like fuel portions in the intake manifold. When the wall film is torn at the inlet valve orifice, large fuel droplets are formed, which lead to the described disadvantages. Even during steady-state engine operation the amount of fuel film entering the combustion chamber in consecutive cycles is not constant but shows cyclic variations, which leads to marked cycle-to-cycle variations of the overall lambda value.

Besides the many requirements already placed on mixture preparation for steady-state conditions, correct fuel metering for transient engine operation is yet another problem.

2.5.1. Fuel Atomization

Except for any special forms of preparation (fuel-film deposits on walls, surface evaporation), the mixture preparation process commences with fuel breakup into individual droplets. In the case of low-viscosity fuels, the minimum energy required for this is determined, primarily, by the newly formed droplet surface and the fuel surface tension [2.97]:

$$E_{min} = \sigma_K \cdot \Delta O_T \qquad\qquad (2.16)$$

E_{min} [J]   ..... minimum energy required for atomization
ΔO_T [m^2]   ..... surface enlargement on account of the breakup process
σ_K [N.m^{-1}] ..... surface tension of the fuel

However, since under real engine conditions, with energy being added, the kinetic energy and internal energy of the fuel are increased substantially more than the energy used for surface enlargement, the actual energy requirement for breaking up the fuel is much greater than the theoretical minimum breakup energy.

A variety of atomization systems are available for breaking up the fuel. According to reference [2.72], atomization systems may be classified as high-speed atomizers and centrifugal atomizers.

Centrifugal atomizers, in which the fuel is broken up by mass forces - in particular centrifugal forces - are not of any importance for spark- ignition engine fuel systems, except for any special forms such as the swirl atomizer.

The mixture formation systems presently used in series-type spark- ignition engines can be classified almost entirely as belonging to the high-speed atomizer types. In such systems, atomization of the coherent mass of fluid takes place by way of surface inertia and viscosity forces which are generated, for example, when a relative motion is created between the fuel to be atomized and the ambient medium.

Two different types of atomization are distinguished in high-speed atomizers, depending on how the required atomization energy is added.

- In the case of airless atomization, the fluid to be atomized is moving, whereas the surrounding gaseous medium is at rest. Gas forces do contribute to fuel breakup, however the velocity and energy of the gas are negligible for breakup into droplets.

- In air-assisted atomization, on the other hand, gas velocity is the parameter which determines the droplet breakup. The fluid to be atomized is nearly quiescent, the atomization energy mainly originates from the kinetic energy of the gas. This type of fuel breakup is found in carburetors, for instance, or, in a similar form, in systems where masking of the inlet valve is used.

In actual mixture formation systems, especially fuel injection systems, the atomization process consists of a superposition of these two breakup mechanisms. Airless atomization normally leads to a primary breakup of the fuel stream. By adding the air velocities occurring in the manifold to the velocities of the droplets formed, a secondary breakup process is caused by air-assisted atomization. Generally, these primary and secondary fuel breakup processes do not only occur consecutively, but also simultaneously. For instance during the induction period, the breakup resulting from airless atomization is strongly influenced by the air velocities in the intake manifold, especially in manifold injection systems.

Both in multi-point and, to a smaller extent, in single-point mixture formation systems, the mixture formation process is governed by marginal conditions which vary widely with time. Depending on the timing of mixture intake in relation to valve timing and thus in relation to the prevailing flow and temperature conditions in the intake manifold, the mixture formation process is the result of various individual, superimposed fuel breakup processes and mechanisms and the resultant wetting of the walls and droplet vaporization.

These rather complex processes of droplet formation can be described by using nondimensional parameters.

The Reynolds number Re

$$Re = \frac{v \cdot d}{\nu_{Fl}} \tag{2.17}$$

v $[m.s^{-1}]$ .....velocity
d $[m]$ .....characteristic nozzle diameter
ν_{Fl} $[m^2.s^{-1}]$ .....kinematic viscosity of the fluid

is significant for the flow profile and, with certain limitations, for turbulence in the nozzle until the fluid is discharged.

The Ohnesorge number Oh

$$O_h = \frac{\eta_{Fl}}{(\sigma \cdot \rho_{Fl} \cdot d)^{0,5}}$$

(2.18)

η_{Fl} [Pa.s^{-1}] .....dynamic viscosity of the fluid
σ [N.m^{-1}] .....surface tension
ρ_{Fl} [kg.m^{-3}] .....density of the fluid
d   [m]   .....characteristic nozzle diameter

describes the physical properties of the fuel which are essential for breaking up the fuel jet and specially includes the effect of fuel viscosity.

When the Weber number We

$$We = \frac{\rho \cdot v_{rel}^2 \cdot d_T}{\sigma}$$

(2.19)

ρ [kg.m^{-3}] .....density
v_{rel} [m.s^{-1}] .....relative fuel/air velocity
d_T [m]   .....droplet diameter

is formed using the density of fuel, it charaterizes the relation of the fluid inertia forces to the surface tension forces. When the Weber number is formed using the density of air, it determines the relation of the pressure forces resulting from the air inertia forces to the surface tension forces. If the breakup is primarily caused by air forces, a Weber number that exceeds a certain critical value is usually taken as a simplified breakup criterion.

2.5.1.1. Airless Atomization

A fuel jet discharged through a nozzle can be caused to break up by many different mechanisms. A critical analysis of possible causes of breakup such as

- Aerodynamic interaction of jet surface and ambient medium;

- Jet turbulence, especially radial velocity components in turbulent nozzle flow;

- Varying velocity profiles, boundary layer accelerations, and boundary layer instabilities at the nozzle exit;

- Cavitation phenomena within the nozzle;

- Fuel pressure variations, and

- Instabilities due to surface tension,

will show that breakup of a spray jet has a single cause only in exceptional cases. In general, jet breakup depends on the conditions at the nozzle exit (exit geometry, jet velocity and turbulence, physical properties of fuel and surrounding medium, etc.) and is the result of complex, super-imposed individual breakup mechanisms.

When disregarding the breakup regimes that are of little importance for combustion engine application, such as the continuous and dripping flow [2.73, 2.74, 2.75], there are basically three breakup regimes of interest here:

- Rayleigh droplet flow (with and without air assistance)

- Aerodynamic or bag breakup (with and without spray mist separation)

- Atomization

In the range of very low exit velocities, the jet breaks up due to axisymmetrical oscillations which, based on initial disturbances due to the effect of surface tension, leads to an instability and breakup of the jet into individual drops greater in diameter than the original jet diameter (Rayleigh instability). The effect of surface tension forces is enhanced by air forces, and this is the main cause of jet breakup.

The importance of the air forces increases with increasing exit velocities. The asymmetrical undulations of the jet, which occur on account of the fuel/air relative velocity, grow faster than the axisymmetrical disturbances and, at critical amplitudes, cause the jet to be torn at the peaks - aerodynamic or bag breakup of the jet occurs (according to Weber-Hänlein [2.77, 2.78]). Here the surface tension acts against the growth of these wavelike disturbances of the jet.

With certain nozzle geometries, the jet surface is torn even before breakup proper by the aerodynamic breakup regime when the exit velocity is increased; a surface disturbance of comparatively high frequency occurs at the nozzle exit which, through a Kelvin-Helmholtz instability [2.79, 2.80, 2.81], **Fig. 2.15**, leads to separation of individual small satellite droplets. This "aerodynamic breakup with spray mist separation" first mentioned in reference [2.82] represents the transition to the range of atomization.

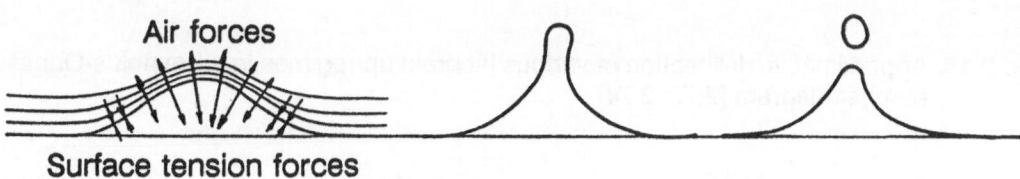

Fig. 2.15. Mechanism of droplet formation according to Kelvin-Helmholtz [2.97]

At very high exit velocities, the jet breaks up - without any apparent regularity immediately upon discharge through the nozzle - into individual droplets whose diameter is small compared to the original jet diameter: this is called atomization.

Atomization is a breakup process under turbulent flow conditions with a wide spectrum of droplet sizes. Especially concerning the area of atomization, somewhat diverging statements are found in the literature as to the theory of the breakup. Under combustion-engine conditions, the predominant mechanism of atomization is aerodynamic interaction between the surface of the jet and the ambient medium. This interaction, which starts with initial disturbances at the nozzle exit, leads to growing unstable waves on the stream surface and finally results in breaking up of the stream. Both, the initial disturbances and the mechanisms which determine them, are very dependent on the exit conditions and have a major influence on atomization.

Generally, these initial disturbances, which are of great importance not only for atomization but also for the aerodynamic and Rayleigh breakup regimes, cannot be determined with accuracy and the Reynolds number gives a much too inadequate characterization of them. Therefore, separation of individual breakup regimes in a Reynolds- Ohnesorge diagram [2.73, 2.74, 2.75] permits only approximative theories as to the type of breakup to be expected; there are no clear delineations but areas of transition, where various breakup regimes may occur simultaneously, **Fig. 2.16**.

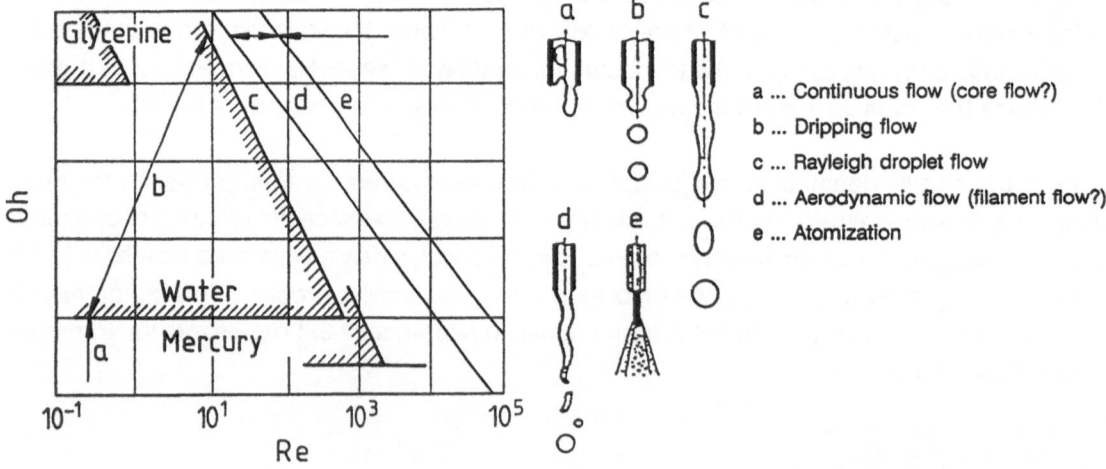

Fig. 2.16. Approximative delineation of various jet breakup regimes in a Reynolds-Ohnesorge-number diagram [2.73, 2.74]

The fact that the initial disturbances cannot be determined adequately leads to the necessity - despite a profound analytical understanding of the breakup of fuel jets discharged through concentric spray holes [2.72, 2.76, 2.77, 2.83, 2.84] - of using empirical factors for analyzing jet breakup.

Predominantly airless atomization of a fuel jet discharged through a concentric spray orifice occurs, for example, in orifice-type injection nozzles. So far this type of injection nozzle has been used only to a small extent for mixture-formation systems in spark-ignition engines, and when it is used it has a metering function rather than an atomizing function.

In the case of pintle-type injection nozzles as commonly used today, the fuel jet is broken up, under the usual operating conditions of multi-point injection systems, in the form of a laminated spray pattern [2.86, 2.87, 2.88, 2.89, 2.90, 2.91, 2.93, 2.94, 2.95, 22.96, 2.97].

In the case of conventional pintle-type injection nozzles, a cone- shaped spray plume, whose thickness decreases toward the leading edge is formed on account of deflection at the pintle immediately after discharge through the annular orifice. Depending on the exit conditions (exit velocities and geometries), this spray plume is characterized by a certain frequency and amplitude spectrum of initial disturbances. Under the influence of the ambient gas, these initial disturbances generate exponentially growing waves on the spray plume. At critical amplitudes, this instability of the plume leads to a breakup into individual ligaments which, under the influence of surface tension and gas forces, rapidly form individual drops.

The breakup of the spray plume may be caused, similar to that of a pencil jet, by two types of oscillation - see **Fig. 2.17.**

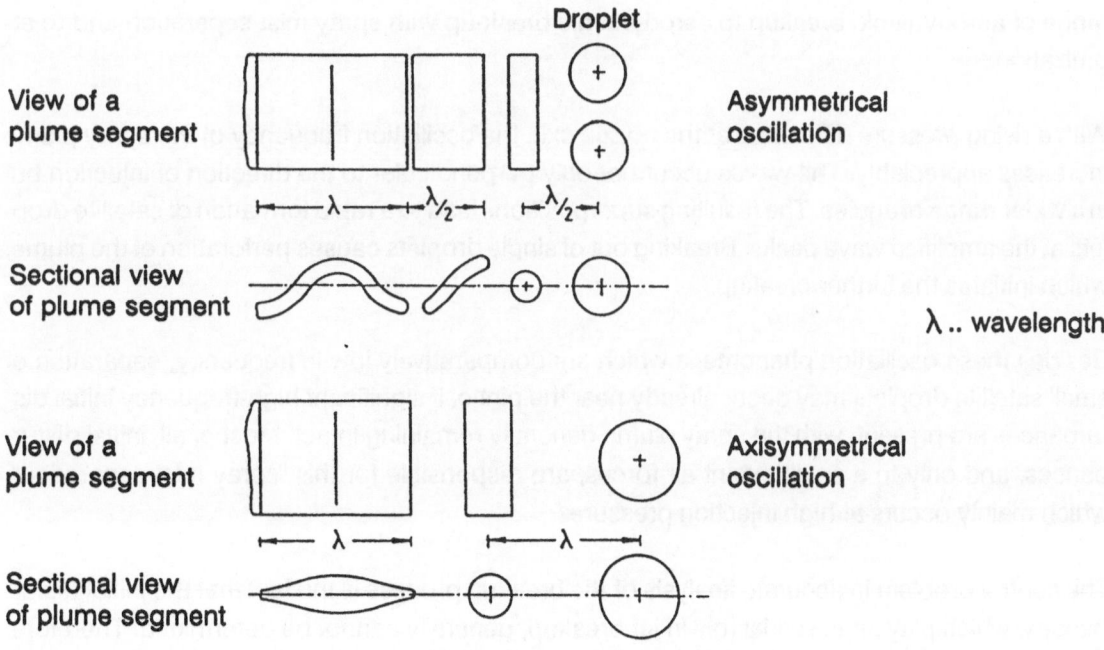

Abb. 2.17. Forms of breakup of a spray plume (idealized) [2.97]

In the case of an asymmetrical oscillation, the spray plume breaks up by tearing at the "wave peaks" with half the oscillation wavelength. An axisymmetrical oscillation, on the other hand, leads to a breakup in the "valleys"; here the breakup wavelength corresponds to the full wavelength and causes correspondingly larger drop diameters.

Besides the conditions within the spray plume, distribution, shape, and frequency of the initial disturbances have a great influence on the type of oscillation generated.

In a steady fuel jet, part of these initial disturbances is caused by the turbulent flow conditions at the annular spray orifice (due to the short nozzle bore, inlet disturbances cannot be expected to vanish even with Re 2300), another part by deflection of the jet at the pintle. Both frequency and amplitude and also local distribution of initial disturbances may be influenced by the prevailing operating conditions (e.g. injection pressure).

Under steady-state conditions, a dominant growth of asymmetrical waves may be deferred from breakup mechanisms of fan jets [2.85, 2.87, 2.88] for the marginal conditions applying here.

In an unsteady fuel jet, due to the strongly varying pressure and flow conditions, the opening process of the nozzle needle causes a marked axisymmetrical oscillation which has a crucial effect on breakup at the tip and a minor effect on breakup in the core of the fuel jet. Although, in the case of laminated spray patterns discussed here, it is more difficult to make a distinction between different breakup regimes than in the case of a pencil jet, a transition is observed from the range of aerodynamic breakup to aerodynamic break-up with spray mist separation and to atomization.

With a rising pressure difference at the nozzle exit, the oscillation frequency of the spray plume increases appreciably. The waves occur not only perpendicular to the direction of injection but in a wider range of angles. The resulting superpositions lead to a rapid formation of satellite droplets at the amplified wave peaks. Breaking out of single droplets causes perforation of the plume, which initiates the further breakup.

Besides these oscillation phenomena which are comparatively low in frequency, separation of small satellite droplets may occur already near the pintle, if significant high-frequency initial disturbances are present, with the spray plume generally remaining intact. Most of all, initial disturbances, and only to a small extent air forces, are responsible for this "spray mist separation", which mainly occurs at high injection pressures.

The central problem in accurate analysis of the breakup process is the fact that the initial disturbances, which play an essential role in jet breakup, generally cannot be determined. Therefore, the breakup process can only be analyzed approximatively and using simplifications.

When looking at a steady spray jet only, mainly asymmetrical oscillations, which are amplified by the effect of gas forces, lead to breakup of the jet. Under the usual operating conditions in manifold injection systems, the portion of small satellite droplets generated by high-frequency fractions of the initial disturbances at short distances from the pintle is comparatively small. The major part of the droplets are formed by being torn off those surface waves which grow fastest under the influence of gas forces. From a spectrum of initial disturbance wavelengths that will always be present, those waves grow fastest which correspond with the oscillation of the spray plume itself. For this growth of wave lengths, which occurs under the influence of surface tension, inertia, and viscosity forces in interaction with an ambient medium of negligible viscosity, complete linear solutions exist for a pencil jet [2.77].

For cone-shaped spray plumes, however, not only the marginal conditions are different but also the growth rate of the amplitudes, with inclusion of viscosity, additionally depends on the varying plume thickness. Therefore, accurate solutions which are valid for a continuous jet cannot be transferred to a cone-shaped spray plume; in the latter case only approximative solutions with empirical coefficients are possible. As in the case of a pencil jet [2.77], also here assumption of a critical ratio of the amplitude of plume oscillation to the amplitude of the initial disturbance seems to be a useful approach [2.80, 2.85, 2.87, 2.97].

The main factors affecting the breakup process are:

- Exit velocity of the spray jet at the nozzle exit;

- Exit geometry (exit cross-section, jet turbulence);

- Fuel properties (surface tension, viscosity, density);

- Air density in the intake manifold.

A decreasing air density may not only effect a decrease of the air forces which are essential for the breakup process but, on account of the reduced absolute pressure, may also lead to boiling of the fuel already at the exit cross section. This may cause foaming of the fuel and result in a complete collapse of the original breakup regimes.

2.5.1.2. Air-Assisted Atomization

If a fluid breaks up in a gas of relatively high velocity, this breakup into individual droplets is mainly determined by the gas velocity, the initial kinetic energy of the fluid having only a minor influence on the formation of droplets. While superposition of such secondary and primary breakup mechanisms (e.g. as in airless atomization) usually occurs in manifold injection systems, a breakup to be classified almost exclusively as air-assisted atomization is found in carburetors.

When looking at a spherical droplet that is inducted into a high- velocity gas stream, deformation of the droplet at the transition point will occur when the dynamic pressure at this point is greater than the internal droplet pressure, which depends on surface tension.

Knowing the internal droplet pressure p_i

$$p_i = \frac{4\sigma}{d_T} \tag{2.20}$$

and dynamic pressure p_{st}

$$p_{st} = \frac{\rho_G \cdot v_{rel}^2}{2} \tag{2.21}$$

ρ_G [kg.m^{-3}] ..... gas density
v_{rel} [m.s^{-1}] ..... relative velocity droplet/gas

a relation for beginning droplet deformation can easily be determined [2.77]:

$$\frac{d_T \cdot \rho_G \cdot v_{rel}^2}{\sigma} = We > 8 \tag{2.22}$$

We   [-]   ..... Weber number

If initial disturbances are present, the droplet will lose its spherical shape even earlier; based on the different drag coefficients in this case, oscillation may occur even at Weber numbers smaller than 8. Beginning droplet deformation, however, is definitely not to be considered as the determining breakup criterion, but is to be seen as an excitation of the droplet to oscillate. Above all viscosity forces can dampen this droplet oscillation even at Weber numbers greater than 8.

Calculation of a critical Weber number, at which droplet breakup occurs, is possible on the basis of a stability analysis of the droplet. The critical Weber number range $5 < We < 20$ observed in practice is also found in theoretical [2.98, 2.99, 2.100, 2.101] and theoretical-empirical works [2.72, 2.73, 2.102], depending on whether viscosity is taken into account or not. The effect of viscosity should be taken into account also in the case of spark-ignition engine fuels of comparatively low viscosity; here it seems useful to define the critical Weber number as a function of the Ohnesorge number [2.102].

A major factor of influence on droplet breakup is not only the magnitude of the relative velocity but also its variation with time. Droplet breakup is affected both by the type of turbulence spectrum in case of turbulent inlet conditions and by the drop initially attaining relative speed.Therefore, rapid changes of velocity lead to appreciably lower critical Weber numbers than in the case of a continuous gas flow [2.101, 2.102, 2.103].

Although fracturing of a droplet in a high- speed gas flow occurs via the intermediate stage of a plume being formed on account of droplet deformation, theoretical analysis of the secondary breakup mechanisms of the cone-shaped spray plumes discharged by multi-point injection nozzles is quite difficult.

Here, the non-negligible plume velocity is superimposed on local velocity components of the gas flow, values for which are only approximatively known. The droplet sizes resulting from superposition of primary and secondary breakup mechanisms are dependent not only on the relative velocity but also on a number of additional marginal conditions and, in general, cannot be determined by theoretical analysis. Therefore, in the area of air-assisted atomization, the empirical-theoretical works prevail which, based on dimensional analyses, use numerical equations for expressing individual parameters affecting the breakup process [2.104, 2.105, 2.106, 2.107, 2.109, 2.110, 2.111, 2.112, 2.113, 2.114, 2.115, 2.116, 2.117].

If the droplet sizes resulting either from a predominantly airless or from a predominantly air-assisted atomization are related to the mean relative velocity between fuel and ambient medium, airless atomization in the range of low relative velocities gives appreciably smaller droplet diameters, see **Fig. 2.18**.

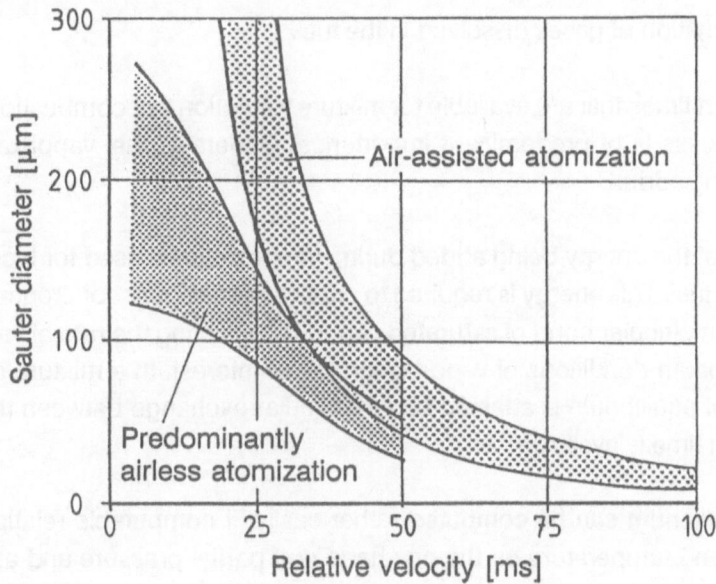

Fig. 2.18. Droplet diameters of a predominantly airless or predominantly air-assisted atomization as a function of the relative velocity fuel/air (for details on Sauter diameter see page 134) [2.97]

In the case of a predominantly airless atomization, as shown here on the example of a multi-point manifold injection system, the air forces resulting from the mean relative velocity of fuel and air are not the sole cause of breakup, but they increase the effect of initial disturbances, which effect a marked breakup of the jet even at low mean relative velocities.

2.5.2. Fuel Vaporization

Breakup of the fuel stream into individual droplets is only the first stage of mixture formation in spark-ignition engines. The transition of the mixture from the liquid to the gaseous phase is of paramount importance for the subsequent combustion process. There are three types of phase transition:

- Evaporation: If there is a sufficiently large free fuel surface, transition to the gaseous phase takes place even below the distillation temperature of the fuel, however at a very slow rate. Depending on the marginal conditions (extraction of fuel vapor, combining of fuel with air, pressure and temperature), the order of magnitude to which this happens is 0.1 to 0.8% evaporated vapor volume per hour [2.118, 2.119].

- Vaporization: If the fuel temperature is higher than the distillation temperature, a much faster transition from the liquid to the gaseous phase takes place, the rate of vaporization being affected by a variety of marginal parameters;

- Segregation of gases dissolved in the fuel.

Owing to the short times that are available for mixture formation in a combustion engine, the fuel vaporization process is of predominant importance. Generally, fuel vaporization takes place when heat is being added.

The largest part of the energy being added during vaporization is used for increasing the internal energy of the fuel. This energy is required to crack the molecular coherence and to generate the much looser molecular bond of saturated vapor. If neglecting the rate of vaporization at this stage, the equilibrium conditions of vaporization are of interest. In a mixture consisting of fuel and air, a state of equilibrium is attained by a molecular exchange between the vapor and the liquid, if sufficient time is available.

This state of equilibrium can be computed rather easily, if computable relationships exist between pressure and temperature on the one hand and partial pressure and amount vaporized on the other, as is the case with chemically uniform, pure substances. In this case, the state of the mixture can be computed for both substances using the ideal gas equation and Dalton's law, according to which the total pressure of the gas mixture equals the sum of partial pressures which each gas alone would have at constant temperature and constant volume. Each pressure has a corresponding temperature at which a fluid vaporizes and vice versa, the pressure generally being called vapor pressure or saturation pressure, the corresponding temperature being called distillation temperature or saturation temperature.

The relation between vapor pressure and saturation temperature is expressed by the equation of the vapor pressure curve. **Fig. 2.19** shows the vapor pressure curves of some hydrocarbons.

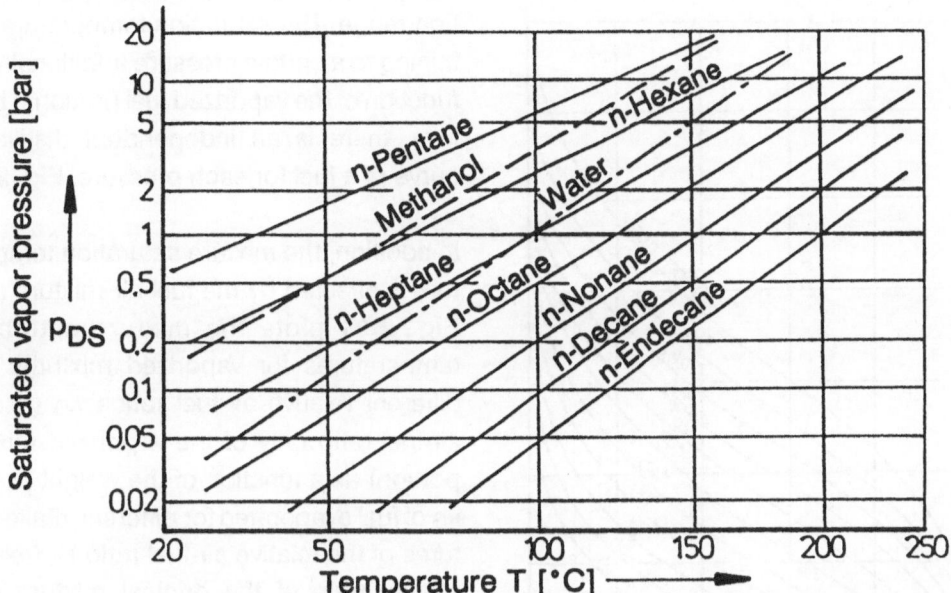

Fig. 2.19. Saturation pressure of various hydrocarbons and water as a function of temperature [2.120, 2.126]

If the mixture contains the maximum possible amount of fuel vapor at a given temperature, i.e. it is saturated, the partial pressure of the fuel is equivalent to its saturation pressure. The corresponding mixture saturation temperature can be seen on the vapor pressure curve. Since, in a mixture of air and fuel vapor, the partial pressures are dependent on the air/fuel mixture ratio at a given total pressure, the mixture ratio has to be taken into account additionally when computing the saturation temperature of the mixture. If the amount of heat required for vaporization is removed from the mixture, a higher mixture temperature at the beginning of fuel vaporization is required in addition. Knowing the heat of vaporization in the given temperature range and the mixture ratio, the degree of cooling of the mixture during fuel vaporization can be determined.

The state of a mixture consisting of an ideal gas (air) and condensable vapor with equal distillation points (hydrocarbon constituent) can be computed also by using the i-x diagram [2.121]. This diagram is useful above all when unsaturated humid air is being mixed with a fuel component and when inclusion of the enthalpies of the liquid and vaporized water portion of air thus would require a complex computation of the mixture temperature. However, the effect of water vapor is small according to reference [2.121], since a reduced degree of cooling of the mixture during fuel vaporization, due to water condensation, is observed only below the dew line of water. If, on account of fuel vaporization, the mixture temperature drops below the freezing point of water, engine operation may be affected by ice formation on the carburetor [2.121, 2.126].

So far only single, chemically uniform fuel components have been discussed. Common spark-ignition engine fuels, however, are mixtures of about 270 different fuel constituents [2.122] with in part widely varying distillation characteristics. Such mixtures therefore do not have a distillation point at a given pressure but a distillation range. The saturation temperature pertaining to a certain pressure additionally is a function of the vaporized fuel portion. Therefore, there is an independent distillation curve of a fuel for each pressure, **Fig. 2.20**.

In addition, the mixture saturation temperature is affected by the fuel-air mixture ratio. **Fig. 2.21** plots the mixture saturation temperatures for vaporized mixtures with different relative air-fuel ratios λ_D (relative air-fuel ratio only of the vaporized mixture portion) as a function of the weight portion κ_G of fuel evaporated for different intake mixtures of the relative air-fuel ratio λ_A (relative air-fuel ratio of the original mixture consisting of a vaporized and a liquid portion) for two engine operating points (full-load manifold pressure 1 bar and part-load manifold pressure 0.4 bar).

In Fig. 2.21 curves for a constant λ_D (broken line) have been added to those for a constant λ_A. Mixture saturation temperatures T_{GS} for a completely vaporized fuel, i.e. for $\lambda_A = \lambda_D$, can be read on the ordinate above $\kappa_G = 100\%$ [2.126].

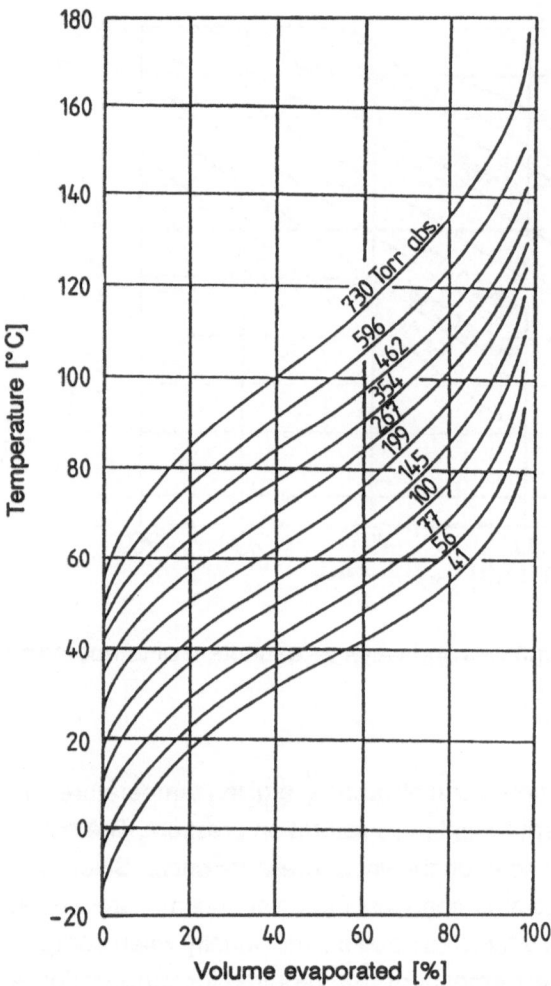

Fig. 2.20. Distillation curves of a spark-ignition-engine fuel at different pressures [2.118]

From this relation, the value the relative air-fuel ratio of the supplied mixture λ_A may have so that a combustible vaporized mixture with a given relative air-fuel ratio λ_D will be produced at a specific mixture saturation temperature, may easily be determined. Unless the amount of heat required for vaporization is added from external sources (preheating of air or fuel, heating of manifold), fuel vaporization will cause an appreciable drop in mixture temperature. In order to obtain a saturated vapor mixture of the desired concentration, the initial temperature of the mixture must in this case be higher by the amount of heat of vaporization dissipated.

The discussion so far was based on the assumption that the time available for mixture formation is sufficiently long for a state of equilibrium of vaporization to be established. From numerous investigations on mixture formation processes in the intake manifold it is clear, however, that such a state of equilibrium is hardly ever attained in actual engine operation. The perfection of mixture preparation is determined not so much by the equilibrium conditions than by the rate of

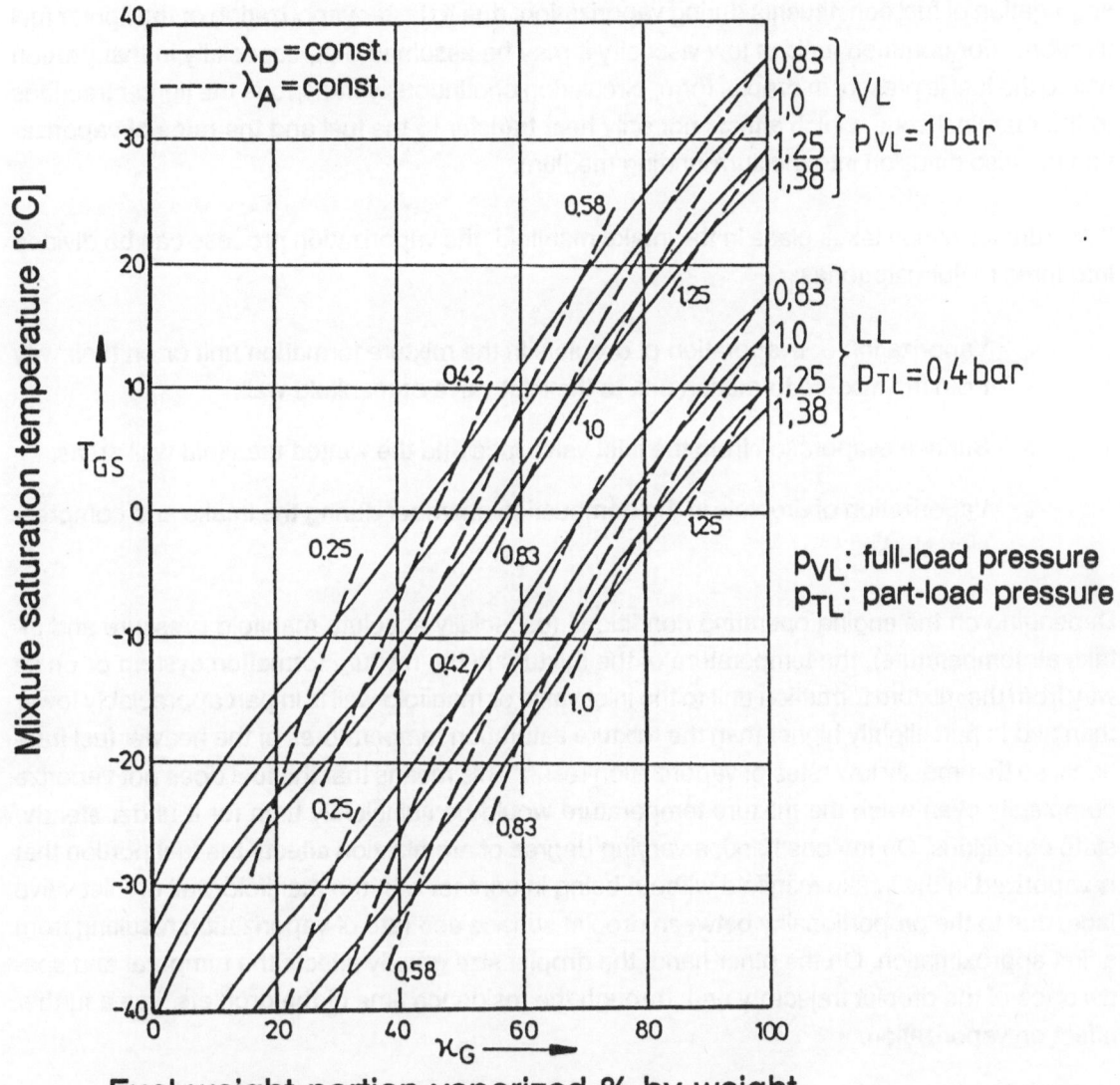

Fuel weight portion vaporized % by weight

Fig. 2.21. Relationship between mixture saturation temperature T_{GS} and weight portion κ_G of fuel evaporated, with the parameters: relative air-fuel ratio λ_A of the supplied mixture and relative air-fuel ratio λ_D of the vaporized mixture at full load (manifold pressure 1 bar) and at idle (manifold pressure 0.4 bar) [2.126]

vaporization. Since the rate of vaporization with a given fuel surface is essentially influenced by the difference between instantaneous partial pressure and partial saturation pressure in the vicinity of the gasoline wall film or droplet, vaporization can be accelerated by raising the temperature of the intake air and/or fuel. In this context, heat transfer to the intake air or fuel becomes very important.

A divergence of the vaporization process from idealized conditions may further be caused by segregation of fuel constituents during vaporization, due to faster vaporization of the lighter fuel fractions. For common fuels of low viscosity it may be assumed that, especially in that portion where the fuel is present in droplet form, circulation continuously transports the lighter fractions to the marginal zone, which affects not only heat transfer to the fuel and the rates of vaporization but also diffusion into the surrounding medium.

If mixture formation takes place in the intake manifold, the vaporization process can be divided into three major categories:

- Vaporization or evaporation of droplets in the mixture formation unit or on their way from the mixture formation unit to the inlet valve or manifold wall;

- Surface evaporation from the inlet valve face and the wetted manifold wall areas;

- Vaporization of droplets in the combustion chamber during the intake and compression strokes.

Depending on the engine operating conditions (especially absolute manifold pressure and intake air temperature), the temperature of the mixture in the mixture formation system or on its way from the mixture formation unit to the inlet valve or manifold wall is in part appreciably lower than and in part slightly higher than the mixture saturation temperatures of the heavier fuel fractions, so that mostly low rates of vaporization result. This means that the fuel does not vaporize completely even when the mixture temperature would be sufficiently high for it under steady-state conditions. On the one hand, a varying degree of atomization affects the fuel portion that is vaporized in the intake manifold without being in contact with the manifold wall or inlet valve face, due to the proportionality between droplet surface and rate of vaporization resulting from a first approximation. On the other hand, the droplet size greatly affects the temporal and spatial trace of the droplet trajectory and, through the residence time of the droplets, has a further effect on vaporization.

A large portion of the fuel droplets, however, do not vaporize while travelling in the intake manifold but impinge on the manifold wall or inlet valve face. Impinging droplets usually coalesce, forming a coherent fuel film. Only in single cases impinging large drops will be shattered. Since these manifold regions, where frequent droplet impingement is to be expected, already have high wall temperatures (inlet valve) or are heated for improved fuel vaporization (manifold heating, hot spot), heat transfer from the hot wall to the fuel becomes of paramount importance. The excessive temperature of the wall as compared to the temperature of the fuel represents the characteristic temperature gradient for this heat transfer. Fuel vaporizes at the highest rate when the surface temperature is about 40-50 $^\circ$C above the distillation temperature of the respective fuel constituents or lies in the range of distillation temperatures of common spark-ignition-engine fuels. With higher wall temperatures, as for instance at the inlet valve face, insulating effects of the fuel vapor causes the rate of vaporization to drop again to a point of maximum droplet lifetime (Leidenfrost point).

Accelerated surface evaporation occurs during valve overlap on account of a blowback of hot residual gas and also during the intake stroke in the turbulent boundary layer of the fuel accumulated on the manifold wall or inlet valve, with remaining liquid fuel portions being further atomized due to the high gas velocities at the inlet valve orifice.

Both, particles torn off the fuel film at the inlet valve orifice and droplets whose residence time in the manifold has not been sufficiently long for complete vaporization, contribute to a substantial fuel portion being inducted into the combustion chamber in droplet form, especially in a cold engine or in the case of injection into the open inlet valve. During the induction phase, vaporization in the combustion chamber basically is a continuation of the vaporization process initiated in the intake manifold, the process being accelerated by charge turbulence and heating of the intake air on the hot combustion chamber walls.

During the compression stroke, however, the rise in mixture temperature and pressure leads to a fundamental change in conditions of vaporization. Since, on the one hand, higher droplet temperatures are required for faster vaporization under the rising compression pressure but, on the other hand, also substantially higher gas temperatures prevail, heat transfer here becomes an essential factor of influence on droplet vaporization. The process of fractional distillation of the fuel droplet initiated already in the intake manifold due to the more rapid vaporization of the lighter fuel fractions, is continued in the combustion chamber, with the differences in the rates of vaporization becoming smaller as pressure and temperature rise [2.123, 2.124]. Since the rise in temperature has a stronger effect during compression than the rise in pressure and since the heat of vaporization of the hydrocarbons decreases appreciably with rising pressure and rising temperature, the conditions of vaporization improve as compression progresses. However, with enhanced vaporization the droplets contain increasing portions of heavier fuel fractions. The faster vaporization of small droplets does not only lead to a limitation of the droplet size spectrum but also to an increase of characteristic droplet diameters with progressing vaporization.

When analyzing the results of numerical simulations [2.125], for one-component fuels, based on simplified approaches for heat transfer and diffusion as a function of operating conditions, critical diameters can be determined for droplets which may be assumed not to have vaporized by the time ignition starts. Especially in the case of injection into the open inlet valve, the presence of fuel droplets at the time of ignition cannot be excluded from theoretical analysis for all engine operating conditions.

2.6. Mixture Flow and Distribution

Since the mechanisms of mixture flow and distribution are different in induction systems with single-point and multi-point mixture formation, the respective processes will be discussed separately for each system.

2.6.1. Mixture Flow and Distribution in Single-Point Mixture Formation Systems

2.6.1.1. General Description

The more or less well prepared multiphase mixture of air, fuel vapor, and fuel droplets which is supplied by the single-point mixture formation system is delivered to the working cylinders via a connecting duct between the mixture formation unit and the cylinder head, the so-called intake manifold. Depending on fuel composition, degree of atomization, manifold pressure, intake air and fuel temperatures, manifold wall temperatures, manifold geometry, and Reynolds number of the individual phases, part of the droplets contained in the mixture will be deposited during this process, which leads to the formation of a fuel film on the wall. This wall film flows at a substantially lower mean flow velocity than the remaining mixture stream travelling to the individual cylinders. Using the conventional designations of multiphase fluid dynamics, the types of flow prevailing in induction systems with single-point mixture formation, depending on the load point and operating conditions, can be described approximatively as vapor or film distribution, **Fig. 2.22.**

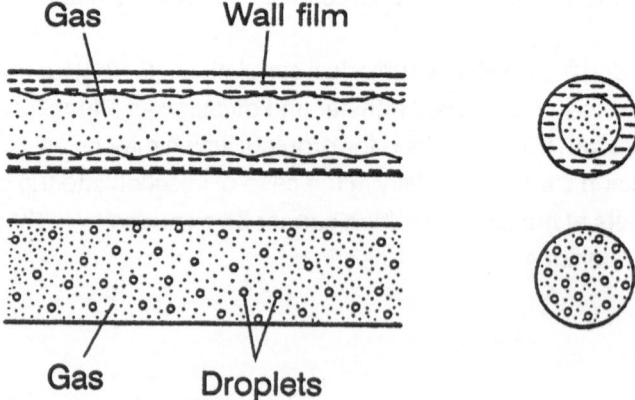

Fig. 2.22. Schematic of vapor and film distribution [2.18]

Whereas a state of equilibrium between the fuel portion being built up on walls as a liquid film on one hand and the fuel portion leaving the wall film in liquid or vaporized form on the other will be gradually established during steady-state engine operation, any change in load point - due to the resultant change of at least one of the above parameters - will lead to a disturbance of this state of equilibrium and thus inevitably to an increase or diminishing of the film-like fuel deposits in the intake manifold. Under transient operating conditions, the mixture therefore varies constantly both in quantity and in composition.

Now, if not only one but several cylinders are supplied with mixture from a single mixture formation unit (single-point mixture formation system), the intake manifold, which is located between cylinder head and mixture formation unit, not only has the function of transporting the mixture but also of distributing it to the various cylinders. It is obvious that distribution of a centrally prepared mixture to, say, four cylinders arranged in line requires elbows and pipe branches and that, in view of the many parameters determining multiphase mixture flow, uniform distribution of all fuel fractions contained in the mixture is an extremely difficult task.

This problem of even mixture distribution has been occupying the minds of engine designers and developers for more than 60 years. If research in the beginning was driven mainly by trying to achieve maximum engine power and operational safety (preventing knock), in the course of the years the center of interest shifted towards minimizing fuel consumption (extension of the lean limit) and finally towards rigorously cutting down on exhaust gas pollutants. Reference [2.17] contains a review of literature, which is an excellent general account of the numerous works published in German and in English until 1965.

The theory of flow and distribution of the individual mixture phases will be discussed in greater detail in the following chapters. The conclusions to be drawn from this for the design and construction of mixture formation systems and intake manifolds will be discussed in the subsequent chapters 4 and 5 headed "Types of Mixture Formation Systems" and "Intake Manifold Design".

2.6.1.2. Air and Fuel Vapor (Gas)

The requirement of rapid flow and uniform distribution of the mixture to the individual cylinders is largely met by the gaseous mixture portions. Investigations on spark-ignition engines operated with gaseous fuels have shown that differences in volumetric efficiency and relative air-fuel ratio λ are generally very small in such engines.

Different volumetric efficiencies may result from non-uniform geometrical intake conditions for the individual cylinders. These differences can be largely avoided by designing the individual intake pipes so as to ensure maximum similarity, i.e. equal shape and dimensions, and by providing identical inlet conditions. In induction systems with single-point mixture formation, the above requirements can be met only to a certain extent in most cases, due to non-gaseous fractions being present in the mixture. Still, the maximum differences in volumetric efficiency in such systems rarely amount to more than 2%.

Varying air-fuel ratios for individual cylinders - which means uneven distribution of the gaseous fuel portion - are usually caused by insufficient blending of combustion gas and intake air at the point of inlet. It is unlikely that segregation of the air and combustion gas or fuel vapor would occur later in the intake manifold, since the differences in density of the two constituents are small. Moreover, secondary flows, which are generated in the pipe elbows and which are super-imposed on the main flow, and swirl motion produced downstream of the throttle valve promote good blending of the charge.

2.6.1.3. Fuel Droplets

If droplets are present in the mixture flow in addition to the gaseous fractions, then segregation in the intake manifold is unavoidable during the reoccurring acceleration and deceleration processes, due to the greatly differing mass inertias of gaseous and liquid portions.

If inertia force F_T of a droplet is set equal to air force F_L acting on the droplet during acceleration, the following equations result:

$$F_T = m_T \cdot b_T = \rho_T \cdot \frac{d_T^3 \cdot \pi}{6} \cdot b_T \tag{2.23}$$

$$F_L = \frac{d_T^2 \cdot \pi}{4} \cdot \zeta_T \cdot \rho_L \cdot \frac{v_{rel}^2}{6} \tag{2.24}$$

m_T [kg]   ..... droplet mass
b_T [m.s^{-2}] ..... droplet acceleration
d_T [m]   ..... droplet diameter
ρ_T [kg.m^{-3}] ..... droplet density
ρ_L [kg.m^{-3}] ..... air density
v_{rel} [m.s^{-1}] ..... relative velocity between droplet and air
ζ_T [-]   ..... drag coefficient of the droplet

and where the empirical equation for the drag coefficient of a sphere can be used to approximatively express ζ:

$$\zeta_T = \frac{24}{Re} + \frac{4}{(Re)^{0,5}} + 0{,}4 \tag{2.25}$$

$$Re = \frac{v_{rel} \cdot d_T \cdot \rho_L}{\eta_L}$$

Re [-]   ..... being the Reynolds number,

then droplet acceleration b_T results from:

$$b_T = \frac{3}{4 \cdot d_T} \cdot \zeta_T \cdot \frac{\rho_L}{\rho_T} \cdot v_{rel}^2 \tag{2.26}$$

When additionally taking into account Equation (2.25) for the drag coefficient ζ, then Equation (2.26) makes evident that acceleration (deceleration) of a droplet moving in an air stream is the greater the higher the relative velocity v_{rel} between droplet and air and the smaller droplet diameter d_T. When the air velocity in the intake manifold is varied, the fuel will follow the air stream the closer with respect to time, the finer it has been atomized. **Fig. 2.23** shows droplet velocities - calculated with a computer program for various droplet diameters - in an intake manifold, based on the assumption that the velocity curve of the air flow is sinoidal and that overlap of the individual inlet valves is 80° crank angle. In Fig. 2.23, the droplet velocity was computed only for engine speed (n = 3000 min^{-1}), whereas **Fig. 2.24** shows the effects of both droplet diameter and engine speed on the droplet travel time.

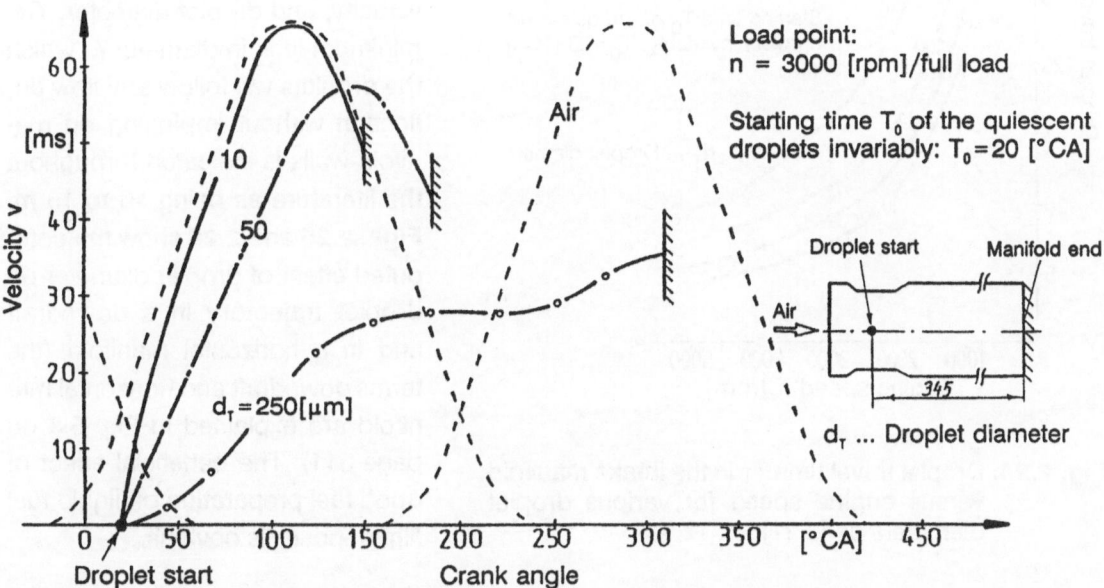

Fig. 2.23. Effect of droplet diameter d_T on droplet velocity v_T versus crank angle, the broken line representing air velocity [2.21]

When Equations (2.23) and (2.24) are now extended to a two- or three- dimensional flow (splitting the forces into the directions x, y, and z) and when buoyant force F_A of the fuel droplets in direction z is included, which yields:

$$F_A = -m_T \cdot g \cdot \left(1 - \frac{\rho_L}{\rho_T} \right) \qquad (2.27)$$

g   [m.s^{-2}] ..... being gravitational acceleration

then the droplet trajectories in the manifold can be computed in addition to the instantaneous

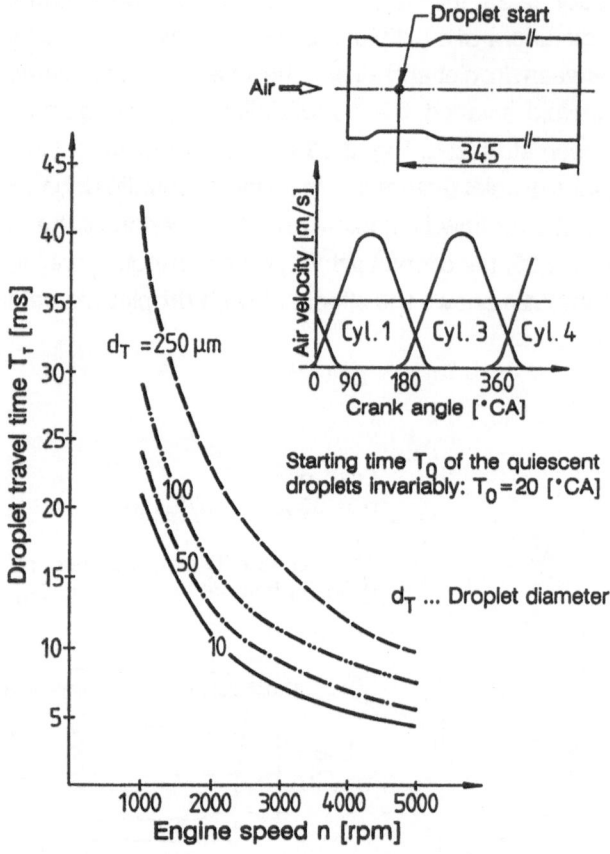

Fig. 2.24. Droplet travel time T_T in the intake manifold versus engine speed for various droplet diameters d_T [2.21]

droplet velocities, if a suitable approach is chosen for the air flow in the intake manifold (e.g. streamline flow). Although the actual processes in the intake manifold can be numerically simulated only in an idealized form, all computations found in the literature [2.21, 2.22, 2.23, 2.24, 2.25, 2.26, 2.27] indicate an appreciable increase in wall film deposits for flow deflections with increasing deflection angle, droplet velocity, and droplet diameter. The minimum limit in diameter at which the droplets will follow any flow deflection without impinging on manifold walls is indicated throughout the literature as being 10 to 15 m. **Figs. 2.25** and **2.26** show the computed effect of droplet diameter on droplet trajectory in a downdraft and in a horizontal manifold (the terms downdraft and horizontal manifold are explained in Fig. 5.3 on page 311). The beneficial effect of good fuel preparation on liquid fuel film deposits is obvious.

While Figs. 2.25 and 2.26 represent computed results, the data shown in **Fig. 2.27** were obtained by a flow bench test which investigated the effect of air velocity at the smallest air funnel cross section - and thus of the degree of atomization - on wall film deposits in an unheated manifold immediately after the carburetor. The test, which was performed on two different induction systems under wide-open throttle conditions, yielded results that show a marked dependence of film deposits on the degree of atomization (deposited fuel portion up to 50%) and thus confirmed the computations.

For maximum uniformity of mixture distribution and minimum wall deposits in the intake manifold, the aim must always be good mixture preparation, i.e. providing a mixture stream with a high portion of vaporized fuel and fuel droplets of small diameter. In single-point mixture formation systems particular attention should be paid to the full-load and near full-load operating range, since in the part-load range mixture formation will usually be satisfactory anyhow, due to the secondary atomization that takes place with a largely closed throttle valve.

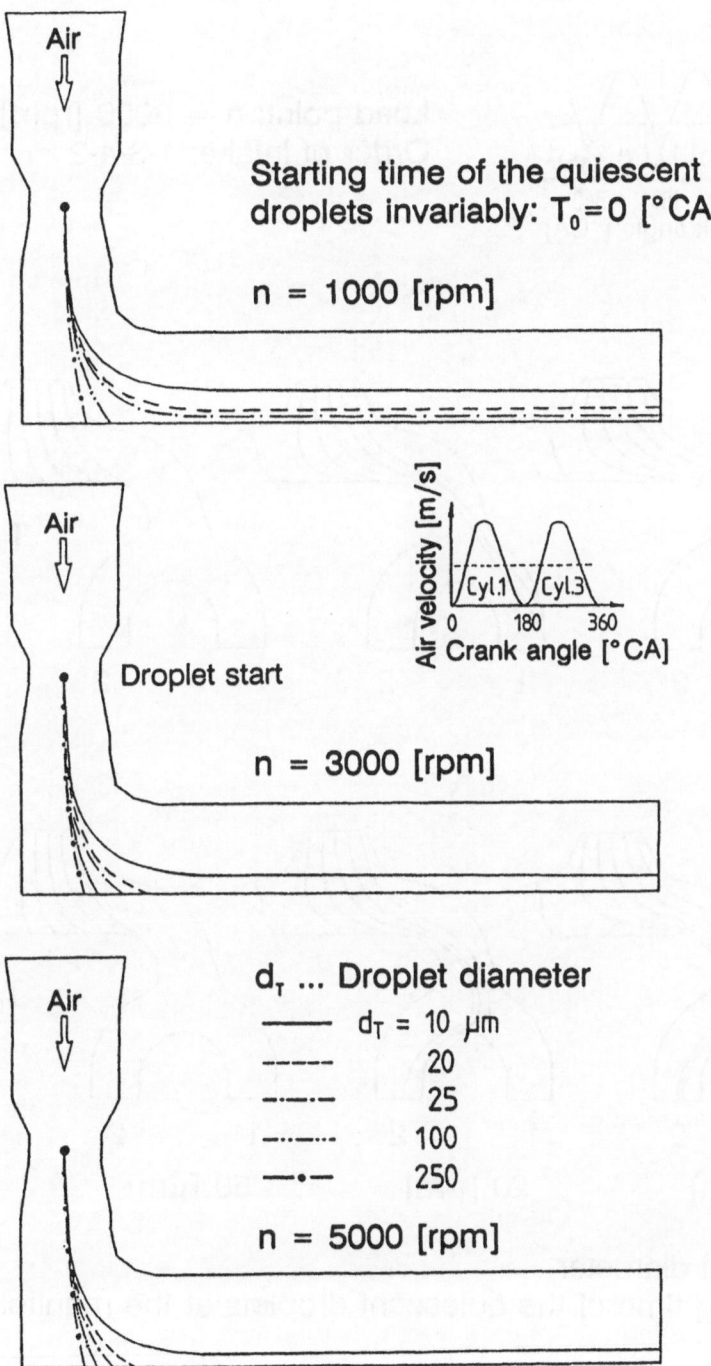

Fig. 2.25. Computed effect of droplet diameter and engine speed on the droplet trajectory in a downdraft manifold under wide-open throttle conditions [2.21]

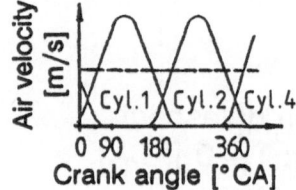

Load point: n = 3000 [rpm]/full load
Order of intake: 1-3-4-2

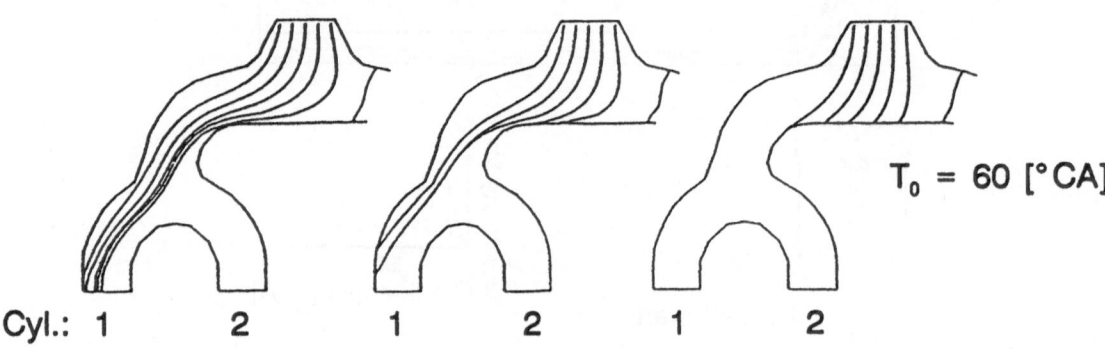

T_0 = 60 [°CA]

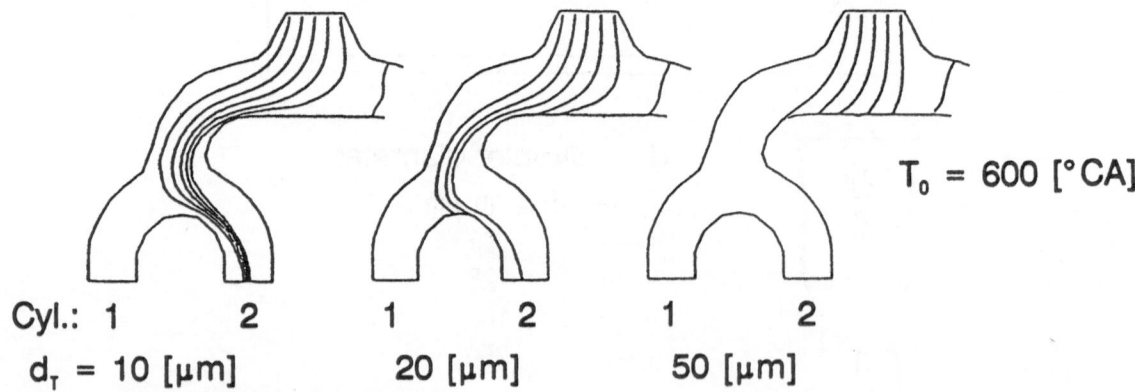

T_0 = 600 [°CA]

d_T ... Droplet diameter
T_0 ... Starting time of the quiescent droplets at the manifold inlet

Fig. 2.26. Computed effect of droplet diameter on droplet trajectory in a siamese horizontal manifold under wide-open throttle conditions [2.21]

Regarding the intake manifold design, the objective is to provide for a flow with least possible and smallest possible deflections (small deflection angles, large radii) and intense heating of those locations where wall film deposits are extensive.

The effect of gravity on the droplet trajectory is negligible, due to the very short residence times of the droplets in the manifold, see Fig. 2.24.

2.6.1.4. Wall Film

As described in the previous chapter, inadequate fuel preparation in combination with a deflection of the mixture stream leads to an impingement of part of the droplets on the manifold walls and consequently to the formation of a liquid fuel film. In accordance with Fig. 2.27, the film-like fuel portion may account for up to one half of the total amount of fuel inducted under wide-open throttle conditions.

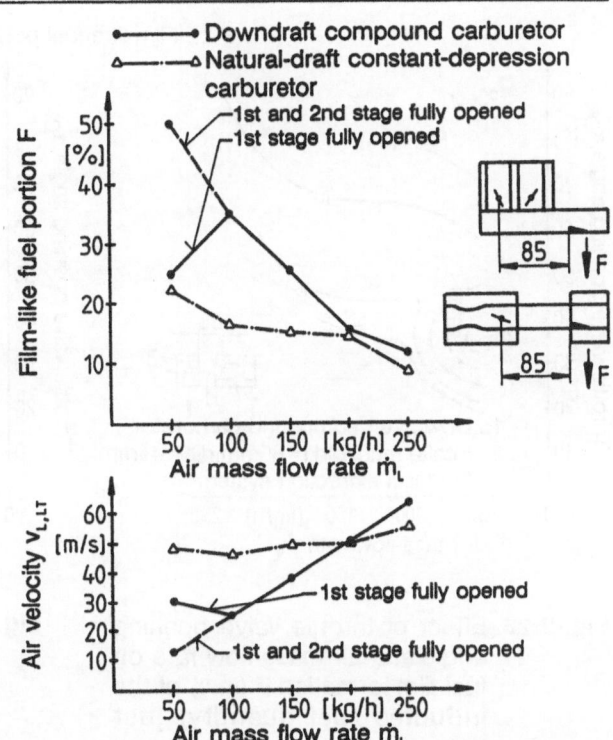

Fig. 2.27. Effect of air mass flow rate corresponding to air velocity at the smallest air funnel cross-section on fuel film deposits for various types of carburetors under wide-open throttle conditions [2.20]

Measurements in straight, unheated single-cylinder horizontal manifolds [2.20, 2.29] have shown that mixture deflection in the manifold does promote the formation of a fuel film but is not a necessary precondition for it. As can be seen from the measurement results charted in **Figs. 2.28** and **2.29**, up to 50% of the fuel contained in the mixture may be deposited on the walls even in absolutely straight manifolds. Studies on the distribution of fuel concentrations across the flow cross section as a function of pipe length [2.30, 2.31, 2.32, 2.33] have shown that leaning-out of the core or, respectively, enrichment of the outer edges of the mixture stream increases with increasing pipe length. The turbulent cross motions occurring in the mixture stream lead to a persistent impingement of the larger droplets near the edges on the surrounding pipe walls. In this context the degree of liquid fall-out is very dependent on the mixture intake conditions at the inlet to the intake manifold. The closing angle of the throttle valve (see **Fig. 2.30**), sharp edges, or other flow obstacles have a prolonged effect on fuel film formation, mixture flow and subsequent mixture distribution. As can be seen from Fig. 2.28, up to 20% of the fuel may be present in the form of a liquid fuel film just downstream of the mixture formation unit not only under wide-open throttle conditions but also in the part-throttle range (directional effect of the throttle valve).

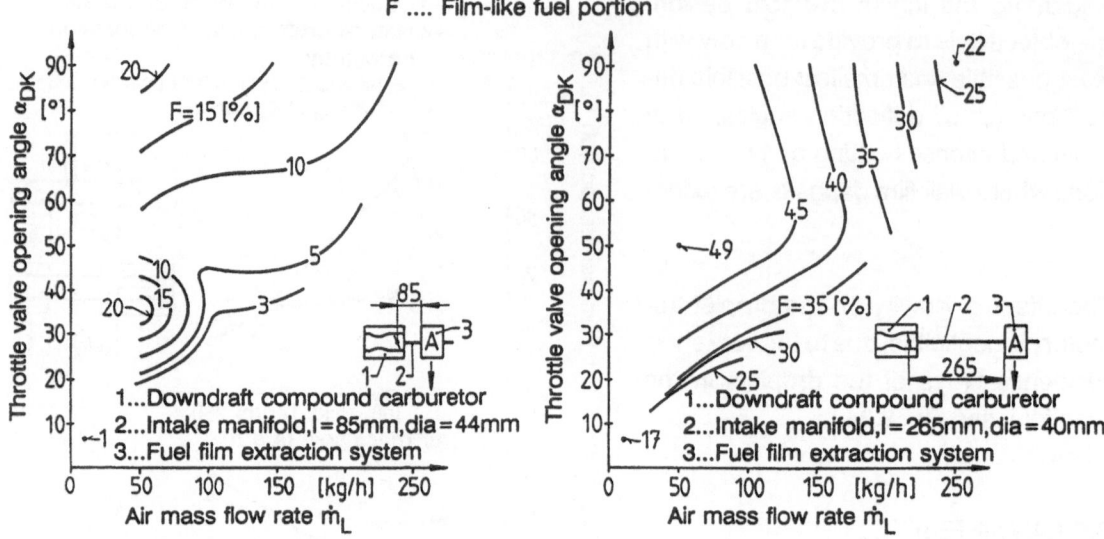

Fig. 2.28. Effect of throttle valve opening angle and air mass flow rate on fuel film formation F (in % of the inducted fuel quantity) just downstream of the carburetor [2.20]

Fig. 2.29. Effect of throttle valve opening angle and air mass flow rate on fuel film formation F (in % of the inducted fuel quantity) in a straight, unheated manifold [2.20]

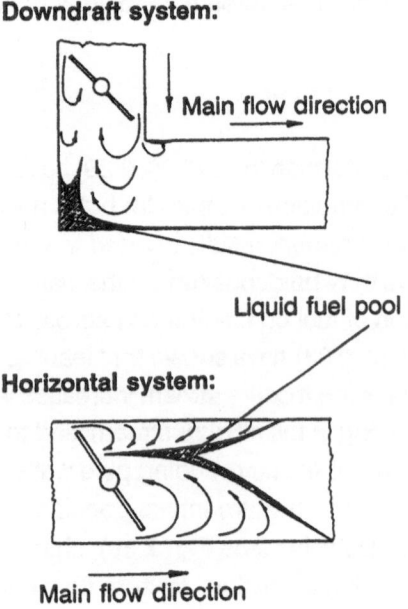

Fig. 2.30. Wall film flow and wall film accumulation just downstream of the throttle valve [2.20]

Under cold-start and warmup engine operating conditions, i.e. when the mixture has not yet been sufficiently preheated, a fuel film may be deposited on the walls even in case of excellent fuel preparation in the mixture formation unit (e.g. complete vaporization). For if the fresh charge in the manifold does not attain mixture saturation temperature, part of the fuel will inevitably condense on the still cold manifold walls until, finally, a state of equilibrium is restored between mixture temperature, mixture pressure (manifold pressure), and new lambda of the remaining gaseous mixture portion. As is evident from **Fig. 2.31**, when taking into account usual lambda values during cold-start and warmup, there is a high probability of the mixture saturation temperature not being attained and thus a wall film being built up throughout the entire engine map.

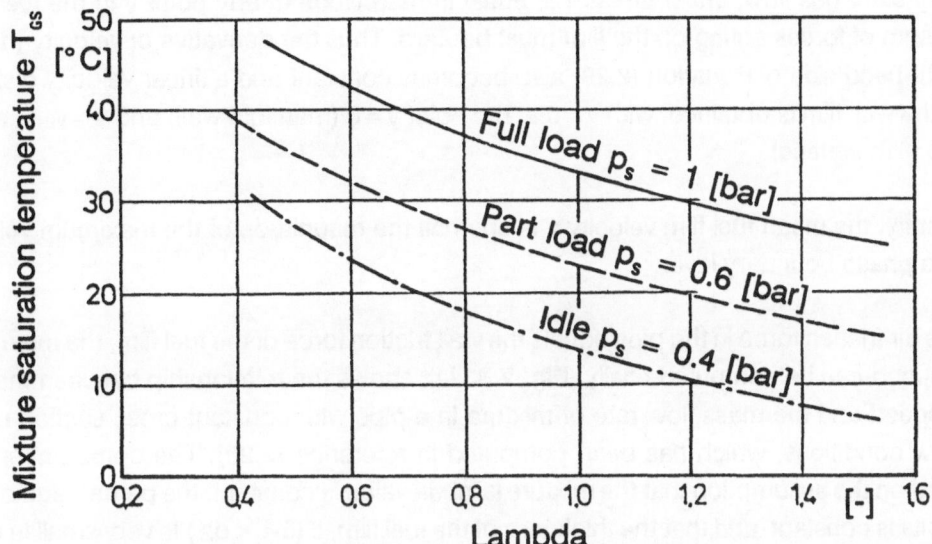

Fig. 2.31. Mixture saturation temperature T_{GS} versus relative air-fuel ratio λ (parameter: manifold pressure p_s) for commercial premium- grade gasoline [2.34]

The gas stream passing over the fuel film surface transports the fuel film, which has accumulated on the manifold walls for different reasons, towards the cylinder head on account of shear stress on the phase boundary area. During this process, the shear stress on the fuel film surface is transferred to the other liquid film layers by internal friction.

For "Newtonian fluids" of laminar flow (e.g. gasoline) the shear stress, τ, in the fluid is proportional to the velocity gradient (dv/dy) and to the dynamic viscosity, see **Fig. 2.32.**

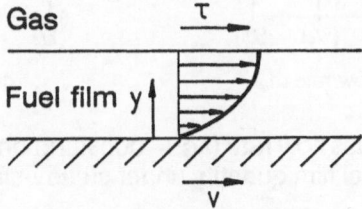

Fig. 2.32. Shear stress and velocity in the liquid fuel film

$$\tau_{Fl.} = \eta_{Fl.} \cdot \frac{dv}{dy} \tag{2.28}$$

$\tau_{Fl.}$ [N.m^{-2}] ..... shear stress (transverse stress) in the fluid
η_F[kg.m^{-1}.s^{-1}] ..... dynamic viscosity of the fluid
v [m.s^{-1}] ..... fuel film velocity at position y
δ [μm] ..... wall film thickness

In a steady-state gas flow, shear stress τ is equal in magnitude at any point y in the fuel film, since the sum of forces acting on the film must be Zero. Thus the derivative of velocity (dv/dy) on the right-hand side of Equation (2.29) also becomes constant and a linear velocity distribution over the wall film is obtained, with v = 0 at the point y = 0 (manifold wall) and v = v_{max} at the point y = δ (film surface).

Consequently, the mean fuel film velocity v is one-half the magnitude of the maximum velocity v_{max} in the phase boundary layer.

Now, if the air friction force in the pipe equals the wall friction force of the fuel film, the mean wall film flow speed can be computed easily. **Fig. 2.33** left shows the relationship between the wall film flow speed and the mass flow rate of mixture in a pipe with constant cross section under steady-flow conditions, which has been computed in reference [2.20]. The computation has been based on the assumption that the mixture lambda value is constant, the percentage of wall film deposits is constant, and that the thickness of the fuel film, δ ($\delta < < d_S$) is very small in comparison to the pipe diameter. Based on the condition of continuity, the overproportional increase of fuel film velocity leads to a decrease in film thickness, which results in the reduction of the accumulated fuel film quantity as shown in Fig. 2.33 right.

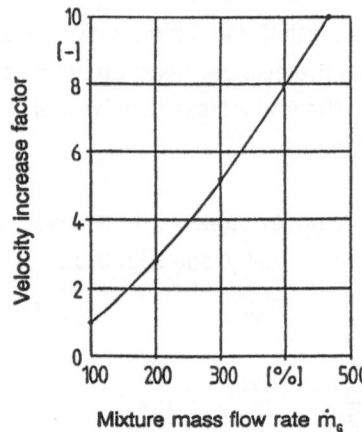 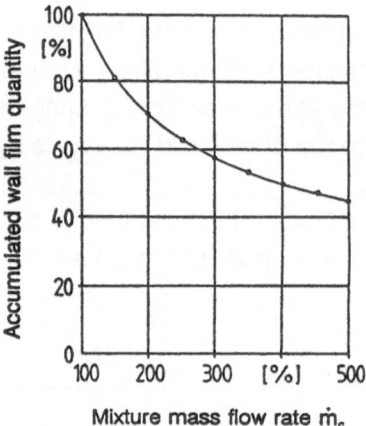

Fig. 2.33. Effect of mixture mass flow rate (λ_G = constant) on the mean fuel film flow speed and the accumulated fuel film quantity under steady-state pipe flow conditions [2.20]

Contrary to the above assumption, the flow conditions prevailing in intake manifolds of spark-ignition engines are not those of steady but of a vigorously pulsating flow. This leads to persistent accelerations and decelerations of the surface flow, resulting in a non-linear velocity distribution over the fuel film. However, the relationships shown in Fig. 2.33 are not fundamentally changed by this, which means that a higher mean gas velocity in the manifold will cause a higher mean wall film velocity and reduced wall film deposits.

In addition to fuel film deposits and fuel film flow on manifold walls, fuel film vaporization of varying intensity, depending on the load point and engine operating conditions, takes place in the intake manifold.

Owing to the turbulent mixture stream passing over the fuel film surface at low fuel partial pressure, a higher wall film portion is able to vaporize than suggested by the distillation curve in Fig. 2.2 for the respective temperature. Since the more volatile fuel fractions vaporize first, an increasing residence time of the fuel film in the intake manifold will lead to an increasing content of heavier fractions in the fuel film due to fractional distillation [2.24, 2.30, 2.31, 2.35, 2.36, 2.37, 2.38]. Owing to the fact that these fractions differ in antiknock quality from the original fuel, uniform mixture distribution has to be ensured not only in quantitative terms but also in qualitative terms (uniform distribution of fuel constituents). Due to vigorous vaporization of the fuel film in a warmed-up manifold, an overall decrease in film thickness in the direction of the cylinder head is observed [2.39].

In addition to transporting the fuel film, the boundary shear stress acting on the fuel film surface also leads to the formation of surface waves, with bits of fuel film being torn off the surface in form of droplets at very high gas velocities.

The influence of gravity on the flow of fuel film is substantially greater, due to the much longer manifold residence time of the particles flowing in the film, than on the flow of droplets [2.20, 2.22, 2.38, 2.39, 2.41, 2.73, 2.109, 2.110]. Especially in the lower full-load range, which is characterized by abundant fuel film accumulation (moderate gas velocity, large film thickness), only slight inclinations of the intake manifold can already effect considerable changes in distribution of the fuel film [2.20, 2.35, 2.67], see **Fig. 2.34**.

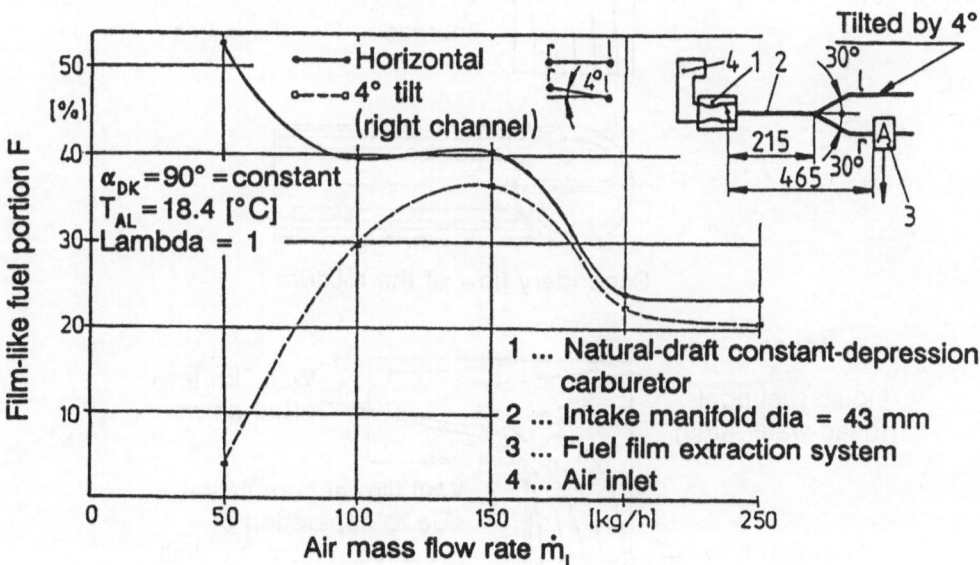

Fig. 2.34. Effect of a slight intake manifold tilt on the fuel film quantity extracted in the right channel [2.20]

In the present study, fuel film was extracted downstream of a manifold branch-off. First, the branch-off was held in a horizontal position and a portion of 52% of the inducted fuel quantity was found to have deposited at an air mass flow rate of 50 kg.h^{-1}. When tilting the channel together with the extraction device, the portion deposited as film dropped to about 4%, which means the fuel film did not follow the 4° ascent but flowed into the other branch-off.

Finally, the flow characteristics of the liquid fuel film in pipe elbows or knees will now be treated in more detail.

First, the flow of pure gas in a pipe will be considered. Owing to wall friction, a flow of pure gas is characterized by a higher flow velocity in the center than near the outer edges. When the flow is deflected in a pipe elbow, the particles flowing in the center tend to move outward due to higher centrifugal forces; this is only possible, however, if a contrary (inward) motion sets in on the side walls. This leads to the generation of eddies which are superimposed on the axial main flow - this is called "secondary flow".

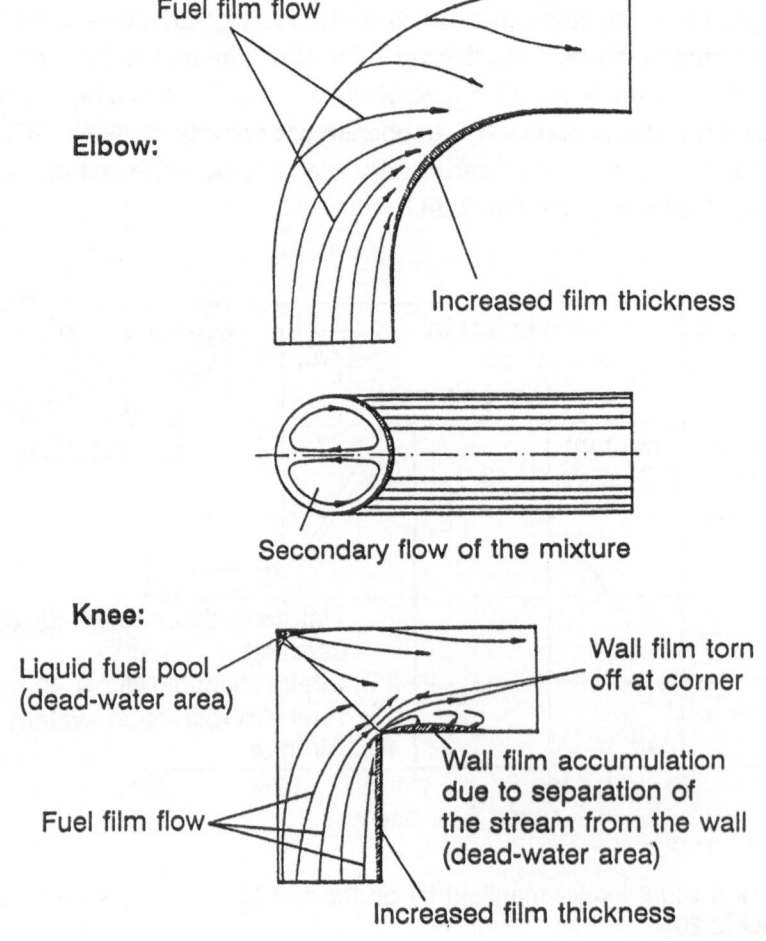

Fig. 2.35. Mixture flow in a pipe elbow (top) or knee (bottom) as a result of secondary flows [2.20]

Now if, instead of a single-phase flow (**Fig. 2.35**), a multiphase fuel stream consisting of gas, droplets, and liquid fuel film is involved, the liquid fuel film accumulated and flowing along the pipe walls is transported to the inside of the bend by the influence of eddies, which leads to an increase in film thickness there. Comprehensive empirical studies [2.20, 2.32, 2.39, 2.40, 2.41, 2.42, 2.43, 2.44] on wall film locations in intake manifolds of spark-ignition engines have confirmed the events described above.

If, again here, droplets are torn off the waves that start to form on the fuel film surface, they first travel to the outside of the bend, then flowing along the top and bottom walls back to the inside.

In respect to pressure distribution across the pipe cross section, a pressure rise can be observed toward the outside of the bend, since the centrifugal forces acting during a motion around a bend have to be taken up by the particles travelling on the outside. According to Bernouilli's equation for a steady, incompressible horizontal flow:

$$p_1 + \frac{\rho \cdot v_1^2}{2} = p_2 + \frac{\rho \cdot v_2^2}{2} = \text{const.} \tag{2.29}$$

velocity differences result, which lead to a deceleration of the stream when it reaches the outside of the bend or, respectively, to an acceleration when it reaches the inside. Immediately after having passed the inside of the bend, the flow is strongly decelerated again, accompanied by a strong pressure rise, which usually causes a separation of the stream from the wall at this point. Separation of the stream from the wall in intake manifolds should be avoided as far as possible, since this - in addition to an increased pressure loss - causes fuel film buildup (the fuel film is not transported) at the area of separation (see Figs. 2.30 and 2.35)

Therefore, the manifold design to be aimed at should principally avoid sudden expansions of the cross section as well as sharp corners or edges.

2.6.2. Mixture Flow and Distribution in a Multi-Point Mixture Formation System

The processes of mixture flow and distribution in induction systems with multi-point mixture formation differ fundamentally from those of single-point mixture formation systems.

The very late addition of fuel to the intake air as compared to single-point mixture formation systems (fuel is injected just before or directly into the inlet valve) leads to the situation that the major part of the intake duct is passed by a stream of pure air. The numerous problems of a multiphase flow, such as phase segregation, uniform distribution of phases, fuel film formation, fuel film buildup on walls, etc., do not occur in this section and consequently there are much more options of designing the intake system with a view to utilizing gas-dynamic supercharging effects.

The object of supercharging is to achieve higher full-load volumetric efficiencies and thus higher mean effective pressures, especially in the lower engine speed range. By providing a higher transmission ratio of the driven axle, equal driving performance can be achieved at lower engine speeds. Consequently, the road resistance curve is displaced to regions of improved fuel consumption in the engine map (see **Fig. 2.36**), which is due mainly to reduced friction in the full-load range and, additionally, to lower charge exchange losses and better conditions of combustion (higher final combustion pressure, decreased residual gas portion) in the part-load range. Savings in specific fuel consumption can be achieved also on account of the fact that numerous engine components such as valve drive, piston, bearings etc. have to be designed only for the lower speed range.

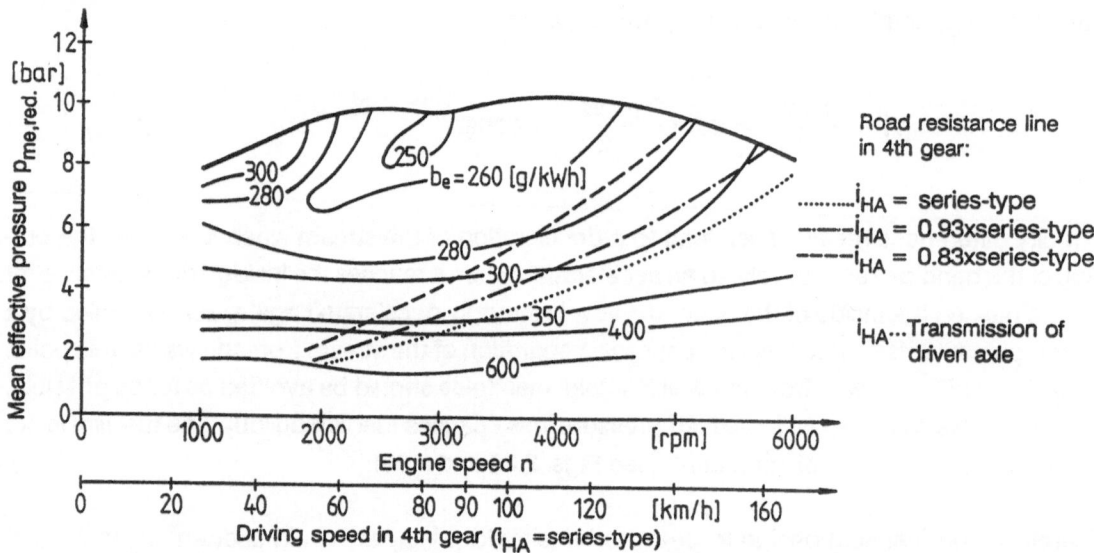

Fig. 2.36. Reduction of the specific fuel consumption of a four- cylinder spark-ignition engine by displacement of the characteristic road resistance line (by modification of the rear axle transmission i_{HA}) to improved combustion map regions [2.56]

There are basically two types of "dynamic supercharging" (that is, no use of a compressor, no utilization of exhaust gas energy) in an internal-combustion engine: Ram pipe supercharging and tuned- intake pipe charging.

The physical principle of ram pipe supercharging is the utilization of the charging effect of pressure waves that travel in intake manifolds from the receiver to the individual cylinders [2.57].

For the desired resonance engine speed, a tank/pipe-system capable of oscillating is designed such that the periodical cylinder intake cycles coincide with the natural oscillation frequency of the tank/pipe-system [2.58]. Based on this resonance, supercharging - in comparison to the situation in a naturally-aspirated engine - of all connected cylinders occurs at resonance engine speed.

The increase in volumetric efficiency and thus in mean effective pressure is limited, however, to a certain engine speed range for both types of dynamic supercharging. When combining the two systems, the increase in mean effective pressure can be extended to a greater speed range [2.59]. The same is true for a combination of tuned-intake pipe supercharging and turbocharging, which leads to a substantial improvement of the torque weakness of turbocharged engines in the lower speed range [2.60, 2.61, 2.62]. Intake system design concepts and computations for ram pipe supercharging and tuned-intake pipe supercharging and the results achievable with such systems are discussed in greater detail in Chapter 5.2 "Intake Manifold Design for Multi-Point Mixture Formation Systems".

The events of mixture flow in the fuel-wetted, short pipe section upstream of the inlet valve are similar to those which occur in intake manifolds of single-point mixture formation systems. Immediately after the fuel has been discharged into the manifold, vaporization of the droplets and wall film commences with varying intensity, depending on the prevailing pressures and temperatures. On account of the small wettable surface area and the high temperatures prevailing near the inlet valve, the fuel portions stored in the form of a liquid film are much smaller than in intake manifolds of single-point mixture formation systems. Fine atomization of the fuel at the injection nozzle also contributes to faster mixture flow and reduced fuel film formation in this case.

Improved mixture preparation can be achieved also by prolonging the very short time available for preparation. This is possible by providing a longer distance between the injection nozzle and the inlet valve. In view of the transient operating conditions of the engine, this however requires a correspondingly stronger acceleration enrichment. A better option is to inject the fuel as early as possible - with individual injection timing for each cylinder - into the still closed valve. Injection into the open inlet valve should generally be avoided on account of the high relative air-fuel ratio gradients which result from droplet vaporization.

Whereas the multiphase mixture flow exhibits similar characteristics to those found in induction systems with single-point mixture formation, mixture distribution is governed by completely different mechanisms. Since in multi-point injection systems, no distribution of the fuel-air mixture occurs in the section between the injection nozzle and the inlet valve, but fuel is metered individually to each cylinder, the quality of mixture distribution depends exclusively on the uniformity of air and fuel metering to the individual cylinders.

Comprehensive investigations on fuel-injected engines have shown [2.63, 2.64] that the main cause of maldistribution of fuel among the individual cylinders are metering errors of the injection nozzles (flow rate tolerances). These errors are greatest at low flow rates (lower load range) and decrease as flow rates increase (higher load), which explains the typical mixture maldistribution in fuel-injected engines in the lower load range (maximum equivalence ratio deviation Δ_{Kmax} between the richest and the leanest cylinder up to 10%), see **Fig. 2.37**.

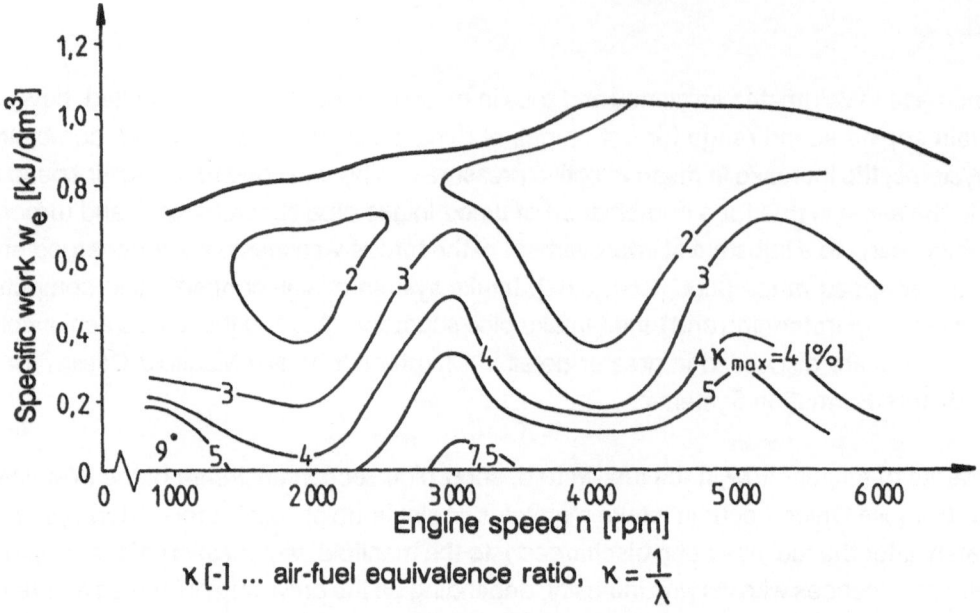

$$\kappa\ [-]\ ...\ \text{air-fuel equivalence ratio,}\quad \kappa = \frac{1}{\lambda}$$

$\Delta\kappa$ [%] .. Deviation of the individual cylinder equivalence
ratio from the calculated mean equivalence ratio
of all four cylinders

Fig. 2.37. Maximum equivalence ratio difference $\Delta\kappa_{max}$ between richest and leanest cylinder in
the map of a fuel-injected engine [2.63]

These metering errors are accentuated (or partly compensated) by pressure variations in the
fuel handling system and, because of different pipe routing, may vary slightly in magnitude from
valve to valve.

Differences in individual volumetric efficiencies begin to have an adverse effect on mixture distribution only at high air mass flow rates; they cause maximum deviations of 3-4% [2.63].

In intermittent multi-point injection systems, there may be yet another cause for mixture maldistribution, when the coordination of injection timing to valve timing differs from cylinder to cylinder. As shown in Fig. 2.38, this leads to fuel injection into the individual ports at different
manifold vacuum pressures. The varying differential pressure at the nozzle causes metering variations of up to 2% [2.63]. A remedy for this is to use a fuel-injection system with individual timing
of injection for each port ("sequential injection"). A detailed explanation of this type of injection
is given in Chapter 4.2.

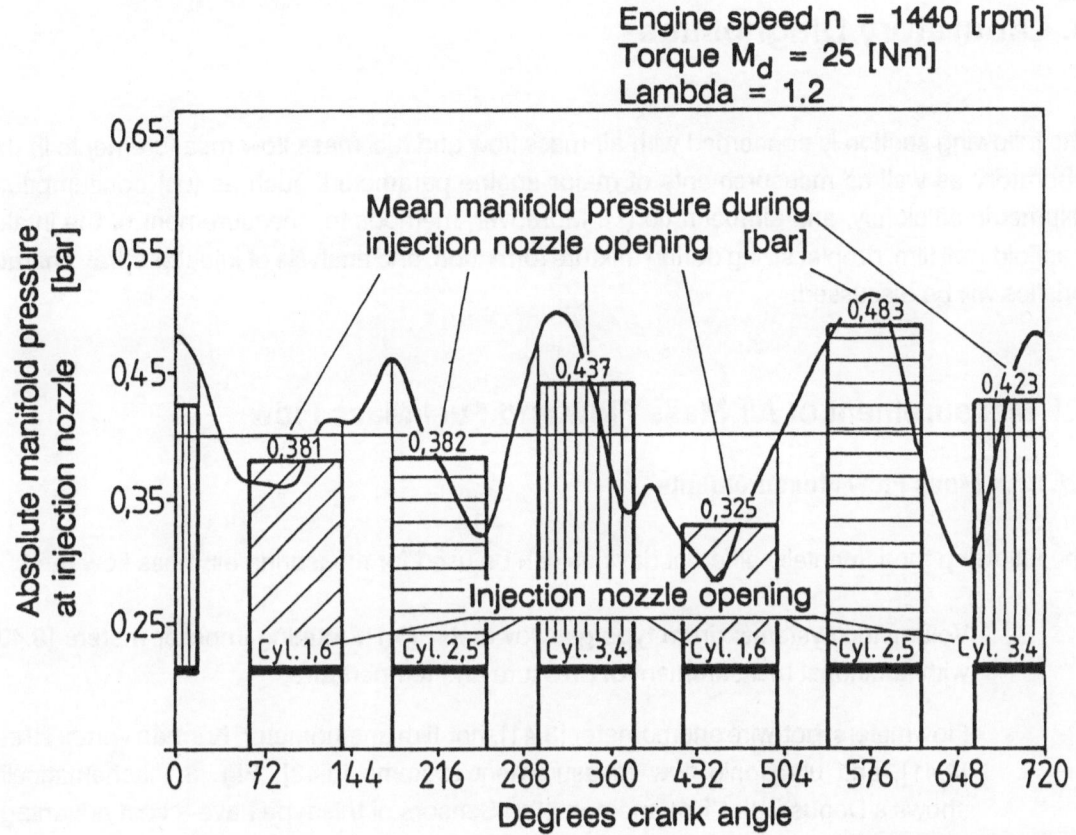

Fig. 2.38. Trace of absolute manifold pressure versus crank angle and graphical presentation of the mean absolute manifold pressures during injection nozzle opening times [2.64]

3. Laboratory Diagnostics

The following section is concerned with air mass flow and fuel mass flow measurements in the laboratory as well as measurements of major engine parameters such as fuel consumption, volumetric efficiency, and air-fuel ratio (λ). Moreover, methods for measurement of the intake manifold wall film, droplet sizing during mixture formation, and analysis of injector spray characteristics will be discussed.

3.1. Measurement of Air Mass Flow and Fuel Mass Flow

3.1.1. Air Mass Flow Measurements

The following fundamentally different devices can be used for measuring air mass flow rates:

- Volumetric systems: drum-type gas flowmeter [3.1], rotating impeller meters [3.40] with additional measurement of pressure and temperature;

- Flowmeters: hot-wire anemometer [3.41], hot-film anemometer, Kármán vortex street [3.41], and ultrasonic flow measurement systems [3.42]. **Fig. 3.1** schematically shows a Degussa hot-film anemometer: Sensors of this type have a cost advantage over hot-wire anemometers, however, they have a slightly less favorable response to changes in flow pattern;

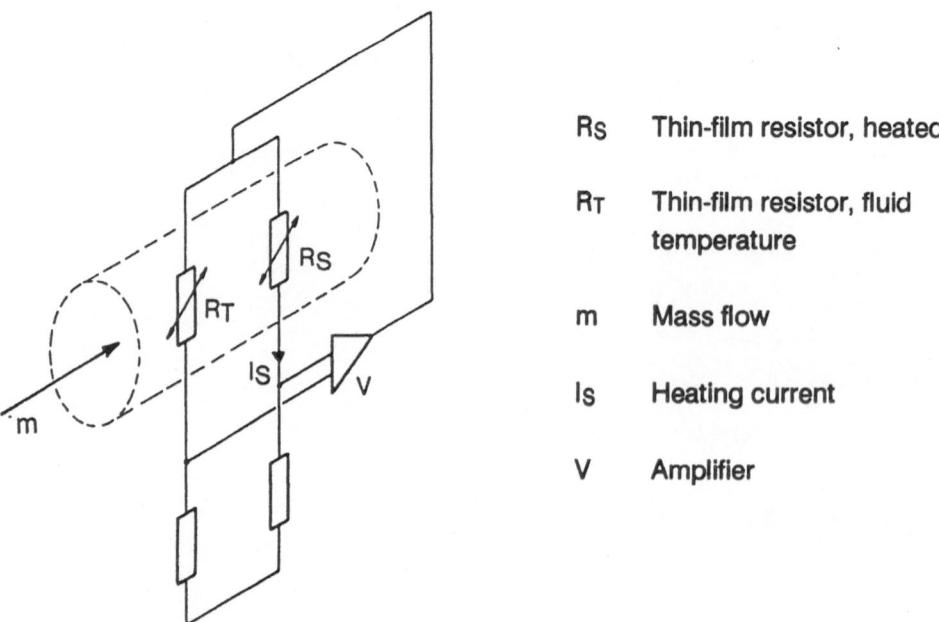

R_S	Thin-film resistor, heated
R_T	Thin-film resistor, fluid temperature
m	Mass flow
I_S	Heating current
V	Amplifier

Fig. 3.1. Schematic diagram of a Degussa hot-film anemometer [3.43]

- Orifice plates: This type of measurement is rarely used in internal combustion engines due to flow losses and pulsations which occur in such systems [3.44];

- Prandtl tubes: These are also rarely used due to the pulsating nature of the flow [3.44].

The main problem in air mass flow rate measurement is the difficulty of determining the absolute value, due to flow pulsations. Most methods, however, do exhibit an excellent reproducibility.

3.1.2. Fuel Mass Flow Measurements

The following methods are commonly used for measuring fuel mass flow rates:

- Gravimetric methods: fuel-weighing devices [3.45];

- Volumetric methods: Seppeler volumeters [3.46] for measuring the flow of a given fuel volume in a period of time to be determined, in some cases automatically by the use of light barriers; flowmeters of the Pierburg-PLU-type shown in **Fig. 3.2**; another option is the measurement of injection nozzle opening times to determine fuel mass flow rates.

1 Gear-type flowmeter
2 Differential-pressure
  sensor
3 Servo-drive
4 Control unit

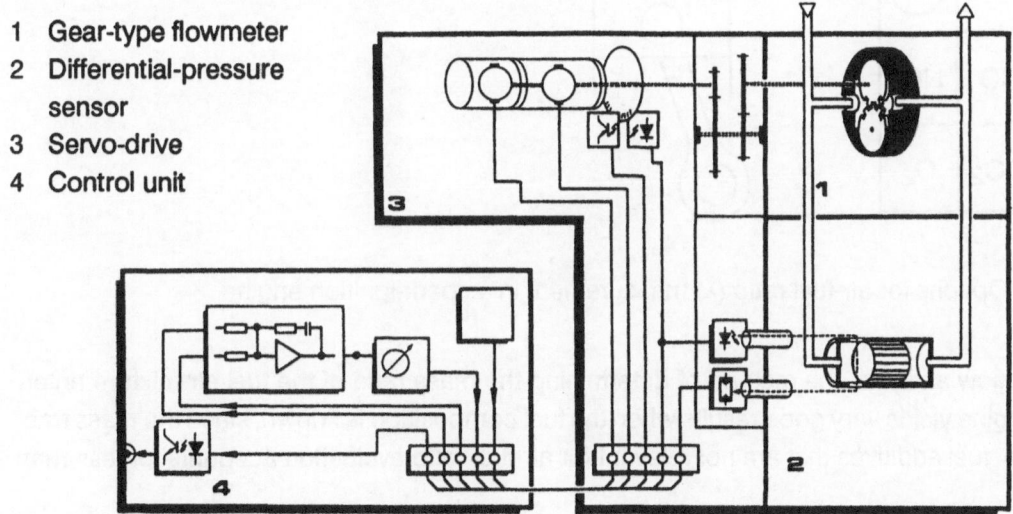

Fig. 3.2. Pierburg PLU flowmeter [3.48]

In the majority, the described techniques of metering fuel mass flow rates are suitable only for steady flow measurements, with the exception of the last method mentioned.

3.2. Determination of the Air-Fuel Ratio (Lambda)

3.2.1. General

The following basic options are available for measurement of the air- fuel ratio λ, which is the central parameter that gives an indication of normal operation of the mixture formation system (carburetor, injector, or gas mixer) - see **Fig. 3.3**:

- Fuel-to-air mass ratio on the upstream side of the engine;

- Balance procedure based on the molar concentrations of the exhaust gas on the downstream side of the engine.

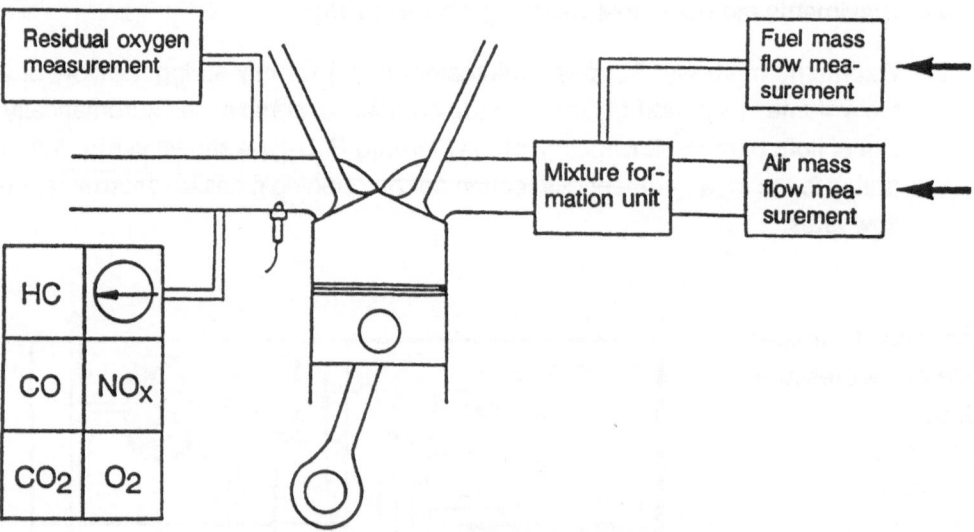

Fig. 3.3. Options for air-fuel ratio (λ) measurement in a spark-ignition engine

In steady-flow analysis, the method of determining the mass ratio of the fuel-air mixture entering the engine yields very good results when the fuel composition is known, since the mass fraction of the fuel additives that are not relevant for air-fuel ratio evaluation accounts for less than 0.5%.

For transient analysis, however, especially for analysis of individual-cylinder air-fuel ratios in multicylinder engines, the air mass flow rate measurement has to be taken near the inlet valve due to oscillations of the inlet air and pulsations in the manifold. Moreover, the blowdown air and fuel masses have to be known in order to be able to determine the mixture composition in the cylinder, and there is also the additional problem of sometimes having to deal with a nonhomogenous mixture distribution due to charge stratification in the combustion chamber.

Therefore, air-fuel ratio measurements are preferably carried out on the downstream side of the engine.

This means that, knowing the mole weights of the substances, the fuel/air mass ratio can be evaluated from the concentrations of the measured exhaust gas constituents and the atom balance procedure covering all constituents. Since it is impossible to measure all exhaust gas constituents with reasonable effort, a variety of methods for computing the air-fuel ratio (λ) has been developed that differ from each other mainly in the number of constituents measured and in the methods used for calculating those constituents that cannot be measured. An essential parameter in this context is the water-gas reaction equilibrium constant K, which permits a temperature-dependent calculation of the exhaust gas constituent H_2, which is difficult to measure, in the rich operating range.

The following are a few examples of air-fuel ratio computation procedures: D'Alleva and Lovell [3.7], Weatherford [3.3], Spindt [3.4], Eltinge [3.5], Lange [3.6], and Brettschneider [3.7]. Among the mentioned methods, the Brettschneider procedure, based on measurement of five exhaust gas constituents (CO_2, CO, O_2, HC, and NO_x) and on the inclusion of oxygenized fuels and intake air humidity, yields the most accurate results.

Generally it holds true that the lower the number of exhaust gas constituents being measured, the more inaccurate the results obtained.

This statement does not hold true for single-constituent air-fuel ratio calculation methods on the basis of post-combustion of the unburned exhaust gas constituents CO, HC, and H_2 with subsequent determination of the residual oxygen content. This category of measurement techniques includes:

- Oxygen sensors incorporated in the exhaust system and

- Single-constituent air-fuel ratio metering instruments such as the so-called "Lambda-meter" developed at the Institute of Internal Combustion Engines and Automotive Engineering at Technical University of Vienna, where the O_2-sensor operates under constant ambient conditions [3.8].

Complete combustion by definition means that free oxygen can be present in the exhaust gas only with excess air. If measurements are to be made also in the rich lambda range ($\lambda < 1$), oxygen or air has to be added to the exhaust gas.

When using oxygen sensors, this addition of oxygen is accomplished by utilizing the oxygen ion pumping effect when applying voltage to a zirconium-dioxide sensor [3.9], which causes O_2-ions from the ambient air to be pumped to the marginal zone or into the space between the diffusion layer and the ZrO_2 ceramic unit. Post-combustion takes place in this marginal zone. When simultaneously measuring the electromotive force of the sensor, an almost constant oxygen concentration in the marginal zone can be adjusted.

In the case of single-constituent air-fuel ratio metering instruments, air is added to the exhaust gas via capillaries or hypercritical flow nozzles. For details refer to reference [3.10].

3.2.2. Accuracy of Different Air-Fuel Ratio Measurement Techniques

3.2.2.1. Atom Balance Procedure

In reference [3.7], Brettschneider offers a detailed error analysis of this multi-constituent measurement technique. Additional sources of error are indicated in **Fig. 3.4**.

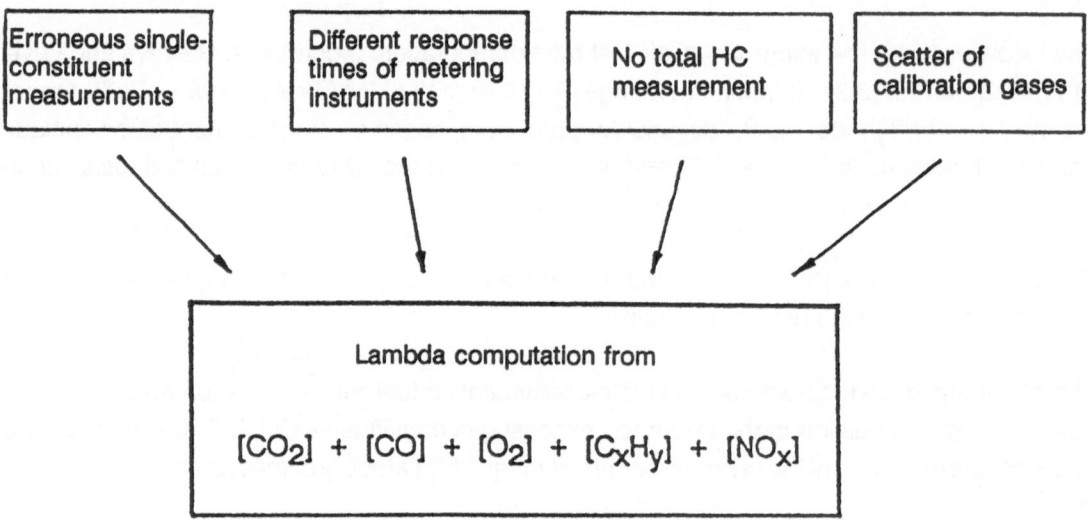

Fig. 3.4. Sources of error in multi-constituent air-fuel ratio measurement procedures

Very briefly, these are:

- Measurement errors and different response times of exhaust gas metering instruments;

- Inaccurate total hydrocarbon measurements;

- Scatter of calibration gases and differing temperatures of calibration gas and analyzed gas.

When paying attention to the mentioned sources of error, high accuracy of measurement can be achieved with this method. An increase of the mean lambda error from 0.5% for $\lambda = 1$ to 1.5% for $\lambda = 1.3$ [3.7] is observed.

Necessary requirements to achieve such high accuracies are precise laboratory metering instruments, heated sample lines, and an accurate total HC measurement. Paramagnetic devices, which are practically free of interference by unburned exhaust gas constituents, should be used for oxygen measurements.

3.2.2.2. Single-Constituent Air-Fuel Ratio Computation

In contrast to oxygen sensors, which measure continuously, lambda closed-loop control units (three-way catalyst converter systems), which basically operate like a switch, are excluded from the discussion here, since they only utilize the voltage increase toward infinity in accordance with Nernst's theorem for oxygen concentrations of zero percent by volume.

Three tasks have to be accomplished in this context:

- Addition of oxygen to the exhaust gas, in order to be able to measure in the rich mixture range;

- Oxidation of the unburned fractions of the exhaust gas constituents CO, HC, and H_2;

- Measurement of the residual oxygen content.

The following factors affect these tasks:

a) When using oxygen sensors, illustrated in **Fig. 3.5**:

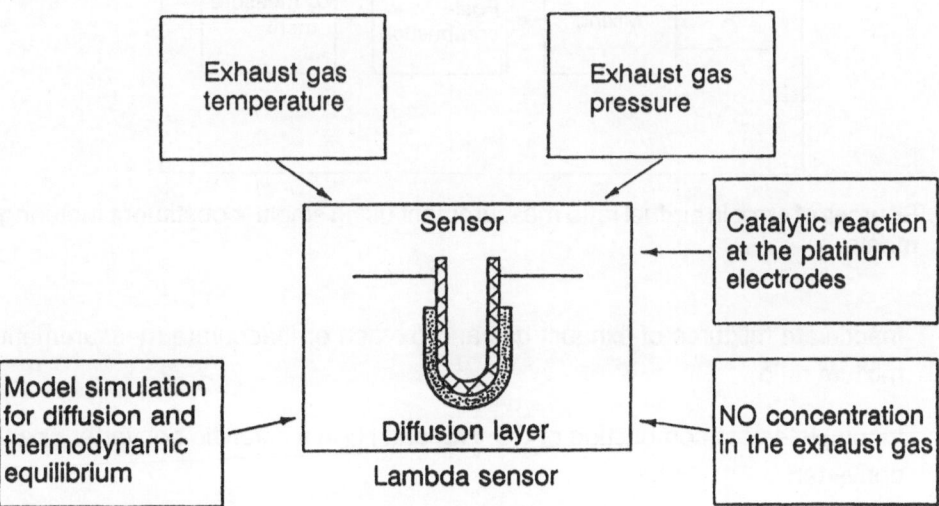

Fig. 3.5. Sources of error in air-fuel ratio (λ) measurements using oxygen sensors

- Temperature variations of the exhaust gas and, in particular, the unknown temperature gradient in the solid electrolyte;

- Pressure variations in the exhaust gas, which influence the electromotive force in the case of voltage sensors and the oxygen ion pumping current in the case of current sensors;

- Diffusion of gas mixtures, which is difficult to determine, and the thermodynamic state of equilibrium which is established between the unburned exhaust gas fractions CO, HC, H_2 and the products of oxidation, CO_2 and H_2O. These factors have to be taken into account by using a strongly simplified model;

- Catalytic reaction at the platinum electrodes of the sensor, which causes a temperature rise of the sensor surface due to post- combustion;

- The unknown NO concentration in the exhaust gas.

b) When using single-constituent air-fuel ratio metering instruments, as illustrated in **Fig. 3.6**:

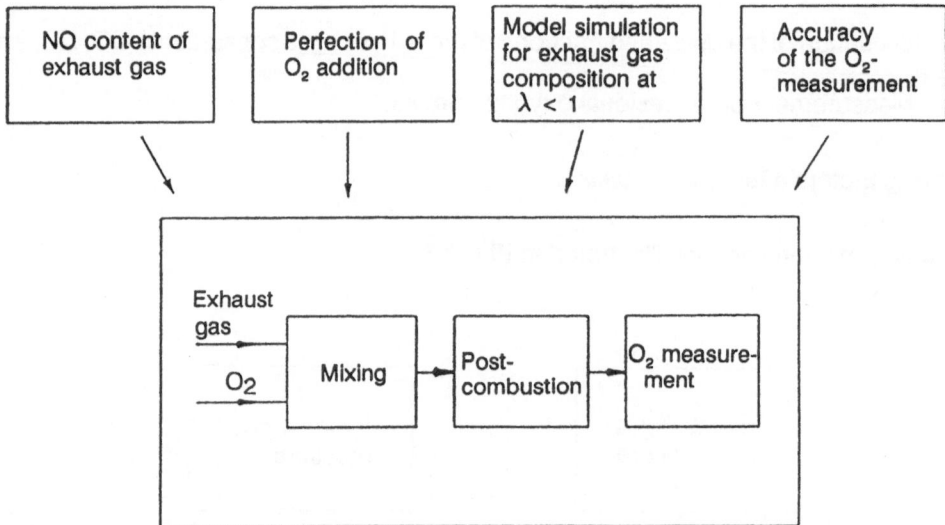

Fig. 3.6. Sources of error in air-fuel ratio measurement using single- constituent metering instruments

- Inaccurate mixtures of exhaust gas and oxygen or inaccurate measurement of the mixture ratio;

- Incomplete post-combustion of CO, HC, and H_2 in a catalytic converter or a thermal converter;

- The accuracy of residual oxygen measurement;

- The chosen model simulation of exhaust gas composition in the rich mixture range, since the volume of the sample gas changes on account of post-combustion of the exhaust gas/oxygen mixture;

- The NO concentration in the exhaust gas, which is unknown in this case.

Both methods of measurement are additionally affected by the hydrogen/carbon ratio of the fuel. However, for residual oxygen measurements with lambda values near one this effect is practically zero. The same applies to the oxygen/carbon ratio. For measurements in the "rich" or, respectively, in the "lean" mixture range, however, these values do have to be taken into account.

Thus, the mean measurement error for the air-fuel ratio λ can be computed according to the law of error propagation.

In the literature, the accuracy of oxygen sensors of this type are indicated to be 1% to 2% error at lambda = 1 and 4% in the range of lambda = 0.8 to 1.5, according to references [3.12 and 3.13] under the condition of a relatively constant exhaust gas temperature. The spread of indicated tolerances for a stoichiometric mixture results from the fact that some combined current/voltage sensors can be operated as pure lambda closed-loop control units, which results in the more accurate measurement of the lean/rich transients.

With single-constituent air-fuel ratio metering instruments, accuracies of approx. 0.5% by volume may be expected for mixture ratio measurements and an accuracy of 0.05% by volume for O_2-measurements, under the condition of constant temperature and constant pressure. An uncertainty of 0.4% by volume has to be included for full-load nitric oxide that has not been measured. Substituting in Equation (3.1) for λ = 1 yields a mean measurement error of 1.5%. If the NO concentration is included as a measured value, accuracy is improved to 0.6% error.

3.2.3. Comparison of Air-Fuel Ratio Measurement Procedures Based on Exhaust Gas Analysis

Table 3.1 compares different methods of determining the air-fuel ratio from the engine exhaust using the criteria: cost, response time, and accuracy of measurement.

	Costs	Response Time [s]	Measurement Error [%]
Five-constituent-measurement	High	1	0,5
Residual oxygen measurement/oxygen sensor	Low	0,05	2
Single-constituent lambdameter + NO-measurement	Average	0,1	0,6

Table 3.1. Comparison of air-fuel ratio measurement techniques based on exhaust gas analysis

When the exhaust gas constituent NO is included in the measurement, an accuracy comparable to that of the multi-constituent method can be achieved by the residual oxygen measurement technique.

3.2.4. Transient Air-Fuel Ratio Measurement

Fig. 3.7 shows the transient air-fuel ratio trace measured with the transient "lambdameter", whose principle of operation is based on the residual oxygen measurement technique [3.14] - response time 65 milliseconds - during a throttle valve step of 500 degrees per second in comparison to injection duration. It clearly shows the short-term engine "lean-out" and the coincidence of control fluctuations around lambda = 1. The lambda curve obtained by measurements with a conventional device has been plotted in the same diagram for comparison. Transient processes cannot be detected with conventional instruments.

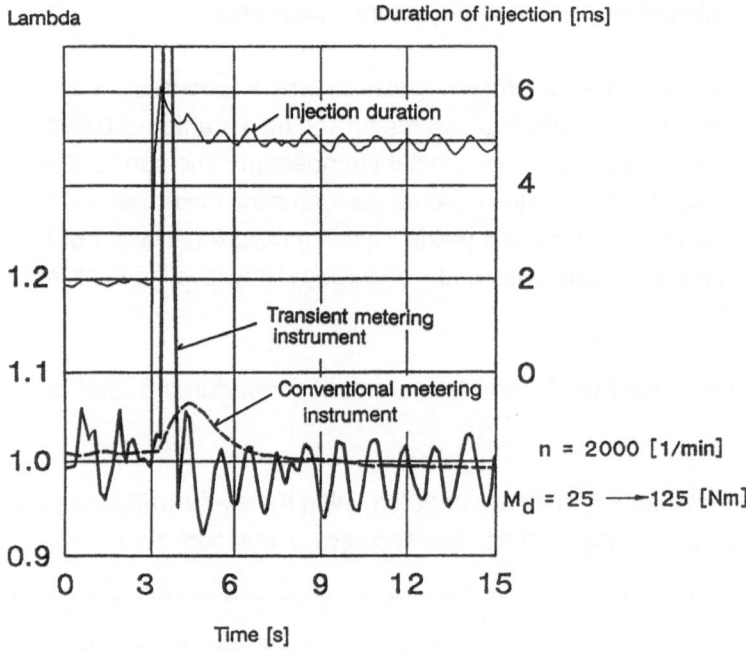

Fig. 3.7. Lambda curves and injection duration at a load increment of 500 [/s] throttle valve speed for various metering instruments [3.37] .

3.2.5. Air-Fuel Ratio Determination in Two-Stroke Engines

In mixture-lubricated two-stroke engines, the lambda value can be determined on-line when using the type of lambdameter developed at the Institute of Internal Combustion Engines and Automotive Engineering at the Technical University of Vienna, whose principle of operation is based on residual oxygen measurement and which incorporates a special- type cyclone oil separator in a hot sampling line.

3.3. Wall Film Measurements

Knowing the quantities of liquid fuel film accumulated on manifold walls under various load and operating conditions is not only important for judging the cold-start, warm-up, and response behavior of an engine, but moreover is the basis for analyzing the causes of mixture maldistribution. A variety of wall film measurement procedures are offered in the literature [3.31, 3.32, 3.33, 3.34, 3.35, 3.36, 3.37, 3.38], all of which have the disadvantage, however, of requiring substantial interference with the geometry of the intake system to be measured.

Therefore, a wall film measurement technique has been developed at the Institute of Internal Combustion Engines and Automotive Engineering at the Technical University of Vienna which permits measurements of the fuel film quantities accumulated in the intake manifold without having to modify the existing intake system and without any additional apparative requirements. The fundamental principle of the procedure is to cause vaporization of the wall film accumulated during the previous load point by a sudden closure of the throttle valve and to determine its quantity by simultaneous analysis of the exhaust gas. The amount of fuel thus vaporized will be equal to the total amount of liquid fuel film deposited in the manifold only when no film-like fuel remains, with the throttle valve closed, which may be assumed on account of the wall temperatures and vacuum pressures at idle [3.30].

Since the above-described exhaust gas analysis, which is conducted in parallel to the wall film vaporization process, is a dynamic metering process, the transmit characteristics of all metering instruments involved have to be determined beforehand. In the computer program used here, this is accomplished by generating a step function of concentration, with delay times and control lags of the individual metering instruments being calculated automatically by a computer on the basis of the characteristic response of an individual device to this step. The mathematical approach to this problem is discussed and a detailed description of the procedure is given in reference [3.30].

The standard use of lambda sensors in production spark-ignition engines today provides the possibility, by suitable choice of the manifold design, of achieving nearly full, on-line compensation of wall film variations in single-point mixture formation systems. One possible technique is described in reference [3.47].

3.4. Droplet Sizing

Accurate determination of droplet sizes is an aspect of central importance in the development of mixture preparation systems. Only accurate quantification of the droplet size distribution will lead to a successful empirical optimization of mixture preparation on one hand and verification of theoretical principles of mixture preparation on the other.

The degree of preparation of a fuel-air mixture is characterized by the following three parameters:

- Portion of fuel evaporated,

- Droplet size distribution

- Droplet number density

Major importance in the judgement of spark-ignition engine mixture preparation systems is attributed to the liquid phase, which is determined by the droplet size distribution and droplet number density distribution.

3.4.1. Droplet Sizing Techniques

Determining size distributions of droplets usually requires a more tedious procedure, due mainly to the inconsistency of liquid drops, than determining size distributions of solid particles. While a variety of methods of nearly any desired accuracy is available for sizing solid particles, only a few solid particle sizing procedures can also be used for liquid drop size measurements. Depending on their basic principle of operation, there are four fundamental types of drop size measurement techniques:

- Collection techniques, which are based on the preservation of droplets for subsequent measurement;

- Fractional separation techniques, by which droplets are separated according to size-dependent physical properties;

- Electrical methods, which use electrical characteristics of drops to determine their size;

- Optical methods, where either images of droplets or optical/physical properties are used for determining droplet sizes.

There are several other principles of measurement in addition to the ones mentioned, which, however, are of little practical importance.

3.4.1.1. Photographic Techniques

The oldest optical methods are those of direct photography [3.11, 3.16, 3.17, 3.18, 3.19, 3.20, 3.21, 3.22]. In addition to conventional cameras with macro- and high-speed photography capabilities, video- based recording systems, which are usually coupled with automatic evaluation units, are being used more and more for recording. The choice of the light source is determined mainly by the droplet velocity: permanent light, electronic flash, stroboscope, spark flash, or pulsed laser are used, depending on the individual case of application. In most cases, pictures are taken using transmitted light methods, that means the droplets are imaged as shadows against an illuminated background.

The main problem of macrophotography is the limitation in depth of field, which may be vast depending on the drop size. This limited depth of field not only makes correct evaluation of individual pictures more difficult but also may require a computational correction of measured results [3.11] for all drop sizes in order to distinguish between identical measurement volumes.

A somewhat improved differentiation of drops located within and outside the focal length of the lens is possible when a coherent laser is used as a light source in the transmitted light technique. By diffraction and interference phenomena of the laser beam, droplets at different distances from the focal plane can be distinguished (**Fig. 3.8**).

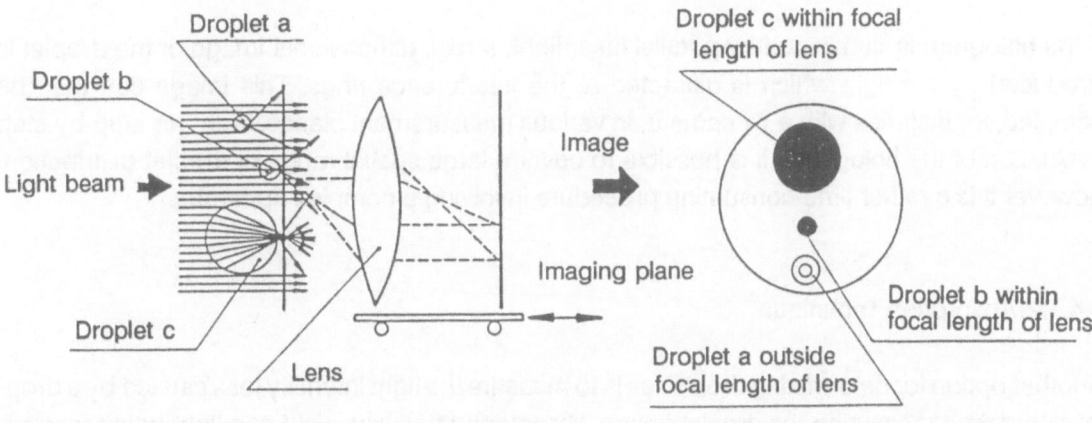

Fig. 3.8. Droplet image in the transmitted light technique with a coherent laser beam in accordance with reference [3.21]

However, since the depth of field is a function of droplet size also in this case, problems of evaluation similar to those of macrophotography arise.

Exact analysis of the entire spatial distribution of drops is possible with holographic techniques [3.18, 3.19, 3.23, 3.24, 3.25]. Pulsed lasers, which give extremely short flashes of coherent light (**Fig. 3.9**) are used for taking pictures of moving drops.

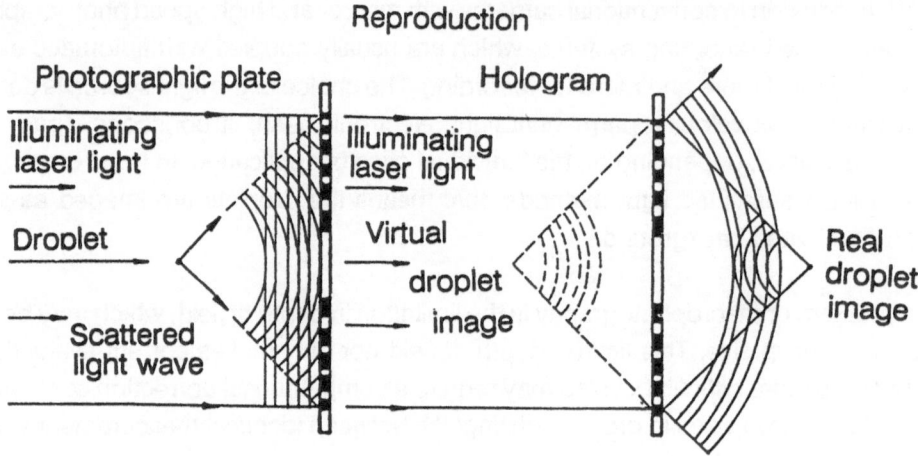

Fig. 3.9. Principle of holographic drop size measurement [3.24]

Behind the droplet, the undisturbed light waves are superimposed on the scattered light waves emitted by the droplet and thus produce interference patterns, which are imaged as a so-called hologram.

If the hologram is illuminated by parallel laser light, a real, dimensional image of the droplet is produced by the light which is diffracted at the interference rings. This image can then be sampled, for instance with a TV camera, in various measurement planes. With this step-by-step evaluation of the holograms it is possible to cover a large spatial range of droplet distribution, however it is a rather time-consuming procedure involving a complex apparatus.

3.4.1.2. Absorption Technique

Another option for determining drop sizes is to measure the light intensity loss caused by a droplet rather than measuring the droplet image. By entering the wave field of a light beam, a droplet causes the field to change by absorbing energy and by scattering light in all directions, both being a function of droplet size.

3.4.1.3. Scattered Light Method

Other commercially used photographic methods of particle and drop size measurement are the various scattered light techniques.

The scattered light generated by a droplet is composed of three fractions, which result from:

- Diffraction on droplets,
- Refraction in droplets, and
- Surface reflection.

The angle-dependent scattered-light energy distribution can be used for sizing fuel droplets. The scattered-light techniques based on this principle can be subdivided into two groups:

- Counting methods, where scattered light evaluations are made separately for individual drops,
- Collective measurements of the droplet population, where the light scattered by a great number of particles is measured and evaluated simultaneously.

The group measurement techniques are presently the most efficient procedures for determining droplet size distributions in spark- ignition engine mixture formation systems, due to the large measurement volumes, wide droplet size and droplet number density ranges covered on one hand and due to the rapidity and accuracy of measured data evaluation on the other. Advantages and disadvantages of these methods are discussed in greater detail in Section 3.4.2.

3.4.2. Theoretical Principles of Scattered-Light Techniques

By entering the wave field of a light beam, a droplet changes the field by absorbing and scattering energy in all directions, both fractions being functions of droplet size. In the forward scattering region, that is at scattering angles that diverge only slightly from the direction of the incident light, the diffracted portion is highly predominant in the fraction of droplets with diameters greater than 1 m (for He-Ne laser light). At short distances from the droplet, this phenomenon is called FRESNEL diffraction, at longer distances it is called FRAUENHOFER diffraction. Diffraction is - based on the wavelike nature of light - the deflection of light resulting from interference of elementary waves which is disturbed by an obstacle.

The diffracted light fraction is imaged via a suitable lens system on concentric circles around the focal point (**Fig. 3.10**), the image being a function of diffraction angle and focal length of the lens only and not of the spatial position of the droplet (see **Fig. 3.11**).

Therefore, the relative velocity between droplet and metering instrument does not affect the scattered-light pattern in the focal plane.

In the case of monodispersions, the droplet size can be determined simply from the zero-point positions of the light intensity distribution, which is a single-valued function of drop size.

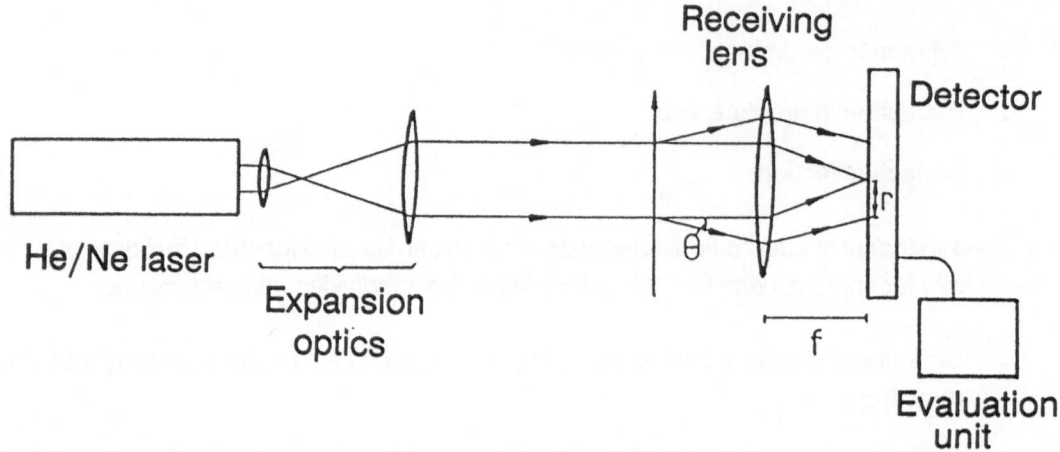

Fig. 3.10. Schematic diagram of a scattered-light metering system for group measurement [3.26]

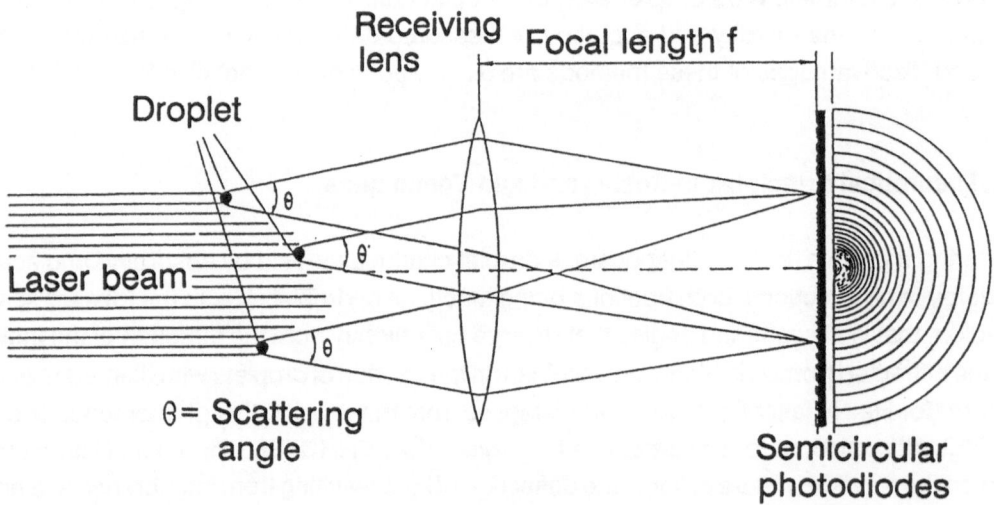

Fig. 3.11. Location-independent image formation of scattered-light profiles through a lens system [3.26]

A much better basis than the radial light intensity distribution for determining size distributions of polydispersions is the light energy distribution across the focal plane (see **Fig. 3.12**).

In contrast to radial light intensity distributions (Fig. 3.12 left picture), the position of the first diffraction maximum is a function also of droplet size in the case of light energy distributions (Fig. 3.12 right picture).

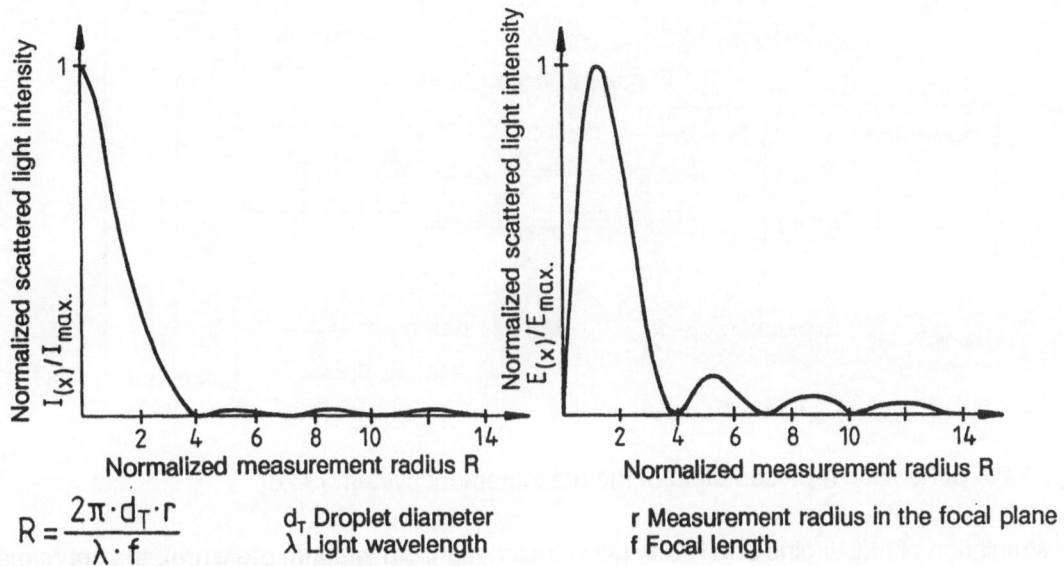

$$R = \frac{2\pi \cdot d_T \cdot r}{\lambda \cdot f}$$

d_T Droplet diameter
λ Light wavelength

r Measurement radius in the focal plane
f Focal length

Fig. 3.12. Dimensionless diffracted light intensity and diffracted light energy distributions for spherical droplets [3.26]

The light energy distribution can be determined by integration of the light intensity distribution over the corresponding area segments in the focal plane.

However, since droplet size distributions with more than one peak may be encountered in mixture formation systems of spark-ignition engines, a model-independent procedure is used to determine such distributions. In this case, the droplet size spectrum is divided into 31 size groups which correspond to the individual measurement rings, and their respective weight portions are varied until a minimum deviation of the measured light energy from the computed light energy is obtained.

The total fuel droplet number density and consequently not only the relative distribution but, with certain limitations, also the number of droplets can be determined on the basis of droplet size distribution, absorption coefficients of individual droplets, and actual light loss of the laser beam.

3.4.3. Arrangement and Apparatus of the Scattered-Light Method

Fig. 3.13 is a schematic diagram of the measurement arrangement. The spray cloud is illuminated by a parallel beam, enlarged to a diameter of 8 mm, of monochromatic coherent light from a 2mW helium-neon laser ($\lambda = 0.6328$ mm).

All scattered light is collected behind the measurement volume by an exchangeable optical receiver system for focal lengths of 63 mm, 100 mm, and 300 mm and reproduced on 31 semicircular photodiodes. The diameters of the individual photodiode rings are chosen so as to ensure good differentiation between individual drop sizes.

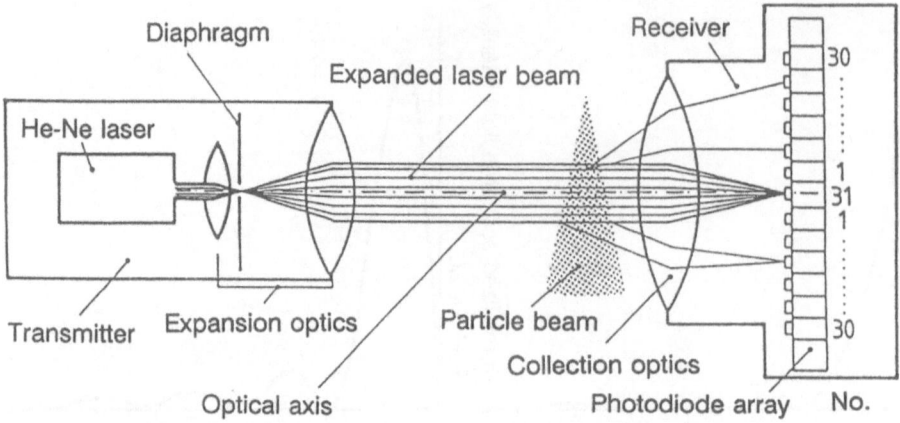

Fig. 3.13. Schematic representation of the measurement system [3.26]

For simulation of real engine conditions (air velocity, manifold vacuum pressure), the entire metering unit is incorporated in a flow bench [3.88]. The required spatial limitation of the volume to be measured had to be optimized first in numerous preliminary trials [3.39, 3.40] to avoid erroneous results.

Fig. 3.14 is a schematic drawing of the metering arrangement adapted for measurement of multipoint injectors in an air flow parallel to the axis.

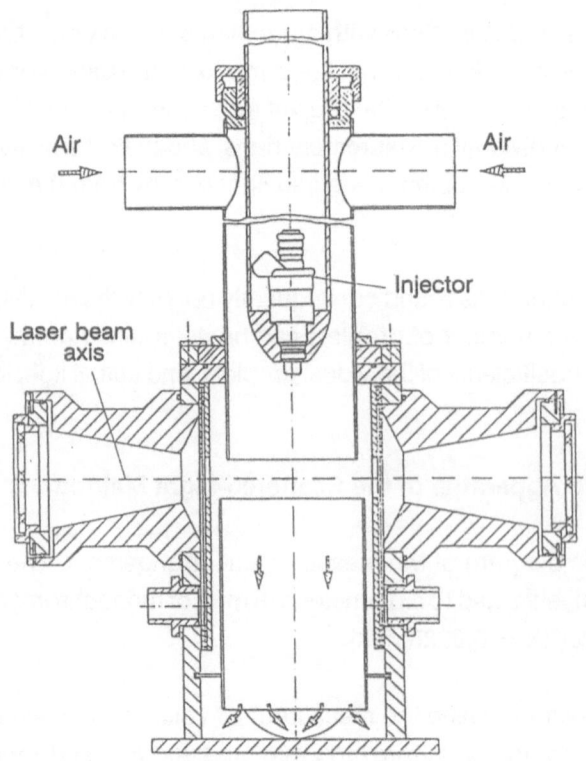

Fig. 3.14. Schematic representation of the metering unit [3.26]

Provisions for suitable alignment of the windows have been incorporated to avoid reflections from the glass panes falling onto the transmitter or receiver lens. Window fouling is prevented by providing an air curtain.

3.4.4. Sources of Error and Measuring Accuracy of the Scattered-Light Method

A successful application of this type of metering system requires good knowledge of potential sources of error as well as the appropriate determination and limitation of measurement conditions.

3.4.4.1. Double Refraction

If, in a dense spray mist, a diffracted ray is incident on another droplet, the ray loses light and is diffracted again. For laser beam absorptions which account for less than 50%, this second diffraction is of little importance. Since, with multi-point injectors, absorption usually is appreciably smaller than this in representative spray regions, the effect of double refraction can generally be neglected both in droplet size measurements and in droplet number density measurements. In case of higher droplet number densities, it is possible to computationally correct a scattered-light energy distribution distorted by double refraction [3.26, 3.27].

3.4.4.2. Nonhomogeneities in the Measurement Beam

Both nonuniform light intensity distributions within the laser beam (negligible when measuring fuel spray jets) and nonhomogeneities in the medium surrounding the droplets can lead to erroneous measurement results.

Both the necessary spatial separation of the measurement volume by glass plates, or panes, and differing density gradients in the air-vapor mixture cause distortions of the measurement signal.

Measurement errors may occur especially when laser beam reflections generated at the glass panes fall back into the transmitting lens (see **Fig. 3.15**).

These effects on the measurement signal can be eliminated almost completely by suitable alignment of the glass panes.

A potential measurement error that is much more difficult to prevent is the effect resulting either from refraction of the laser beam at the phase boundaries between fuel vapor and air or from the high density gradients in the fuel vapor. This may lead to generation of scattered- light profiles similar to those of large drops. Especially with low air velocities and with air pressures greater

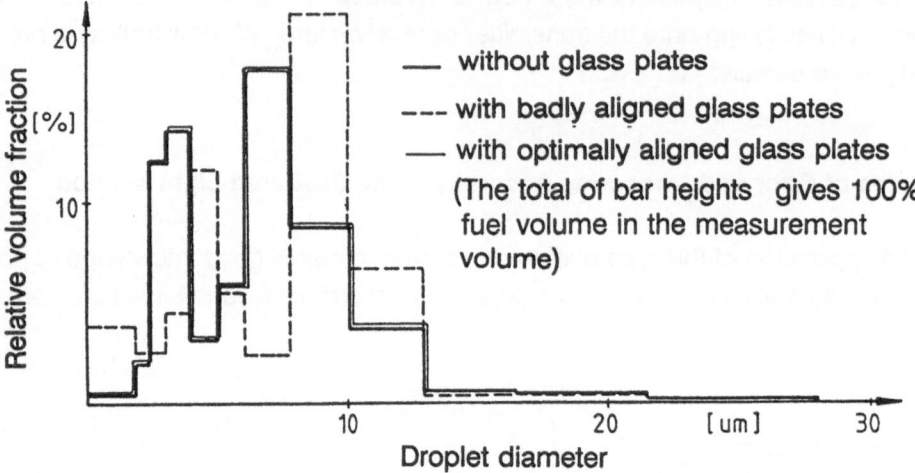

Fig. 3.15. Effect of laser beam reflections on the measurement result [3.26]

than atmospheric pressure, this may lead to substantial distortions of the light energy distribution and consequently to erroneous results.

With measurement conditions as chosen here (i.e. air velocity 2 m/s, absolute manifold pressure atmospheric pressure; extraction of fuel vapor), this effect is also negligible. **Fig. 3.16** shows potential measurement errors caused by refraction of the laser beam when injecting a completely vaporized fuel in comparison to a typical measurement result.

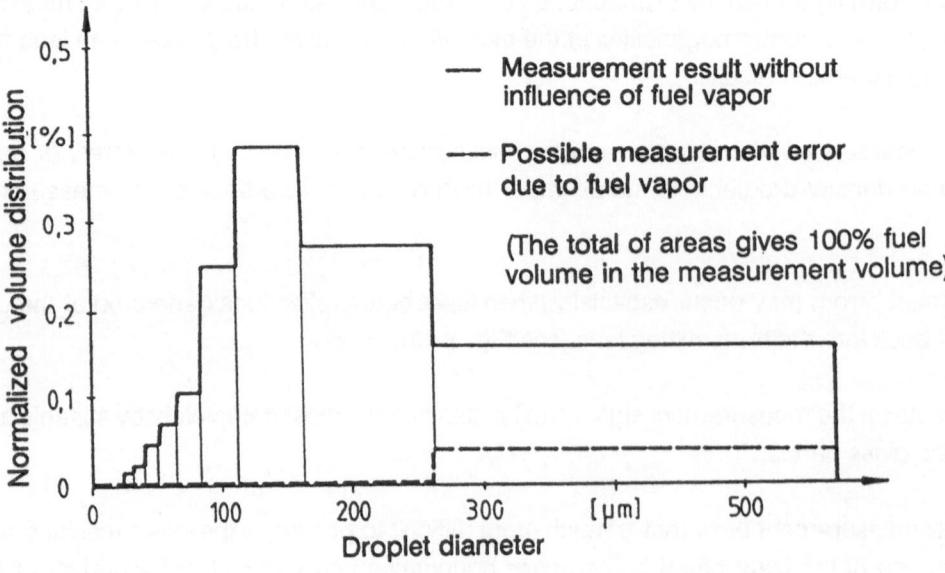

Fig. 3.16. Measurement error caused by fuel vapor under representative measurement conditions [3.26]

3.4.4.3. Evaluation Method

The iterative evaluation of droplet size distributions from scattered- light energy distributions requires an a priori assumption of a droplet size distribution that seems realistic. The assumption can be established either by exact mathematical distribution functions or by dividing the droplet size distribution into size classes. This does not affect measurement results in the case of size distributions with only one peak, since a wide range of possible distributions can be covered by varying the location and shape parameters of the respective distribution models (Rosin-Rammler, Gaussian distribution, and logarithmic Gaussian distribution).

Dividing the droplet size distribution into size classes, which is necessary when analyzing size distributions with more than one peak, is not done at random but is based on the differentiability between scattered-light enery distributions of individual drop sizes. With very small droplets, this yields a high resolution. However, when classifying drop sizes larger than 160 m diameter the resolution will be comparatively coarse.

When graphically presenting droplet size distributions of common multi-point injectors, dividing the continuous spectrum into size classes usually does not mean any significant loss of information (see **Fig. 3.17**).

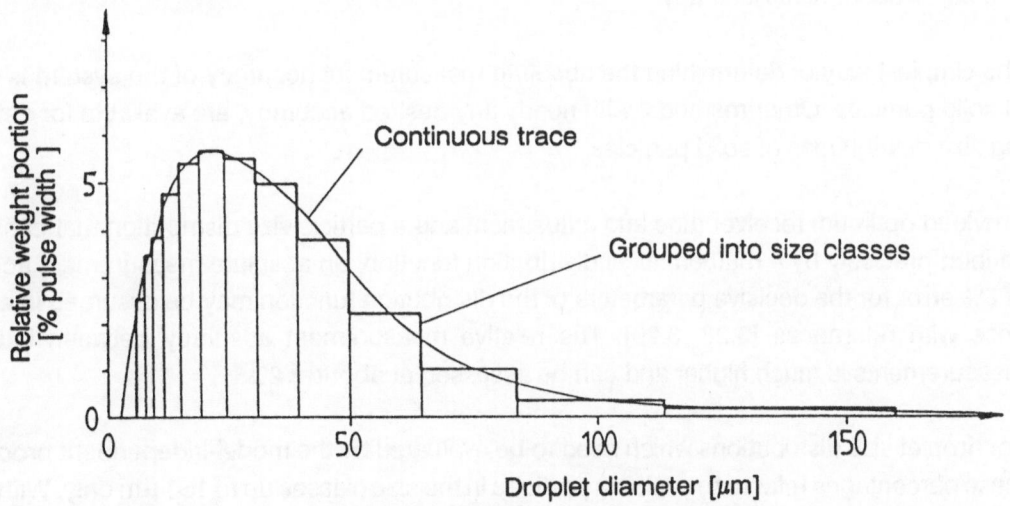

Fig. 3.17. Division of a continuous droplet size spectrum into individual size classes [3.26]

If characteristic diameters are computed from this size-class distribution function, however, substantial errors in the characteristic diameters may result due to round-off errors in the range of small droplet sizes and due to the coarse resolution in the range of large drop diameters. While the measurement error is of a magnitude of 10% for droplet size distributions whose maxima are within the range of 100 to 150 μm, it is not permissible to determine characteristic diameters from the size-class distribution function in the case of very coarse droplet size distributions. Smoothing of the size-class distribution functions leads to improved accuracy, however this requires a very careful choice of smoothing algorithms, especially in the case of droplet size distributions with several peaks.

3.4.4.4. Other Influences on Measurement Results

Erroneous measurement results may originate from the selected measurement location. If the entire measurement volume lies within the focal length of the lens, the influence of the distance between the collector lens and the measurement volume is negligible. If droplets are present outside the focal length of the lens, image formation of individual scattered-light portions on the photodiode array either fails to take place at all or leads to distorsions.

Differing droplet velocities do not distort the measurement result, but they have an appreciable effect on the droplet size distribution. The droplet size distribution is determined by group measurement on the droplet population in a measurement volume large in size compared to the droplet diameter. This results in a direct relation between droplet velocity and number of droplets in the measurement volume. In case of different droplet velocities within a fuel spray jet, the slower droplets are overrated. This effect is observed particularly in dripping flows.

3.4.4.5. Measurement Accuracy

The simplest way of determining the absolute measurement accuracy of the system is by use of solid particles. Other methods, with nearly any desired accuracy, are available for determining size distributions of solid particles.

Provided optimum receiver tune and adjustment and a particle size distribution that can be described precisely by a mathematical distribution function, an absolute measurement accuracy of 5% error for the decisive parameters of the distribution function may be assumed in accordance with references [3.28, 3.29]. The relative measurement accuracy between individual measurements is much higher and can be assessed at about ±2%.

For droplet size distributions which need to be evaluated by the model-independent procedure, these percentages refer to the volume portions in the size classes up to 150 μm only. With larger drop diameters greater measurement tolerances may well be the case due to the coarse resolution in this diameter range and due to measurement errors, which, however, have a comparatively small effect on characteristic diameters.

Therefore, the scattered-light method discussed here exhibits an excellent accuracy, especially in respect of relative measurements. Provided appropriate use of the measurement and evaluation system, a relative accuracy of +/-3% may be assumed, with the exception of very coarse sprays (pencil-jet nozzle).

3.4.5. Options for Presenting Steady-State Droplet Size Distributions

In steady-state analysis of a fuel spray, the spray mist is characterized by the size distribution of the droplets on one hand and by their spatial concentration on the other, the droplet number density being considered as the entire percentual volume portion of liquid fuel in the measurement volume. The droplet size distribution can be presented both as a cumulative and a volume distribution (**Fig. 3.18**).

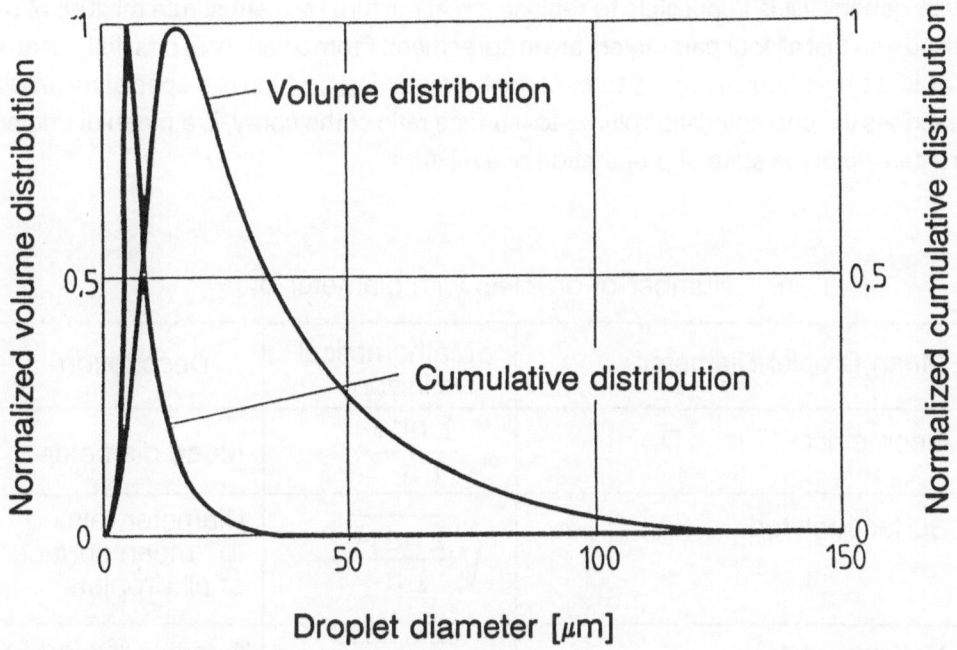

Fig. 3.18. Traces of a droplet size distribution [3.26]

The cumulative distribution, that is the representation of the number of droplets of a specific diameter, conveys the visual impression of a spray.

The volume distribution, which includes the total volume of droplets of a specific diameter, on the other hand, is a much better indication of the state of preparation of the mixture, which is a relevant factor in combustion processes. With this type of presentation, large individual droplets, which only marginally affect a cumulative distribution, are also included in relation to their volume portion.

Besides the distribution function of droplet sizes, there is the option of using a characteristic diameter which describes a "mean droplet size" in an alternative mixture constisting of droplets all equal in size, for depicting such a size distribution approximatively.

This alternative mixture should closely match the real droplet-air mixture in the following parameters:

- Number of droplets

- Sum of diameters

- Total surface of droplets, and

- Total volume of droplets.

However, generally it is impossible to replace a real mixture by a substitute mixture of droplets equal in size so that all four parameters are in agreement. From a variety of possible mean diameters (**Table 3.2**), particularly the "Sauter Mean Diameter D_{32}", which is a special mean diameter that describes the characteristic volume-to- surface ratio of the spray, is a mean diameter which is characteristic of the state of preparation of a mixture.

n ... Number of droplets with diameter d_T

Mean Droplet Diameter	Mathematical Definition	Description
Geometrical $\qquad$ D_{10}	$\dfrac{\Sigma\, n d_T}{\Sigma\, n}$	Mean diameter
Surface-related $\qquad$ D_{20}	$\sqrt{\dfrac{\Sigma\, n d_T^2}{\Sigma\, n}}$	Diameter related to mean surface of all droplets
Volume-related $\qquad$ D_{30}	$\sqrt[3]{\dfrac{\Sigma\, n d_T^3}{\Sigma\, n}}$	Diameter related to mean volume (weight) of all droplets
Volume/surface-related (Sauter Mean Diameter) $\qquad$ D_{32}	$\dfrac{\Sigma\, n d_T^3}{\Sigma\, n d_T^2}$	Diameter related to mean volume/ surface ratio of all droplets

Table 3.2. Mathematical definitions of mean droplet diameters [3.26]

However, since nominally equal Sauter Mean Diameters may result from greatly varying distributions (**Fig. 3.19**), especially in the case of droplet size distributions with several peaks, there is no way around looking at both the Sauter Mean Diameter and the shape and location of the distribution function, if representative comparisons of droplet size distributions are to be made.

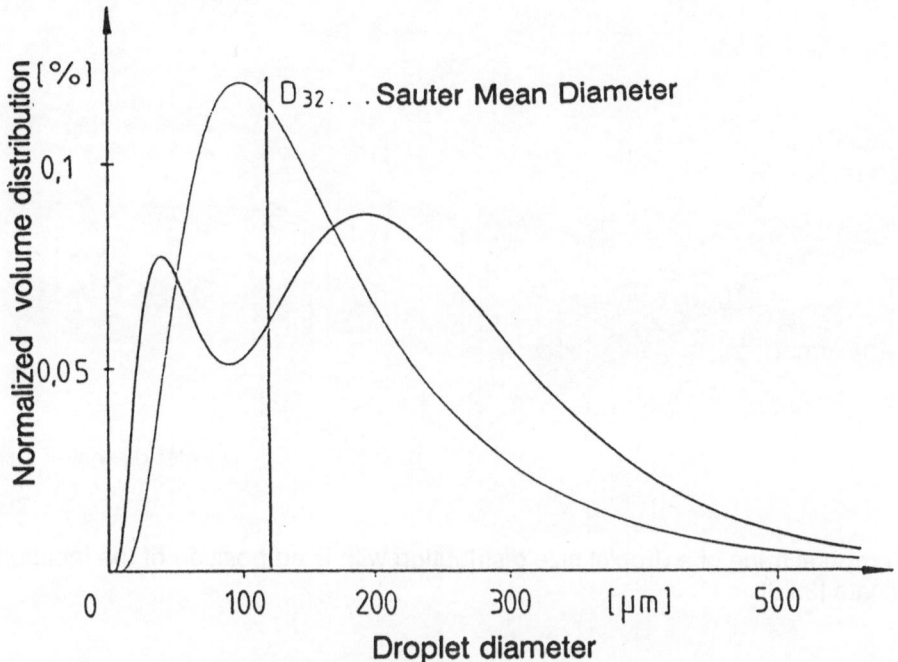

Fig. 3.19. Different droplet size distributions with same Sauter Mean Diameter D$_{32}$ [3.26]

3.4.6. Presentation of Transient Droplet Size Distributions

If temporal variations of a droplet size distribution are not cyclic variations but small-size random fluctuations, as is the case with carburetor systems in many operating ranges, then the droplet size distribution can be described representatively by a mean value of an appropriately high number of individual measurements.

In intermittent mixture formation systems, on the other hand, temporally and spatially highly transient droplet size distributions are common. In this case it is necessary to include the entire temporal and/or spatial trace of the droplet size distribution. A complete presentation of such transient size distributions would require six coordinates:

- 1 Time coordinate

- 3 Space coordinates

- 2 Coordinates for droplet size distribution (droplet size and droplet number density)

The temporal trace of the droplet size distribution can be presented intelligibly in the following graphical form, with the location coordinate being fixed in position (**Fig. 3.20**).

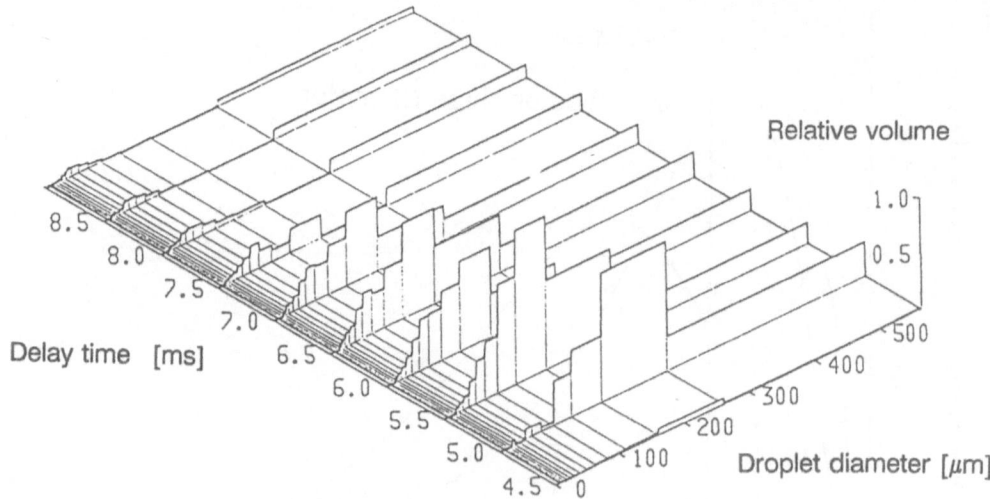

Fig. 3.20. Temporal trace of a droplet size distribution with fixed position of the location coordinate [3.26]

In this presentation the entire spray event is depicted at a spatially representative point, in close temporal gradation, including both droplet size distribution and droplet number density, which yields a three-dimensional representation of the temporal trace of the droplet size distribution.

One possibility of obtaining a clear comparison of different transient droplet size distributions is to describe such size distributions by the traces of Sauter Mean Diameter D_{32} and of droplet number density SC - **Fig. 3.21.**

The top picture of the illustration shows the trace of the Sauter Mean Diameter, the bottom picture plots the droplet number density versus delay time after opening of the injection nozzle.

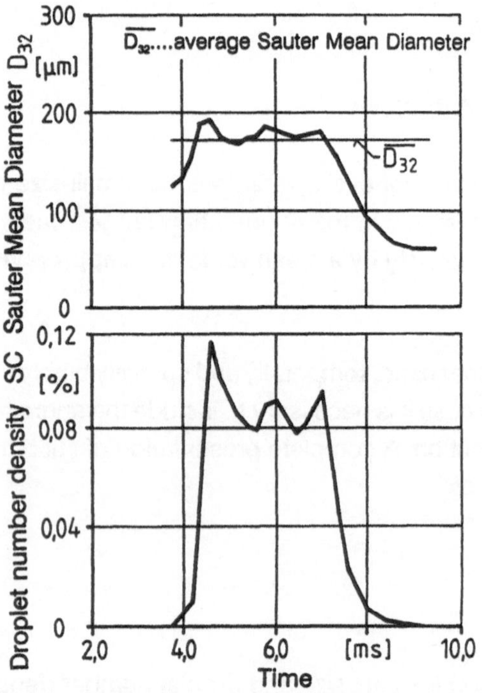

Fig. 3.21. Sauter Mean Diameter D_{32} and droplet number density as a function of delay time after opening of the injection nozzle [3.26]

3.5. Measurement of Injector Spray Characteristics

In the development of new injection nozzles, great importance has to be attributed to finding out what fuel quantities are supplied at certain points of the jet cross-section. A test bench for analysis of spray characteristics [3.39] developed at the Institute of Internal Combustion Engines and Automotive Engineering at the Technical University of Vienna can be used for analyzing spray characteristics produced by injectors, in certain cases automated procedures are used.

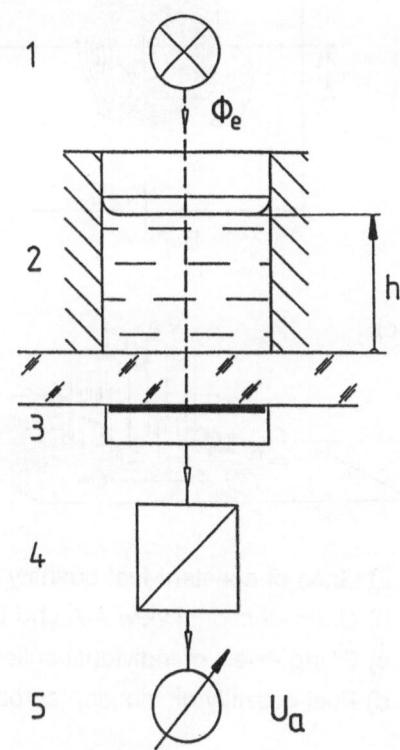

In the test bench arrangement, the injector is mounted on a vertically moveable platform, the fuel being sprayed onto a collector plate with collection bores.

For measurement of the fuel volume in the bores, a non-contact optical system is used which is based on the principle of radiation being influenced when passing through a medium (**Fig. 3.22**).

A low-voltage halogen lamp with adapted optics is used as sensor. Silicon photodiodes function as receivers, their major benefit being rapidity and linearity over wide ranges of radiation strengths.

The measurement arrangement used here converts the geometrical parameter, i.e. fuel level h in the bore, to an electrical parameter, U_a, which is recorded and processed by a computer.

The measurement procedure and evaluation of measured data can be controlled by the user interactively with the computer through a program specifically designed for this purpose. Injection parameters such as frequency and pulse duration, for example, can be varied as desired, making it possible to determine their influence on fuel quantity distribution. The measurement data evaluated by the computer as well as nozzle and injection parameters can be stored in the computer memory.

In addition to evaluating the quantitative fuel distribution according to sectors and circular rings, a variety

1 Transmitter
2 Fuel head
3 Receiver
4 Amplifier
5 Indicating instrument
Φ_e Optical radiation
h Filling level
U_a Voltage

Fig. 3.22. Principle of operation of the test bench for spray pattern analysis

of other options are available for evaluation - **Fig. 3.23** shows a few examples.

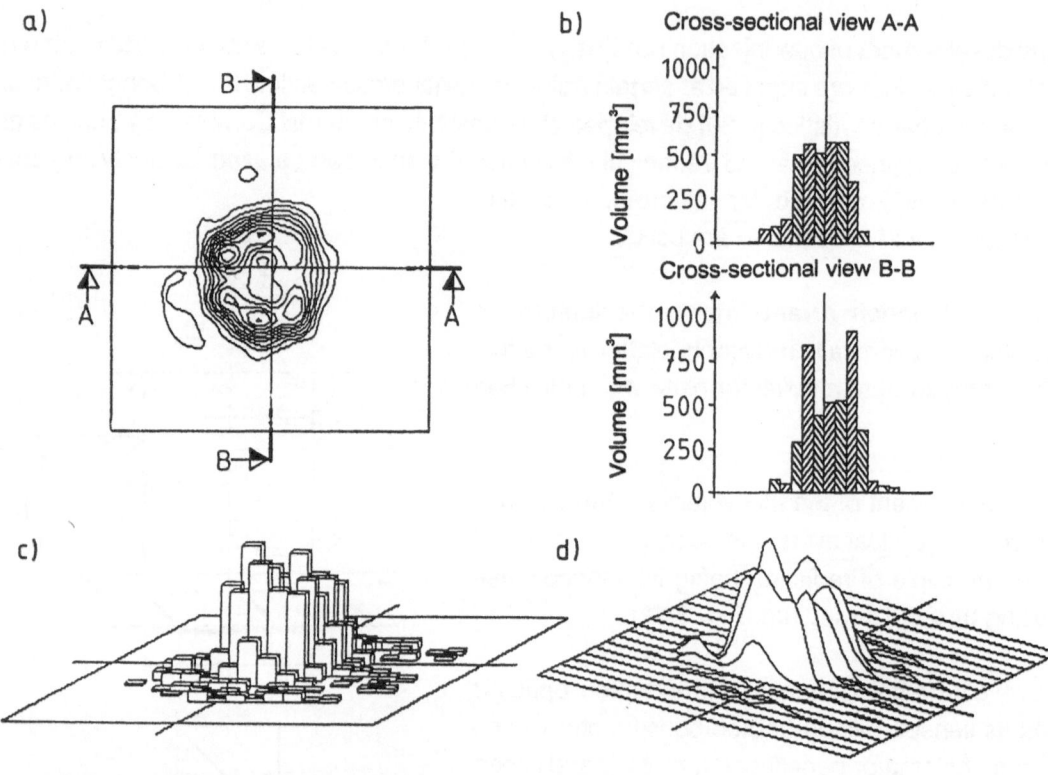

a) Lines of constant fuel quantity per area
b) Cross-sectional view A-A and B-B of a)
c) Filling levels of individual collection volumes
d) Fuel quantity distribution across the area

Fig. 3.23. Spray characteristics of an electromagnetic pintle-type injector in different graphical presentations

4. Types of Mixture Formation Systems

Mixture formation systems are units which feed fuel to the air flowing into the engine in accordance with the requirements described in the previous chapters.

They come as carburetors or injection systems (broken down by function), or single- or multi-point injection systems (broken down by their arrangement within the engine) although we occasionally find systems which do not fit in any of the categories.

4.1. Single-Point Mixture Formation Systems

A single-point mixture formation system is arranged centrally at the engine. It feeds the fuel to the air flowing into the cylinders at a single point. In other words, all cylinders are served by a single system, which incidentally can be either a carburetor or an injection system.

The single-point arrangement is standard and practical in carburetors, while the multi-point arrangement is the system of choice for injection systems (cf. Chapter 4.2).

With injection systems, the only reason for preferring single- point arrangements is because they reduce costs.

Carburetors are usually of the single-point type because, being rather complex affairs consisting of several units, they are difficult to synchronize, i.e. to adjust to identical throughputs of air and fuel.

It is also more difficult to adjust changes in the air throughput equally for all cylinders using parallel throttle valves instead of just one.

Carburetors furthermore take up much space, require sophisticated driving rods, are heavier, difficult to service and in total generate additional costs.

A multi-carburetor system consequently is no longer considered practical. Things were different at a time when no suitable injection system was available for high- performance engines: in order to reduce throttle losses in the manifold baffles, engineers used several carburetors per engine - up to one carburetor for each cylinder - to achieve maximum outputs. The negative effects appeared negligible at the time.

4.1.1. The Carburetor

A carburetor basically is a device in which the air flowing into the engine:

- Generates a vacuum in a constriction which sucks the fuel from the float chamber and adds it to the air;

- Is throttled in a throttle valve in accordance with engine requirements.

A carburetor thus consists of three principal components:

- A constriction (venturi tube),

- A float chamber to maintain constant fuel level, linked to the venturi through a connecting pipe,

- A throttling device, usually in the form of a throttle valve.

This rather simple type of carburetor, however, does not meet all the practical requirements, so that some additional units are needed.

Basic set-up of a carburetor

An operable carburetor thus consists of the following components:

- Venturi;

- Unit to keep the fuel at a constant level: usually made up of a float and float needle valve;

- Nozzle system consisting of a main fuel nozzle, mixing tube and air adjusting nozzle;

- Idling system to determine the composition of the idling mixture through idling air nozzle and idling fuel nozzle, and its volume through the idling mixture regulating screw;

- Bypass system for transition between idling and main systems;

- Accelerating pump;

- Enrichment unit; and

- Starter.

As already described, each speed and load point has its specific optimum air-fuel ratio (λ). Its value depends on the primary requirement for the relevant range: at full load, the driver usually wants maximum output, i.e. a rich combustion mixture with no consideration of consumption or emission of pollutants; at part loads, the emphasis is more on low consumption and emission levels. In the controlled three-way catalytic converter concept the main concern is maintaining a stoichiometric air-fuel ratio (λ).

In order to meet requirements for the engine and exhaust gas cleaning system, it is thus necessary to have several supplementary units which will be described below.

4.1.1.1. Basic Equations

The general carburetor equation is derived from the equations for air and fuel flow. Let us start with the basic carburetor components as illustrated in **Fig. 4.1**. From Bernoulli's equation we get the following correlations for air flow m_L through the carburetor:

$$m_L = \alpha_L \cdot A_L \cdot \rho_L \cdot v_L = \alpha_L \cdot A_L \cdot \rho_L \cdot (2 \cdot \Delta p_L / \rho_L)^{0,5}$$
$$= \alpha_L \cdot A_L \cdot (2 \cdot \rho_L \cdot \Delta p_L)^{0,5} \qquad (4.1)$$

and for fuel flow m_K:

$$m_K = \alpha_K \cdot A_K \cdot (2 \cdot \rho_K \cdot \Delta p_K)^{0,5} \qquad (4.2)$$

where:

m_L, m_K [kg.s^{-1}]	.....	flow volumes for air/fuel
α_L, α_K [-]	.....	flow indices for air/fuel
A_L, A_K [m^2]	.....	flow cross-sections for air/fuel
v_L, v_K [m.s^{-1}]	.....	flow rates for air/fuel
ρ_L, ρ_K [kg.m^{-3}]	.....	density of air/fuel
Δp_L, Δp_K [N.m^{-2}]	.....	differential pressure when air flows through the venturi/when fuel flows through the fuel nozzle

We can assume that the differential pressures are approximately identical for air and fuel flowing through the carburetor, i.e. its calibration cross-sections (venturi and main fuel nozzle), so that we have:

$$\Delta p_L = \Delta p_K = \Delta p \qquad (4.3)$$

deriving for the mixture ratio M:

$$M = \frac{m_L}{m_K} = \frac{\alpha_L \cdot A_L}{\alpha_K \cdot A_K} \cdot \left(\frac{\rho_L}{\rho_K} \right)^{0,5} \qquad (4.4)$$

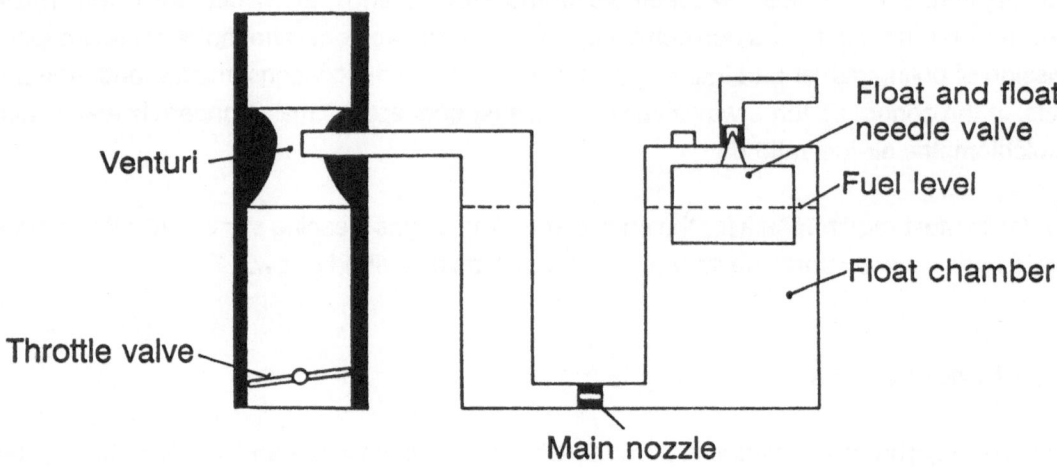

Fig. 4.1. Basic carburetor components

The cross-sections A_L and A_K are constant so that this simple type of carburetor would supply a constant mixture ratio for the entire flow area provided that the flow indices α_L, α_K and the air and fuel densities remain constant.

Actually the only parameters to remain constant are fuel density and air flow index (venturi), while air density and fuel flow index (main fuel nozzle) vary with the throughput (**Fig. 4.2**). The flow index at the main fuel nozzle increases with the throughput, while the density of the intake air declines to accommodate the higher vacuum required for delivery.

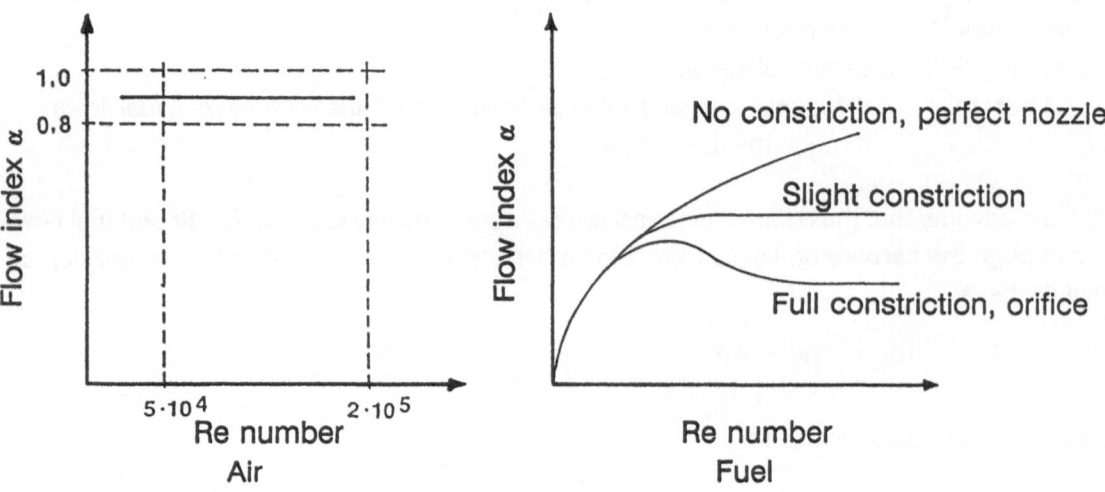

Fig. 4.2. Variations in flow indices [4.8]

Without going into the calculation of flow indices (which is described in full details in [4.1, 4.2]), it is sufficient to note that the flow index depends mainly on two interrelated factors, namely viscosity and inertia of the fluid. In general it can be said that with low Reynolds' numbers viscosity is high compared to inertia and vice versa. When the nozzle throughflow is characterized predominantly by viscosity, the flow index will increase with increasing Re numbers; when inertia is the primary factor, the flow index will decline with increasing Re numbers, provided that the orifices or nozzles are not perfectly round, as a result of the inertia acting at the constriction. In practice the two forces will be found to overlap. **Fig. 4.3** provides an example of a flow index graph measured at a customary carburetor fuel nozzle [4.8].

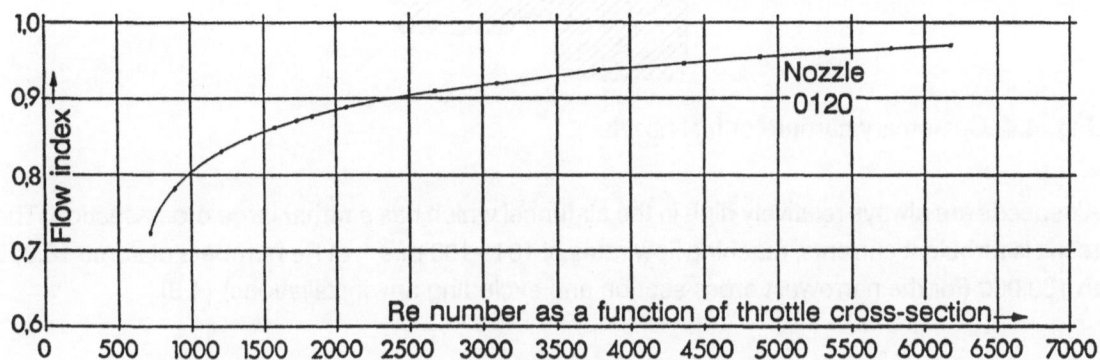

Fig. 4.3. Flow index at a customary carburetor fuel nozzle [4.8]

In situations characterised by pulsating flows, orifices etc. can be calculated on a quasi-steady basis [4.3], provided that the pulsating parameter

$$\Phi = \frac{v_m \cdot t}{d} \tag{4.5}$$

where:

v_m	[m.s^{-1}]	..... mean flow rate;
t	[s]	..... pulsation period;
d	[m]	..... diameter of carburetor nozzle;

is not less than several multiples of 1.

The flow index calculated for the respective steady flow then applies for the relevant time of pulsation. Recent research into advanced carburetors has found that the fuel throughput depends on the vehicle speed: at low speeds, the throughput is higher at pulsating flows than at steady flows and vice versa [4.18].

Carburetors usually have a high pulsation parameter, except when operating in overrun, where the fuel flow rate is low and pulsation periods are short, and where the parameter may be reduced to a few multiples of 1.

Fig. 4.4 shows a cross-section of a customary fuel nozzle.

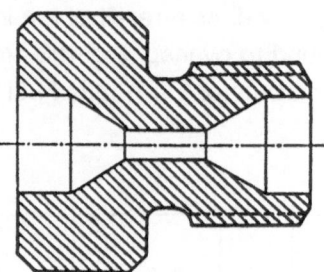

Fig. 4.4. Customary carburetor fuel nozzle

Air speeds are always relatively high in the air funnel which has a rather large cross-section. The result is turbulent currents, reaching flow rates of 10 to 100 m.s^{-1}, at Re numbers of some 10,000 to 180,000 (for the narrowest cross-section and excluding any installations) [4.8].

4.1.1.2. Basic Carburetor Systems

4.1.1.2.1. Air Funnel, Throttle Valve

Flow conditions in air funnels at unobstructed steady flow have been studied in detail. For a survey of the research literature see [4.4]. With regard to practical application in carburetors, it is of primary interest to find air funnel shapes with minimum pressure loss and maximum vacuum at their narrowest point. The pressure loss is the combined loss from friction and separation losses.

The overall pressure loss is lowest at diffuser angles of approx. 7° to 12°, depending on the Re number, where sharp angles have high Re numbers. According to [4.4], energy utilization is best when diffusers are enlarged to almost the separation angle. With regard to diffuser efficiency and inflow conditions see the studies by [4.5, 4.6].

The practical result varies greatly depending on installations in the carburetor such as a fuel outlet or enrichment tube and because of the fact that fuel is added. Another consideration is the low height required for carburetors which usually limits the diffuser length.

It is necessary to know the pressure distribution in an air funnel for a range of air velocities, firstly because pressure changes caused by the velocity may vary in individual parts of the air funnel so that it is possible to influence fuel delivery through increasing throughput, and secondly because a minor local change in the fuel outlet design (such as occurs from manufacturing tolerances) may be sufficient to cause major functional deviations in some parts of the air funnel. The location of the fuel outlet within the air funnel is thus of significant importance for the carburetor function.

Fig. 4.5 shows an air funnel used to measure pressure distribution at various throughputs.

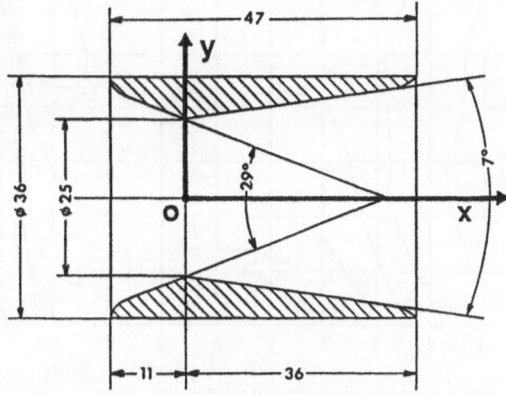

Fig. 4.5. Test air funnel [4.8]

Fig. 4.6 plots the pressure distribution for the radial and axial positions while **Fig. 4.7** charts the isobars in the air funnel.

Fig. 4.8 illustrates the flow at a throttle valve as a function of its angle.

In view of the parameters acting on it, and as described in the previous chapter, it is obvious that carburetors require a number of auxiliary systems to ensure a satisfactory mixture ratio at all operating conditions.

Fig. 4.9 shows vacuums in the manifold and air funnel.

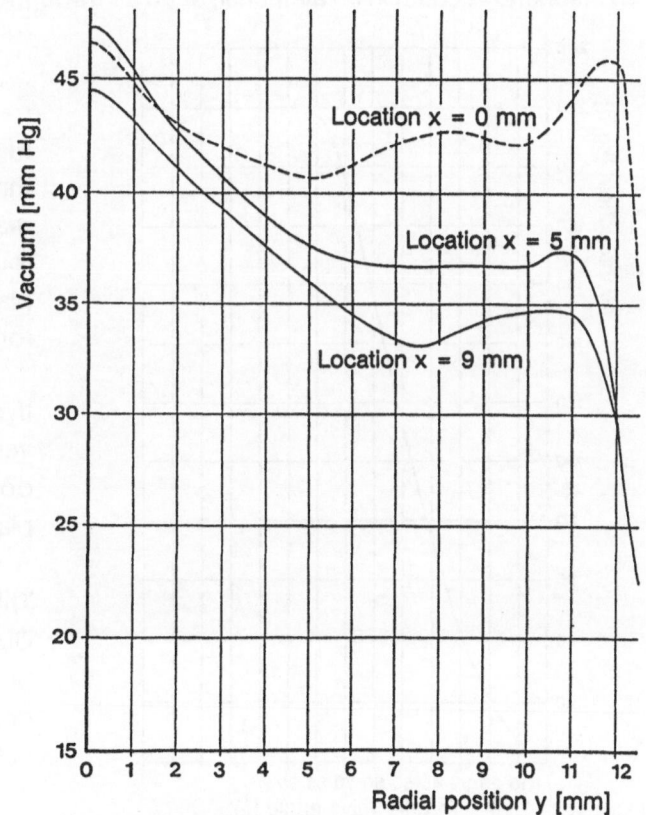

Location x and radial position y as per Fig. 4.5

Fig. 4.6. Trace of pressure in an air funnel in terms of its radial position, at an air throughput of 37.5 l/s [4.7]

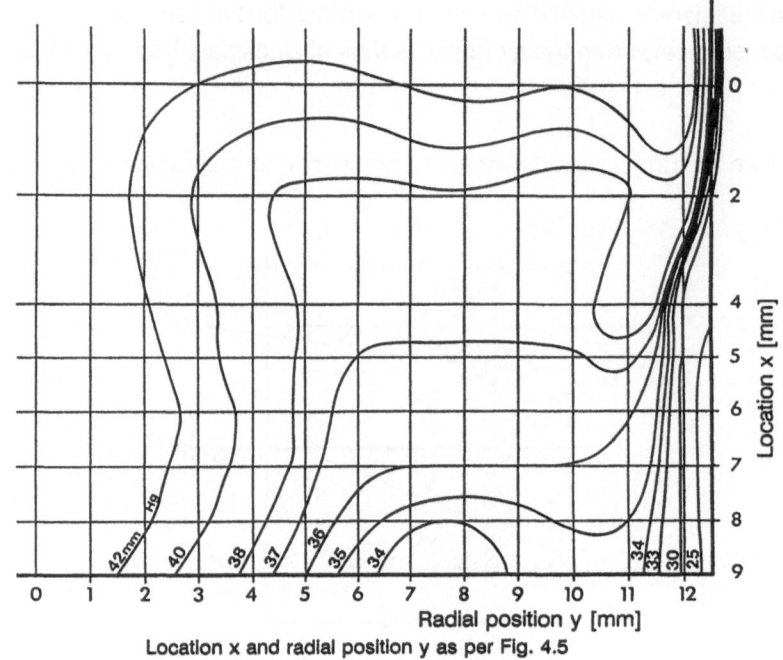

Fig. 4.7. Isobaric vacuum in an air funnel, at an air throughput of 37.5 l/s [4.7]

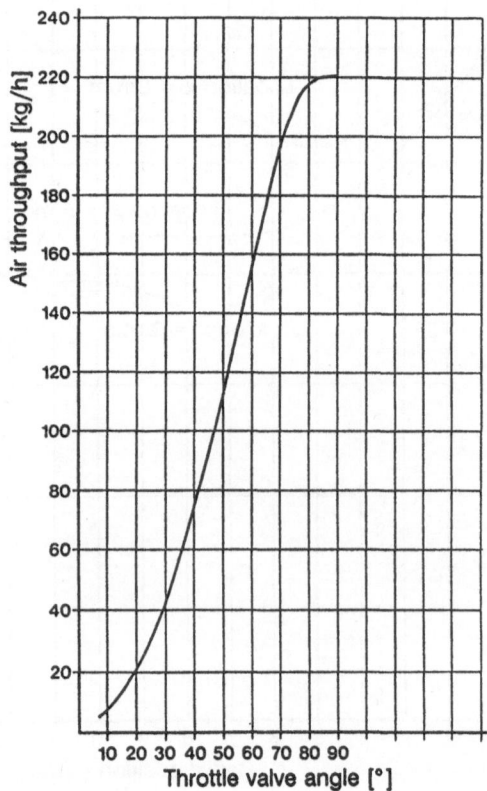

Fig. 4.10 provides graphs for identical throttle valve positions and identical manifold vacuums of a four-cylinder four-stroke Otto engine with a displacement of 2 l and a single carburetor [4.8].

In order to improve the vacuum in the venturi, engine designers often use double venturis, i.e. two venturis placed in sequence.

This system was studied in detail by Dutta [4.10] with regard to its flow and pressure conditions and its optimum design. His findings are reproduced below.

Fig. 4.8. Flow at a throttle valve in terms of its position; differential pressure: 100 mm Hg, throttle valve diameter: 30 mm [4.8]

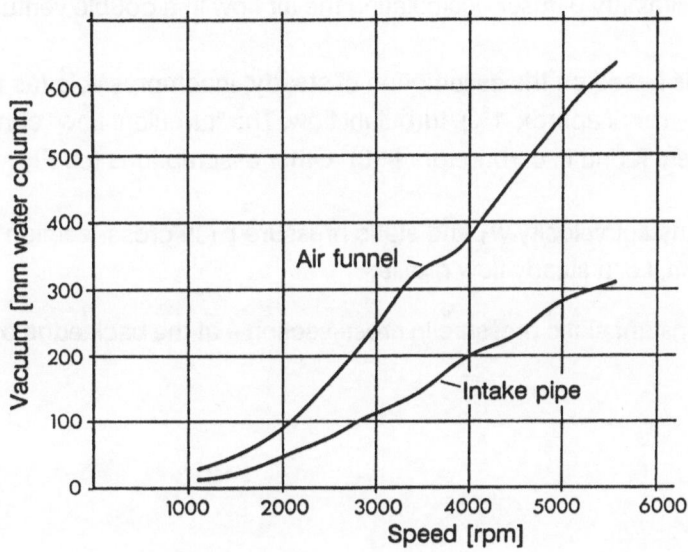

Fig. 4.9. Vacuums in the manifold and air funnel [4.8]

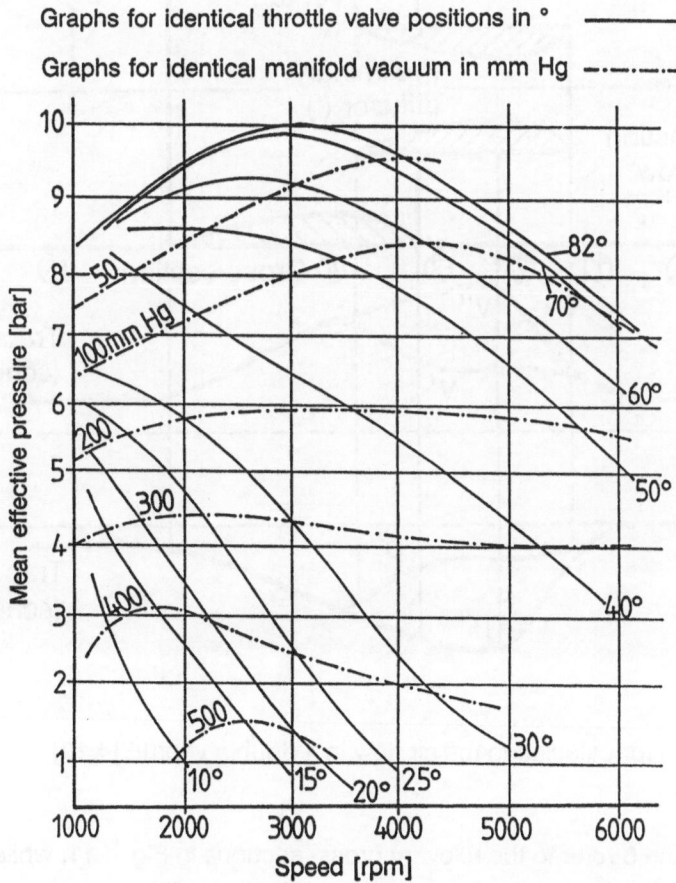

Fig. 4.10. Characteristic map with graphs for identical throttle valve positions and identical manifold vacuums for a four-cylinder four-stroke Otto engine with a displacement of 2 l and a single carburetor Solex 40 PDSI [4.8]

Action of the preliminary diffuser - calculating the air flow in a double venturi:

The calculation is based on the assumption of steady, incompressible (as pointed out by [4.9] this includes an error of approx. 1%), turbulent flow. The "turbulent flow" condition is always met by the Re numbers found in carburetors [4.8]. Other assumptions (cf. **Fig. 4.11**):

- a) Constant velocity w_1 and static pressure p_1 in cross- section 1 across the cross-section, i.e. a steady flow profile;

- b) Constant static pressure in cross-section 4 at the back edge of the preliminary diffuser.

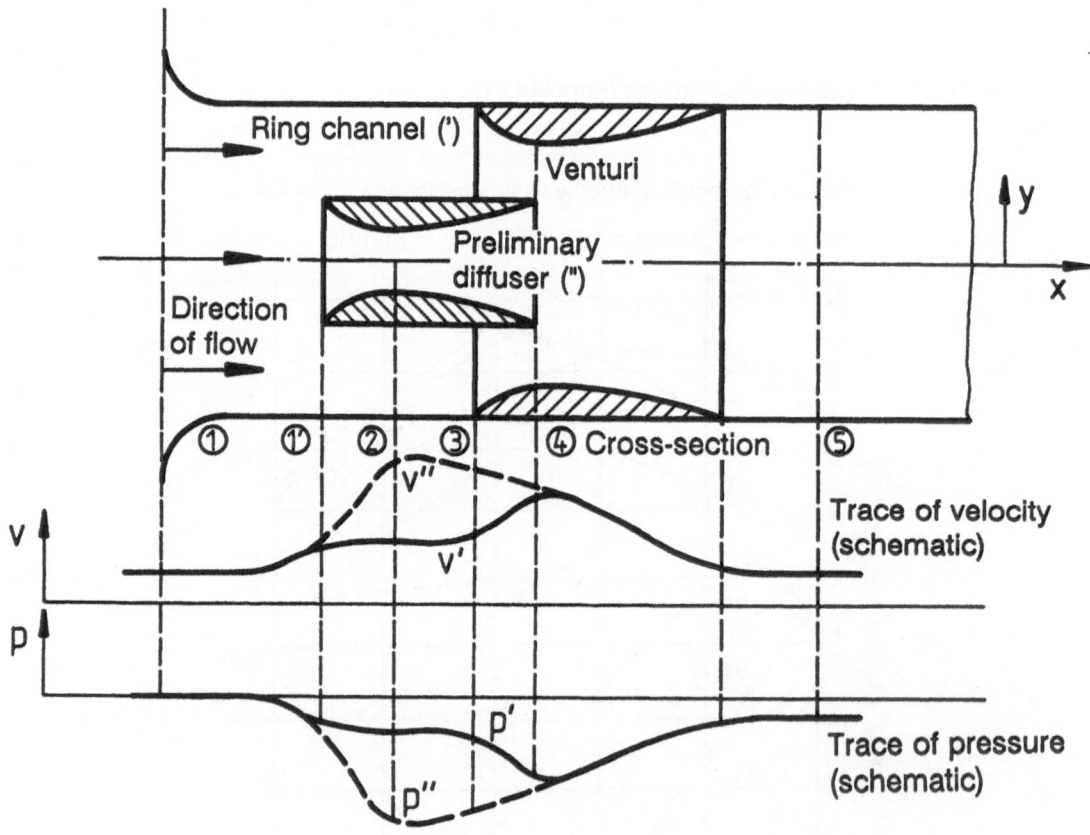

Fig. 4.11. Diagram for calculating the air flow in a double venturi [4.10]

Indices 1, 2, 3, 4 and 5 refer to the relevant cross- sections in Fig. 4.11, while the indices ' and " indicate the ring channel with the air funnel and preliminary diffuser.

The air flow is divided into two streams. The outer stream flows through the ring channel (indicated by ') and is accelerated up to the narrowest cross-section of the air funnel. The inner stream flows through the preliminary diffuser (indicated by "), where it is accelerated, only to be decelerated before it leaves the preliminary diffuser. It is assumed that the velocity at the preliminary diffuser outlet is identical to that of the outer stream. This assumption is made in order to have a genuine mathematical fix for calculating the resistance curve, and is necessary for the next calculations.

Equations:

When drawing up equations the first consideration must be how to do the actual calculations and how to express flow losses. Such losses may be represented as an additive value, or alternatively as multiplicative values (efficiency). Both methods lead to the same result. We are primarily interested in the static pressure at various points of the flow channels, and particularly in the narrowest cross-section of the preliminary diffuser (2").

Pressures and velocities in terms of pressure losses [4.10]:

For the air funnel we have:

$$p_{LT} + p_5 + \frac{\rho_L}{2} \cdot v_5^2 = p_4 + \frac{\rho_L}{2} \cdot v_4^2 \qquad (4.6)$$

where:

p_{LT} $[N.m^{-2}]$ ..... pressure loss in the air funnel.

Taking: $v_5 \cdot A_5 = v_4 \cdot A_4$

we have:

$$p_4 = p_5 + \frac{\rho_L}{2} \cdot v_5^2 \cdot \left[1 - \left(\frac{A_5}{A_4} \right)^2 \right] + p_{LT} \qquad (4.7)$$

For the preliminary diffuser we have:

$$p_2'' + \frac{\rho_L}{2} \cdot v_2''^2 = p_4 + \frac{\rho_L}{2} \cdot v_4^2 + p_{VZ} \qquad (4.8)$$

where:

p_{VZ} $[N.m^{-2}]$ ..... pressure loss in the preliminary diffuser.

Taking: $v_4 \cdot A_4'' = v_2'' \cdot A_2''$

we have:

$$p_2'' = p_4 - \frac{\rho_L}{2} \cdot v_5^2 \cdot \left(\frac{A_5}{A_4} \right)^2 \cdot \left[\left(\frac{A_4''}{A_2''} \right)^2 - 1 \right] + p_{VZ} \tag{4.9}$$

Adding equation (4.7) to (4.9) we get:

$$p_2'' = p_5 - \frac{\rho_L}{2} \cdot v_5^2 \cdot \left[\left(\frac{A_5}{A_4} \cdot \frac{A_4''}{A_2''} \right)^2 - 1 \right] + p_{LT} + p_{VZ} \tag{4.10}$$

For external air intake we have:

$$p_1 = p_5 + \frac{\rho_L}{2} \cdot v_5^2 + p_{LT} \tag{4.11}$$

By adding equation (4.11) to equation (4.10) we get:

$$p_2'' = p_1 - \frac{\rho_L}{2} \cdot v_5^2 \cdot \left(\frac{A_5}{A_4} \cdot \frac{A_4''}{A_2''} \right)^2 + p_{VZ} \tag{4.12}$$

Special case:

If only one preliminary diffuser is available:

$$p_2'' = p_1 - \frac{\rho_L}{2} \cdot v_5^2 \cdot \left(\frac{A_4''}{A_2''} \right)^2 + p_{VZ} \tag{4.13}$$

If only one air funnel is available:

$$p_4 = p_1 - \frac{\rho_L}{2} \cdot v_5^2 \cdot \left(\frac{A_5}{A_4} \right)^2 \tag{4.14}$$

Pressures and velocities in terms of efficiency

Flow separation in the diffuser parts of the air funnel and preliminary diffuser results in pressure losses, i.e. the conversion of energy into heat.

Using diffuser efficiency η_{LT} for the air funnel and η_{VZ} for the preliminary diffuser, we have:

$$p_4 - p_2'' = \eta_{VZ} \cdot \frac{\rho_L}{2} \cdot \left(v_2''^2 - v_4^2 \right)$$

$$\tag{4.15}$$

$$p_5 - p_4 = \eta_{LT} \cdot \frac{\rho_L}{2} \cdot \left(v_4^2 - v_5^2 \right)$$

together with $\quad v_2'' \cdot A_2'' = v_4'' \cdot A_4''$ and $v_4 \cdot A_4 = v_5 \cdot A_5$ (4.16)

we get:

$$p_4 - p_2'' = \eta_{VZ} \cdot \frac{\rho_L}{2} \cdot v_5^2 \cdot \left[\left(\frac{A_5}{A_4} \cdot \frac{A_4''}{A_2''} \right)^2 - \frac{A_5^2}{A_4^2} \right]$$

(4.17)

$$p_5 - p_4 = \eta_{LT} \cdot \frac{\rho_L}{2} \cdot v_5^2 \cdot \left(\frac{A_5^2}{A_4^2} - 1 \right)$$

For a connex between the two expressions we have:

$$p_{VZ} = (1 - \eta_{VZ}) \cdot \frac{\rho_L}{2} \cdot v_5^2 \cdot \left(\frac{A_5}{A_4} \right)^2 \cdot \left[\left(\frac{A_4''}{A_2''} \right)^2 - 1 \right]$$

(4.18)

$$p_{LT} = (1 - \eta_{LT}) \cdot \frac{\rho_L}{2} \cdot v_5^2 \cdot \left(\frac{A_5^2}{A_4^2} - 1 \right)$$

Looking at this term from the energy point of view and reducing energy conversion E (losses at the preliminary diffuser and air funnel) to a minimum, we arrive, after some mathematical transformations, at the main result:

$$\frac{A_4'}{A_2''} = 3{,}12 \approx 3{,}0$$

(4.19)

The ratio between the available ring area of the air funnel and the available preliminary diffuser area thus should be approximately 3:1, as illustrated in **Fig. 4.12**.

Once we have this key result, it is possible to determine the other values as well.

We find that, when diffuser efficiences for the air funnel and preliminary diffuser are identical, it is preferable to increase

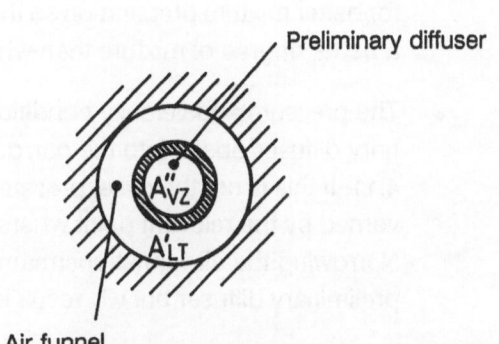

Preliminary diffuser

Air funnel

Fig. 4.12. View of the air funnel and preliminary diffuser cross-section

$$p_1 - p_2'' \quad \text{bzw.} \quad v_2''^2 - v_4^2 \quad \text{or} \quad \frac{A_4''^2}{A_2''^2}$$

$$\text{instead of}$$

$$p_1 - p_4 \quad \text{bzw.} \quad v_4^2 - v_5^2 \quad \text{or} \quad \frac{A_5^2}{A_4^2}$$

In the former expressions, energy (loss) conversion is lower by a factor of A_4''/A_4 than in the latter. In other words, it is better to increase air velocity (and generate a vacuum) in the preliminary diffuser than in the air funnel.

Energy conversion is lower by a factor of A_4''/A_4 with just a preliminary diffuser than with just an air funnel.

Deductions

In summary we may thus state [4.10]:

- When using a preliminary diffuser of the venturi type, the preliminary diffuser achieves a higher air velocity and higher vacuum than is the case in an air funnel. Mixture preparation is improved.

- When using a preliminary diffuser of the cylinder type, air velocity and vacuum are identical in the preliminary diffuser and air funnel, but air velocity is available for a longer period which can be expected to result in a better mixture.

- When the air funnel is constricted appropriately without using a preliminary diffuser, we can achieve the same air velocity and the same vacuum as would be possible with a preliminary diffuser, except that air throttling is higher.

- Provided that the throttle valve does not obstruct any attempt to avoid wall wetting, and provided further that the preliminary diffuser does not itself collect large drops from wall wetting, it is preferable to have a preliminary diffuser because it provides for better mixture preparation so that the air-fuel mixture entering the air funnel shows a better degree of mixture than when it leaves an outlet.

- The pressure and velocity conditions described above occur only when the preliminary diffuser opens into the narrowest point of the air funnel, as is illustrated in Fig. 4.11. If this is not the case, pressure conditions in the preliminary diffuser will be governed by the relevant point where the preliminary diffuser opens into the air funnel. Narrowing the air funnel upstream of that point will not increase the vacuum in the preliminary diffuser but will result in throttle losses.

- It also appears that wall wetting will be increased when the mixture from the preliminary diffuser enters a delayed flow since delayed flows tend to generate backflows more than accelerated flows. If it is not possible to run the preliminary diffuser into the narrowest cross-section of the air funnel it should be designed to open into the air funnel upstream, rather than downstream, of that point.

Dutta [4.10] also made detailed tests to check and supplement his calculations:

Fig. 4.13 indicates the vacuum at the preliminary diffuser and air funnel as a function of air throughput.

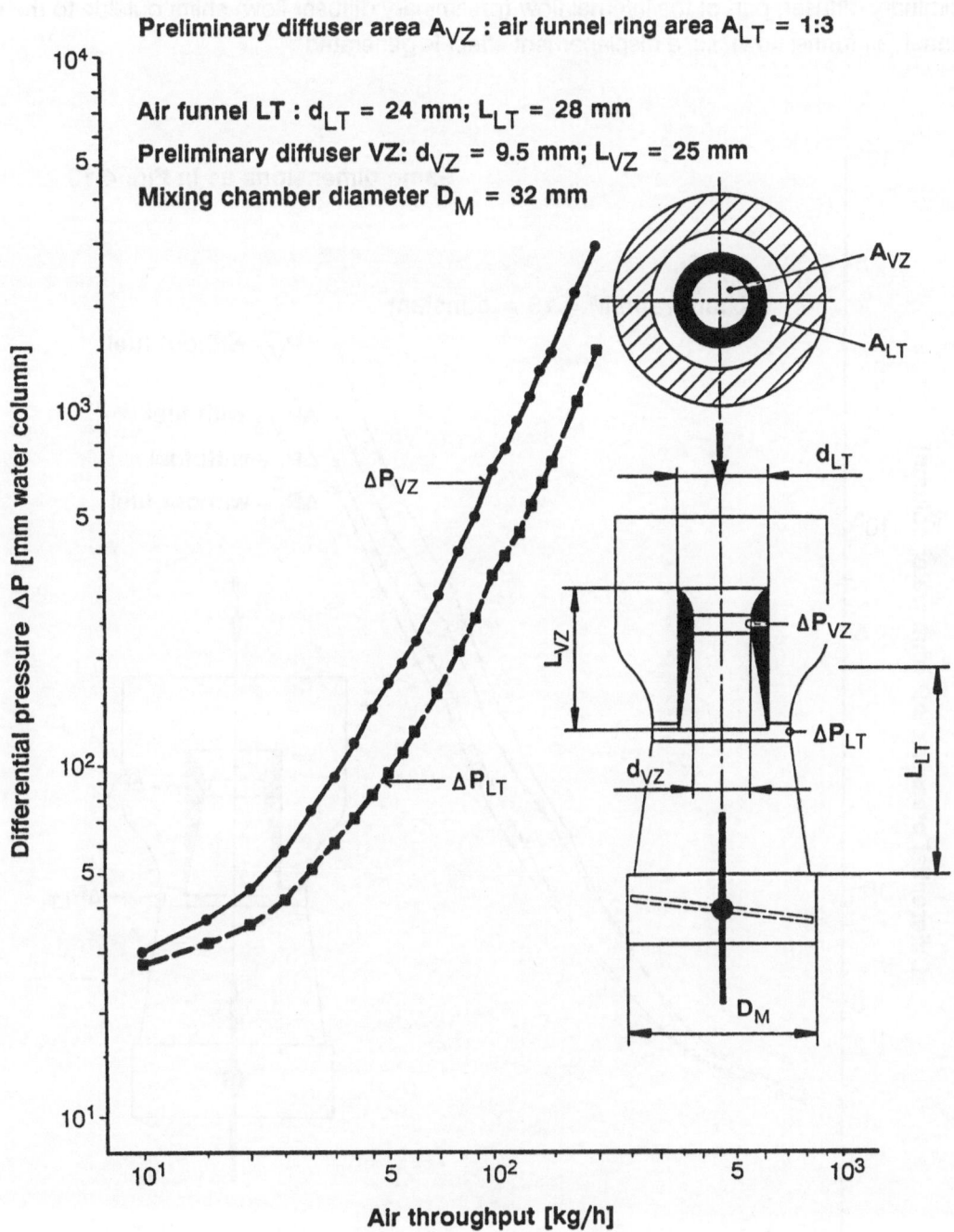

Fig. 4. 13. Vacuum testing at the preliminary diffuser and air funnel when the throttle valve is open/closed [4.10]; measurements were made without fuel

Fig. 4.14 supplies vacuum figures for the preliminary diffuser and air funnel with and without fuel. Measurements using fuel found lower vacuums in the preliminary diffuser than measurements without fuel; the result, according to Dutta, of the energy expenditure required to transport the fuel. This applies only to the preliminary diffuser while vacuums in the air funnel are higher with fuel than when measured without fuel. Dutta assumes that when fuel is added to the preliminary diffuser, part of the internal flow (preliminary diffuser flow) shifts outside to the ring channel (air funnel flow), i.e. a displacement effect is generated.

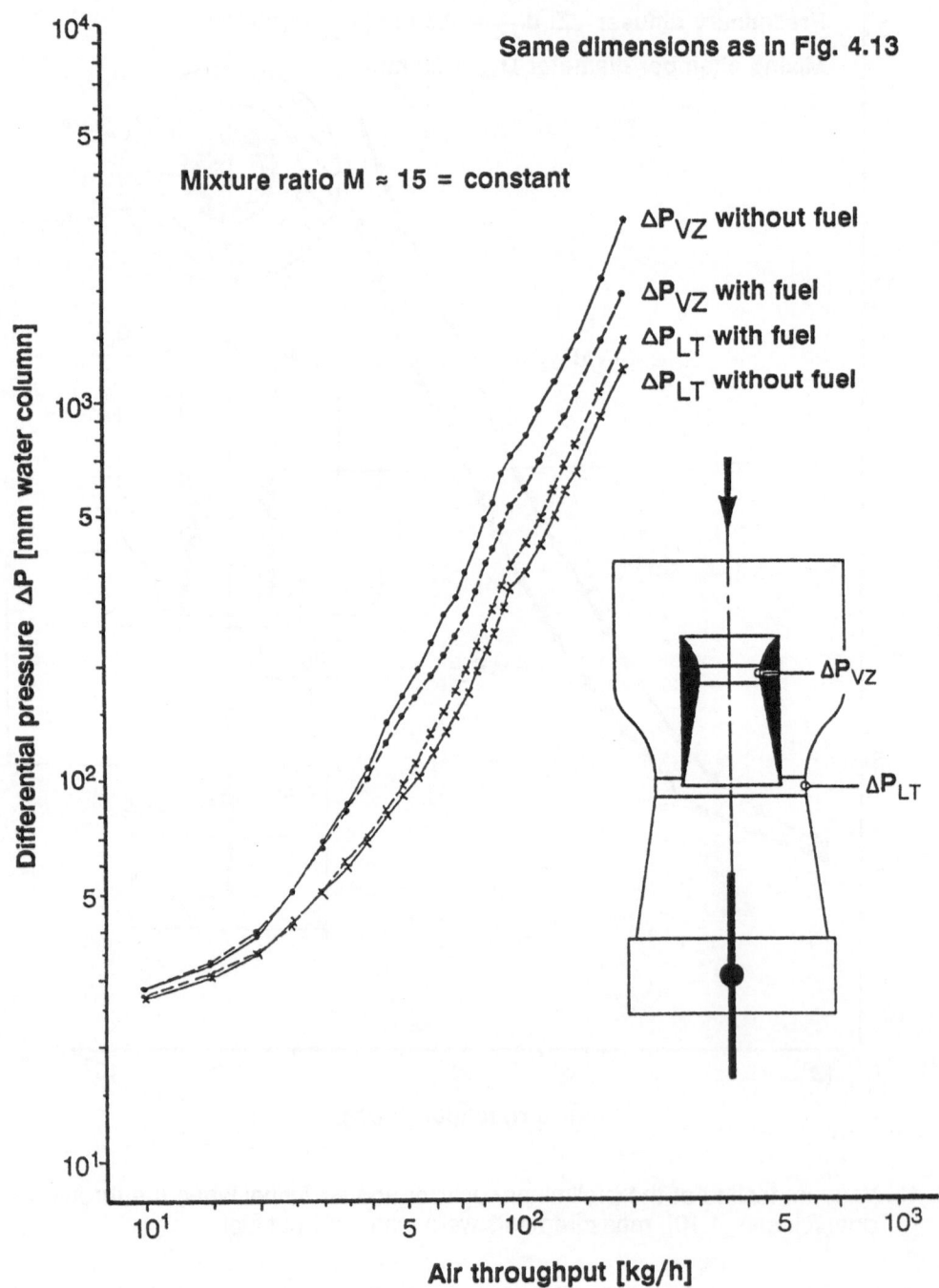

Fig. 4. 14. Vacuum testing at the preliminary diffuser and air funnel with and without fuel [4.10]

Fig. 4.15 charts vacuum graphs in the preliminary diffuser for various mixture ratios in terms of the air throughput. No significant changes can be found at the lower throughput range, but vacuums increase with increasing mixture ratios when the throughput is higher.

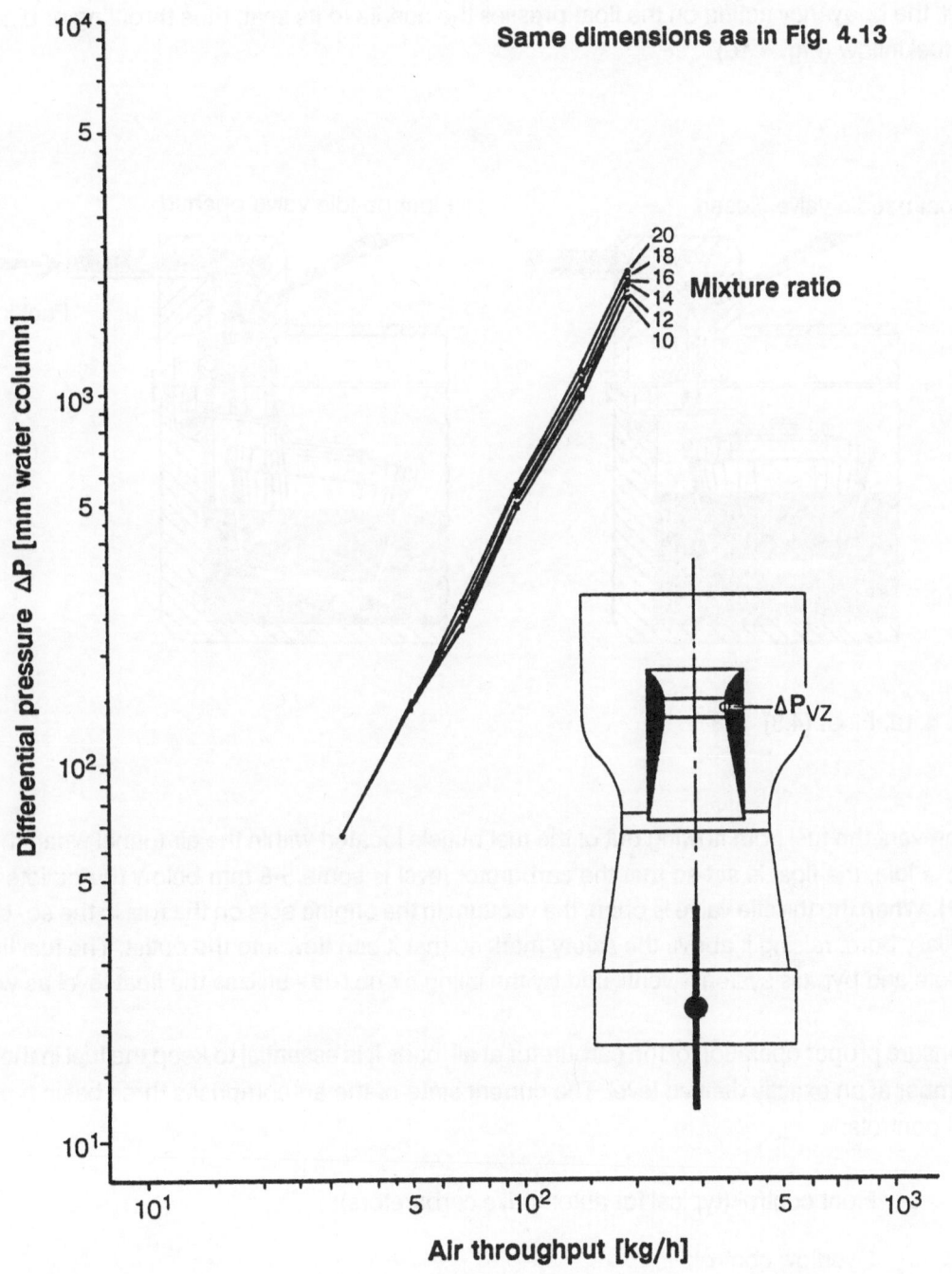

Fig. 4. 15. Vacuum at the preliminary diffuser for different mixture ratios of kg air/kg fuel [4.10]

4.1.1.2.2. Devices to Control the Fuel Level

Such devices, usually consisting of a float and float needle valve in automotive carburetors, are designed to keep the fuel in the carburetor at a constant level. When the fuel has reached a given level, the buoyancy acting on the float presses the needle to its seat, thus throttling or blocking the fuel inflow (**Fig. 4.16**).

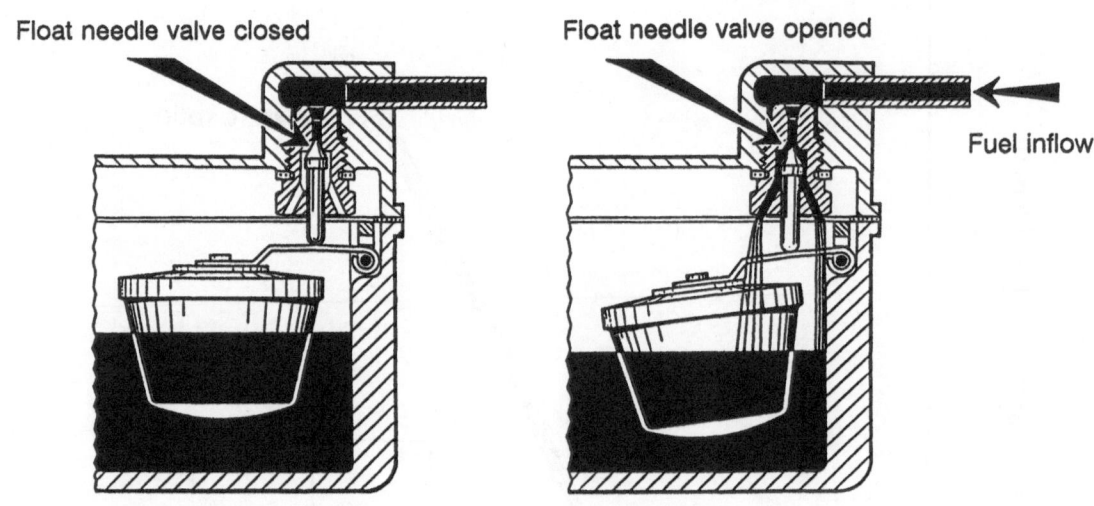

Float needle valve closed   Float needle valve opened

Fuel inflow

Fig. 4.16. Float [4.8]

To prevent the fuel from flowing out of the fuel outlets located within the air funnel when the engine is idle, the float is set so that the carburetor level is some 5-8 mm below the outlets (**Fig. 4.21**). When the throttle valve is open, the vacuum in the engine acts on the fuel in the so-called auxiliary bore, raising it above the safety mark so that it can flow into the outlet. The fuel line to the idle and bypass system - ventilated by the idling air nozzle - utilizes the float level as well.

To ensure proper operation of the carburetor at all loads it is essential to keep the fuel in the float chamber at an exactly defined level. The current state-of-the-art comprises three basic types of level controls:

- Float control (typical for automotive carburetors);
- Overflow control;
- Diaphragm control.

For a proper design of the level control system it is first necessary to define the conditions to be met:

- Constant level at the auxiliary bore and main nozzle;

- Prevention of level variations, such as are caused by fuel inflow, fuel delivery, hysteresis, vibrations, etc.;

- Stability in curves;

- Acceleration and deceleration stability, although it is usually acceptable that the level in the auxiliary bore rises with acceleration and drops with deceleration;

- No impairment of operating conditions when the vehicle is inclined longitudinally or laterally (in particular when reversing out of a basement garage).

The carburetor should be fitted so that the float chamber is up front when viewed in the direction of driving. The result is that the mixture tends to be richer during acceleration or when the vehicle is tilted upward which in turn improves performance (**Fig. 4.17**).

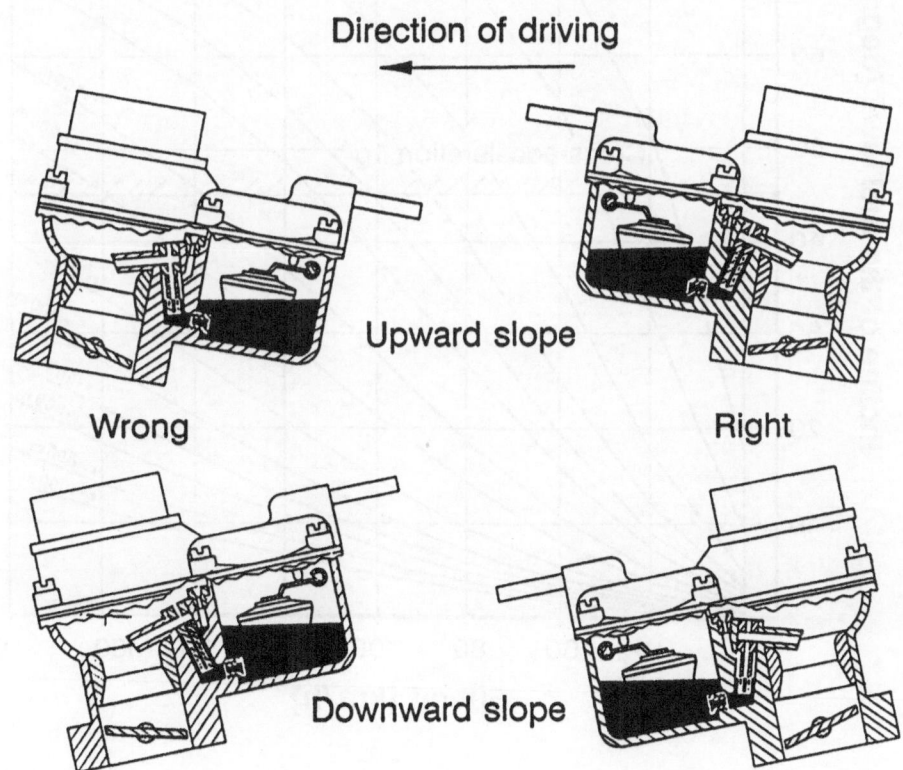

Fig. 4.17. Float chamber arrangement [4.8]

To avoid the float system from being affected by curves, the float chamber and float are placed symmetrically, as viewed in the direction of driving, to the fuel outlet. For the same reason it is recommended to keep the distance between the mixture outlet and float center as small as possible. By these measures it is possible to keep to a minimum the impact of variations in the fuel level position in curves on the fuel delivery (**Fig. 4.18**). Other factors to take into account when designing float systems: linking the float at the level of the float's center of gravity to prevent the centrifugal force from acting on the float in a curve; and a transmission ratio between float lift and float needle valve lift to ensure that a low float lift produces a high float needle valve lift. Such measures combine to keep the fuel at a more or less constant level at all driving conditions [4.12].

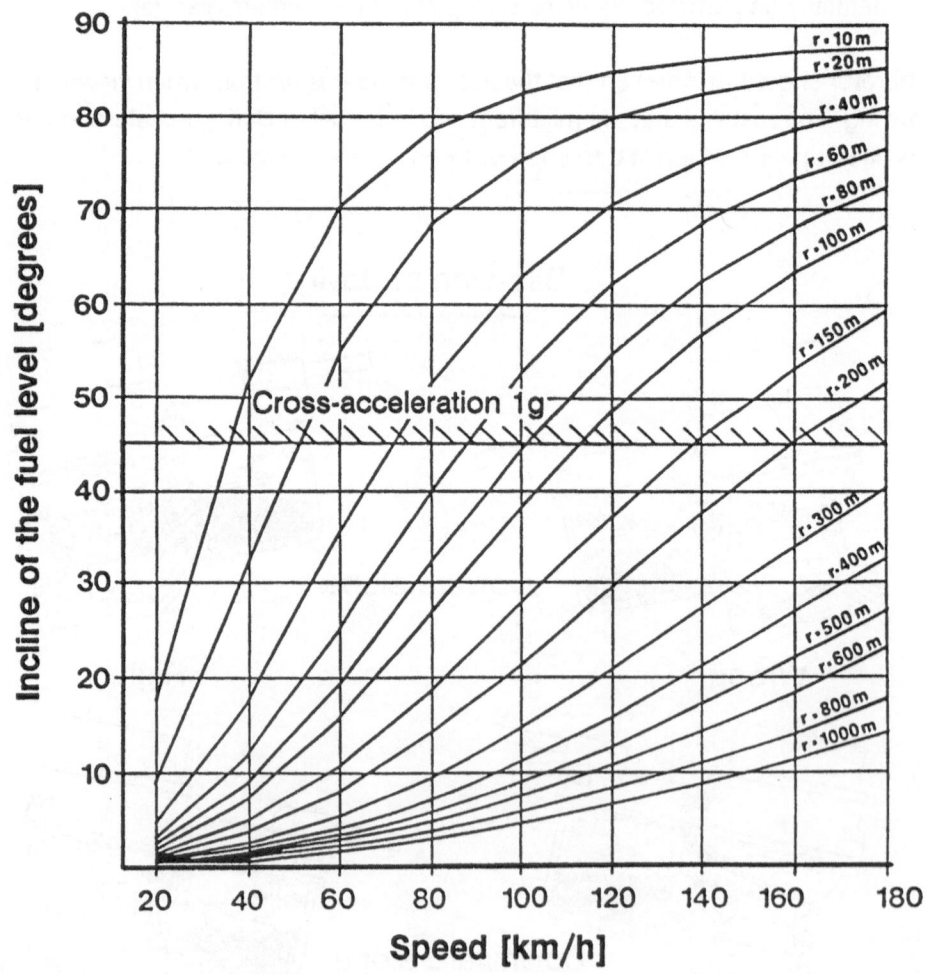

Fig. 4.18. Incline of the fuel level at different curve radii in terms of the speed [4.8]

Carburetors fitted with an overflow pipe to keep the fuel at a constant level are particularly insensitive to curves, tilts and acceleration. They are provided with a separate float chamber (**Fig. 4.19**) or alternatively a special fuel return pump to feed the overflowing fuel back into the tank.

The former have a fuel pump (on the right-hand side in Fig. 4.19) to pump the fuel into a float chamber placed underneath the carburetors. From there the fuel is fed by a second pump to the carburetors each of which are fitted with an overflow pipe to run the excess fuel back into the float chamber. The result is a highly constant fuel level in the carburetor.

Diaphragm systems to control the fuel level are rarely used in automotive carburetors. They are normally installed in small carburetors, such as are fitted into logging saws, as their controlling action, while independent of the carburetor position, is relatively inaccurate.

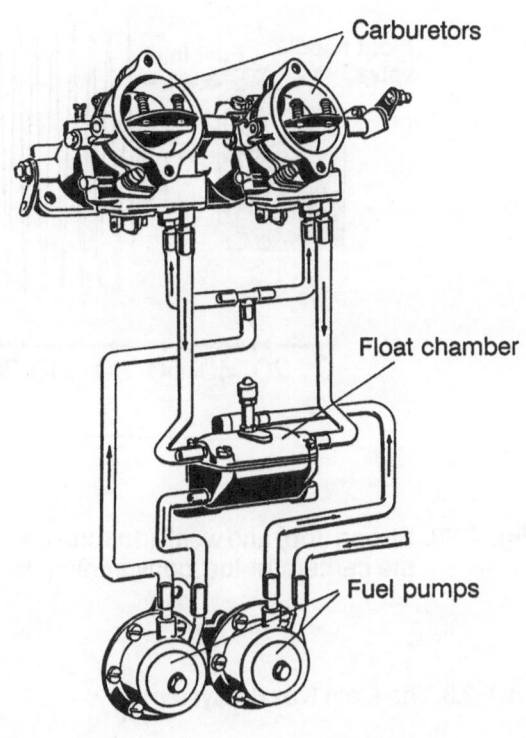

Fig. 4.19. Carburetors with separate float chamber [4.8]

A problem affecting not so much practical operation but calibration of carburetors is the pendulum action of the float. With the carburetor float constituting a mass, and the float joint and valve needle generating friction, the float will not normally be in a position that corresponds to the mean fuel delivery.

It will oscillate between a position of excess fuel delivery and one of insufficient fuel delivery. This inherent movement nevertheless is so negligible that it may be virtually ignored.

The movement becomes stronger when fuel flows from the float needle valve onto the float or float joint. Possible design measures to prevent this pendulum action are placing the float needle valve underneath the fuel level or laterally to the float. In those cases where the float needle valve must be placed atop, the float can be stabilized by putting runoff sheets from the float joint to below the fuel level.

Fig. 4.20 illustrates the oscillation of a float/float needle valve system.

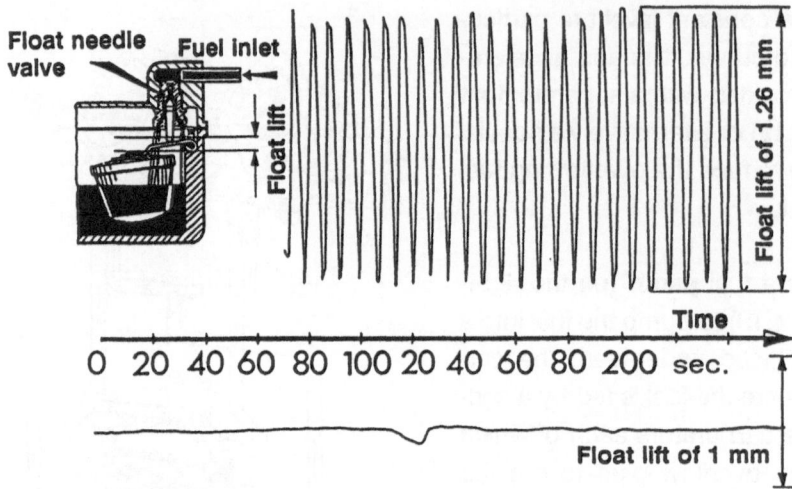

Fig. 4.20. Strong (top) and weak (bottom) oscillations of the float/float needle valve system when the carburetor fuel consumption is constant [4.8]

4.1.1.2.3. The Main Nozzle System

The composition of the combustion mixture is determined by the following carburetor components:

the main nozzle to supply the basic fuel quantity; the air funnel to meter the basic air quantity; and the compensating air nozzle to act, together with the mixing pipe, as a compensator by adding more air as the throughput increases, to prevent excessive enrichment of the mixture (**Fig. 4.21**).

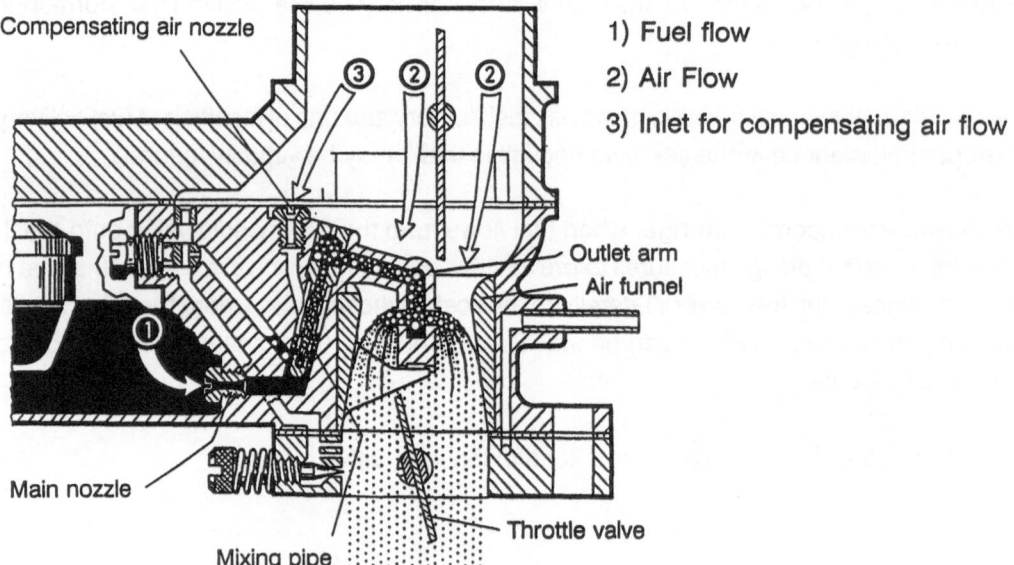

1) Fuel flow

2) Air Flow

3) Inlet for compensating air flow

Fig. 4.21. Main metering system for a downdraft carburetor [4.8]

In this type of carburetor, the compensating air nozzle is placed next to the air funnel. The vacuum in the air funnel causes fuel and air to be sucked off through the bores of the mixing pipe and outlet arm. **Fig. 4.22** sketches the arrangement of a compensating air nozzle and mixing pipe.

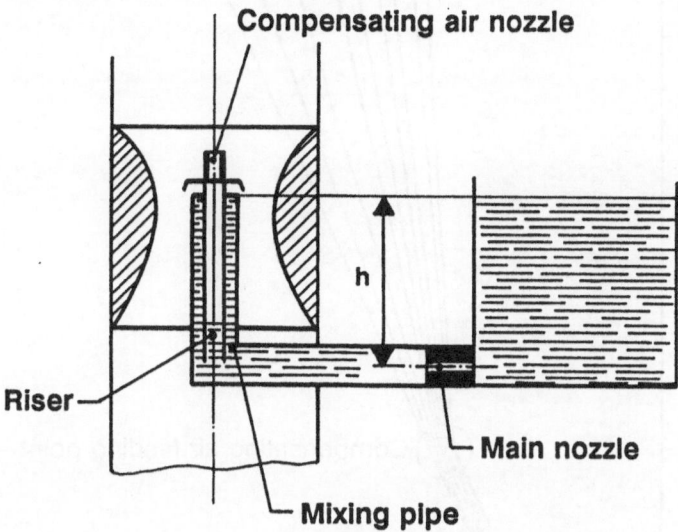

Fig. 4.22. Schematic arrangement of the compensating air nozzle [4.8]

The mixing pipe in the illustration is simplified by showing only one outlet bore. The fuel level in the mixing pipe drops when the vacuum rises in the air funnel. When it has dropped by a factor h, compensating air flows into the riser where it forms an emulsion of fuel and air. By this action, a flow is generated in the compensating air nozzle which reduces the differential pressure at the main nozzle that causes fuel to be delivered so that further fuel deliveries are reduced relative to the increase of the air funnel vacuum.

The effect of various sized compensating air nozzles on the differential pressure at the main nozzle is given in **Fig. 4.23**.

The action of the compensating air nozzle was calculated mathematically by Linzer [4.13]. On the assumption of friction-free flow and neglecting air compressibility, [4.13] the following equations define the mixture ratio of a carburetor provided with a compensating air nozzle and mixing pipe outlet:

$$\frac{m_L}{m_K} = \frac{A_L \cdot \alpha_L \cdot K_1}{A_K \cdot \alpha_K} \cdot [\frac{\gamma_L}{\gamma_K} \cdot \frac{\Delta p_{LT}}{(p_A - p_2) + h \cdot \gamma_K}]^{0,5} \qquad (4.20)$$

$$K_1 = (1 - \frac{2 \cdot A_K}{A_2} + \frac{2 \cdot A_K^2}{A_2^2})^{0,5} \qquad (4.21)$$

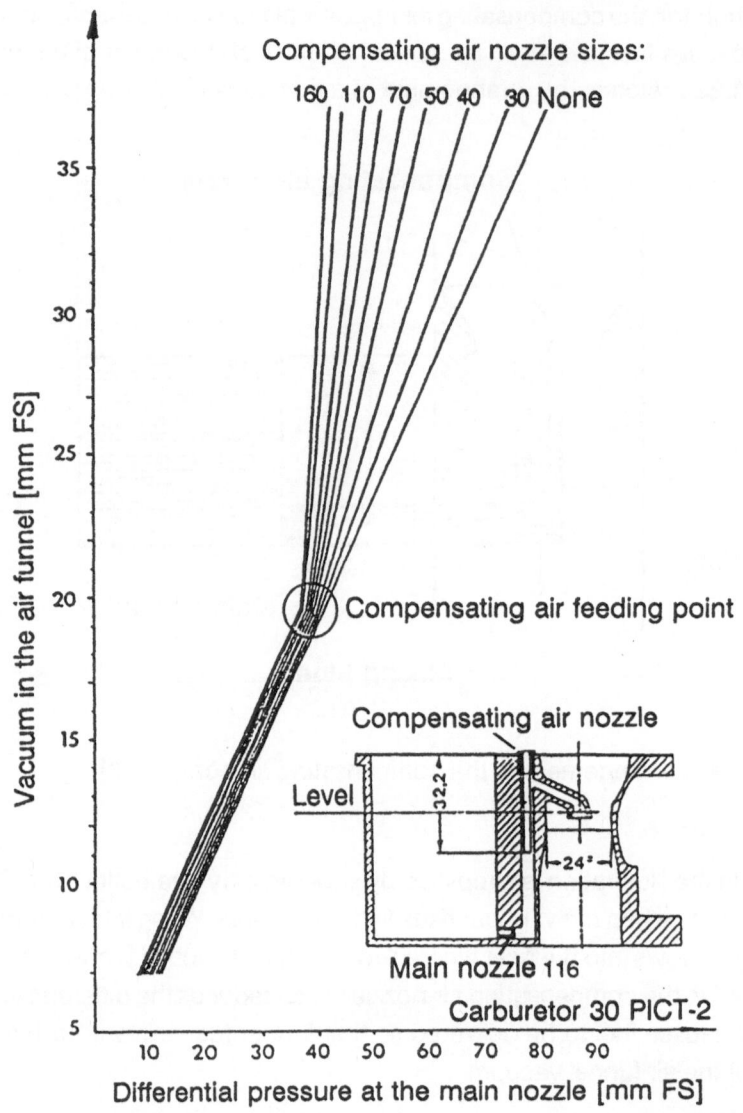

Fig. 4.23. Effect of the compensating air nozzle and mixing pipe on the differential pressure at the main nozzle (FS = liquid column) [4.8]

where:

α_{KD}, α_{LT} [-]   ..... flow indices for the fuel nozzle/air funnel

A_{KD}, A_{LT} [m²]   ..... passages of the fuel nozzle/air funnel

γ_L, γ_K [kg.m⁻³] ..... relative density of air/fuel

h   [m]   ..... distance between the compensating air inlet bore and the float chamber
                                   fuel level, i.e. immersion depth of the mixing pipe

Δp_{LT} [N.m⁻²]   ..... vacuum in the air funnel

p_A [N.m⁻²]   ..... external pressure

p_{KD} [N.m⁻²]   ..... pressure at the fuel nozzle

A_2 [m²]   ..... cross-sectional area of the fuel inlet line

The relation between (p_A - p_2) and Δp_{LT}, provided that compensating air is fed and that $A_1 = A_2 = A_3$, is derived from:

$$\Delta p_{LT} = (p_A - p_2) \cdot \left\{ 1 + \frac{2 \cdot K_3^2 \cdot \alpha_{LK} \cdot A_{LK}}{K_1 \cdot K_2 \cdot \alpha_{KD} \cdot A_{KD}} \cdot \left(\frac{h \cdot \gamma_K}{p_A - p_2} + 1 \right)^{0,5} \cdot \right.$$

$$\left. \left[\left(\frac{\gamma_K}{\gamma_L} \right)^{0,5} + \left(\frac{\gamma_L}{\gamma_K} \right)^{0,5} + 2 \cdot \frac{K_3^2 \cdot \alpha_{LK}^2 \cdot A_{LK}^2}{K_2^2 \cdot \alpha_{KD} \cdot A_{KD}^2} + \frac{h \cdot \gamma_m}{p_A - p_2} \right\} \right. \tag{4.22}$$

When

$$K_2 = \left(1 - \frac{2 \cdot \alpha_{LK} \cdot A_{LK}}{A_3} + \frac{2 \cdot \alpha_{LK}^2 \cdot A_{LK}^2}{A_3^2} \right)^{0,5} \tag{4.23}$$

(where:

$A_{LK}, a_{LK} \, [m^2, -]$ ..... surface/flow index of the compensating air nozzle
$A_3 \quad [m^2]$ ..... surface of the compensating air channel
$A_1 \quad [m^2]$ ..... surface of the fuel channel

as in Fig. 4.22.

$$K_3 = \frac{\alpha_K \cdot A_{KD}}{A_3} \tag{4.24}$$

γ_m is the mean relative density of fuel and air achieved after mixture.

$$\gamma_m = \frac{\gamma_K + \dfrac{K_1 \cdot \alpha_{LK} \cdot A_{LK} \cdot \gamma_L}{K_2 \cdot \alpha_K \cdot A_{KD}} \cdot \left\{ \dfrac{1}{\gamma_L \cdot [h / (p_A - p_2)] + \gamma_L / \gamma_K} \right\}^{0,5}}{1 + \dfrac{K_1 \cdot \alpha_{LK} \cdot A_{LK}}{K_2 \cdot \alpha_K \cdot A_{KD}} \cdot \left\{ \dfrac{1}{\gamma_L \cdot [h / (p_A - p_2)] + \gamma_L / \gamma_K} \right\}^{0,5}} \tag{4.25}$$

As long as no compensating air enters the mixing pipe we have:

$$\Delta p_{LT} = (p_A - p_2) + h \cdot \gamma_K \tag{4.26}$$

Fig. 4.24 outlines the mixture ratios found at a carburetor main system when various sized compensating air nozzles are used and when none is used. It establishes clearly that the use of a compensating air nozzle results in a leaner mixture.

Fig. 4.25 and **4.26** demonstrate the effects of different main nozzles and compensating air nozzles for the carburetor in general.

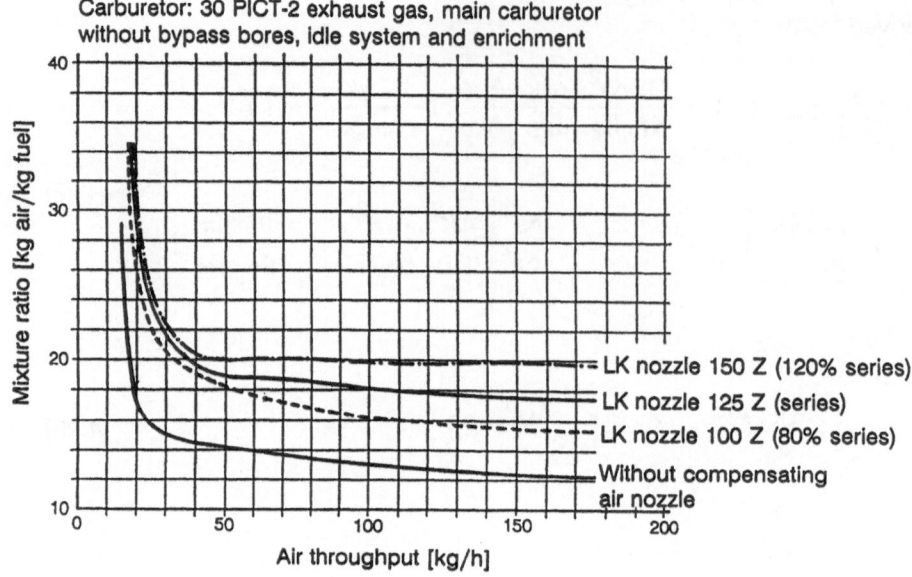

Fig. 4.24. Mixture ratio of a carburetor main system using various sized compensating air nozzles and using no nozzle [4.8]

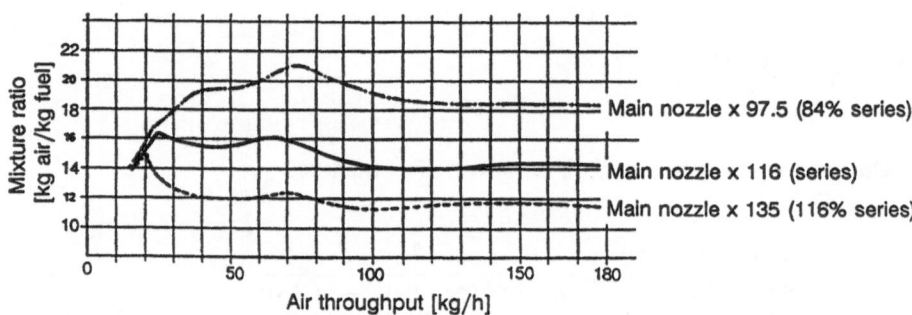

Fig. 4.25. Effect of different main nozzles on the mixture ratio in a Solex 30 PICT-2 carburetor [4.8]

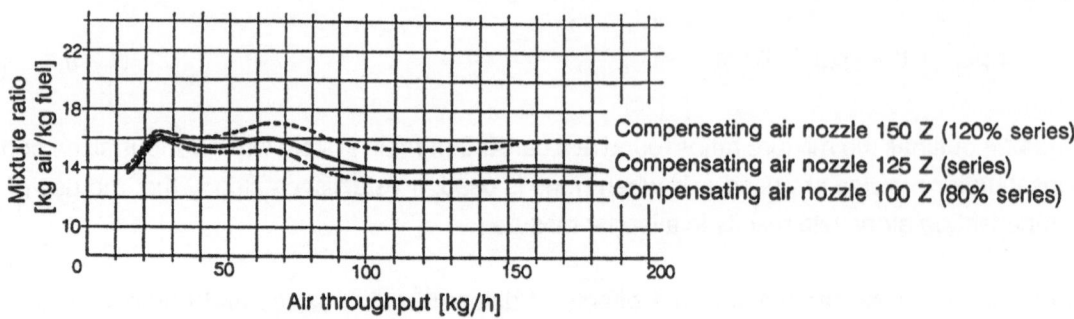

Fig. 4.26. Effect of different compensating air nozzles on the mixture ratio in a Solex 30 PICT-2 carburetor [4.8]

4.1.1.2.4. The Idling System

When the throttle valve is almost closed, the vacuum in the air funnel drops enough to stop the fuel from flowing out through the main system. A second nozzle system is thus required to keep the engine running in idle shift (**Fig. 4.27**).

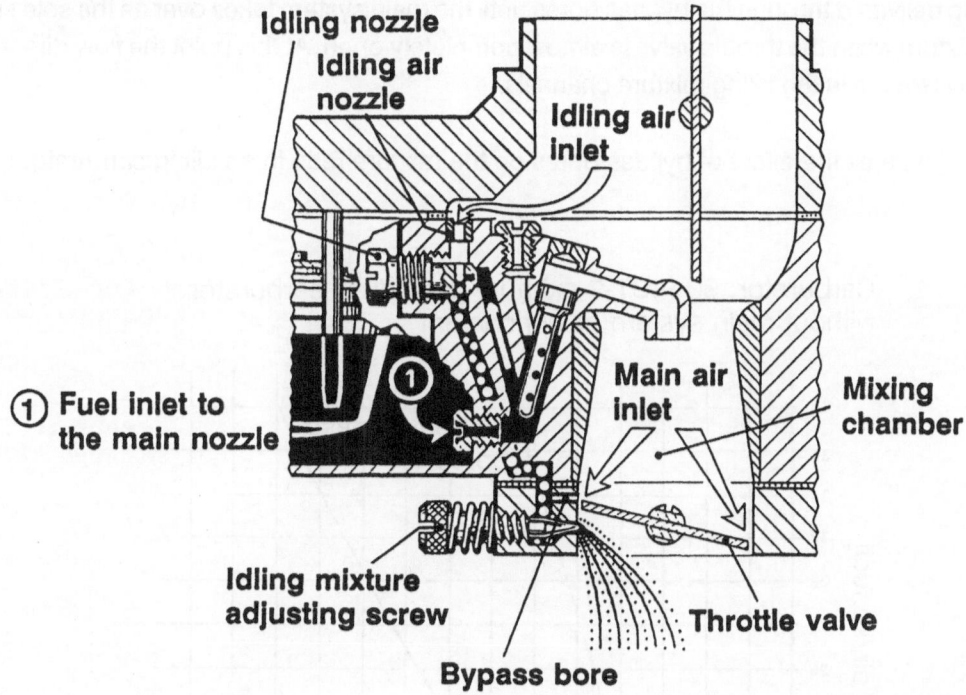

Fig. 4.27. Idling system [4.8]

The system, generally called "idling system," has its outlet underneath the throttle valve. Here too it is important to supply the engine with the proper ratio of fuel and air. This is done on the one hand by the idling fuel nozzle and idling air nozzle, where dosing is by the idling mixture adjusting screw, and by a setting of the throttle valve that allows an air volume to flow through the throttle valve gap which can be adjusted to regulate the idling speed.

The idling system operates up to a speed of some 60 km/h in engines of about 1.5 liters of cubic capacity, and even higher speeds in larger engines, so that its proper adjustment is an important factor in fuel consumption.

4.1.1.2.5. The Bypass System

"Bypasses" are bores between the mixing chamber and a fuel- carrying channel near the throttle valve which are covered by the throttle valve when it opens.

The bypass system, sketched schematically in the form of a single bore above the throttle valve in Fig. 4.27, provides for initial ventilation of the idling system. When the throttle valve opens it stops ventilation and more mixture is delivered. When the throttle valve continues to open mixture is also delivered through the bypass bores until the main system takes over as the sole supplier of mixture when the throttle valve is almost completely open. At this point the flow direction will usually reverse in the idling mixture channels.

Fig. 4.28 illustrates the effect of bypass bores on the mixture ratio in an idling carburetor system.

Carburetor: 30 PICT-2 exhaust gas, idling carburetor
without main system or enrichment

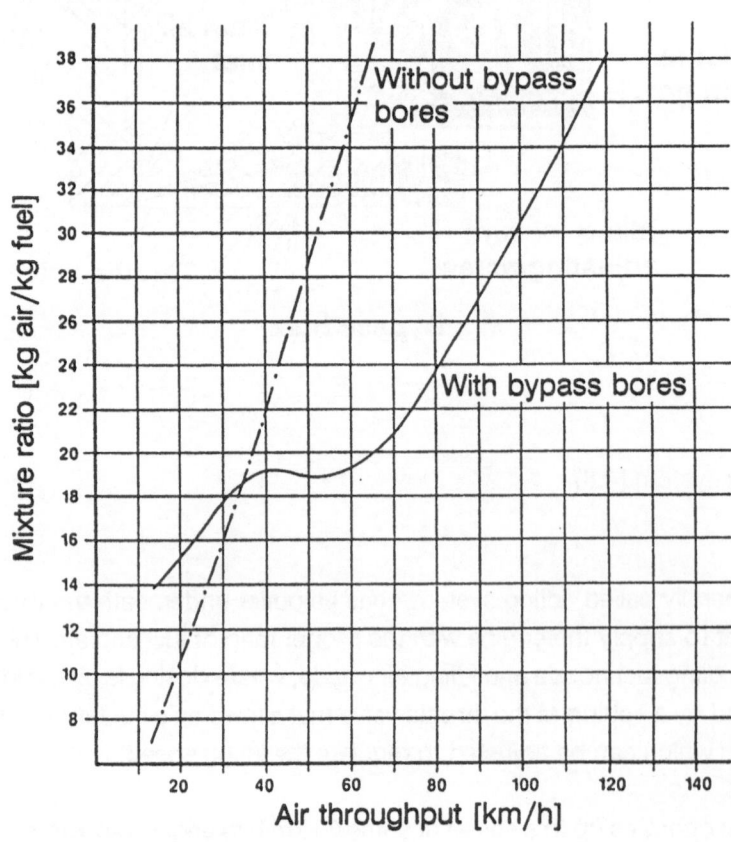

Fig. 4.28. Mixture ratio in an idling carburetor system, with and without bypass bores [4.8]

Fig. 4.29 and **4.30** demonstrate the pressure traces at no load, part load and full load for two idling systems [4.14]. In Fig. 4.29 the idling air inlet is positioned in the upper part of the idling system, while in Fig. 4.30 it is slightly below the narrowest point of the air funnel. The measurements were carried out at a carburetor test bench. After setting a specific manifold vacuum and corresponding air throughput, pressures were taken at four points of the idling system. The measuring points, marked 1-5 in Figures 4.29 and 4.30, were located as follows:

- Measuring point 1: in the throttle valve component next to the idling mixture outlet;

- Measuring point 2: in the idling channel above the bypass bores;

- Measuring point 3: in the idling channel near the idling fuel nozzle, which in this case is designed as an immersion nozzle;

- Measuring point 4: in the auxiliary bore next to the idling fuel nozzle;

- Measuring point 5: on the air inlet side of the idling air nozzle.

Now for a more detailed explanation of Fig. 4.29:

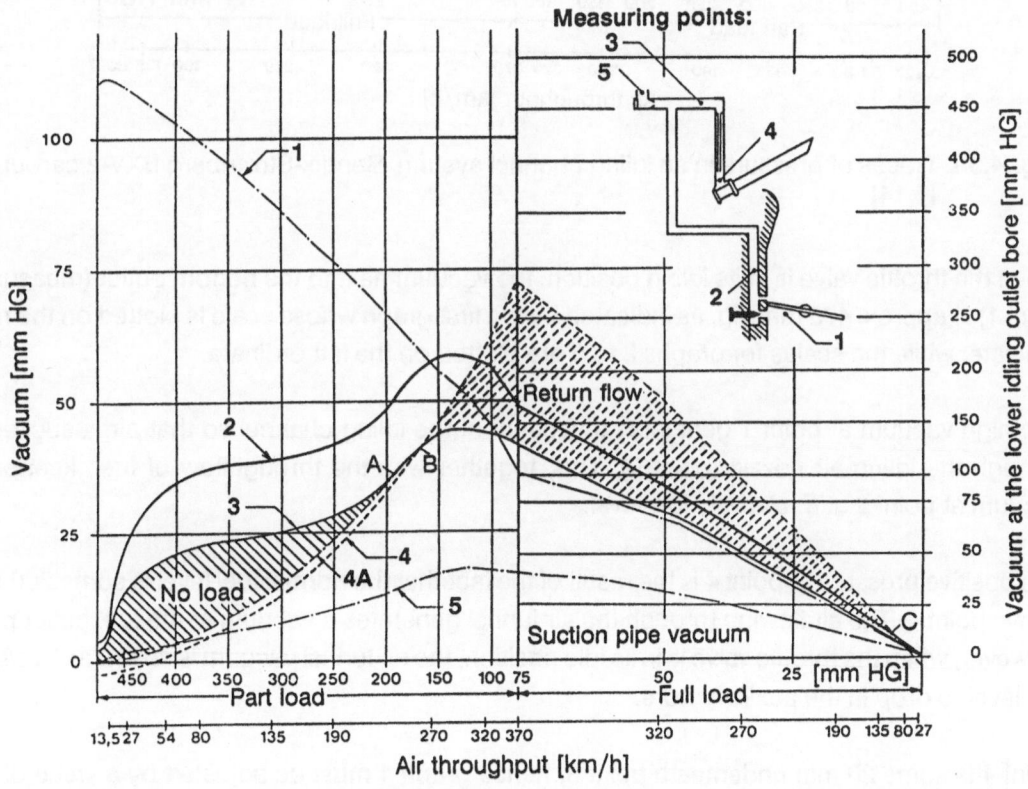

Fig. 4.29. Traces of pressure in an idling channel system. Bendix-Stromberg BXV-2 carburetor [4.14]

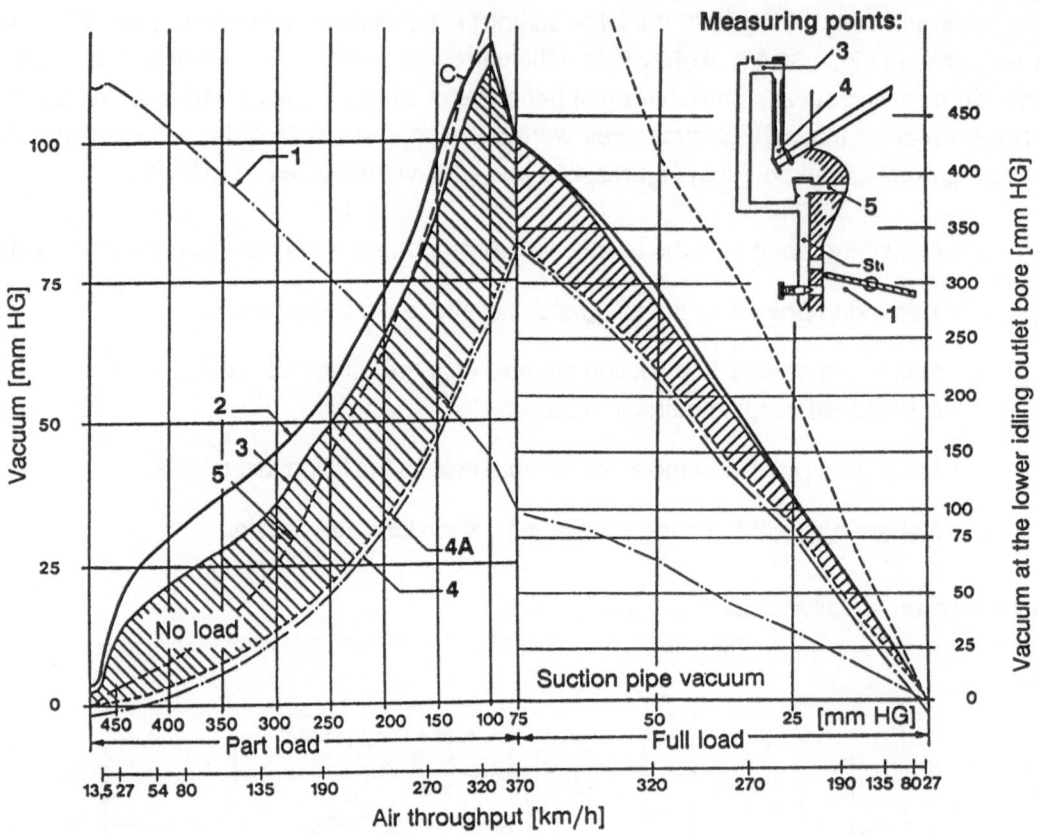

Fig. 4.30. Traces of pressure in an idling channel system. Bendix-Stromberg BXV-2 carburetor [4.14]

When the throttle valve is in its idling position, the vacuum next to the bottom outlet (measuring point 1) is approx. 475 mm Hg, as indicated by the first graph whose scale is plotted on the right ordinate, while the scales for graphs 2 to 5 are plotted on the left ordinate.

The high vacuum at point 1 generates a vacuum at the idling channel so that air is sucked in through the idling air nozzle (point 5). This, together with the throughflow of fuel, keeps the vacuum at point 2 at a relatively low level.

The positive pressure at point 4 is the result of the fact that the normal fuel level is some 200 mm above point 4. The air flowing through the air funnel generates a vacuum at the fuel outlet pipe. However, when the throttle valve is in its idle position, the air funnel vacuum is so low as to cause the level to drop in the auxiliary bore.

Point 4 is some 20 mm underneath point 3, hence graph 4 must be adjusted by a value of approx. 20 mm liquid column to achieve a pressure drop comparable to other measuring points, which in turn results in fuel flowing through the idling system. Graph 4, adjusted by 20 mm LC, is plotted as graph 4A.

When the throttle valve is opened from its idling position, the bypass bore shifts from low vacuum to high vacuum, changing in character from idling channel ventilation bore to idling mixture outlet bore, as demonstrated by the sharp ascent of graph 2.

With the throttle valve opening, the fuel level in the auxiliary bore drops initially until a vacuum is generated in the air funnel, which raises the level so that fuel is delivered through the idling channels as well as through the fuel outlet arm.

The vacuum at point 4 rises with the opening of the throttle valve (graph 4A) until it is similar to the vacuum at point 3 (measuring point A in Fig. 4.29). At this point fuel delivery through the idling system is stopped; the channel between 3 and 4 is filled with fuel. When the air flow continues to increase, the fuel level between 3 and 4 is gradually decreased until the channel is free of fuel at measuring point B in Fig. 4.29. Graph 4A blends into graph 4. From measuring point B until the fully opened throttle valve, the vacuum is higher at point 4 than at 3, so that air is sucked through the idling air nozzle into the fuel outlet arm. The other idling channels are also drained of fuel.

This reversal of the flow direction and drainage of the fuel channel has two consequences: better emulsification of the fuel delivered through the fuel outlet arm with the air delivered through the idling system, which in turn results in a leaner mixture.

When the throttle valve is suddenly changed from full-load to no-load position the flow direction is reversed and a short delay occurs in the fuel delivery for the no-load condition until the channels have been filled with fuel. This may cause some jerking of the vehicle.

To prevent this, the idling air nozzle may be arranged as shown in Fig. 4.30 where it is placed some 3 mm underneath the narrowest point of the air funnel. By this means the flow in the idling channels is not reversed until the point where the idling air nozzle runs into the upper part of the channels, as illustrated from the trace of pressure in Fig. 4.30. The result is a relatively short section of the channel which is to be filled when the throttle valve closes (measuring point 2) and where the flow is reversed, which section ventilates the air funnel bore via the bypass bore starting from measuring point C in Fig. 4.30.

Starting at this air throughput no more air is delivered through the idling air nozzle which delivers fuel and air into the air funnel instead.

The practical setup depends on the application requirements; designers can choose from a wide range of variations to customize the system to the respective demands.

Fig. 4.31 shows the effect of the throttle valve position on the mixture ratio when the throttle valve covers the bypass bores. Smooth transition of the engine from no load to normal operation generally depends on the arrangement of the bypass bores.

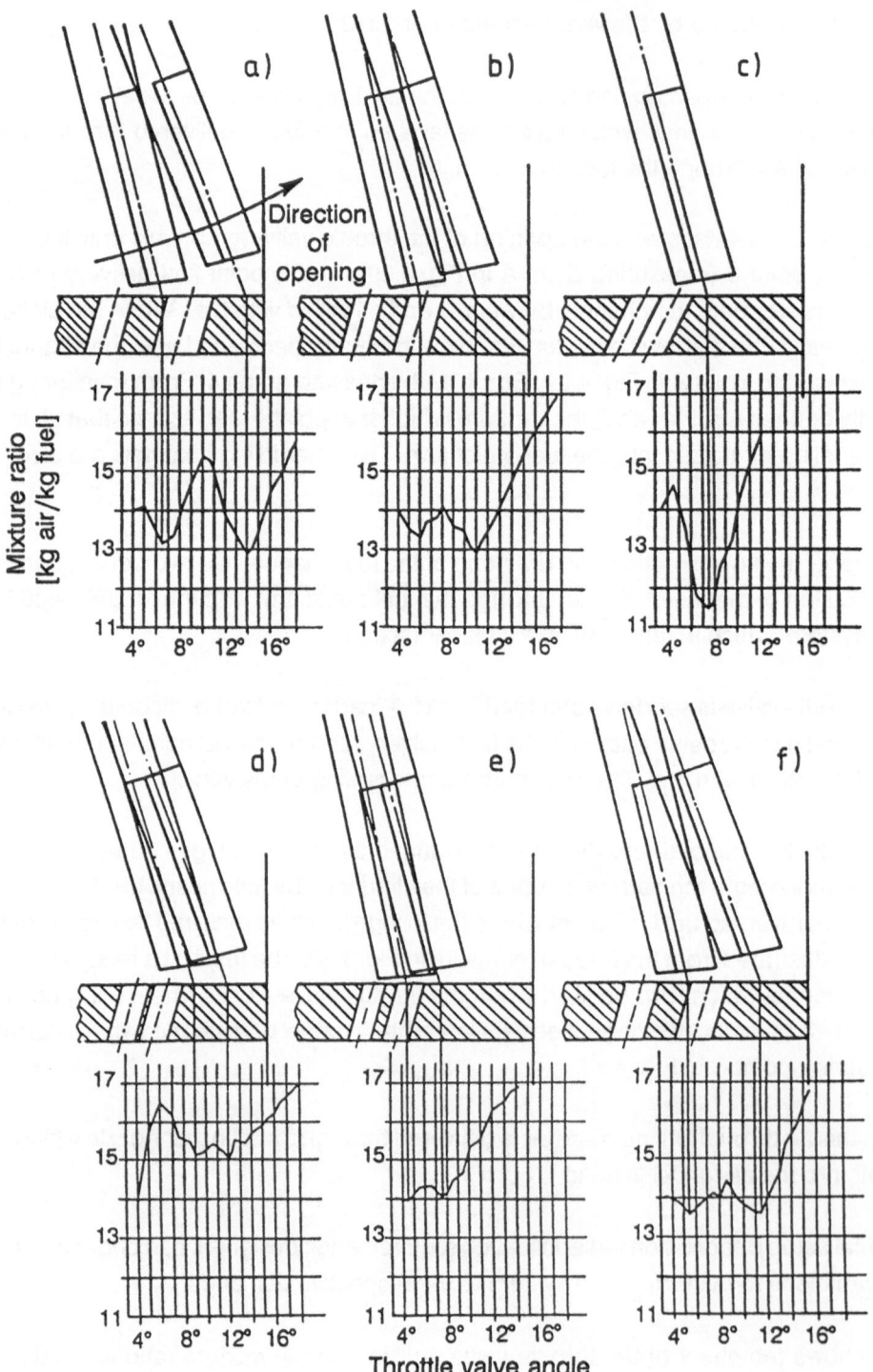

Fig. 4.31. Effect of the throttle valve position on the mixture ratio when the throttle valve covers the bypass bores [4.8]

Enrichment of the mixture is strongest when a bypass bore is fully covered by the throttle valve so that the manifold vacuum can exert its full effect on the bore. The leanness of the mixture between two bypass bores is proportional to the distance between the bores.

The further away the bypass bore is from the bottom edge of the throttle valve (when measured in its closed position) the higher must be the throttle valve angle and with it the air throughput to produce enrichment.

Enrichment improves with larger bypass bores, while the point of maximum enrichment remains the same (same air throughput/same throttle valve angle). When two bypass bores are enlarged we get two enrichment points; the leanness between the two bypass bores remains the same absolutely. Such a change would have approximately the same effect as enlarging the idling fuel nozzle.

From Fig. 4.31a we see considerable variations in the mixture ratios when the throttle valve covers the bypass bores. The bores are too far away from each other so that the mixture between them is of a significantly leaner nature.

Fig. 4.31b demonstrates that the situation improves when the bores are narrowed and spaced more closely.

When two bypass bores are too close together they act as if they were a single large bore (Fig. 4.31c), causing considerable enrichment when the throttle valve covers them. The situation is remedied by spacing them more widely (Figures 4.31d and 4.31e).

The design shown in 4.31e is faulty in that the mixture becomes lean already at a throttle valve angle of $7.5°$. By spacing the bores more widely and enlarging them (Fig. 4.31f) the mixture ratio could be maintained at a throttle valve angle of $14°$ to approx. $12°$. This design has already reached the limits of spacing, as indicated by the slightly leaner mixture beginning at an angle of $8.5°$.

The bypass bores may be replaced by a single slot which permits gradual adjustment of the mixture ratio at the transition between idle and main carburetor system in line with engine requirements. An example for a T-shaped slot is given in 4.1.1.5.

4.1.1.2.6. The Starting and Warm-up System

A slight vacuum is generated during starting, as already noted above. It is, however, not sufficient to trigger the main nozzle system or to supply the engine with an adequate quantity of mixture through the idling system, in particular since most of the mixture is precipitated as a deposit on the manifold walls. Accordingly, it is necessary to use auxiliary devices for which we have two basic alternatives:

The first option is illustrated in **Fig. 4.32**: here the air funnel is placed so as to ensure maximum vacuum, by adding a second throttle valve (the choke or initial throttle) between the air funnel and air filter which is closed during starting. The main throttle valve must be opened slightly to permit the vacuum to build up in the air funnel and mixture outlet.

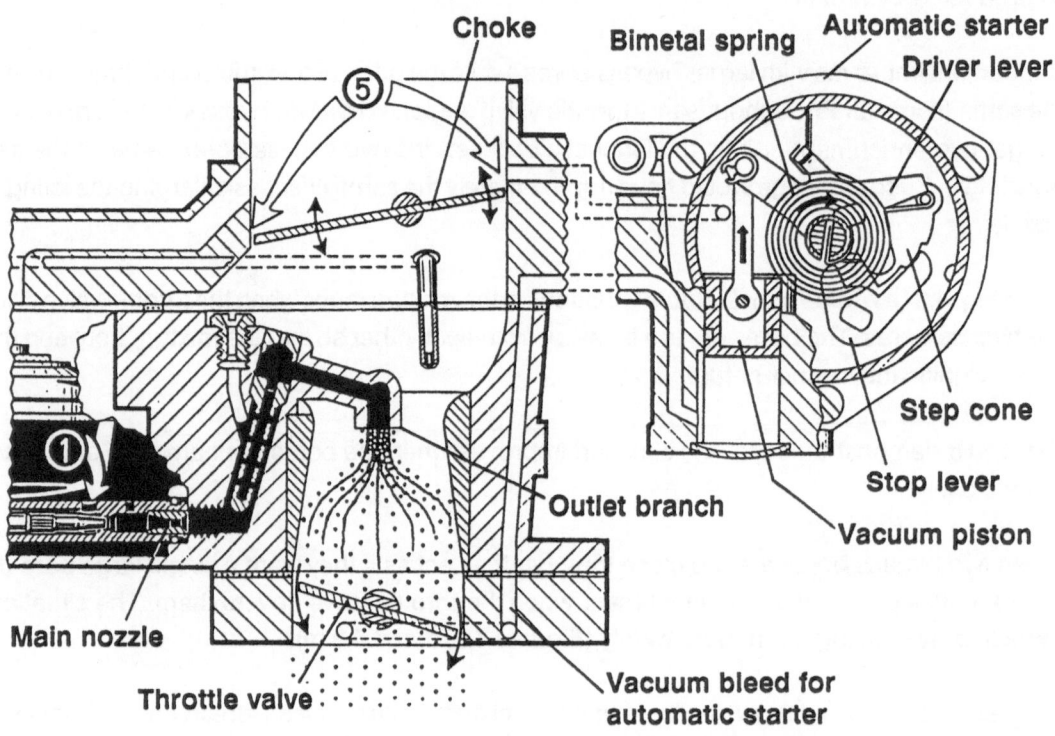

Fig. 4.32. Choke and automatic starter during cold start: (1) mixture inlet, (5) starter air inlet [4.8]

The choke is actuated manually via a pull wire or through an automatic device. In an automatic starter, such as shown in **Fig. 4.33**, the choke shaft is tensioned by a spiral-shaped bimetal spring which responds to differences in temperature. With a cold engine, the choke is partially closed because the bimetal spring forces it into its closed position during cooling.

When the bimetal spring heats up, it loses its closing power so that the choke opens, uncovering the air inlet when the engine has reached its normal operating temperature. The bimetal spring is heated electrically, by the cooling water, or by the engine exhaust. Opening of the choke is facilitated by the eccentric position of its shaft in the air inlet piece which means that the blades are of unequal size.

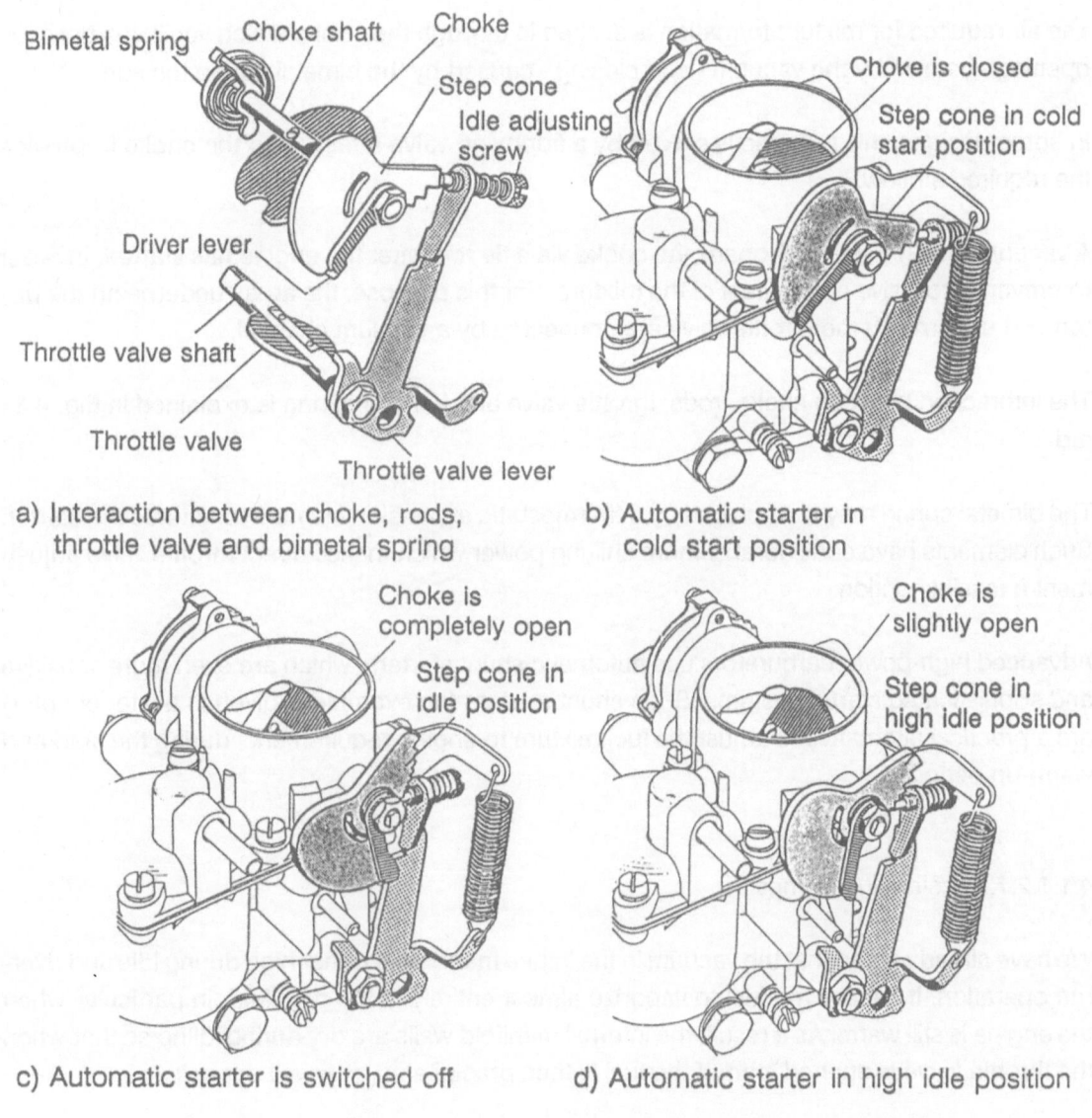

a) Interaction between choke, rods, throttle valve and bimetal spring

b) Automatic starter in cold start position

c) Automatic starter is switched off

d) Automatic starter in high idle position

Fig. 4.33. Automatic starter [4.8]

While the choke is in its closed position, the throttle valve is slightly open. This is ensured by the driver lever. Rigidly connected to the choke shaft, it lifts the freely movable step cone when the choke closes so that its catch engages with the idle adjustment screw at the throttle valve lever.

When the choke reaches its maximum closed position the idle adjustment screw assumes its top-most step cone position which in turn causes the throttle valve to open slightly. As a result, more air can flow in and the vacuum generated during starting can act all the way to below the choke, improving the start triggering of the main nozzle system.

The air required for mixture formation is sucked in through the choke which vacillates between opening - caused by the vacuum - and closing - caused by the bimetal spring tension.

In some designs, this purpose is served by a floppying valve attached to the choke to provide the required air flow.

A vacuum piston (Fig. 4.32) opens the choke via a tie rod, after the engine has started, in order to prevent excessive enrichment of the mixture. For this purpose, the areas underneath the piston and underneath the throttle valve are connected by a vacuum channel.

The interaction between choke, rods, throttle valve and bimetal spring is explained in Fig. 4.33 a-d.

The bimetal spring may be replaced by a thermostatic expansion element to actuate the starter. Such elements have considerably more shifting power which makes direct throttle valve adjustment a feasible option.

Advanced high-power carburetors use automatic shunt starters, which are even more sensitive and sophisticated starter systems. Such shunt starters (an example is given in a later chapter) are a practical alternative to adjust the fuel mixture to engine requirements during the start and warm-up period.

4.1.1.2.7. Accelerating Pumps

We have already noted that the vacuum in the intake manifold is rather high during idle and over-run operation. It causes the fuel to vaporize almost entirely in the manifold, in particular when the engine is still warm. As a result, the internal manifold walls are dry during idling so that when the throttle valve is opened, part of the fuel is then precipitated as a wall deposit.

Furthermore, the air flow accelerates more rapidly than the fuel flow in the pipelines. Accordingly, the mixture would become leaner exactly when it should be enriched in order to achieve maximum performance.

To avoid this, an accelerating pump injects additional fuel when the driver steps on the gas, usually 1-1.5 cm^3 for each movement of the accelerator. If no such pump were provided the vehicle would be sluggish in its response to acceleration, which would be felt as a disturbing jerk if it exceeds about 3/100 seconds.

The accelerating pump (**Fig. 4.34**) is actuated mechanically in modern carburetors, where it is connected to the throttle valve shaft by a rod.

Earlier versions of carburetors feature pneumatically actuated accelerating pumps. The working area of the pump is filled with fuel sucked in from the float chamber through the suction valve. In its neutral position, the pump diaphragm is pressed against the pump lever by the diaphragm spring. When the throttle valve opens, the movement is transmitted via a lever which presses the diaphragm inside, so that fuel is injected into the mixing chamber through an injection pipe and ball valve. The ball valve in the pump inlet prevents the fuel from flowing back into the float chamber during the pump pressure stroke. The ball valve in the pump outlet, on the other hand, prevents air from entering the pump system during the pump intake stroke. Injection volume and timing can be adjusted to the engine requirements.

Another type of accelerating pump uses a piston instead of a diaphragm.

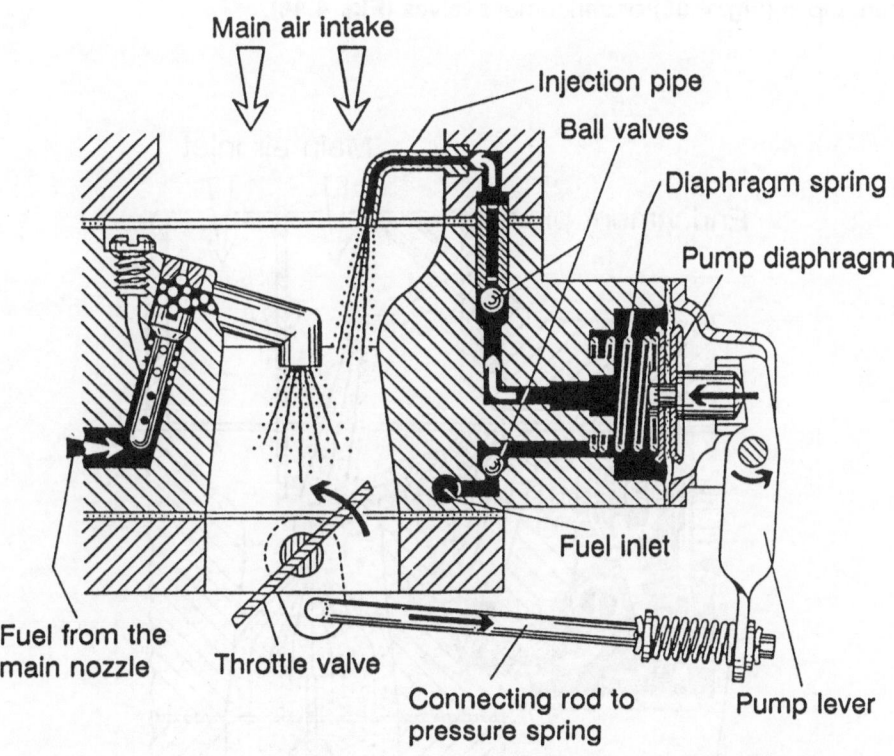

Fig. 4.34. Accelerating pump [4.8]

4.1.1.2.8. Devices to Control Fuel Enrichment

It has already been noted that while air-fuel mixtures ignite only within narrowly defined limits it is still possible to make an ignitable mixture "richer" or "leaner." The effect of the fuel composition on the output, consumption, and exhaust gas was discussed in Chapters 1 and 2. Added to this is the fact that a rich mixture requires more heat in the cylinders to vaporize the fuel than is the case with a lean mixture. Also, the flame propagation speed in the cylinder is at its highest at an air deficiency of some 10%, and drops rapidly with leaner mixtures, so that the composition of the mixture is a key factor for the operating behavior of an engine.

In the range of part loads it is more economical to use lean mixtures while rich mixtures are preferred for full load operation, for the reasons outlined above. Enrichment systems are a way of fine-tuning the carburetor to the desired engine characteristics. The main nozzle is usually designed to handle part loads so that the mixture can be enriched for full load operation, either by enrichment pipes (**Fig. 4.35**) or enrichment valves (**Fig. 4.36**).

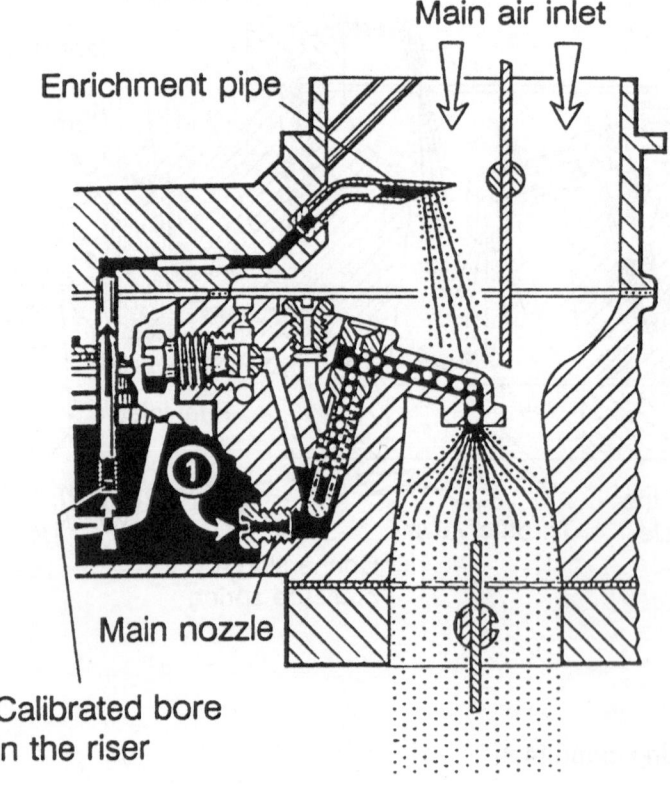

Fig. 4.35. Enrichment system featuring an enrichment pipe [4.8]

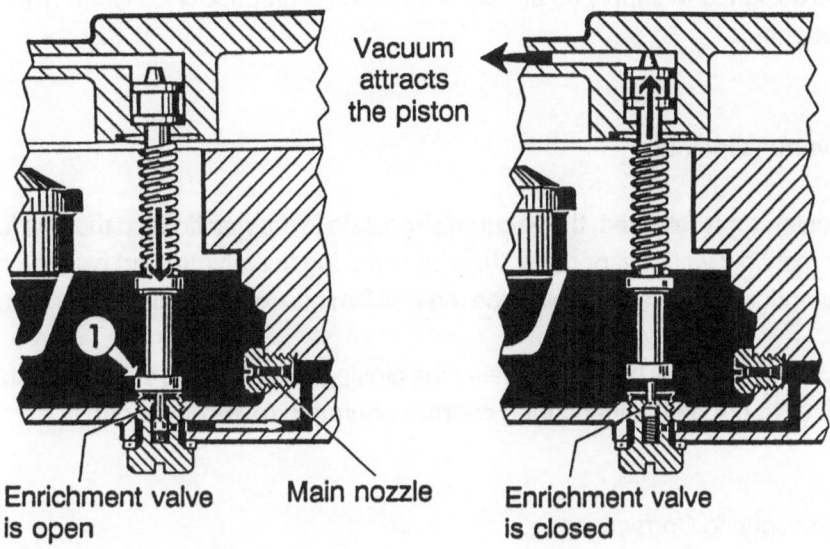

Vacuum attracts the piston

①

Enrichment valve is open   Main nozzle   Enrichment valve is closed

Fig. 4.36. Enrichment system featuring valve control through a vacuum piston [4.8]

4.1.1.2.9. Auxiliary Mixture Systems

These systems have been designed specifically to reduce harmful emissions. The throttle valve should always be in the same position vis-à-vis the bypass bores and ignition vacuum extraction bores even when the mixture volume required to maintain no-load operation differs because of different friction forces. To ensure this engineers use an auxiliary mixture system in the form of a small auxiliary carburetor which operates in parallel with the main carburetor and which supplies the different mixture volumes (**Fig. 4.37**).

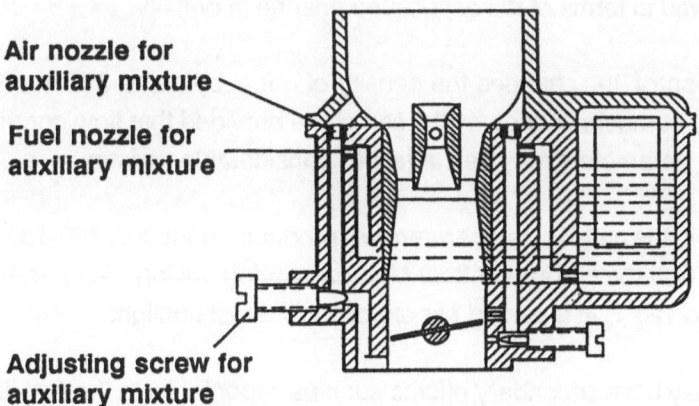

Air nozzle for auxiliary mixture

Fuel nozzle for auxiliary mixture

Adjusting screw for auxiliary mixture

Fig. 4.37. Auxiliary mixture system [4.8]

The system is designed to supply an air-fuel ratio which is practically constant over the relevant range of throughput.

4.1.1.2.10. Overrun Operation

When the accelerator is released, the vacuum shoots up in the manifold so that the fuel film flowing along the manifold walls evaporates, thereby enriching the mixture and causing higher emissions. Consumption increases as well since the wall film has to be rebuilt during re-acceleration.

For this reason devices may be used to delay the closing of the throttle valve (throttle valve closing dampers) or to cut off the fuel during overrun operation (overrun cutoff).

4.1.1.2.11. Atmospheric Corrections

Variations in the physical condition of the air are important for the engine with regard to three factors: temperature, pressure and humidity. Its composition is approximately similar up to an altitude of some 10,000 meters, due to the constant mixture caused by the unlevel ground, wind, and rain.

The weather produces variations in the air pressure of not more than about 30 mm Hg or 8%, while local densities may vary by more than 10%. Altitudes may intensify the effect so that any such changes may have a considerable impact on fuel mixture composition and, as a consequence, on output, consumption, and emissions (cf. **Table 4.1**).

For this reason, national and international reference states have been defined to characterize internal combustion engine performance (cf. Chapter 2.1).

With regard to the "air" component of the fuel mixture, changes in pressure, temperature, and humidity may be viewed in terms of the associated change in density.

A temperature variation of 40° changes the density of a fuel by some 3% and its viscosity by some 30-40%. The Re numbers change by the same rate provided that flow conditions are identical. As a result, flow parameters may be subject to considerable variations.

Variations of 20-40°C in the ambient temperature will produce variations of 50-100°C underneath the hood and 40- 75°C in the carburetor float chamber during driving. Temperatures under the hood may reach up to 120°C in a parked car exposed to direct sunlight.

High temperatures may have secondary effects such as vapor locks in the fuel lines or a build-up of fuel vapor in the carburetor, which in turn may produce an enrichment of the mixture.

In general, the effect of changes in the atmospheric conditions on fuel mixture composition is primarily expressed by the change in the air density. The effect on carburetors is calculated on the basis of the general carburetor equation (4.1.1.1).

When disregarding air expansion and changes in the flow indices, the change in the mixture ratio is proportional to the root of the change in air density. This is a preliminary approximation only which does not include the many factors which have an impact on carburetors in practice. But the figure at least gives an idea of the dimensions involved.

Location	Altitude [m]	Atmospheric Pressure [mm Hg]
Düsseldorf	40	app. 760
Munich	520	app. 715
St.-Gotthard-Pass (CH)	2110	app. 590
Grossglockner mountain road	2500-2700	app. 570-550
Denver (U.S.)	1600	app. 630
Mexico City (Mexico)	2300	app. 580
Tunnel underneath Loveland-Pass (U.S.)	3400	app. 505
Koi-Tesek route (Pamir S.U.)	4700	app. 450
Sary-Tasch route (Pamir S.U.)	3800	app. 490

Table. 4.1. Selected altitudes and mean atmospheric pressures

Some physical and chemical properties of the fuel change with the temperature, in particular density, surface tension and viscosity.

Any change in fuel density affects the mixture ratio, according to the general carburetor equation. It also changes the float chamber level owing to the change in float buoyancy.

The fuel surface tension affects metering and preparation of the fuel mixture.

Variations in fuel viscosity may affect composition because, provided that the pressure differential at the nozzle is constant, the Re numbers and with them the flow indices will be changed (**Fig.** 4.2 and **4.38**).

Fig. 4.39 charts the enrichment values as a function of altitude, as measured on a carburetor test bench combined with an altitude cabin.

Values calculated for carburetors with fixed and variable air funnel cross-sections are plotted as well. The theoretical values show excellent consistency with those collected from testing.

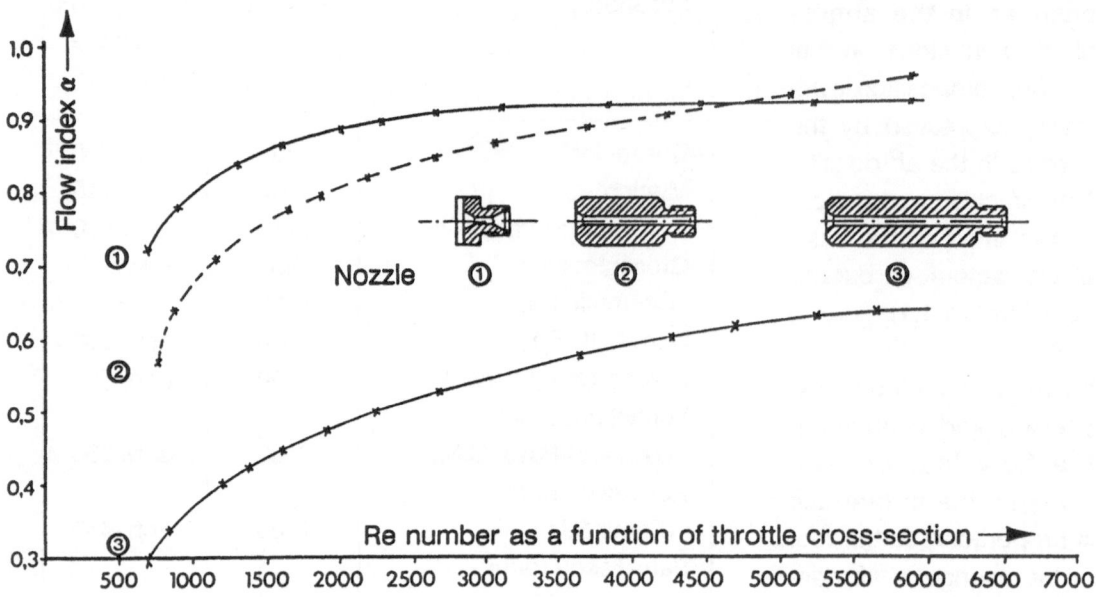

Fig. 4.38. Flow indices for carburetor nozzles (nozzle 1: standard carburetor nozzle) [4.8]

Enrichment devices can double the natural enrichment of a simple carburetor because they respond already at a lower altitude. When the mixture is enriched through a tube running from the float chamber into the suction line, the device responds earlier because the vacuum increases with decreasing air density when the air throughput volume remains constant.

In carburetors fitted with an enrichment valve which is actuated when the vacuum in the suction line deteriorates, altitude-caused actuation occurs earlier because the throttle valve must be opened wider with decreasing air density, at a constant air volume, which decreases the vacuum in the manifold.

Fig. 4.40 illustrates the effect of an enrichment system on mixture enrichment in the carburetor in terms of altitude. The measurements involved part loads at constant air throughput. The throttle valve was opened progressively with increasing altitude.

The scope of the effect of enrichment on performance, consumption, and emission as compared to the basic mixture depends on the air-fuel ratio of that basic mixture. In general it differs at each load point in the diagram.

In a carburetor that has no enrichment system and where the basic air-fuel ratio is $\lambda = 1.1$, an enrichment of 16% such as occurs on the Grossglockner mountain road (which spans an altitude difference of 2,000 meters) will raise the mean pressure by almost 10%, which counteracts

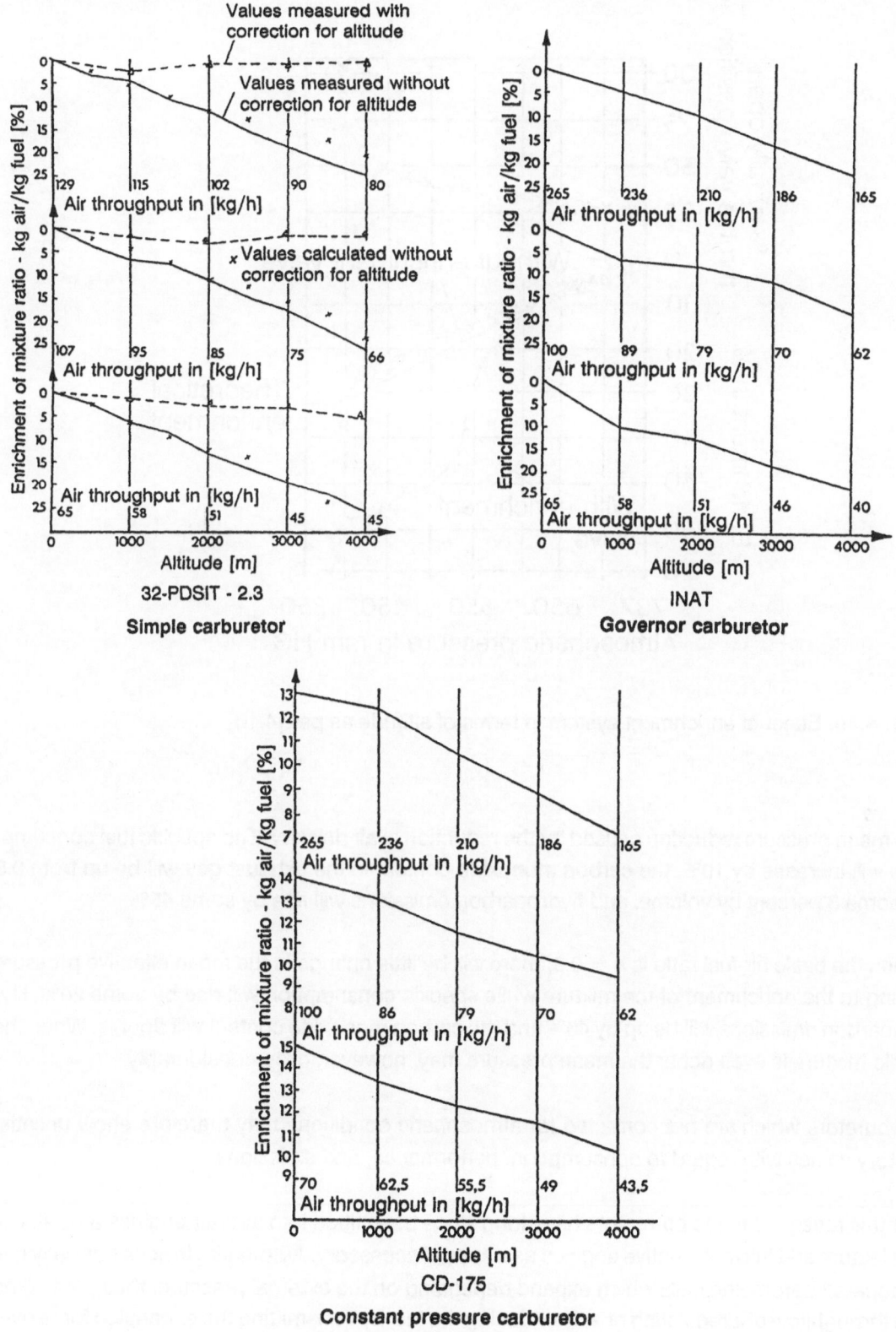

Fig. 4.39. Enrichment of carburetors in terms of altitude [4.8]

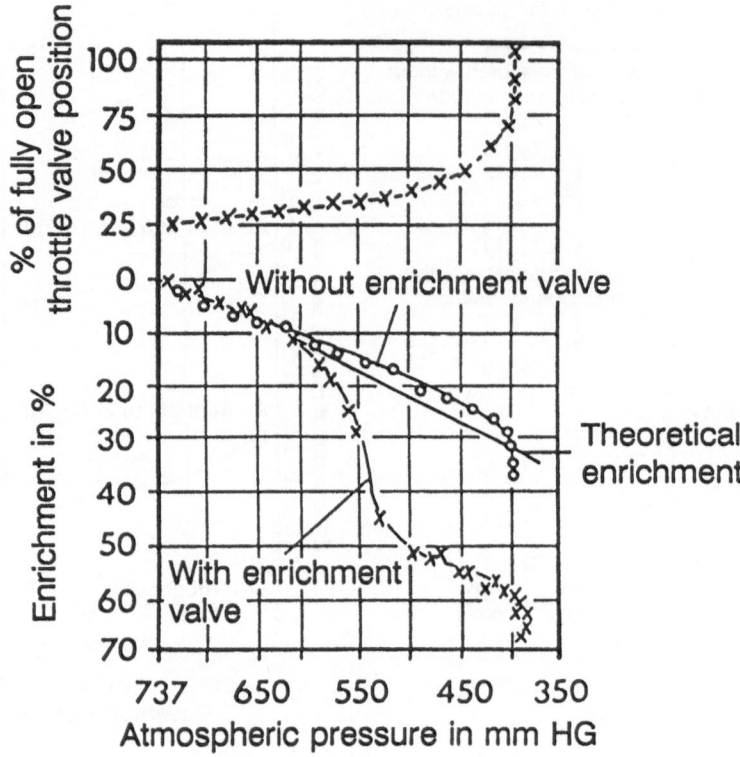

Fig. 4.40. Effect of enrichment system in terms of altitude as per [4.16]

the mean pressure reduction caused by the reduction in air delivery. The specific fuel consumption will increase by 10%, the carbon monoxide content in the exhaust gas will be up from 0.5 to some 3 percent by volume, and hydrocarbon emissions will rise by some 45%.

When the basic air-fuel ratio is $\lambda = 0.9$, there will be little change in the mean effective pressure owing to the enrichment of the mixture while specific consumption will rise by some 20%. Hydrocarbon emissions will be up by 40% and the carbon monoxide content will double. When the basic mixture is even richer the mean pressure may, however, drop considerably.

Carburetors which are not corrected for atmospheric conditions may therefore show unsatisfactory values with regard to consumption, performance, and emissions.

For this reason, altitude correctors have long since been fitted into aircraft engines as a standard feature and into automotive engines as a special accessory. Automatic devices basically use evacuated barometric cells which expand depending on the external pressure, thus controlling the throughflow of a regulating nozzle or dosing the fuel by transmitting the expansion force over a lever system.

The regulating nozzles, which are typical for automotive engines, are used for dosing the fuel, the compensating air, float chamber ventilation, or the air volume bypassing the air funnel to reduce the vacuum. It is also possible to change the ignition timing in terms of the altitude correction.

Fig. 4.41 illustrates the basic function of correction devices. A highly evacuated barometric cell will respond to pressure only while less thoroughly evacuated cells will also react to temperature.

Maximum engine performance depends basically on the volume of oxygen, i.e. on the quantity of air supplied, so that an altitude corrector obviously cannot prevent the natural decline occuring as a more or less linear function of the decrease in air density, but it can at least provide the best possible mixture ratio for any particular engine condition.

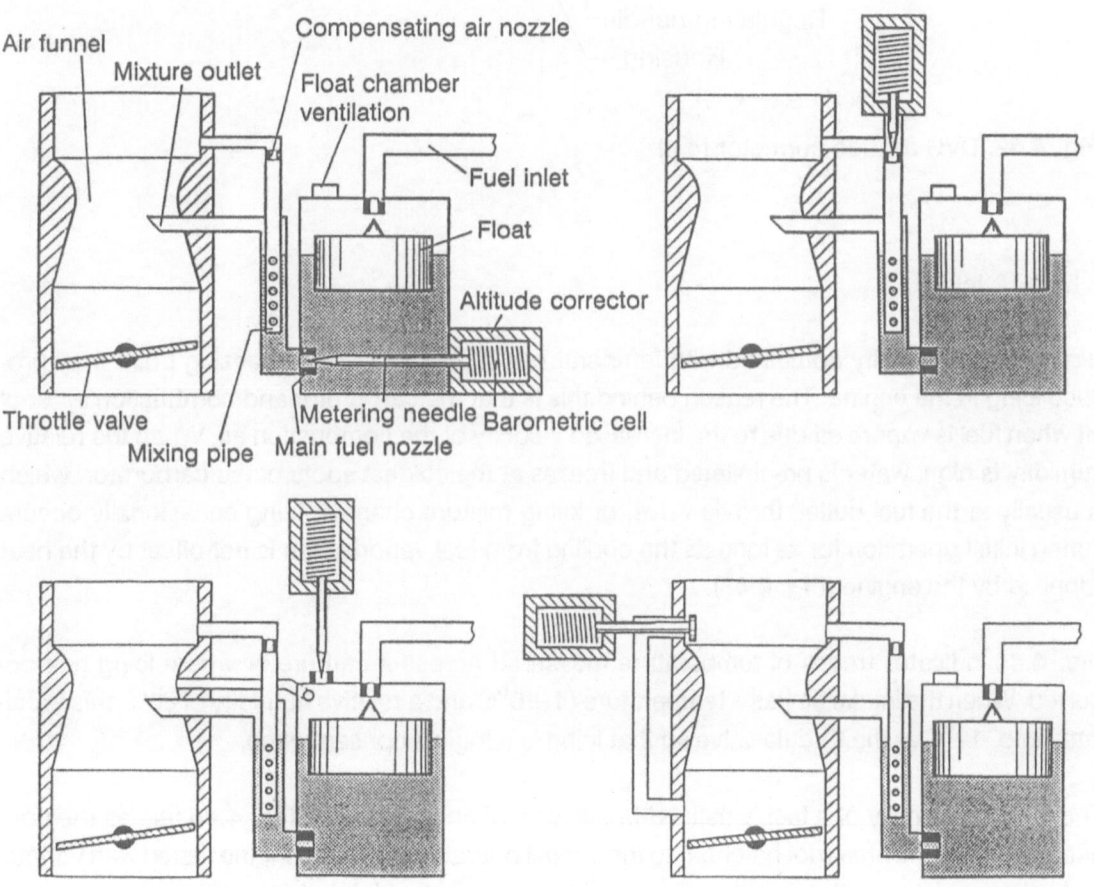

Fig. 4.41. Altitude correction by acting on the main nozzle (upper left); compensating air nozzle (upper right); float chamber pressure (bottom left); bypass venturi (bottom right) [4.8 and 4.17]

Fig. 4.42 illustrates an altitude corrector as had been fitted into the VW 1600 Automatic, which acts on the main nozzle.

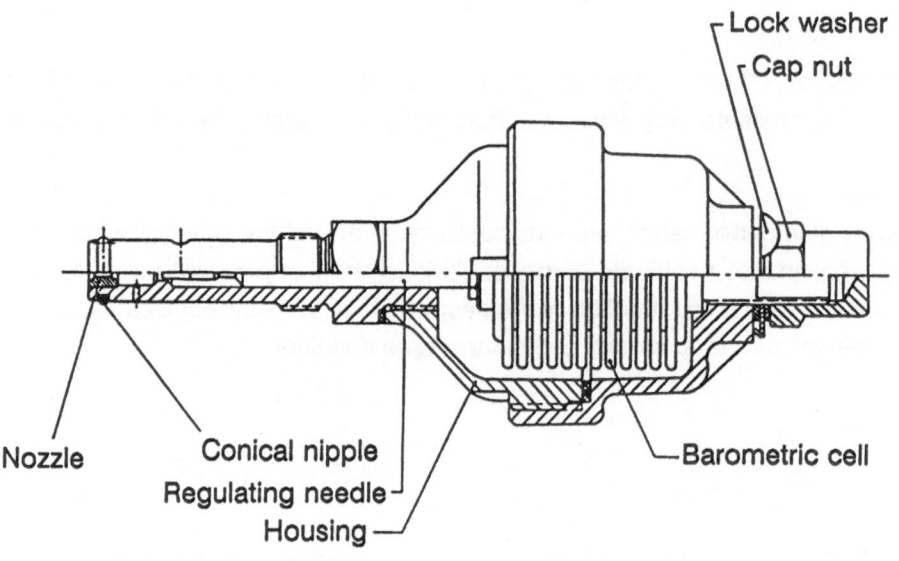

Fig. 4.42. DVG altitude corrector [4.8]

4.1.1.2.12. Icing

High relative humidity combined with temperatures slightly above the freezing point may produce icing in the engine. The reason behind this is that the carburetor and combustion air cool off when fuel is vaporized due to the increased velocity of the combustion air. When the relative humidity is high, water is precipitated and freezes at the coldest spots of the carburetor, which is usually at the fuel outlet, throttle valve, or idling mixture channel. Icing occasionally occurs during initial operation for as long as the cooling from fuel vaporization is not offset by the heat supplied by the engine (**Fig. 4.43**).

Fig. 4.44 indicates traces of temperature measured across a carburetor where icing has occurred. When the intake air has a temperature of +6°C and a relative humidity of 80%, this translates into -12°C at the throttle valve so that icing is a logical consequence.

The icing propensity of a fuel is defined in the form of an icing index. **Fig. 4.45** relates the non-affected operating area (not hatched) to the overall operating area. An engine tested with various fuels showed icing at the throttle valve and idling channel which in turn caused the engine to stop while idling.

Carburetor icing can be prevented by supplying heat, although precautions have to be taken to avoid the generation of vapor locks.

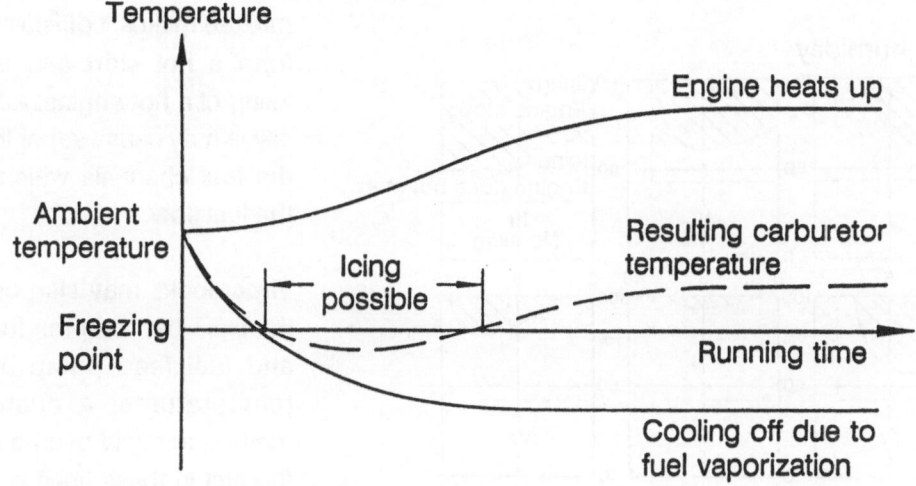

Fig. 4.43. Carburetor temperature as a function of engine running time [4.8]

Practical solutions are supplying heat to the carburetor throttle valve parts or - the most usual course today - using anti-icing additives with the fuel.

Heating the intake air - as is common in many carburetor engines - reduces the icing propensity because of increased fuel vaporization, and has the added advantage of improving mixture distribution.

4.1.1.2.13. Vapor Locks

When a hot engine is started, particularly in summer when ambient temperatures are very high, problems may occur when vapor locks are generated in the carburetor.

When the carburetor heats up, the low- boiling fractions are released and enter the manifold. The resulting

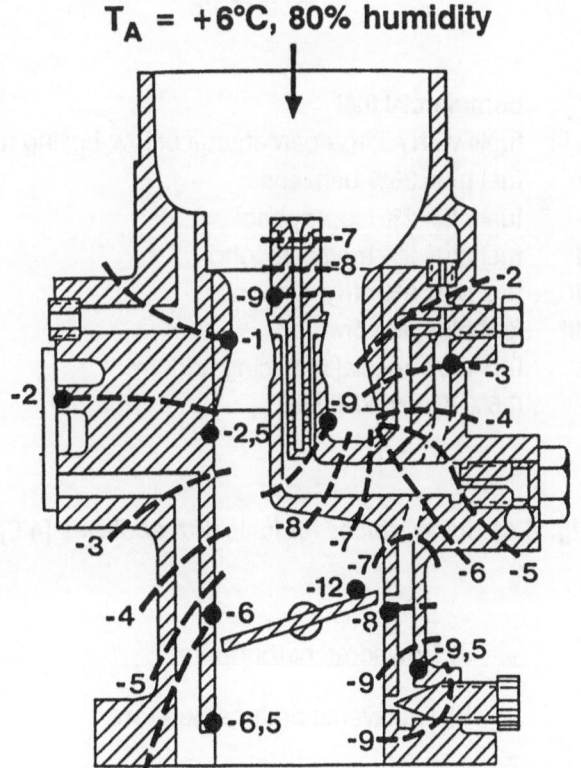

Fig. 4.44. Carburetor temperatures when icing occurs [4.8]

Relative humidity

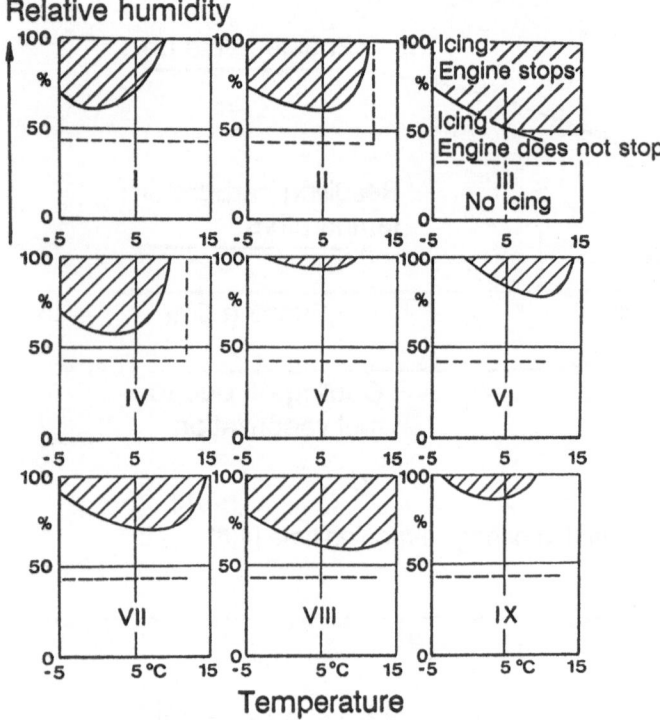

I   commercial fuel
II, III   fuels with a large percentage of low-boiling fractions
IV   fuel plus 36% benzene
V   fuel plus 2% isopropanol
VI   fuel plus 2% methyl alcohol
VII   fuel plus 2% ethyl alcohol
VIII   fuel plus acetone
IX   fuel plus special anti-icing additive:
           0,5% Kerofluid PGM.

Fig. 4.45. Icing indices for fuels with additives [4.8]

excessive enrichment of the fuel mixture makes it difficult to perform a hot start and impairs idling of a hot engine. Additionally, it may cause vapor locks in the fuel channels which block the fuel flow.

Vapor locks may also occur in the lines between the fuel tank and fuel feed pump at high temperatures. A customary method to avoid overheating of the fuel in these lines is to fit a fuel return valve at the carburetor which diverts some of the fuel upstream of the carburetor through a return line into the tank when the engine is idling, or running at part load. This is an effective means to prevent excessive heating of the fuel, and resulting vapor locks.

4.1.1.3. Carburetor Types

We distinguish the following types of carburetors, by their arrangement of the manifold within the engine and by the direction of the suction flow within the carburetor (**Fig. 4.46**):

- a) Downdraft carburetors;

- b) Transverse draft carburetors;

- c) Horizontal draft carburetors;

- d) Rising draft carburetors.

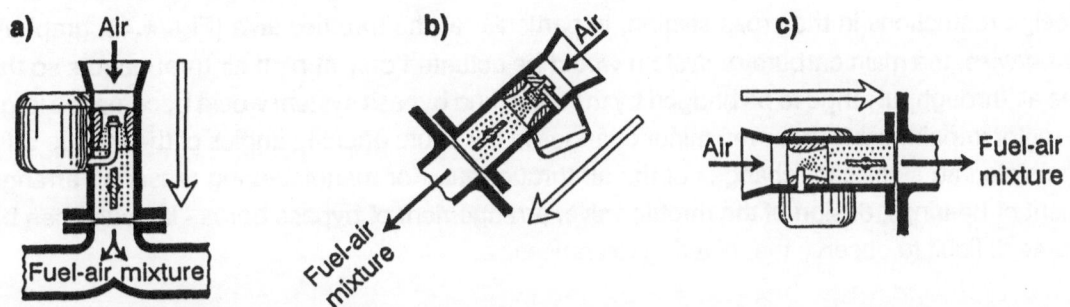

Fig. 4.46. Carburetor types: a) downdraft carburetor, b) transverse draft carburetor, c) horizontal draft carburetor

Rising draft carburetors are no longer in use.

Breaking down carburetors by the number and design of mixing chambers we distinguish (**Fig. 4.47**):

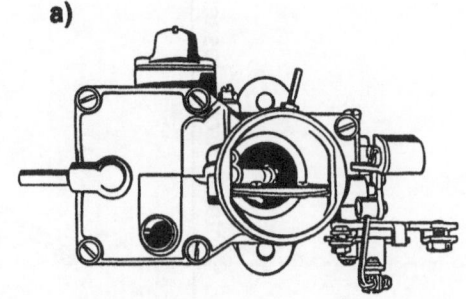

- a) Single-barrel carburetors for a single manifold;

- b) Double-barrel carburetors for separate manifolds;

- c) Compound carburetors for a single manifold with successive stages.

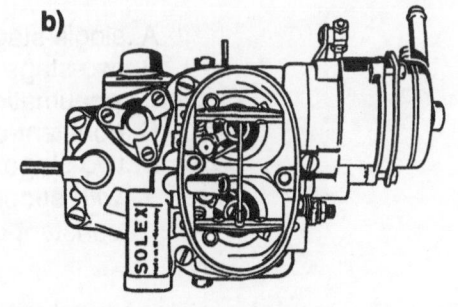

Compound carburetors have a second stage (i.e. a second carburetor) which is connected when a preset air throughput has been reached. Connection may be of the mechanical type, governed by the position of the first-stage throttle valve (mechanical control), or by the first-stage vacuum (pneumatic control).

When comparing carburetor types in terms of throttle losses we find:

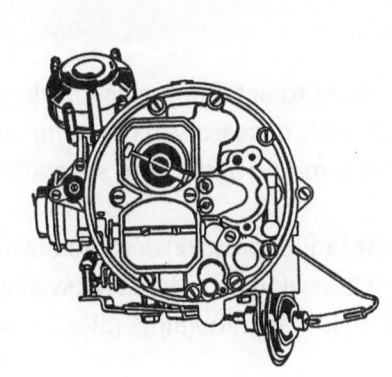

Fig. 4.47. Carburetor types:
   a) Single-barrel carburetor for a single manifold,
   b) Double-barrel carburetor for separate manifolds,
   c) Compound carburetor

Single-stage carburetors suffer considerable throttle losses at high throughputs because of the design restrictions in the cross-section, in particular at the throttle valve (**Fig. 4.48**, graph A). Otherwise, the main carburetor system would be actuated only at high air throughputs so that the air throughput range to be bridged by the idling and bypass system would become too large. A major impairment is that even minor changes in the acute opening angles of the throttle valve translate into significant changes of the air throughput. For manufacturing reasons - arrangement of bearings, design of the throttle valve, arrangement of bypass bores - it would then become difficult to observe the metering tolerances.

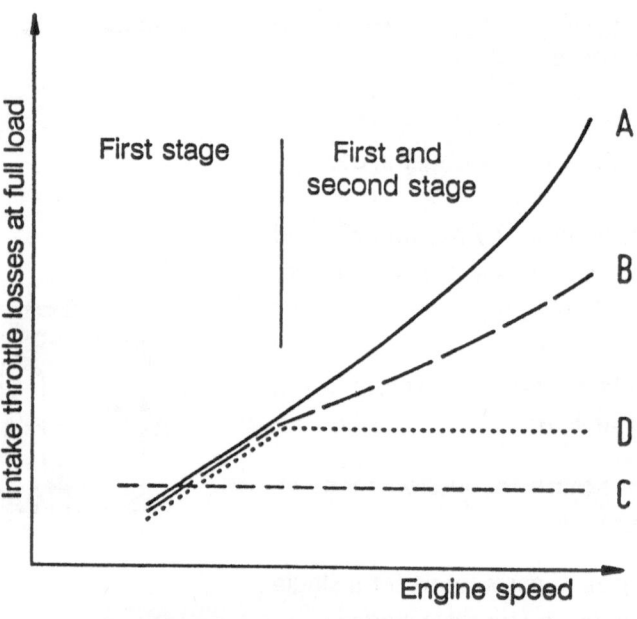

A..single-stage carburetor
B..two-stage (compound) carburetor with
    pneumatic second-stage control
C..constant-depression carburetor
D..two-stage (compound) carburetor
    with second stage designed as a
    constant-depression carburetor

Fig. 4.48. Throttle losses in carburetors

In order to achieve precise mixing and low throttle losses in engines that have a wide speed range it is necessary to use compound carburetors with a relatively narrow primary stage for accurate mixing and a wide secondary stage to reduce throttle losses (Fig. 4.48, graph B).

Constant-depression carburetors are relatively simple in their design. They usually show excellent transition behavior and low throttle losses in the high full-load speed range (Fig. 4.48, graph C). At low air throughput rates, i.e. when the air flow sensor (piston or air flap) shows little movement, it is, however, difficult to achieve accurate fuel mixing since the air flow sensor governing the fuel metering is not always in the same position due to friction.

From the above it follows that the best solution for engines with a wide speed range is a compound carburetor where the second stage is designed as a constant-depression carburetor (Fig. 4.48, graph D). The second-stage throttle valve may then be opened mechanically, governed by the vacuum at a specified spot in the carburetor.

4.1.1.4. Constant-Depression Carburetors

Constant-depression carburetors (**Fig. 4.49**) have air funnels which widen or narrow their flow cross-section depending on the air throughput so that an approximately constant vacuum is generated in the mixing chamber.

A bore in the piston connects the mixing chamber with the area above the diaphragm. When the throttle valve opens, the vacuum rises in the mixing chamber in response to the higher air throughput, lifting the piston above the diaphragm and pressing it against a piston spring until the pressure generated by the vacuum is in equilibrium with the spring tension.

The fuel flows through the float needle valve, float chamber, nozzle holder, and needle jet into the mixing chamber. It is metered through the needle jet whose needle is attached to the piston.

This type of carburetor does not require any acceleration pump or even - in some designs - a special idling system. When the throttle valve is opened suddenly, the inertia of the piston causes a short-term vacuum in the mixing chamber and thus short-term enrichment which has the same effect as an acceleration pump.

The vacuum is relatively high when the throttle valve is in its idle position so that the idling fuel is sucked out of the fuel nozzle. A slight disadvantage of this design is that idling adjustment must be done by changing the fuel outlet space, i.e. by moving the needle seat. For this reason, the idle setting also affects fuel supply at higher throughputs. This type of carburetor thus has on occasion been fitted with special idling systems (**Fig. 4.50**).

Constant-depression carburetors (as well as other systems) use needle jets which may be a source of problems because the throughflow volume depends on the needle's eccentricity in the nozzle.

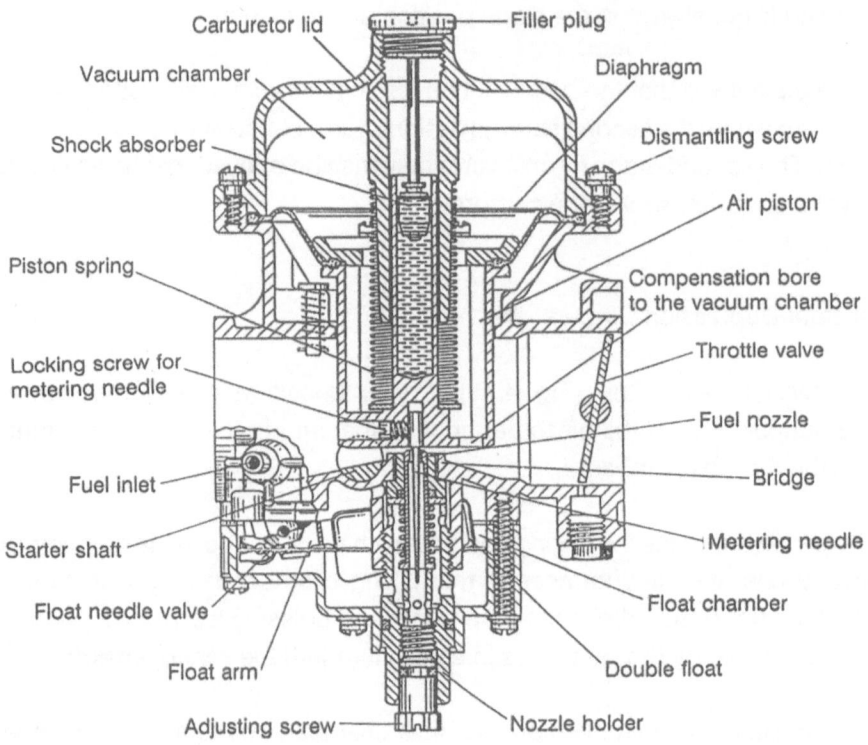

Fig. 4.49. Constant-depression carburetor, Zenith-Stromberg- CD, schematic cross-section [4.8]

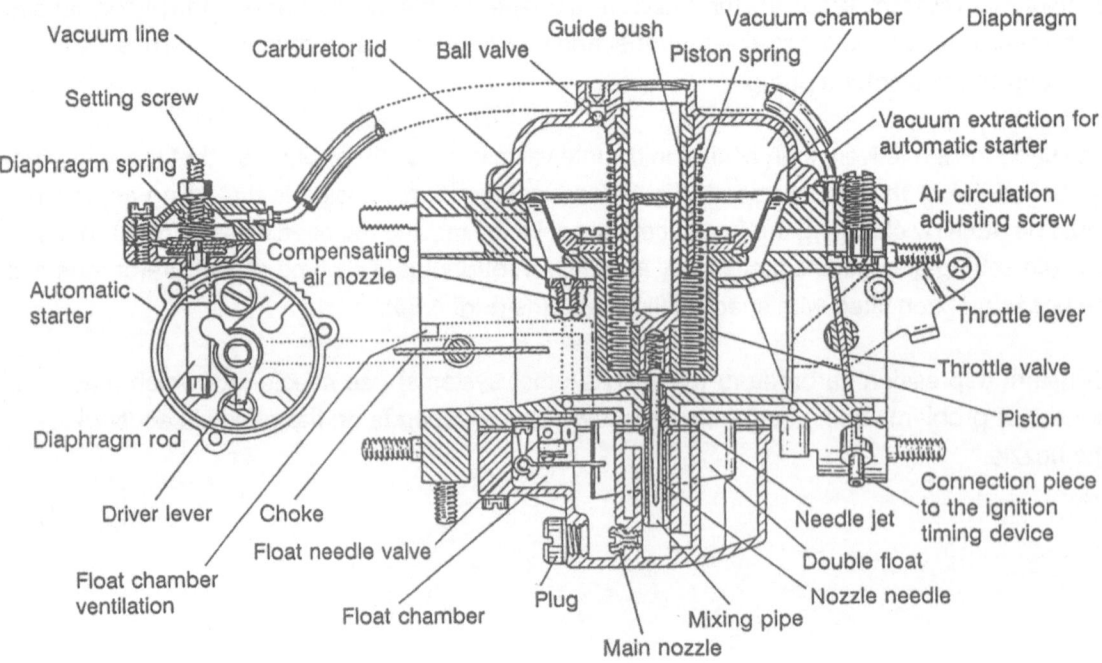

Fig. 4.50. Constant-depression carburetor with separate idling system and starter device, Solex 34 SWT [4.8]

From **Fig. 4.51** it can be seen that the flow is subject to greater variations when the eccentricity is changed in the front (when viewed in the flow direction) than in the back of the nozzle.

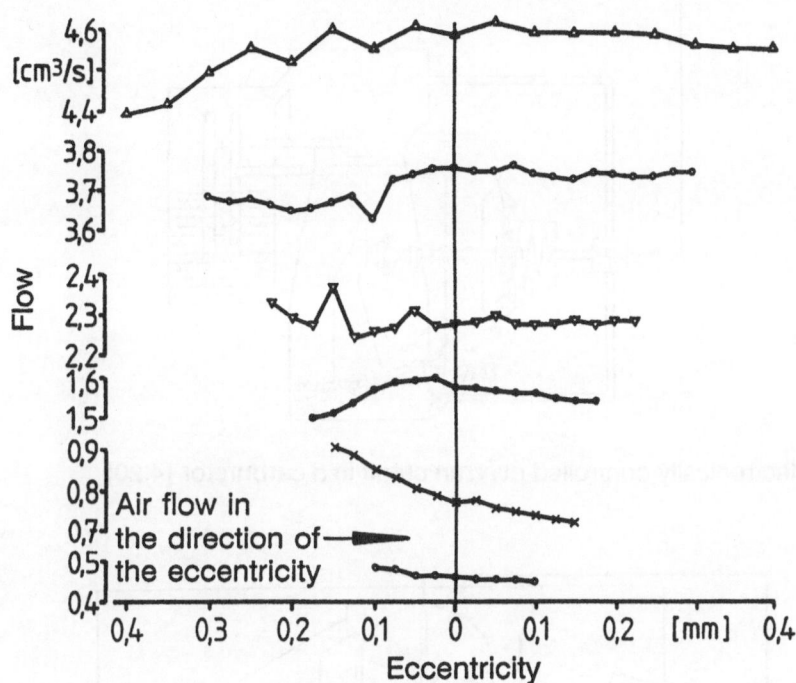

Fig. 4.51. Flow across the needle eccentricity [4.8]

Consequently it is preferable to place the nozzle needle so that it contacts the back wall to prevent any change in the throughflow by accidental shifts in the eccentricity.

4.1.1.5. Carburetors with Electronic or Closed-Loop Control

Electronically controlled carburetors are a new development in the field. The concept is based on the premise of eliminating the multitude of systems required for a high- performance carburetor and electronically controlling and governing fuel mixture metering in a simple and basic carburetor. In **Fig. 4.52** electronic control is restricted to overrun cutoff. **Fig. 4.53** is an outline of the basic Ecotronic system developed by Pierburg where all the major auxiliary functions are controlled electronically.

Based on a relatively simple carburetor with idling, bypass, and main systems, its electronic control extends to idling speed, start and warm-up, acceleration enrichment, and overrun cutoff. An electronic controller provides data on temperature, throttle valve position and speed, and on the

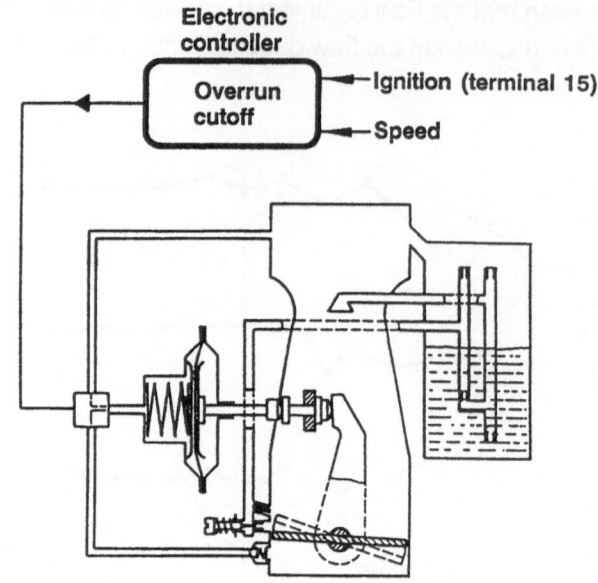

Fig. 4.52. Electronically controlled overrun cutoff in a carburetor [4.20]

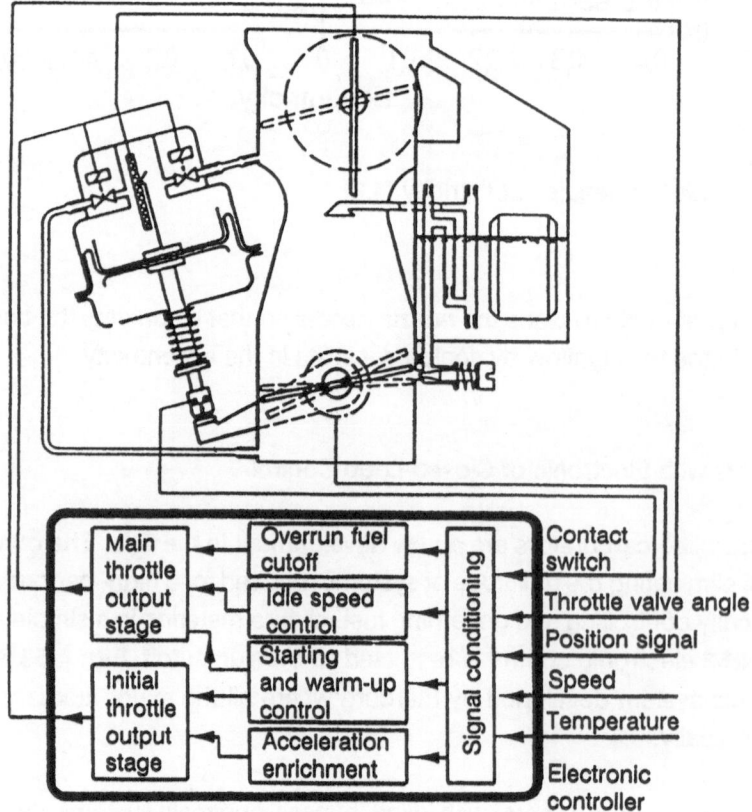

Fig. 4.53. Electronically controlled Ecotronic carburetor by Pierburg [4.20]

basis of these data, the positions of the initial throttle valve and throttle valve adjuster, to achieve specific fuel mixture enrichment rates or mixture throughputs.

Fig. 4.54 illustrates an extended electronic-control carburetor which facilitates the closed-loop control concept. A lambda probe placed in the exhaust line measures the partial oxygen pressure and regulates the mixture formation system to ensure a stoichiometric air-fuel ratio at all times (lambda = one).

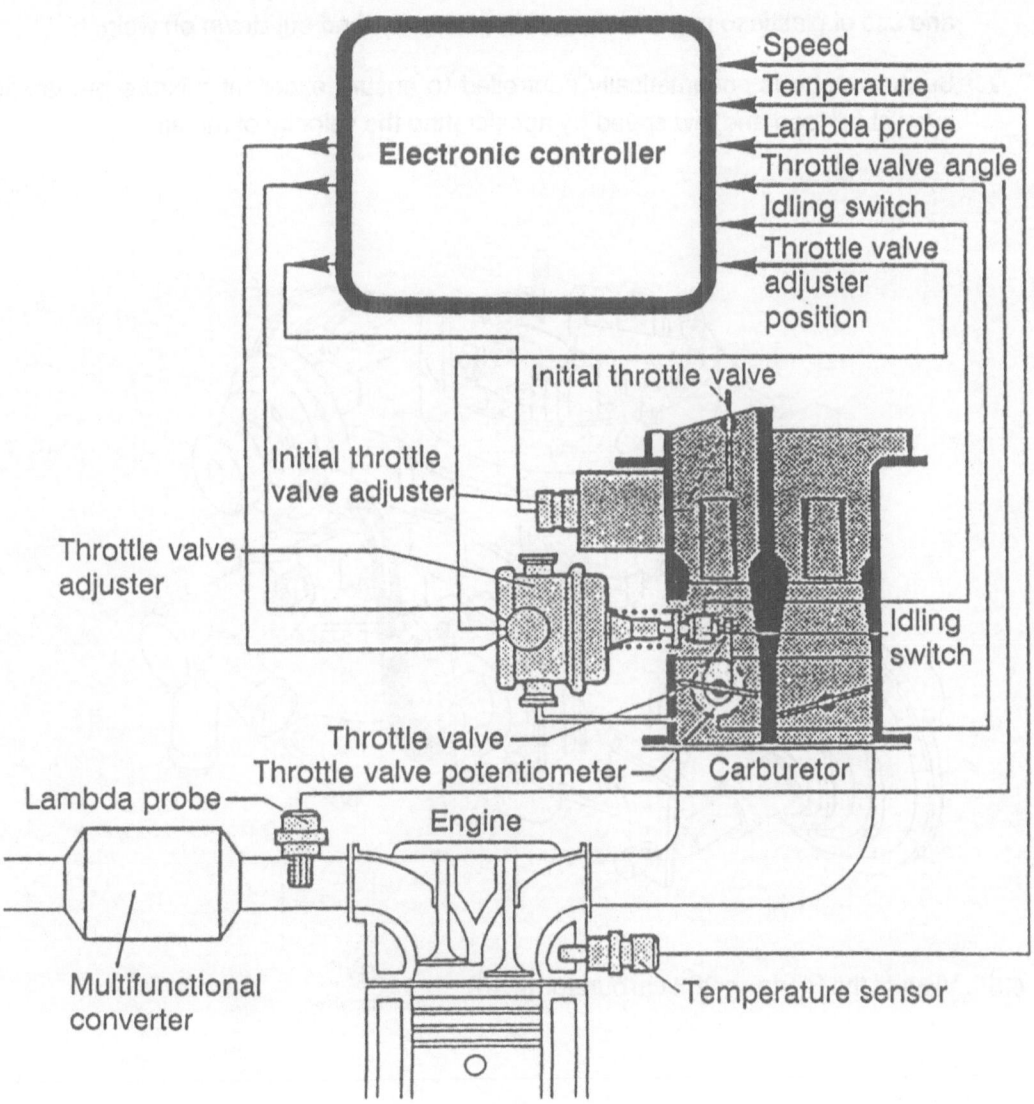

Fig. 4.54. Lambda-controlled electronic carburetor by Pierburg [4.40]

4.1.1.6. Carburetor Design Examples

4.1.1.6.1. Pierburg Compound Carburetor 2 E [4.15]

Fig. 4.55 shows the carburetor and fully automated starter. Its special design features are:

- Immersed nozzle systems to ensure good hot-start behavior;

- Use of aluminum for the float chamber casing, carburetor top, starter body and cover, and use of plastic to improve corrosion protection and cut down on weight;

- Second stage is pneumatically controlled to ensure excellent mixture preparation even at full load and low speed by accelerating the velocity of the air.

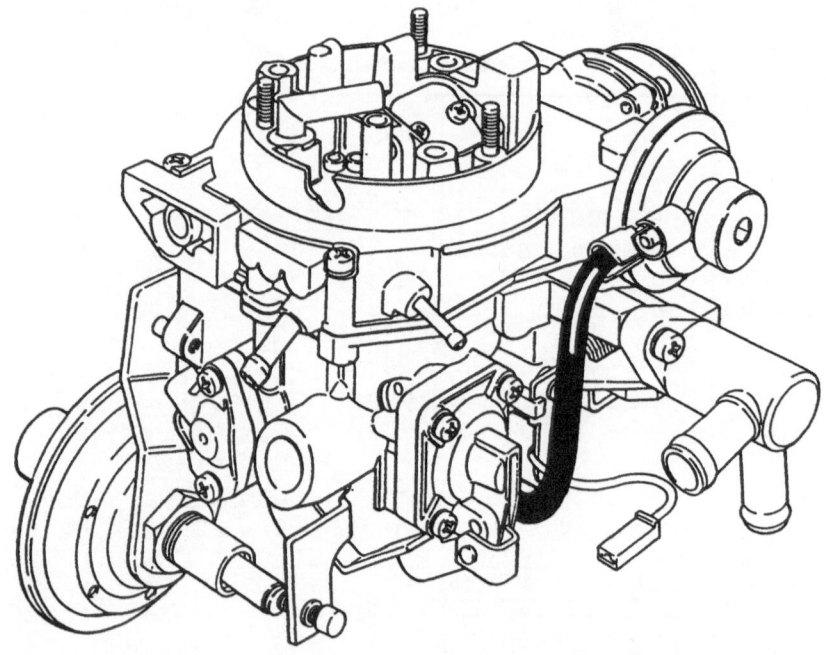

Fig. 4.55. View of the Pierburg 2E2 carburetor [4.15]

Starter devices

The 2E carburetor accomodates the following starter devices:

- Manual starter 2E1;

- Fully automated starter 2E2;

- Conventional automatic starter 2E3.

The manual and conventional automatic starters have already been described in a previous chapter, so that we concentrate on the fully automated starter 2E2. This device automatically adjusts all changes in the fuel-air ratio and filling necessary for starting to the engine's respective operating temperature:

Adjusting the Mixture Ratio

Acting through a bimetal spring, an initial throttle increases the delivery pressure on the main system of the first carburetor stage, thereby changing the mixture ratio for the cold engine.

While the engine runs up, the mixture becomes leaner through a pulldown device.

The opening angle of the initial throttle is adjusted to engine requirements by multi-step electric and/or coolant heating of the bimetal spring. Contrary to the setup in conventional starters, the bimetal spring does not change the engine filling via a starter cone pulley and connecting rods.

Adjusting the Cold-Start Air Throughput

In order to smooth cold starts and optimize cold-start fuel consumption, the components adjusting the cold start mixture ratio and the cold-start air requirement operate separately and independently on a fully automated scale.

The higher throughput for cold starts is ensured by the carburetor throttle valve setting. It is actuated by a mechanism illustrated in **Fig. 4.56**.

A pneumatic adjuster, acting against the resistance of the throttle valve reset spring (not shown in this figure), opens the throttle valve of the first carburetor stage to the position required to ensure starting at a low temperature.

The vacuum building up in the manifold after the engine has started will then cause the diaphragm to retract the adjuster to its hot idle run position in line with a specific control function.

Depending on the engine operating temperature, the air throughput is increased by adjusting the throttle valve in line with a run-up function which is controlled through the entire engine warm-up range up to the hot idle run air requirement by a coolant-heated thermostatic expansion element.

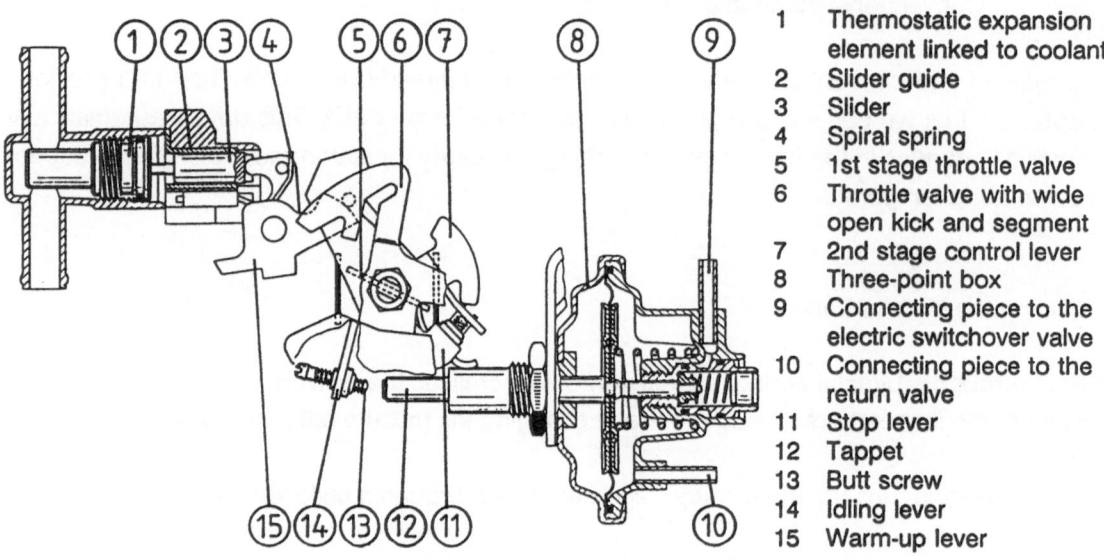

1   Thermostatic expansion element linked to coolant
2   Slider guide
3   Slider
4   Spiral spring
5   1st stage throttle valve
6   Throttle valve with wide open kick and segment
7   2nd stage control lever
8   Three-point box
9   Connecting piece to the electric switchover valve
10 Connecting piece to the return valve
11 Stop lever
12 Tappet
13 Butt screw
14 Idling lever
15 Warm-up lever

Fig. 4.56. Throttle valve control for cold start and warm-up in the fully automated 2E2 carburetor [4.15]

By separating the devices to adjust the mixture ratio and the air throughput during cold start it is possible to adapt both parameters accurately to the engine requirements, taking into account the different warm-up curves for coolants and engine oil.

Idling System

When the design provides for a split into several idling systems, instabilities are possible at very low idling fuel throughput rates. For this reason the auxiliary Pierburg mixture system to achieve constant CO rates over different idling air requirements is not used in this unit.

Instead, a slot-type idling and transition system outlet has been chosen to ensure the same adjusting benefits. In the throttle valve idling position, the lower part of the slot reaches into the mixing chamber below the throttle valve so that it can feed the fuel for idling requirements.

When the slot is properly adjusted the preset idling mixture ratio will not change even at different idling air requirements (**Fig. 4.57**).

According to Pierburg, the use of transition outlet bores does not produce a similarly satisfactory solution in larger series.

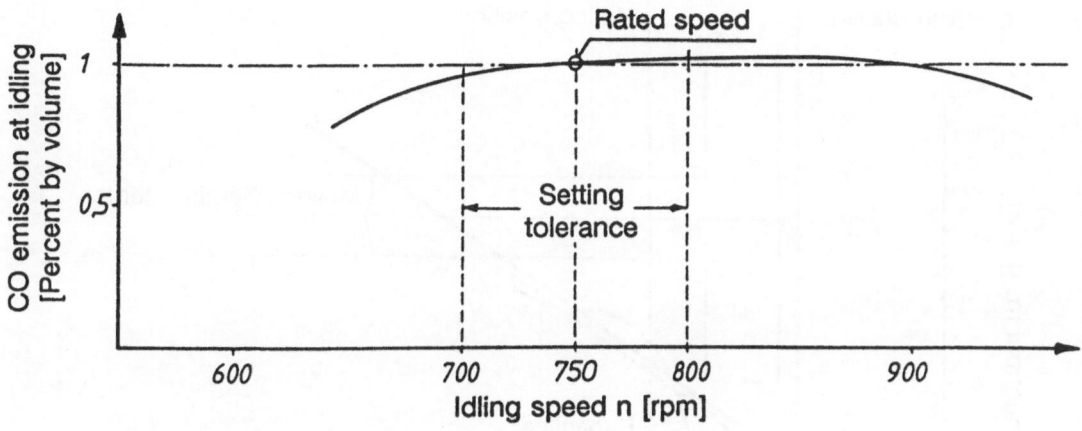

Fig. 4.57. Composition of the idling mixture for speed corrections [4.15]

When using fuel cutoff in the overrun together with low idling throughputs, the slot is shaped like an inverted 'T.' For idling and overrun cutoff, the carburetor throttle valve is closed to a value lower than the idling position. The entire slot moves to a position above the throttle valve and thus stops supplying fuel. The T slot and optimized channel design of the idling system ensure a quick return to the idling mixture ratio when the system is restarted after the overrun stage (**Fig. 4.58**). The arrangement of bypass bores, on the other hand, did not produce the desired effect, as is indicated by the figure.

Fuel Cutoff in the Overrun

During overrun operation the fuel supply is cut off in many designs to save on fuel consumption and, to some extent, to reduce emissions. In the 2E carburetor this is done through the throttle valve (**Fig. 4.59**).

The throttle valve adjuster switches the main throttle valve of the first stage between idling and overrun. The vacuum applied via the connecting line to the adjuster diaphragm is reduced via the positioning valve. The regulating valve position is adjusted in accordance with the desired idling speed. The ventilation to produce the idling position can be switched off by the reversing valve. When the speed controller sends the appropriate signal (e.g. n > 1400 1/min), voltage is applied to the reversing valve which then closes. The lack of ventilation makes the diaphragm of the throttle valve adjuster move into the overrun position. The throttle valve, no longer kept open because the accelerator is released, then goes into overrun position, moved by the throttle valve closing springs.

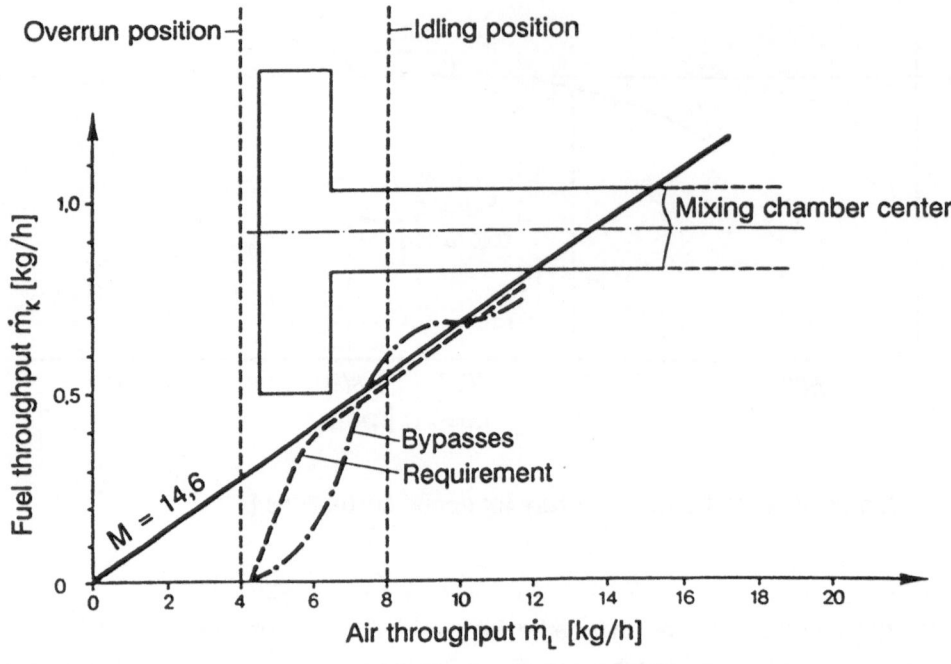

Fig. 4.58. T slot to select the idling system during overrun cutoff [4.45]

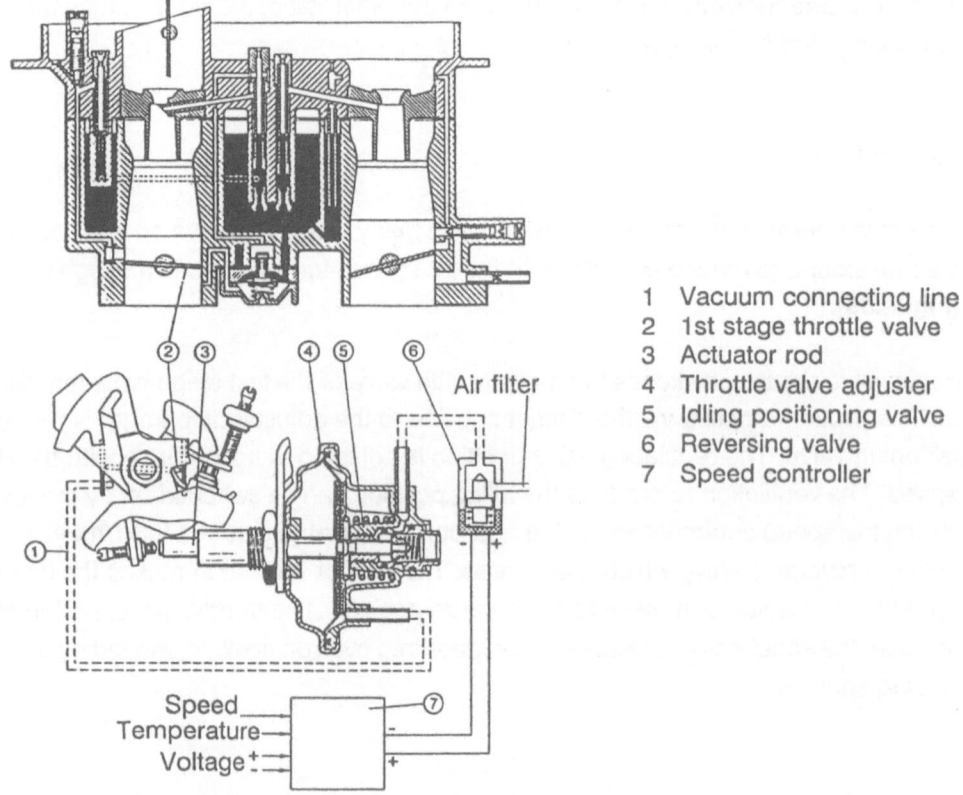

1	Vacuum connecting line
2	1st stage throttle valve
3	Actuator rod
4	Throttle valve adjuster
5	Idling positioning valve
6	Reversing valve
7	Speed controller

Fig. 4.59. Diagram of throttle-valve-controlled overrun cutoff [4.15]

When the engine, now working in thrust, drops below the so- called restart speed, a signal is sent to the reversing valve to reventilate the vacuum area. As a result, the throttle valve adjuster spring returns the main throttle into the idling position within a few milliseconds.

Accelerating Pump

Fig. 4.60 demonstrates the diaphragm-actuated accelerating pump.

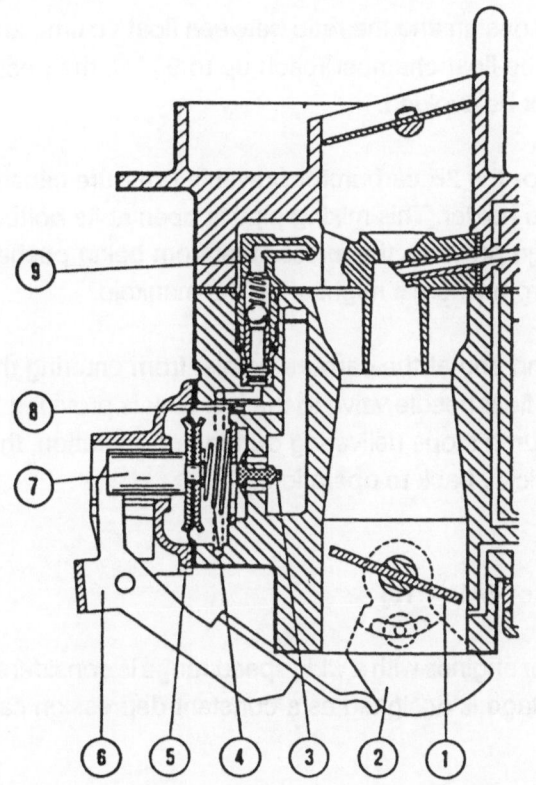

1	1st stage throttle valve
2	Cam
3	Pump suction valve
4	Pump spring
5	Diaphragm/diaphragm disk
6	Transmission lever
7	Diaphragm tappet
8	Return nozzle
9	Pump outlet pipe

Fig. 4.60. Diaphragm-actuated accelerating pump [4.15]

The diaphragm tappet contains a spring which is preloaded when the throttle valve is actuated and which guides the pump diaphragm.

The pump suction valve is of the elastomer type. It is slightly preloaded through a retaining spring, and encounters almost no resistance in the suction direction, where it is sealed by an external lip on a cast surface.

At its outlet side, the accelerating pump system has a pressure valve. A spring-loaded ball valve prevents air from entering the pump system during the intake stroke which otherwise might lead to a reduction in the injection volume.

Hot Start, Hot Run-up

In order to ensure proper restart of the vehicle when it has been switched off after operation, designers have provided for a very small float chamber in the 2E carburetor to keep the rate of vaporizing fuel as low as possible.

Key factors for hot idling are the kinematic float design and the ratio between float volume and float weight. Considering that temperatures in the float chamber reach up to 90 °C, the possibility of freshly fed fuel suddenly foaming cannot be avoided.

A hot start mixing pipe developed specifically for the 2E carburetor prevents pressure caused by vapor bubbles from building up in the nozzle holder. This mixing pipe is open at its bottom and provided with an external collar. This design prevents the excess fuel from being pushed through the mixture outlet by the gas bubbles from where it might enter the manifold.

To prevent the gases generated in the fuel pump and at the carburetor inlet from entering the float chamber, a return line from the carburetor float needle valve to the fuel tank is provided in the 2E carburetor. In the case where the fuel pump stops delivering during hot operation, this return line can assist in returning the system quickly back to operation.

4.1.1.6.2. Pierburg Double Compound Carburetor 4A1 [4.19]

As already discussed, the optimum carburetor for engines with a wide speed range is considered to be a compound carburetor whose second stage is designed as a constant-depression carburetor.

Considering further that uniform mixture distribution is difficult to achieve for more than four cylinders per carburetor, while on the other hand two-carburetor systems are prone to synchronization problems as well as difficult to maintain, it is recommended to use double compound carburetors with a constant-depression-type second stage for six- and eight-cylinder engines, provided that carburetors should be chosen at all. An example of the latter type of carburetor is the Pierburg 4A1 carburetor (**Fig. 4.61**) which features the following advantages:

- Maximum possible accuracy in metering for part loads;

- Minimum possible throttling in the upper full load range;

- Compact construction;

- Easy to maintain;

- Can be used for various engine sizes owing to the constant-depression principle of the second stage.

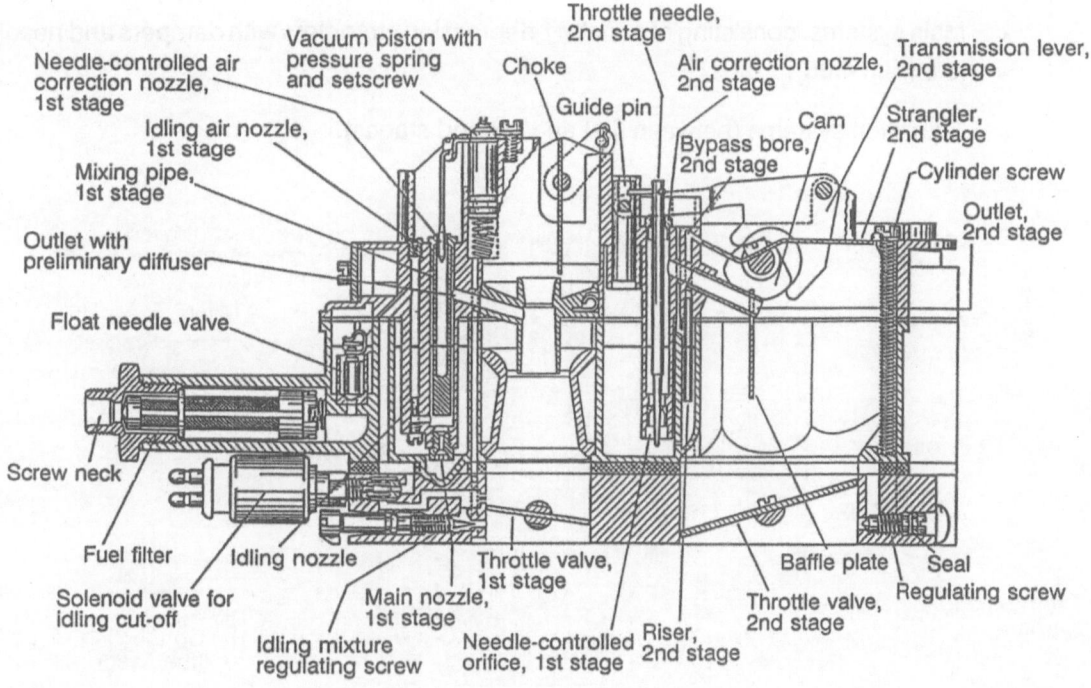

Fig. 4.61. Pierburg double compound carburetor 4A1 [4.19]

The complex setup of the carburetor contains the following systems:

First stage:

- Idling systems (**Fig. 4.62**);

- Bypass systems (Fig. 4.62);

- Main system (**Fig. 4.63**) (2 - air inlet, 3 - compensation air);

- Compensation air systems with vacuum-controlled needle jets (Fig. 4.62);

- Acceleration systems (**Fig. 4.64**);

- Idling cutoff valves;

- Chokes for cold-start enrichment;

- Shunt starters to increase idling mixture during warm-up;

- Idling speed governor.

Second stage:

- Main systems, consisting of constant-depression stranglers with dampers and needle jets (**Fig. 4.65**);

- Transition systems (between first and second stages).

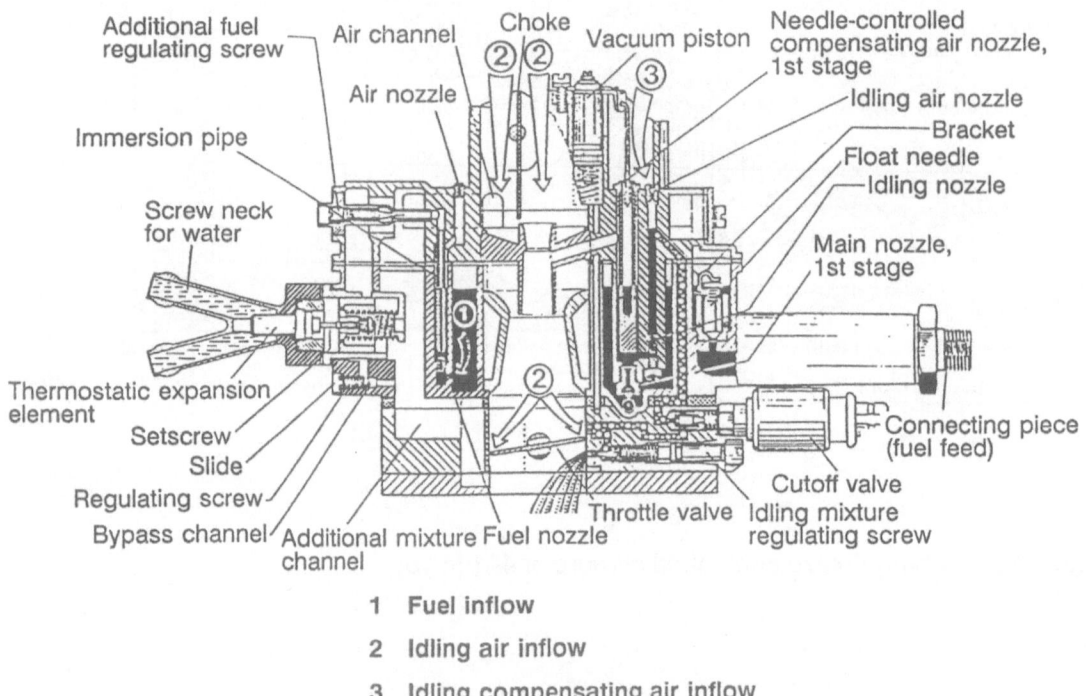

1   Fuel inflow

2   Idling air inflow

3   Idling compensating air inflow

Fig. 4.62. Idling system of the Pierburg double compound carburetor 4A1

This type of carburetor includes some 20 different systems, depending on its design. Obviously it is no easy task to develop and harmonize such a unit. Nevertheless, this most complex and sophisticated of mechanical carburetors proved its worth for many years in the six-cylinder engines used by Daimler-Benz and BMW. Today, fuel injection systems have replaced these highly complicated carburetors.

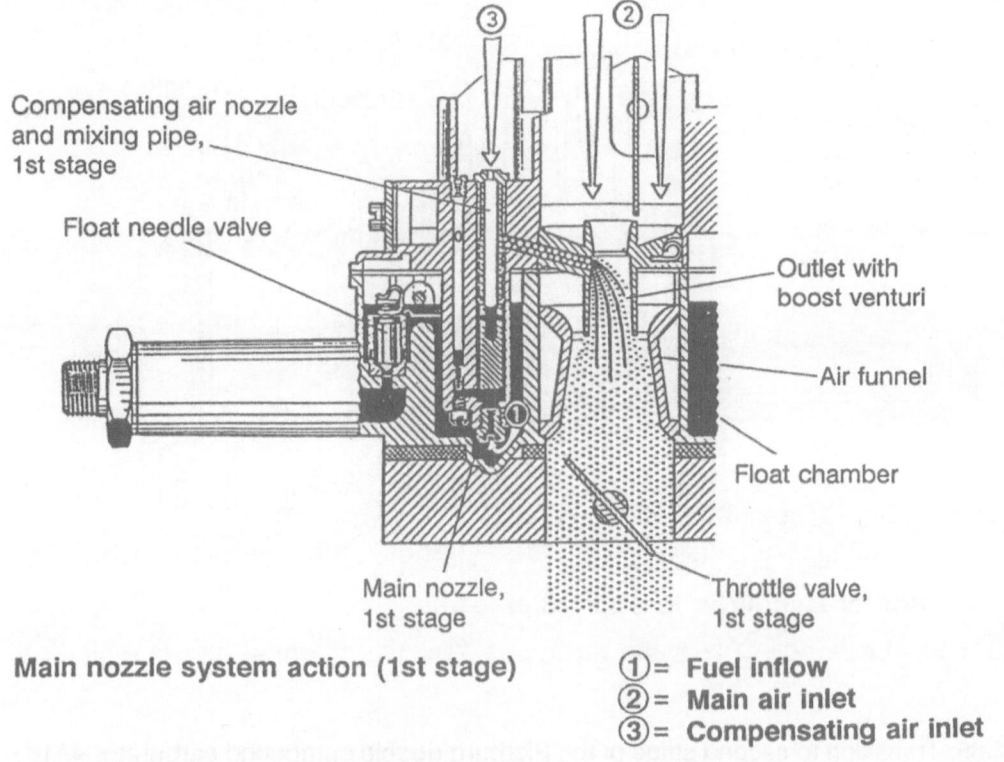

Compensating air nozzle and mixing pipe, 1st stage

Float needle valve

Outlet with boost venturi

Air funnel

Float chamber

Main nozzle, 1st stage

Throttle valve, 1st stage

Main nozzle system action (1st stage)

① = **Fuel inflow**
② = **Main air inlet**
③ = **Compensating air inlet**

Fig. 4.63. Main nozzle system (first stage) of the Pierburg double compound carburetor 4A1

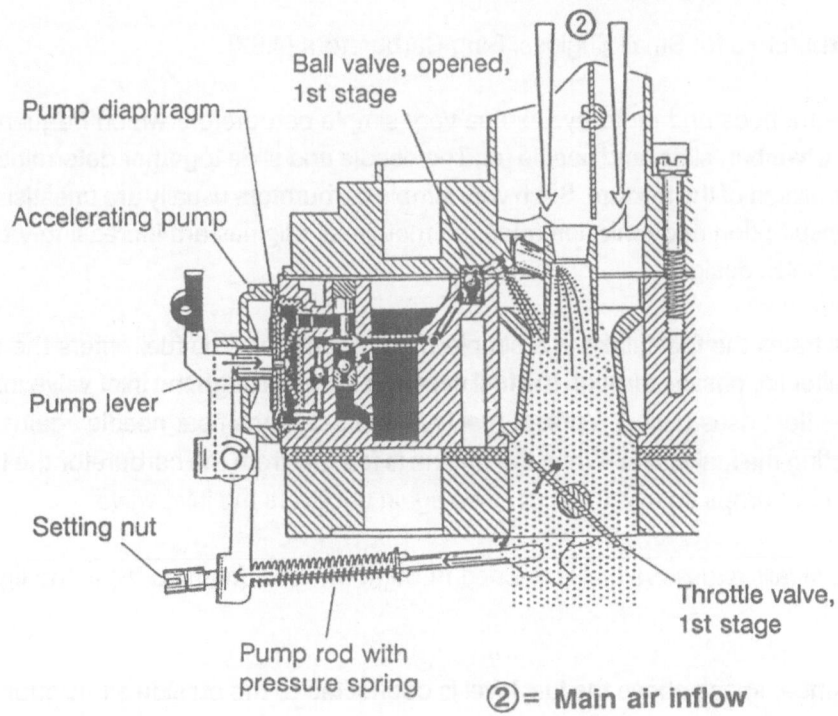

Ball valve, opened, 1st stage

Pump diaphragm

Accelerating pump

Pump lever

Setting nut

Pump rod with pressure spring

Throttle valve, 1st stage

② = **Main air inflow**

Fig. 4.64. Accelerating pump system of the Pierburg double compound carburetor 4A1

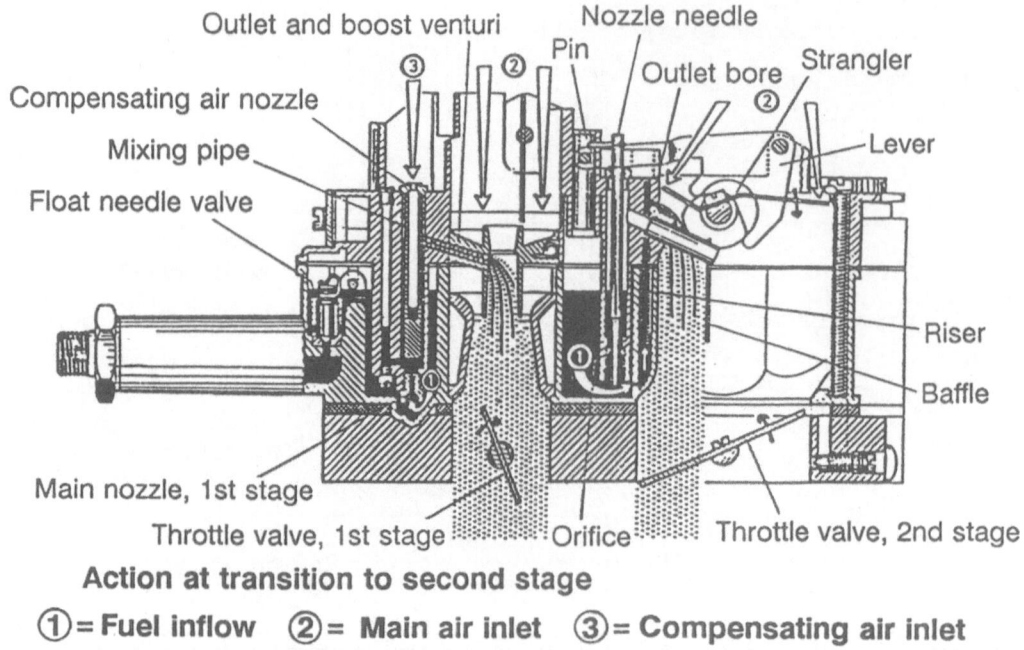

Outlet and boost venturi — ③ ② Pin — Nozzle needle — Outlet bore — Strangler ②

Compensating air nozzle

Mixing pipe

Float needle valve

Lever

Riser

Baffle

Main nozzle, 1st stage

Throttle valve, 1st stage — Orifice — Throttle valve, 2nd stage

Action at transition to second stage

① = **Fuel inflow**   ② = **Main air inlet**   ③ = **Compensating air inlet**

Fig. 4.65. Transition to second stage of the Pierburg double compound carburetor 4A1

4.1.1.6.3. Carburetors for Small Engines, Bing Carburetors [4.57]

Small-volume mopeds and motorcycles use very simple carburetors which frequently consist of just a float chamber, slide, and needle jet. The needle and slide together determine the quantity and composition of the mixture. Such very simple carburetors usually are unsatisfactory with regard to consumption and emission rates, so that small engines are increasingly being fitted with more complex designs.

Fig. 4.66 illustrates the fuel intake in a simple Bing carburetor. The fuel enters the carburetor through the filter lid, passes through the fuel filter and flows through the inlet valve into the float chamber. The float rises until at a preset fuel level it presses the float needle against the valve seat, interrupting the fuel intake. When the engine is fed fuel from the carburetor the fuel level in the float chamber drops so that the float once again uncovers the inlet valve.

If the inlet valve fails excessive fuel is drained through the tube (marked "R" in the figure) in the float cap.

The float chamber space above the fuel level is connected to the outside air through bore "E."

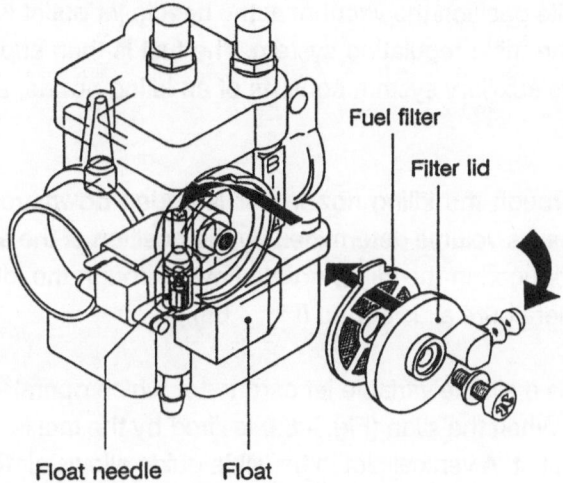

Fig. 4.66. Fuel supply in a simple Bing carburetor [4.57]

Engine performance, i.e. the quantity of mixture sucked in by the engine, is regulated by the passage cross-section uncovered by the gas slide (**Fig. 4.67**). The slide is lifted by a tackle against the pressure exerted by the restoring spring. The air flow generates a vacuum in the passage which sucks the fuel from the float chamber through the nozzle system. The fuel flows through the main nozzle and needle nozzle where it is first mixed with air coming through the compensating air channel and cross-bores in the needle jet (Fig. 4.67).

When the engine runs at part load, the fuel flow to the carburetor passage is throttled by a jet needle which is connected to the slide and which plunges into the needle jet.

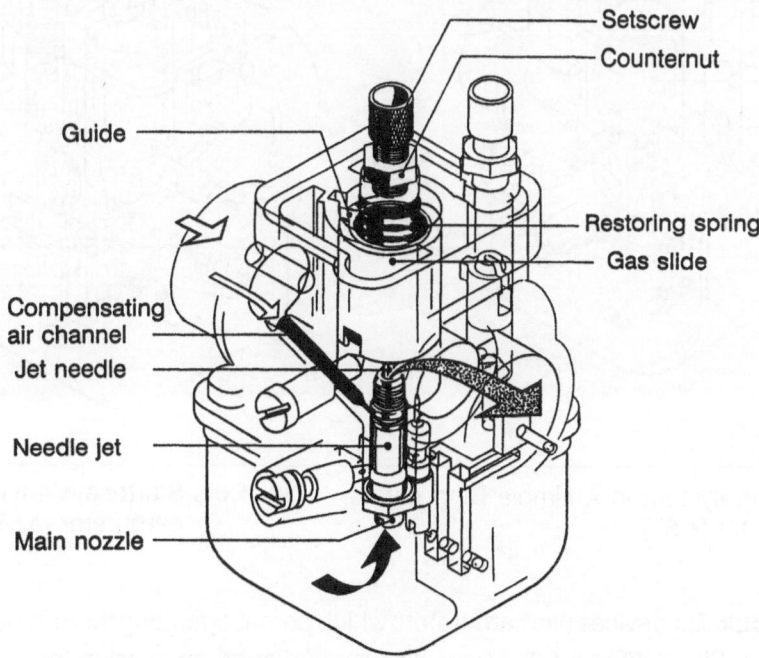

Fig. 4.67. Normal operation system in a simple Bing carburetor [4.57]

With the engine in its idle position the vacuum at the needle jet outlet is so low that no fuel can be delivered through the main regulating system. The fuel is then supplied through the idling system (**Fig. 4.68**). This auxiliary system consists of an idling nozzle, air regulating screw and idling outlet bore.

Here, the fuel flows through the idling nozzle. Air is added downstream of the nozzle bore, through the cross-bores, its volume determined by the position of the air regulating screw. The preliminary mixture produced in the idling nozzle flows through the idling outlet bore into the carburetor passage where pure air is added (Fig. 4.68).

The starter carburetor is a simple variable jet carburetor which operates in parallel to the main carburetor (**Fig. 4.69**). When the slide (Fig. 4.69) is lifted by the tackle against the spring pressure it opens the fuel outlet. A vertical slot in the slide guide allows air to enter from the steadying chamber which is supplied through a bore at the filter side. The preliminary mixture produced in the starter carburetor flows through bore "A" into the carburetor passage (Fig. 4.69). The fuel volume for the starter carburetor is measured by the start nozzle. When the starter carburetor is closed the aerated antechamber will have the same fuel level as the float chamber. When the engine is started with the starter carburetor open, the fuel from the antechamber is sucked off first to ensure cold-start enrichment, followed by a quantity of fuel which can pass through the starter nozzle [4.57].

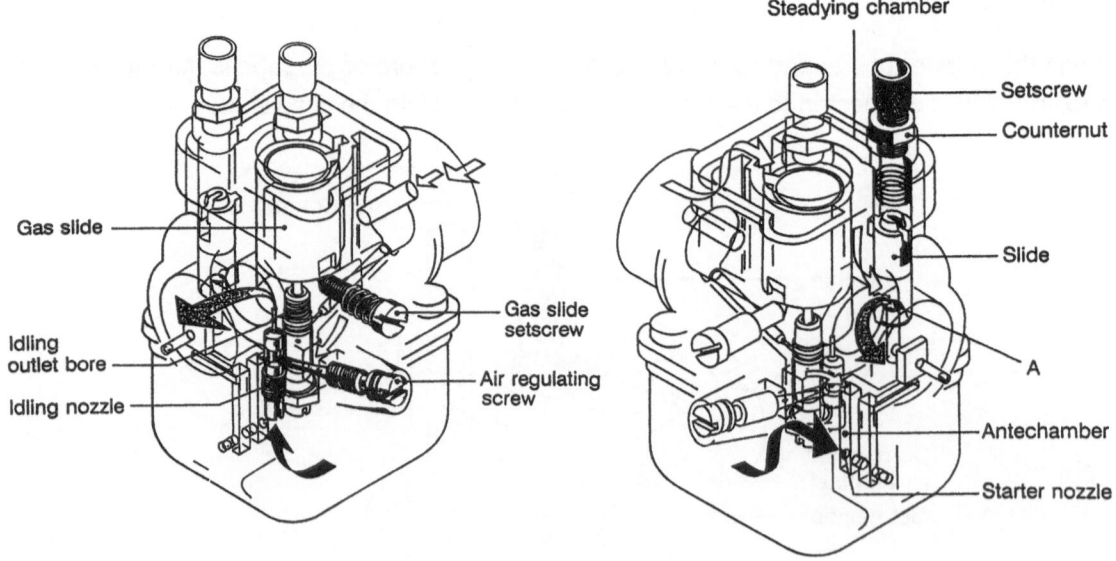

Fig. 4.68. Idling system in a simple Bing carburetor [4.57]

Fig. 4.69. Starter system in a simple Bing carburetor [4.57]

Chainsaws and similar devices use carburetors which permit operating them in slanting or "head-down" positions. **Fig. 4.70** and **4.71** show this type of diaphragm carburetor.

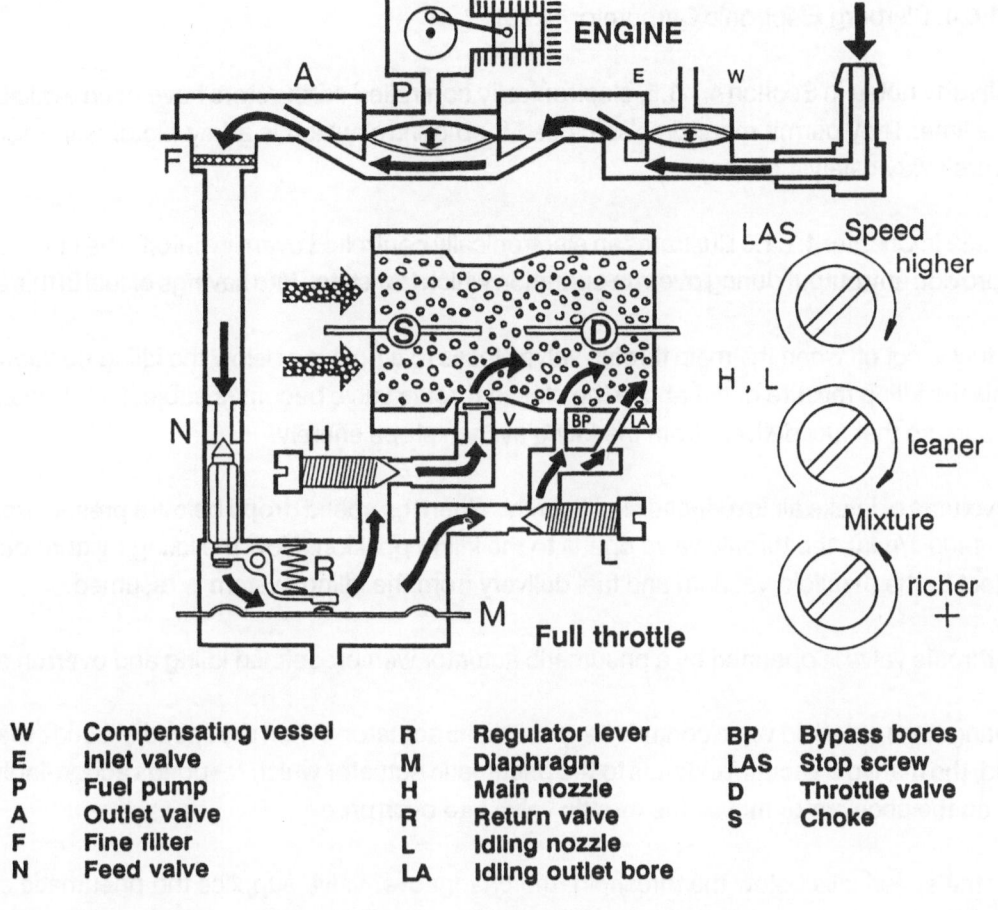

W	Compensating vessel	**R**	Regulator lever	**BP**	Bypass bores
E	Inlet valve	**M**	Diaphragm	**LAS**	Stop screw
P	Fuel pump	**H**	Main nozzle	**D**	Throttle valve
A	Outlet valve	**R**	Return valve	**S**	Choke
F	Fine filter	**L**	Idling nozzle		
N	Feed valve	**LA**	Idling outlet bore		

Fig. 4.70. Full load system of the Tillotson diaphragm carburetor [4.58]

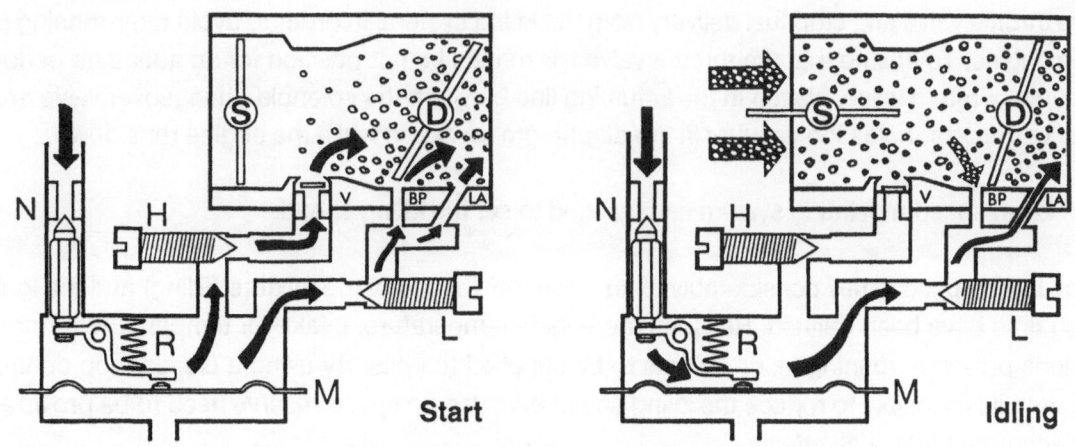

Fig. 4.71. Idling system of the Tillotson diaphragm carburetor [4.58]

4.1.1.6.4. Pierburg Electronic Carburetor [4.20]

As already noted in Section 4.1.1.5, electronically controlled carburetors have been available for some time. They permit more precise closed-loop control which is a key factor with regard to the three-way catalyst concept.

Fig. 4.52 in Chapter 4.1.1.5 illustrates an electronically controlled overrun cutoff. The engine need not provide any output during overrun operation which translates into savings of fuel in this stage.

The fuel is cut off when the main throttle valve closes to a position below the idling position. Due to this the idling mixture outlet above the closed throttle valve becomes subject to atmospheric pressure so that fuel delivery from the idling system stops entirely.

The volume of intake air is reduced significantly. When the speed drops below a preset threshold (e.g. 1400 1/min), the throttle valve opens to the idling position. Now the idling mixture outlet is subject to the manifold vacuum and fuel delivery from the idling system is resumed.

The throttle valve is operated by a pneumatic actuator with predefined idling and overrun stops.

A changeover solenoid valve controls the pneumatic actuator. When the speed exceeds the threshold, the manifold vacuum extends to the pneumatic actuator which responds accordingly. Releasing the accelerator moves the throttle valve into overrun.

After the speed falls below the threshold, the changeover valve supplies the pneumatic actuator with atmospheric pressure so that it opens the throttle valve to its idling position, using a pressure spring.

When the engine is switched off the manifold vacuum expands to the diaphragm box to close the throttle valve and stop fuel delivery from the idling system, in order to avoid after-running of the engine. To ensure that the throttle valve will remain in that position for an adequate period of time, a return valve, placed in the actuating line between the solenoid changeover valve and the carburetor, holds the vacuum in the diaphragm box even while the engine runs down.

An idling speed regulating system can be fitted to set the idling speed:

The idling speed varies considerably even when the volume of the mixture (filling) and the ignition time have been defined. Reasons are engine temperature, intake air temperature, atmospheric pressure, running-in, or outputs to be supplied to units. By using a closed-loop control system it is possible to reduce the idling speed because no speed reserve need to be provided to accommodate such effects.

Closed-loop filling control is provided by the throttle valve (**Fig. 4.72**). An electro-pneumatically operated throttle valve adjuster varies its idle position. This adjuster contains two solenoid valves to transmit the control pressures (atmospheric pressure and manifold pressure) and a flat potentiometer to signal the stroke position of the throttle valve adjuster.

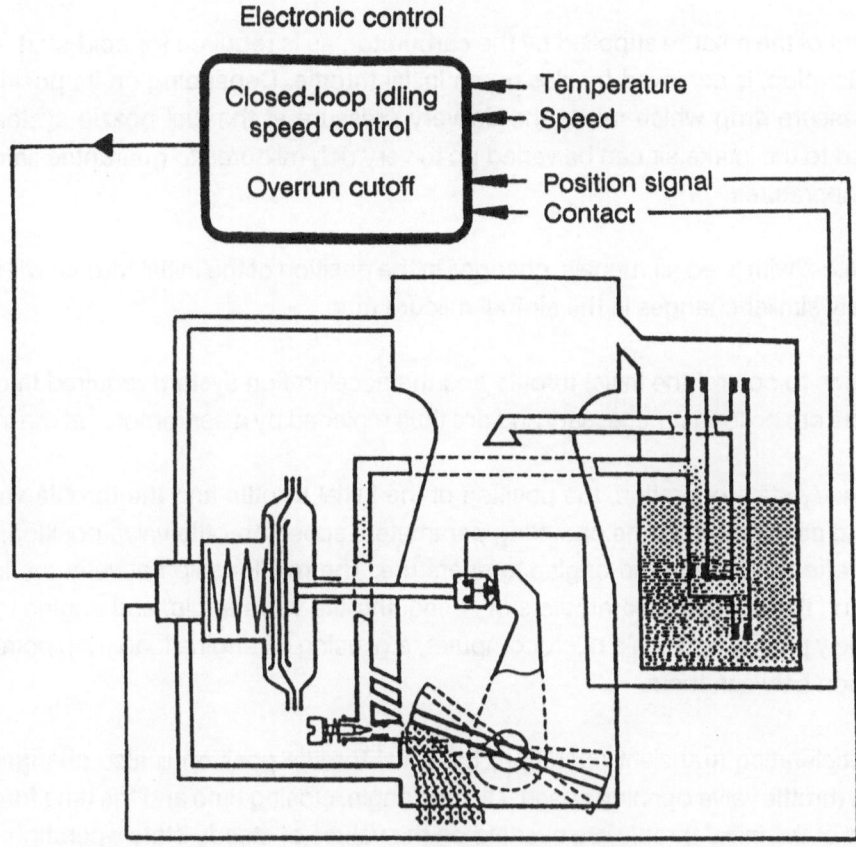

Fig. 4.72. Diagram of electronic closed-loop idling speed control in a carburetor, combined with overrun cutoff [4.20]

During idling, the actual engine speed is constantly compared with the desired rated value. The deviation is fed into a PI closed-loop control algorithm to calculate the required change in the throttle valve position.

When the closed-loop control scope is adequately dimensioned (throttle valve opening angle of approx. 25°) it is possible for the closed-loop idling speed control actuator to regulate fillings from low-temperature starts (wide throttle valve opening) to overrun cutoff (closed throttle valve).

Other carburetor functions that can be electronically controlled are start, warm-up, and acceleration.

In this concept, the basic carburetor systems are adapted to the required mixture ratio at steady-state operation with a warmed-up engine. It is ensured that the vehicle can be operated even when the electronic control should fail.

Enrichment of the mixture supplied by the carburetor, as is required for cold-start, warming up, and acceleration, is achieved by closing an initial throttle. Depending on its position, it generates a pressure drop which raises the delivery pressure at the fuel nozzle systems. The fuel volume fed to the intake air can be varied up to very rich mixtures to guarantee safe starts even at low temperatures.

In carburetors with fixed air funnels, changes in the position of the initial throttle will produce approximately similar changes in the air-fuel mixture ratio.

The systems to control the initial throttle and the acceleration system required in conventional carburetors are no longer necessary and are thus replaced by a servomotor at the initial throttle.

During steady-state operation, the position of the initial throttle and the throttle valve stop are determined as functions of the operating parameters speed, throttle valve position, engine running time after cold start, and engine temperature. The relationship between the initial throttle position and the operating parameters of speed, throttle valve angle, and engine temperatures can be freely programmed in a microcomputer, e.g. using 64 engine load map points and linear interpolation between them.

During acceleration (transient operation) the initial throttle position is also changed as a function of the throttle valve opening speed. Closing angle, closing time and the time function during reopening of the initial throttle are overlaid on the values of steady-state operation.

The initial throttle position is calculated by a microcomputer as a function of the above operating parameters, using additional freely programmable map points with linear interpolation between them.

Rapid adjustment and accuracy of the initial throttle servomotor permit quick and differentiated variations in enrichments to accommodate the above operating parameters. The control system allows precise adjustment of the mixture ratio, especially in cold engines and during acceleration.

4.1.2. Single-Point Injection

In single-point injection systems the carburetor is replaced by a central injection unit consisting of a throttle valve, injector, and transducer. Generally, one injector is sufficient. Fuel distribution to the cylinders is through the manifold, same as in carburetor engines. Single-point injection systems are also known under their company names (e.g. "Bosch Mono-Jetronic") or as central fuel injection (CFI) systems. Their first wide-spread use was in the U.S., in lamba closed-loop controlled systems where they gradually replaced the lambda closed-loop controlled carburetors.

Ford introduced a central fuel injection (CFI) system for V8 engines in 1979. In this system, the carburetor is replaced by the CFI unit, which consists of two injectors for four cylinders each and a pressure regulator. The fuel pump, situated in the fuel tank, at approx. 2.5 bars generates a relatively high pressure (so that the system is now known as a "high-pressure system"). The manifold pressure serves as the main actuating variable.

Chrysler, too, uses a manifold pressure-controlled high-pressure system for four-cylinder engines (**Fig. 4.73**). The CFI unit includes an injector above the throttle valve whose injection angle has been widened.

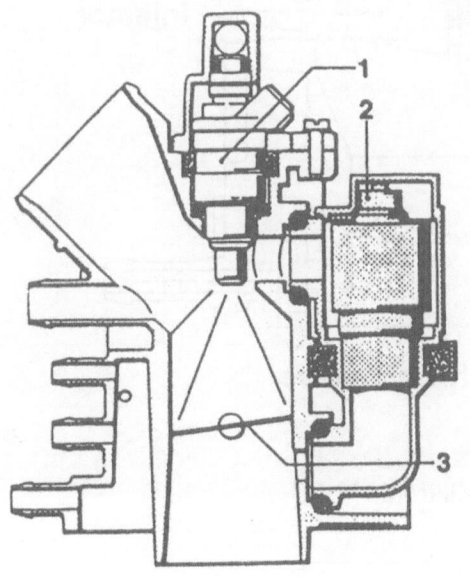

Single-point fuel injection unit for Chrysler
1 Injector, 2 Idle speed actuator, 3 Throttle valve

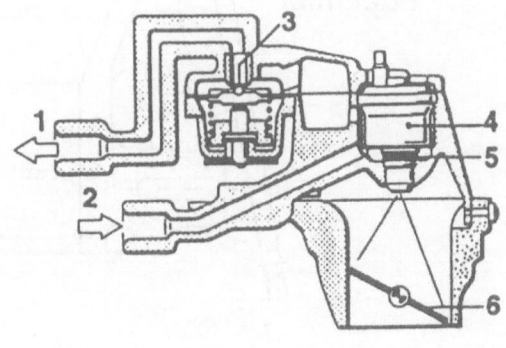

TBI-System of General Motors
1 Fuel outlet, 2 Fuel inlet, 3 Fuel-pressure regulator, 4 Injector, 5 Filter, 6 Throttle valve

Fig. 4.73. Single-point fuel injection systems in the U.S. [4.21]

General Motors has for some time been using a low-pressure system called Throttle Body Injection (TBI) (Fig. 4.73). Its low pressure of 0.7 bars allows the use of a flow pump made of inexpensive plastic components rather than a relatively accurate volume pump as would be required for high pressure. The low pressure does not prevent vapor generation during cutoff which in turn would affect hot-starting. The injector is therefore refilled quickly by a flushing switch. A swirl unit ensures adequate mixture preparation at low pressure.

4.1.2.1. Bosch Mono-Jetronic

The centerpiece of the central injection unit (**Fig. 4.74**) is its solenoid-operated injector which directs a - usually cone-shaped - spray to the throttle valve.

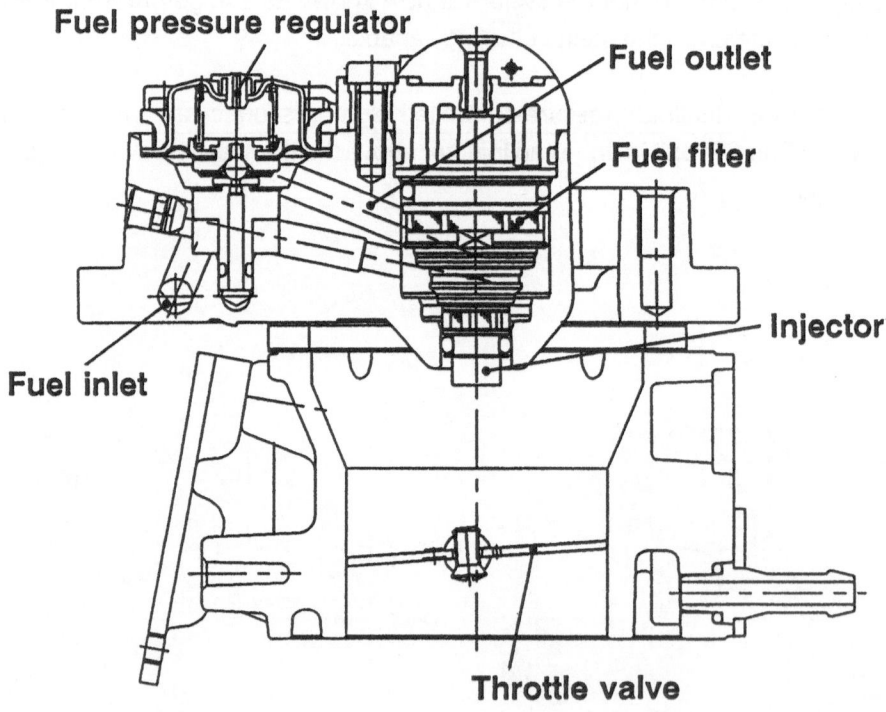

Fig. 4.74. Central injection unit in the Bosch Mono-Jetronic [4.11]

At full load, atomization is ensured by the relative speed of the air and fuel sprays, at part loads it is achieved by the high velocity of the air at the throttle valve gap.

The injector is triggered in time with the ignition pulses. A constant fuel stream through the injector effectively prevents the formation of vapor locks.

Fuel is supplied by an electric fuel pump, usually located in the fuel tank, which delivers the fuel via a fuel filter to the central injection unit. There, a pressure regulator maintains the differential pressure at the injector measuring point at a constant level of approx. 1 bar regardless of the quantity of injected fuel.

Acquisition of operating data:

The ignition system supplies the speed signal to the electronic control unit (ECU). The quantity of air is determined by the throttle valve position as registered by the throttle valve potentiometer. The latter also detects the "idle" and "full load" conditions which are relevant for overrun fuel cutoff and full load enrichment.

Temperature sensors register the engine and air temperatures and signal them to the ECU.

Higher injection quantities, as are necessary for cold start or warming-up, are achieved by lengthening the injection period.

The idle speed is reduced and stabilized by an idle control system: a servomotor positions the throttle valve and feeds air to the engine as appropriate to achieve the desired idle speed as determined by the output requirements to supply auxiliary units, temperature influences, wear, etc.

Fuel enrichment is also necessary (same as in carburetor engines) when the engine is accelerated. The ECU detects acceleration from the throttle valve potentiometer signal and enriches the mixture depending on the engine temperature and throttle valve movements. The same applies vice versa for leaner mixtures when the engine decelerates.

Supplementary functions provided by the system are fuel cutoff when a preset speed has been reached or lambda closed- loop control. In the latter, the injection volume is controlled so that a ratio of lambda = 1 will be maintained at all operating conditions.

Fig. 4.75 gives a functional diagram of the lambda closed-loop controlled Bosch Mono- Jetronic.

Fig. 4.76 provides an overview of the lambda closed-loop controlled Mono-Jetronic.

Fig. 4.77 Shows the Mono-Jetronic solenoid injector. The types of injectors will be discussed in a later chapter.

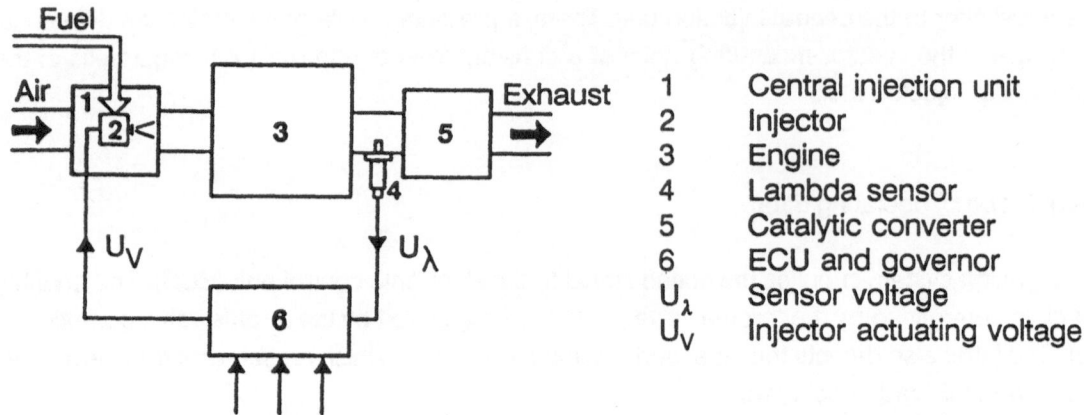

Fig. 4.75. Functional diagram of a lambda closed-loop controlled Mono-Jetronic [4.22]

1	Central injection unit
2	Injector
3	Engine
4	Lambda sensor
5	Catalytic converter
6	ECU and governor
U_λ	Sensor voltage
U_V	Injector actuating voltage

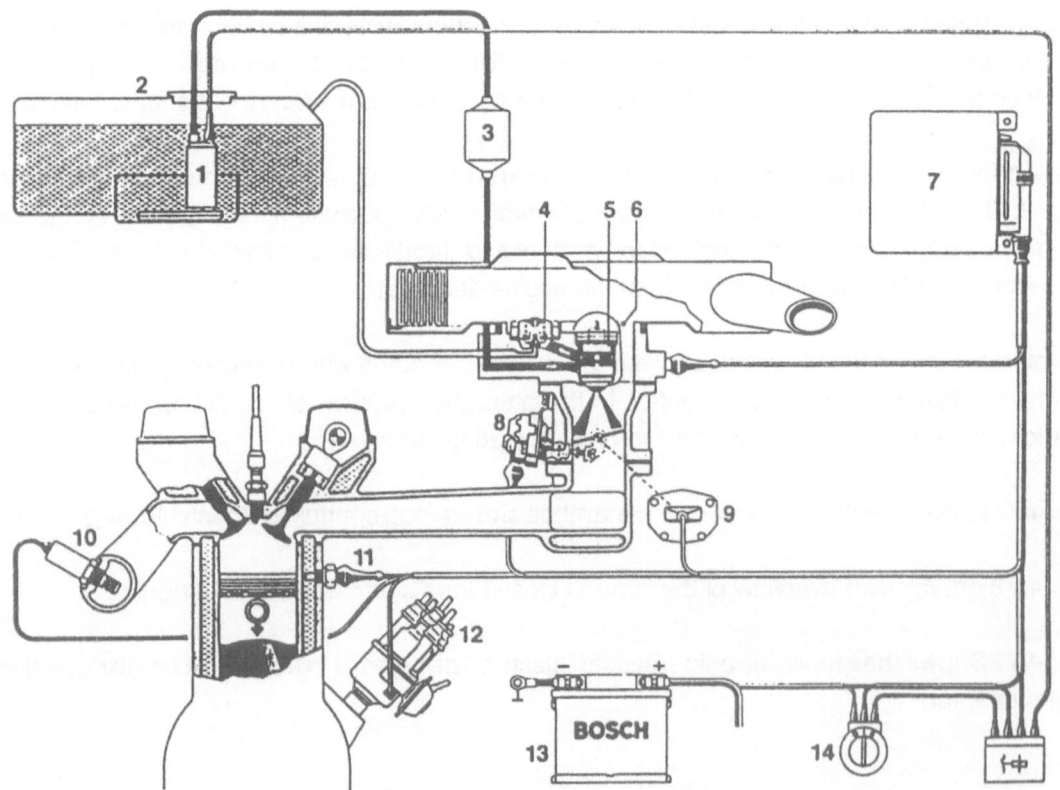

1 Electric fuel pump, 2 Fuel tank, 3 Fuel filter, 4 Fuel-pressure regulator, 5 Injector, 6 Air-temperature sensor, 7 Control unit, 8 Throttle-valve actuator, 9 Throttle-valve potentiometer, 10 Lambda sensor, 11 Engine-temperature sensor, 12 Ignition distributor, 13 Battery, 14 Ignition/starting switch

Fig. 4.76. Diagram of a Bosch lambda closed-loop controlled Mono-Jetronic [4.22]

4.1.2.2. Nissan Single-Point Injection [4.23]

Fig. 4.78 provides an overview of the system, whose main components are an injection body, control unit, and several sensors.

The sensors detect engine operating conditions and signal them to the control unit whose microprocessor then determines the appropriate fuel quantity to be injected. Signals from the lambda sensor located in the exhaust are processed to check and if necessary adjust the mixture ratio. The control unit is provided with a self-test function.

The injection body (**Fig. 4.79**) contains a hot-wire airflow meter in the bypass, an injector, pressure regulator, throttle valve sensor, auxiliary air system, and the throttle chamber with the throttle valve.

The platinum hot wire meters the mass flow rate without being affected by fluctuations in atmospheric pressure or temperatures.

It is, however, affected by air pulsations so that it is located in the bypass where such variations are less severe.

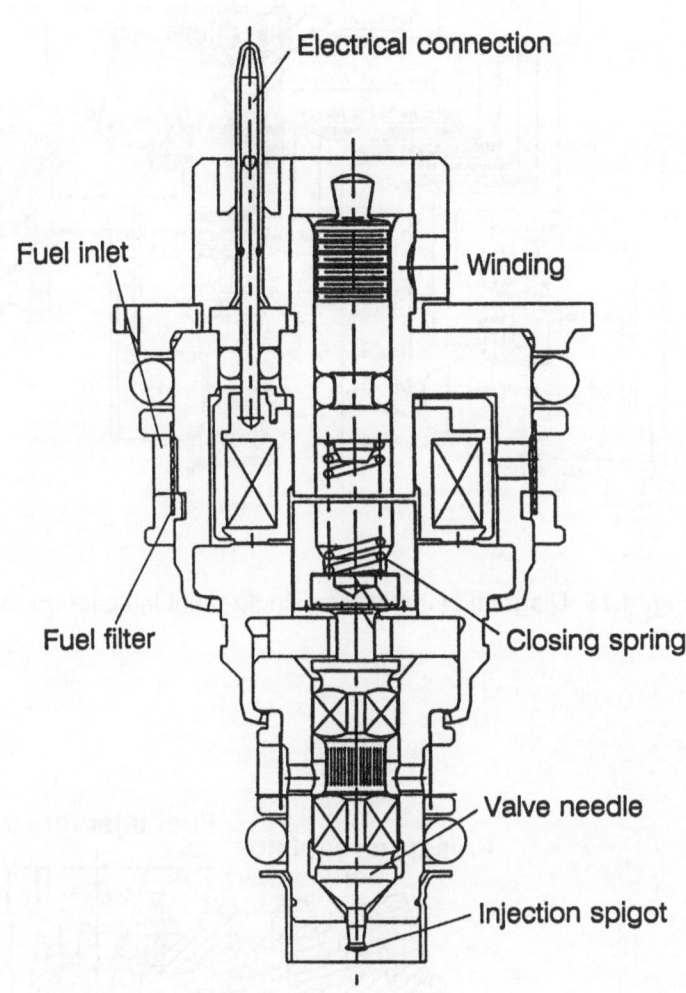

Fig. 4.77.Bosch solenoid injector [4.11]

When the throttle valve is fully opened so that pulsations are very strong, the hot wire signals are processed to provide a weighted average, in order to preclude faulty indications.

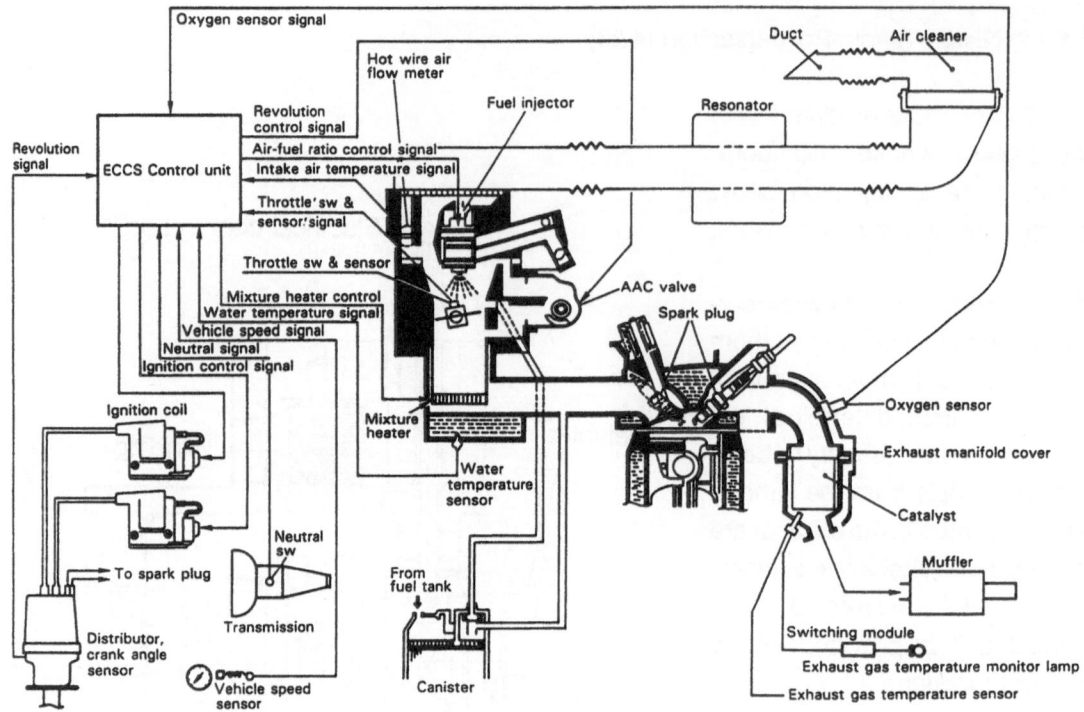

Fig. 4.78. Diagram of the Nissan single-point injection EI system [4.23]

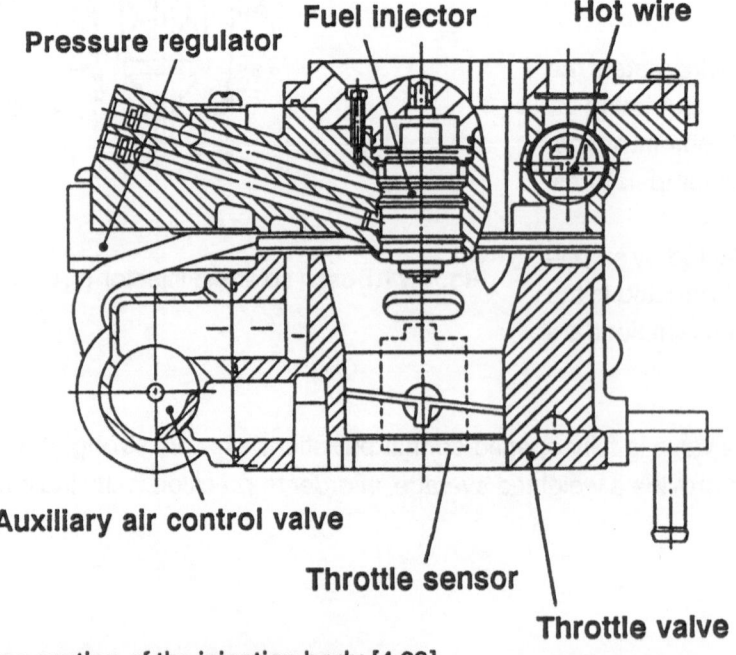

Fig. 4.79. Cross-section of the injection body [4.23]

The hot wire is heated up to a high temperature after switching off the ignition, to burn off any dust deposits.

Such deposits would reduce the heat-emitting surface of the wire and would thus cut down the current necessary to keep the wire at a constant temperature. The result would be an error in the airflow rate signal. **Fig. 4.80** indicates the effect of high-temperature heating of the hot wire with regard to the measuring error.

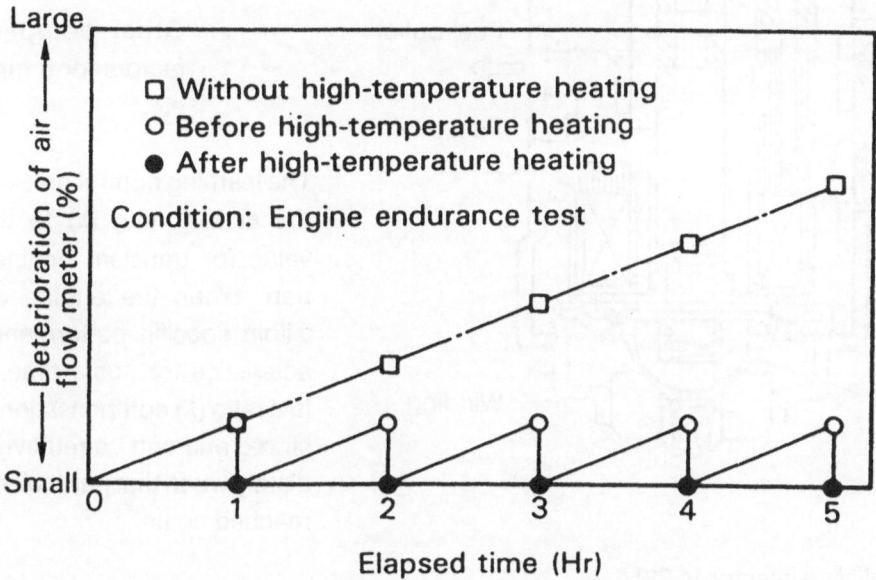

Fig. 4.80. Effect of high-temperature heating of the hot wire [4.23]

The solenoid injector used here is shown in **Fig. 4.81**. The metering orifice is closed by a ball. Fuel pressure is 1 bar, steady fuel flow rate is 600 $cm^3.min^{-1}$.

The injector is constantly flushed by the fuel, to keep it cool. Its orifice is spiral-shaped to give the spray a swirling movement.

The injection pulse signal is applied to the injector in synchronization with an ignition timing signal twice per engine revolution (four-cylinder engine).

The injection quantity depends on the following factors:

- Basic fuel quantity, as determined by the engine speed and volume of intake air;

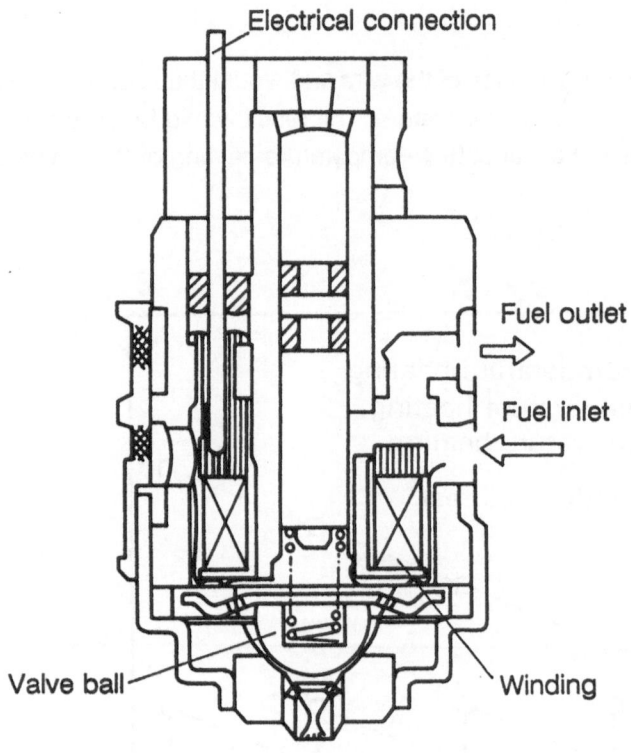

Fig. 4.81. Nissan injector [4.23]

- Air-fuel ratio compensation for lambda = 1 closed-loop control, except for warming-up, deceleration, and full load;

- Compensation factor for acceleration and deceleration;

- Other compensation factors for temperature, etc.

The learning control quickly adapts the air-fuel ratio (λ) to the rated value for transient engine operation. When the engine operates within specific performance characteristics for some time, the air-fuel ratio (λ) compensation factor is stored and can be retrieved immediately when that particular range is reached again.

4.1.2.3. Opel Multec Central Injection Unit [4.38]

Fig. 4.82 provides a functional diagram of the Opel Multec central injection unit.

In this system, the fuel pump is located in the tank. It is designed as a two-turbine pump with a capacity of 80-100 l/h that operates at 0.75 bars and 13.5 V. This capacity exceeds the engine requirements so that the excess fuel flows back into the tank through a return line. In this way it is possible to avoid fuel overheating and vapor locks.

For the fuel pump design see **Fig. 4.83**.

It consists of an electric motor and two turbine wheels placed in a common housing and flushed by fuel, which in turn cools the motor. An inductor and eight-pole collector drives the turbine wheels placed on the inductor axis. The fuel is sucked in at turbine housing 9, and flows through the housing and return valve 5.

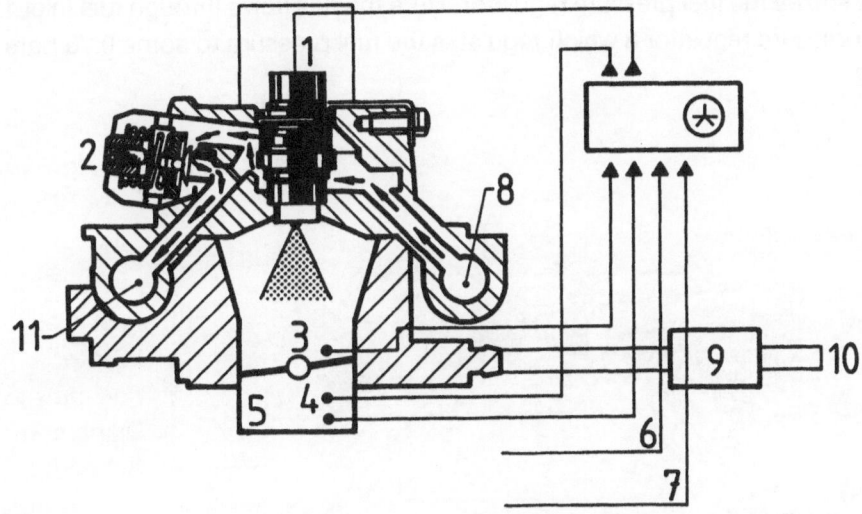

1	Injector	7	Lambda-Sensor
2	System pressure regulator	8	Fuel inlet
3	Throttle valve potentiometer	9	Vaporization monitoring system
4	Idle fill step motor	10	Tank ventilation
5	Intake manifold pressure sensor	11	Fuel return line to the tank
6	Coolant Temperature sensor		

Fig. 4.82. Functional diagram of an Opel Multec central injection unit [4.38]

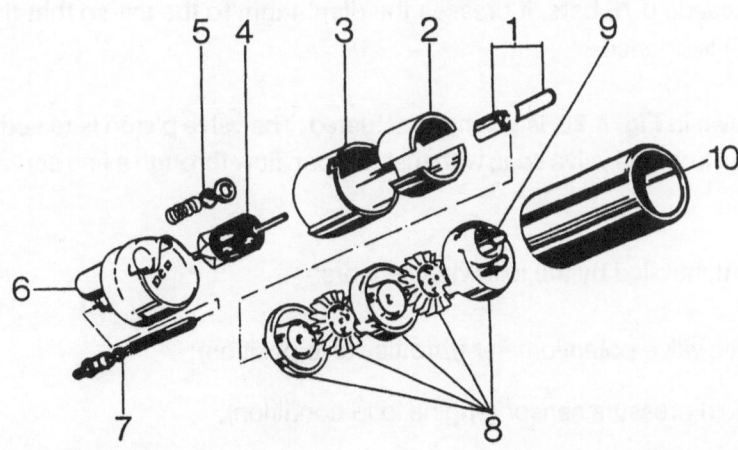

1	Attachment of permanent magnets	6	Delivery outlet, pressure side
2	Permanent magnets	7	Carbon brushes
3	Support of permanent magnets	8	Turbine wheels and seats
4	Inductor	9	Delivery inlet, suction side
5	Return valve	10	Pump housing

Fig. 4.83. Fuel pump in the Opel Multec central injection unit [4.38]

Fig. 4.84 shows the fuel pressure regulator. Here the fuel flows through fuel inlet 1 to injector 2 and fuel pressure regulator 3 which regulates the fuel pressure to some 0.75 bars upstream of injector 2.

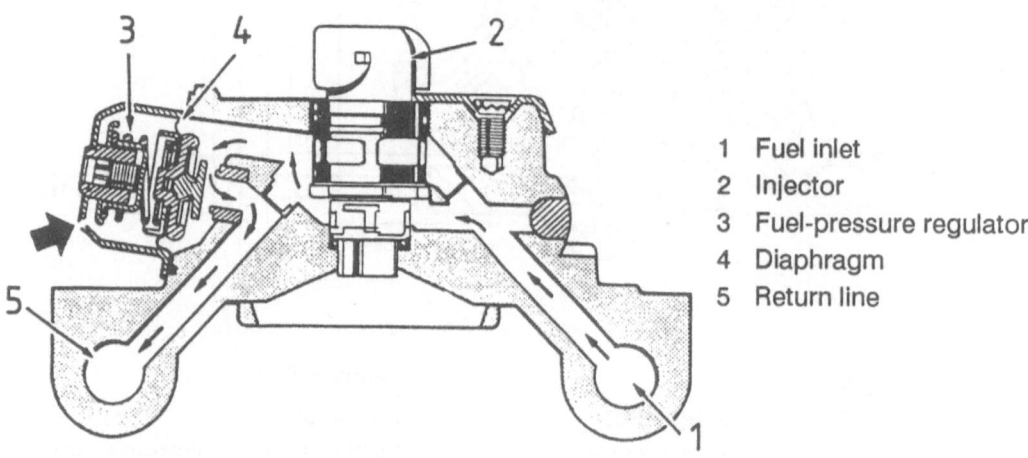

1	Fuel inlet
2	Injector
3	Fuel-pressure regulator
4	Diaphragm
5	Return line

Fig. 4.84. Fuel pressure regulator in an Opel Multec central injection unit [4.38]

The fuel pressure regulator consists of a spring chamber and fuel chamber. The two chambers are separated by a diaphragm (4). On its left side, the diaphragm is tensioned by the atmospheric pressure and pressure spring, while the fuel pressure applies on its right side. When the fuel pressure exceeds 0.75 bars, it presses the diaphragm to the left so that the relief valve to the return line (5) is opened.

The injector, shown in **Fig. 4.85**, is solenoid-actuated. The valve piston is raised so that the pretensioned ball lifts from the valve seat; with that fuel can flow through a fine screen into the spray diffuser.

Fuel metering is controlled by the following sensors:

- Throttle valve potentiometer (throttle valve position);

- Manifold pressure sensor (engine load condition);

- Temperature sensor (coolant);

- Odometer frequency transmitter (to inform the idle speed control whether or not the vehicle moves and to calculate the vehicle speed);

- Lambda sensor, unheated ($\lambda = 1$ for three-way catalytic converter);

- Ignition distributor (speed reference signals).

1 Magnetic pole
2 Lifting cylinder and stop
3 Magnetic coil
4 Housing of magnetic coil
5 Vapor lock filter
6 Inductor ring
7 Inductor
8 Fuel inlet filter
9 Valve housing
10 Diffuser
11 Director
12 Valve seat
13 Valve spring seat
14 Valve inductor spring
15 Guide ring
16 Setscrew, in fixed position
17 Valve piston

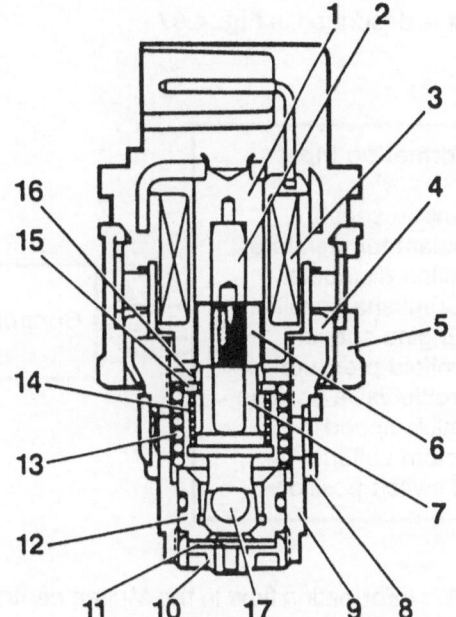

Fig. 4.85. Injector [4.38]

The control unit shown in **Fig. 4.86** processes all input signals, comparing them with the recorded data and selecting the actuators.

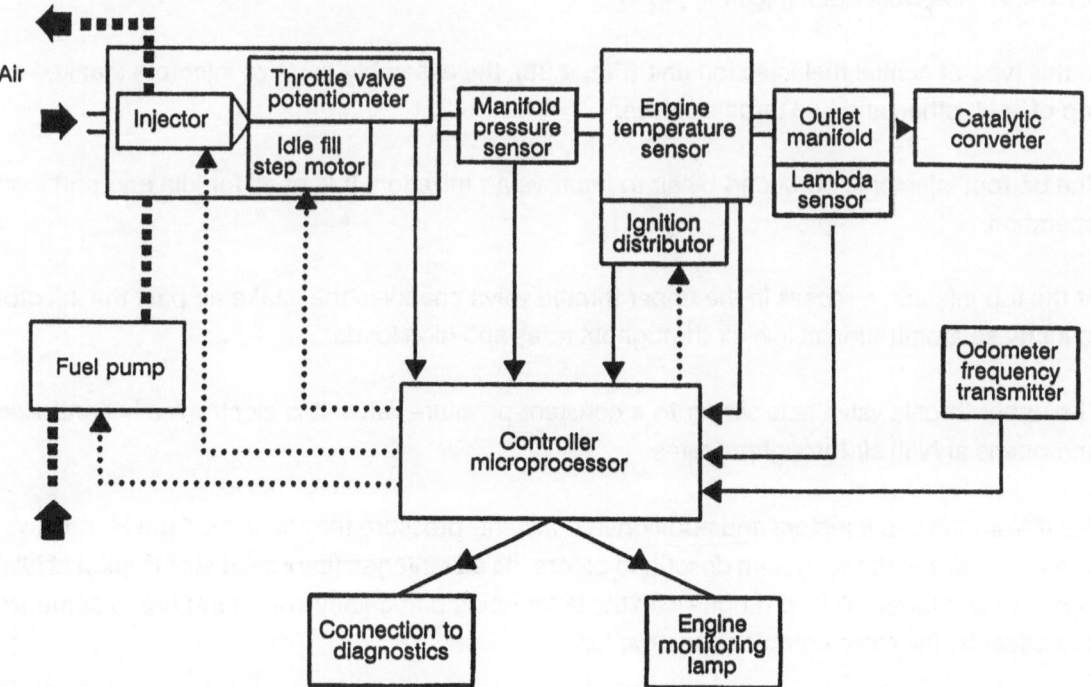

Fig. 4.86. Diagram of the control unit [4.38]

The information flow within closed-loop control, i.e. a ratio of $\lambda = 1$, in the Multec central injection unit is described in **Fig. 4.87**.

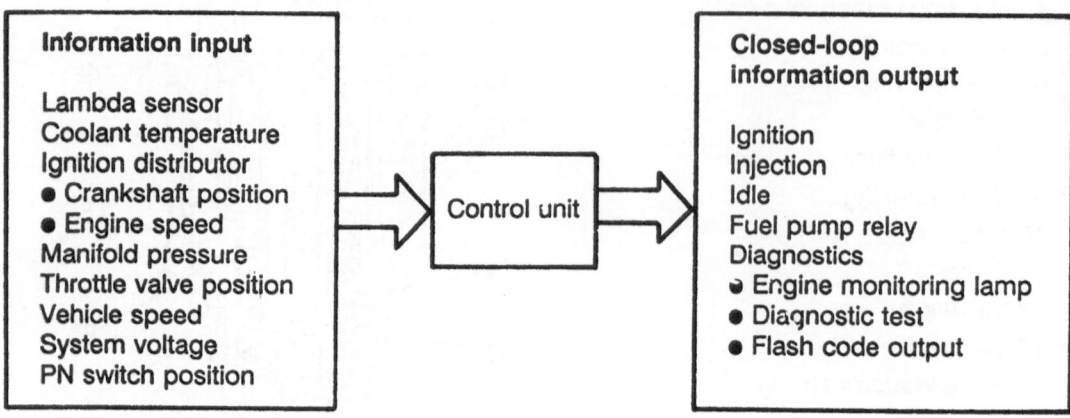

Fig. 4.87. Information flow in the Multec central injection unit [4.38]

When the loop is open, the lambda sensor does not provide any information, i.e. the preset value of $\lambda = 1$ is not polled. This is the case in the cold phase up to a coolant temperature of 20 °C.

4.1.2.4. Honda Dual-Point Injection [4.24]

In this type of central fuel injection unit (**Fig. 4.88**), the assembly has two injectors stacked on top of each other, and two throttle valves.

The bottom injector is shrouded by air to improve atomization. It is used for idle and part load operation.

At the top injector, a recess in the upper throttle valve channels the intake air past the injector to increase atomization at low air throughput rates and high loads.

The upper throttle valve acts similar to a constant-pressure valve. It is electronically controlled and opens at high air throughput rates.

Apart from the two injectors and additional "constant- pressure throttle valve," the Honda system works similar to the system described before. Its advantages (improved atomization at idle, part load and lower full load range - with the latter being particularly important) are to some extent offset by the more complex construction.

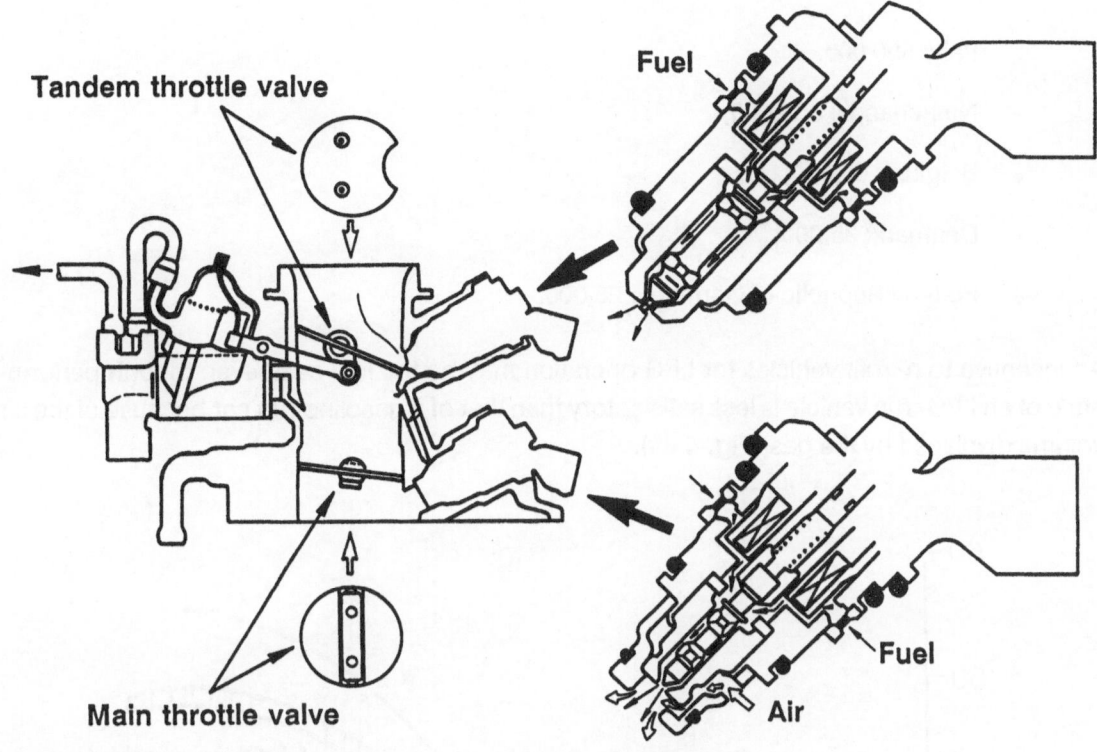

Fig. 4.88. Honda dual-point injection system [4.24]

4.1.3. LPG Mixer Units

Engines run with liquid petroleum gas, i.e. a mixture of propane and butane, require LPG mixer units with vaporizers to prepare the fuel mixture.

LPG liquifies at ambient temperature and low pressure (4-12 bars), so that it can be filled and carried in the vehicle in its liquified state. For a detailed description of its properties see Chapter 2.2.3.3.

Disadvantages of LPG operation are the relatively large high- pressure container which is usually retrofitted in the trunk space and the risk of explosion existing even when a small leakage occurs in a closed space.

Vehicles are usually fitted to run with either gasoline or LPG as the network of LPG filling stations may not yet be very dense in some countries. Thus, 20% of the service stations in the Netherlands or Denmark sell LPG, while other countries offer few or no such stations.

A breakdown of LPG-operated vehicles by the end of 1981:

- Italy: 550,000;

- Netherlands: 380,000;

- Belgium: 60,000;

- Denmark: 35,000;

- Federal Republic of Germany: 23,000.

An incentive to retrofit vehicles for LPG operation may be the low cost, even though performance of an LPG- run vehicle is less satisfactory than that of a gasoline-run car because of the air volume displaced by the gas (**Fig. 4.89**).

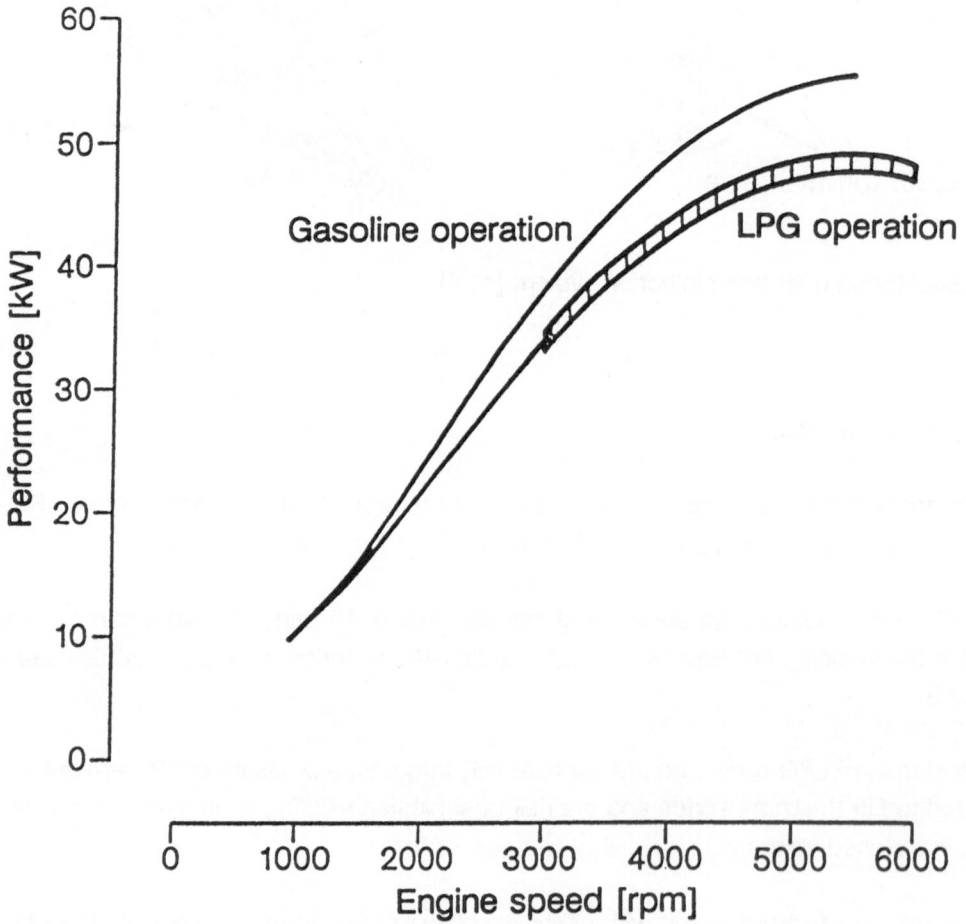

Fig. 4.89. Performance comparison between gasoline and LPG operation at full load [4.25]

An LPG unit that runs on LPG and gasoline alternatively consists of the following main components (**Fig. 4.90**):

- LPG tank;

- Gasoline tank;

- Shut-off valves for LPG and gasoline;

- LPG vaporizer pressure regulator;

- Carburetor and LPG mixer unit.

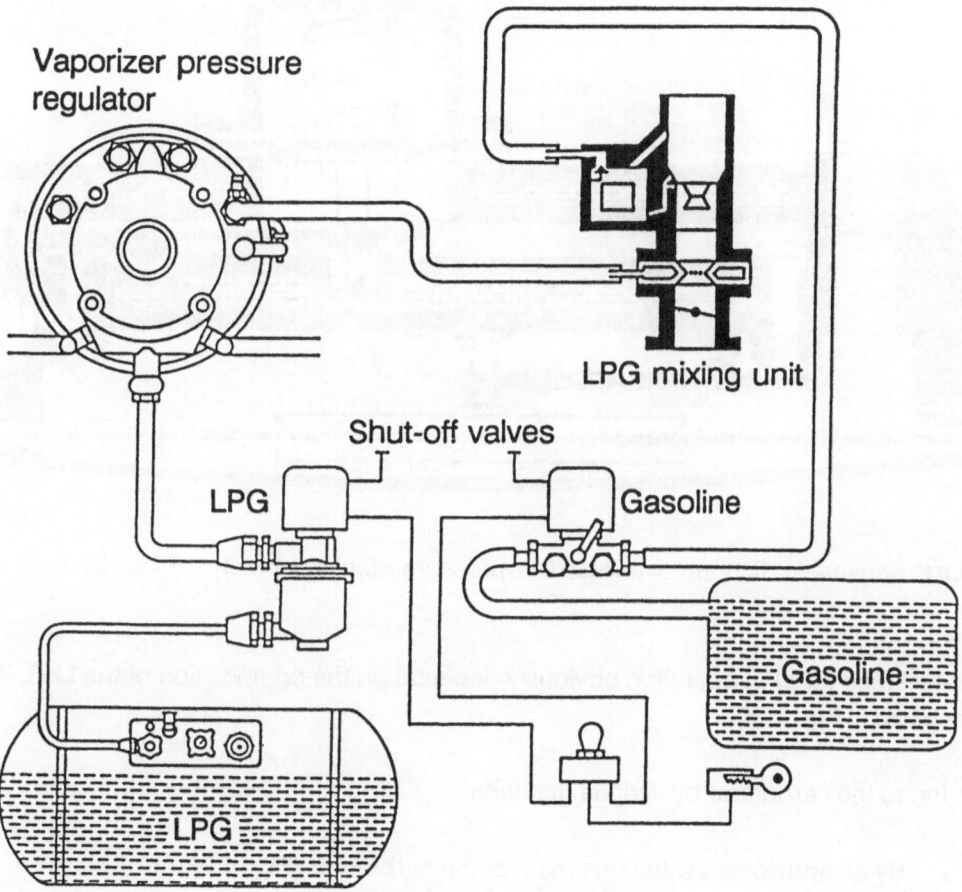

Fig. 4.90. Components of a dual unit [4.25]

The LPG flows under its own pressure from the high-pressure tank via the filter shut-off valve into the vaporizer- pressure regulator. Here the liquid gas is vaporized and its pressure reduced. The heat necessary to do this is supplied by the engine cooling circuit. The gas flows into the mixer unit via a gas hose where it is mixed with air to form a homogeneous gas/air mixture.

Depending on the type, the mixer unit may be installed either upstream or downstream of the gasoline carburetor. When the ignition is switched off, either the solenoid LPG shutoff valve closes during LPG operation, or the gasoline shutoff valve between the gasoline pump and the carburetor closes when the engine is running on gasoline.

The driver selects either LPG or gasoline operation via a selector switch mounted on the instrument panel.

A vaporizer-pressure regulator is illustrated in **Fig. 4.91**.

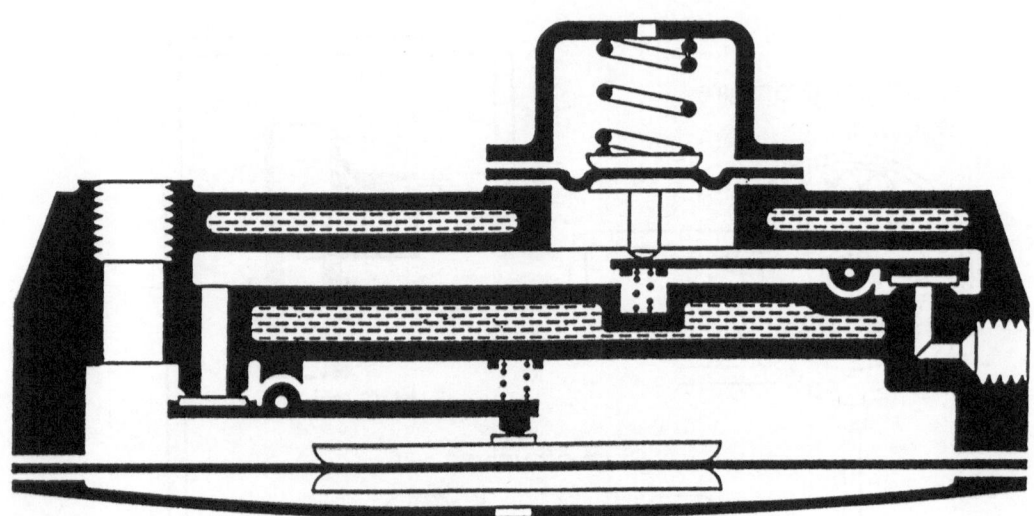

Fig. 4.91. Schematic diagram of a vaporizer-pressure regulator [4.25]

The air-fuel ratio for LPG operation obviously depends on the composition of the LPG. The ratios are given in **Fig. 4.92**.

LPG is fed to the carburetor by various methods:

- By an additional venturi upstream of the carburetor (**Fig. 4.93**);

- In the carburetor air funnel (**Fig. 4.94**);

- At the undercut air funnel (**Fig. 4.95**);

- Downstream of the air funnel (**Fig. 4.96**);

- Downstream of the throttle valve (**Fig. 4.97**).

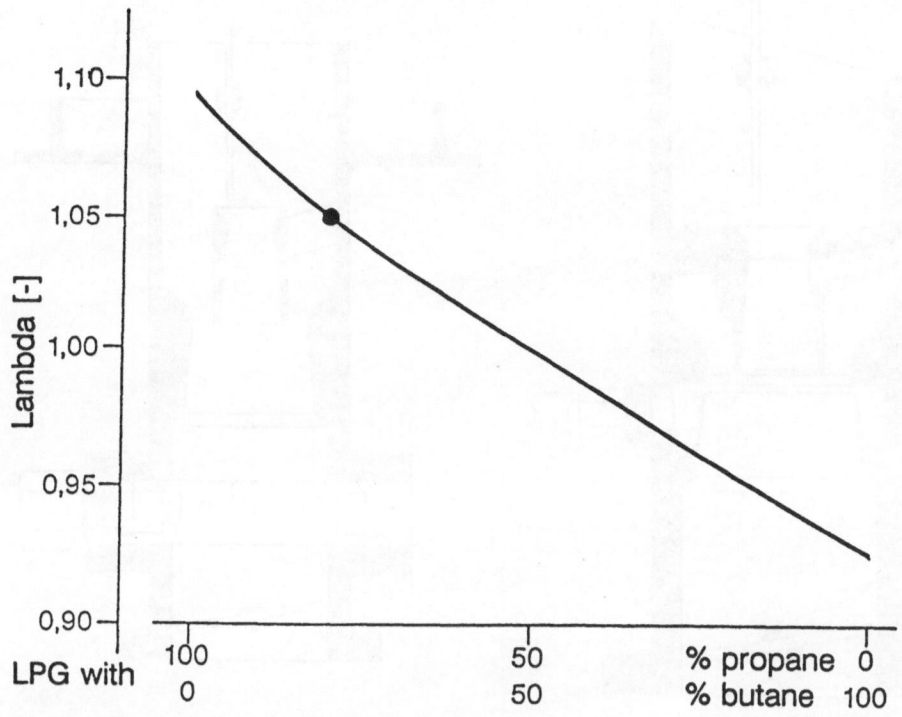

Fig. 4.92. Air-fuel ratio as a function of gas composition [4.25]

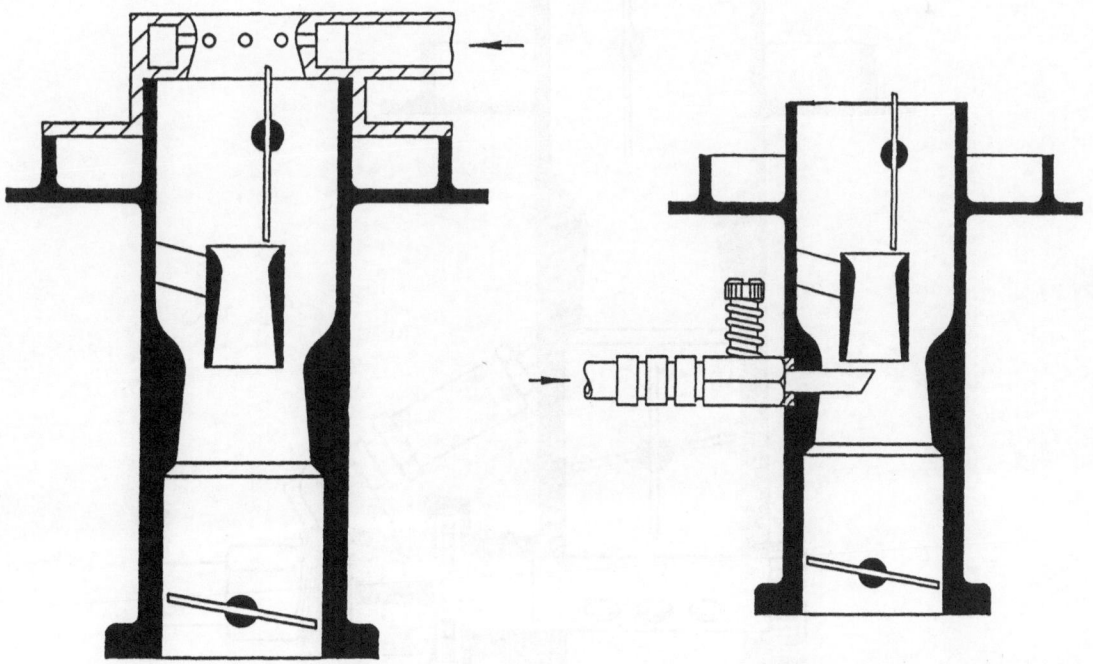

Fig. 4.93. LPG supply at an additional venturi upstream of the carburetor
[4.25]

Fig. 4.94. LPG supply by utilizing the carburetor air funnel [4.25]

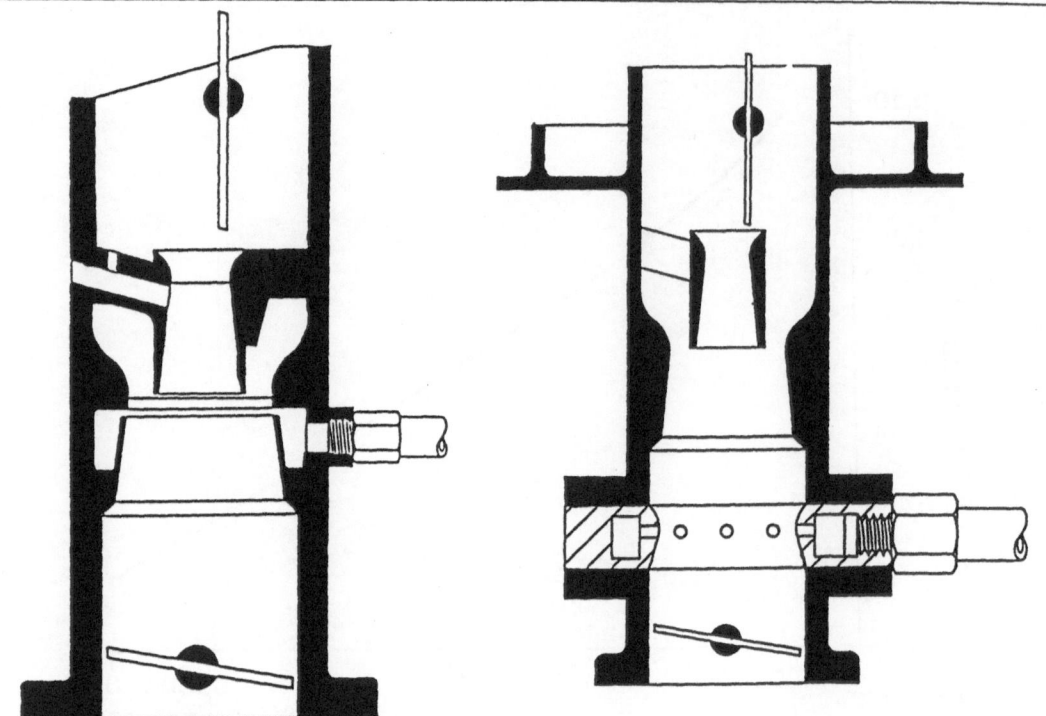

Fig. 4.95. LPG supply at the undercut air funnel [4.25]

Fig. 4.96. LPG supply at the additional venturi downstream of the carburetor air funnel [4.25]

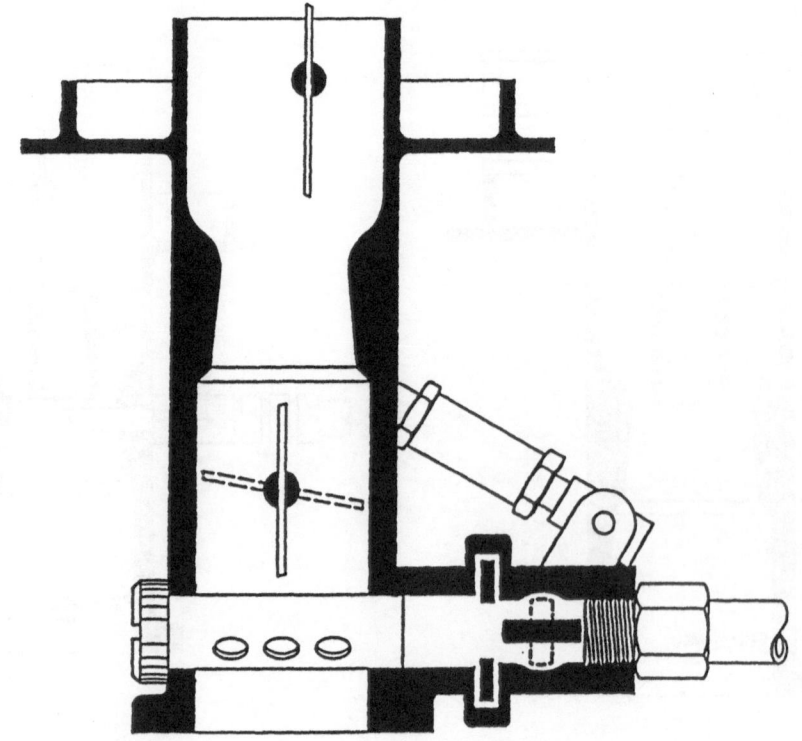

Fig. 4.97. LPG supply downstream of the throttle valve (Century system) [4.25]

Regardless of the method, it is important to ensure that carburetor pressure rates during gasoline operations are not affected, i.e. that the characteristic carburetor parameters are not changed.

Fig. 4.98 outlines LPG supply in an injection engine.

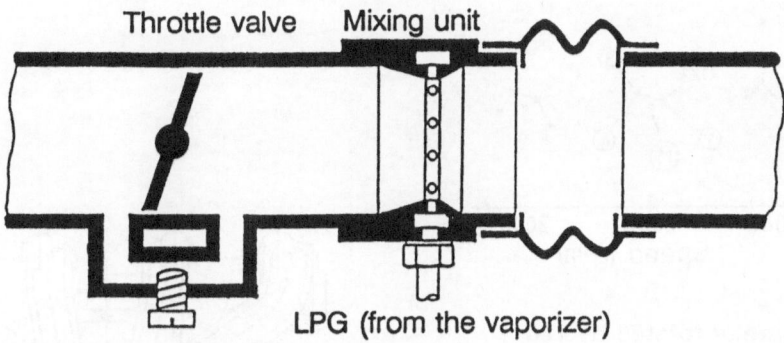

Fig. 4.98. LPG supply in an injection engine [4.25]

4.1.4. The Geometric Design of Mixture Formation Systems and Its Effect on Mixture Distribution

4.1.4.1. General

Mixture formation systems have two main tasks: mixture preparation and metering. As already described in detail in Chapter 2, the design of the mixture formation system thus constantly and indirectly affects the mixture distribution and formation of a wall film in the downstream manifold. Generally it may be said that intensive atomization and vaporization of the fuel and its complete mixture with the air reduce the risk of droplet deposits at the manifold wall, which in turn facilitates even distribution of the mixture to the cylinders.

In addition to this indirect effect of the design on mixture distribution we also find a direct influence, which is often neglected in practice. Downdraft systems in particular may display significant maldistributions caused by the design of the mixture formation system [4.59, 4.60, 4.61, 4.65, 4.66, 4.67, 4.26, 4.27, 4.28, 4.29, 4.30, 4.31, 4.32, 4.33, 4.34, 4.35, 4.36], which usually cannot be rectified by the manifold design. This includes distribution biased towards one half of the manifold in symmetrically designed manifolds, which frequently exceeds symmetric maldistribution in favor of either the internal or external pair of cylinders.

Where the mixture formation system flange is symmetric, design-related maldistributions can most easily be detected by rotating the manifold-fitted system by 180 degrees (**Fig. 4.99**). While the mixture distribution cannot be expected to be mirror-reversed, in view of the different inflow conditions, the effect on the distribution is usually very well verified [4.26, 4.28, 4.31, 4.59, 4.60, 4.61]. Inbalances in the mixture distribution are expressed by deviation Δκ (single cylinder air-fuel equivalence ratio κ - reciprocal value of the air-fuel ratio λ) from the mean equivalence ratio of all cylinders; cf. 5.1.

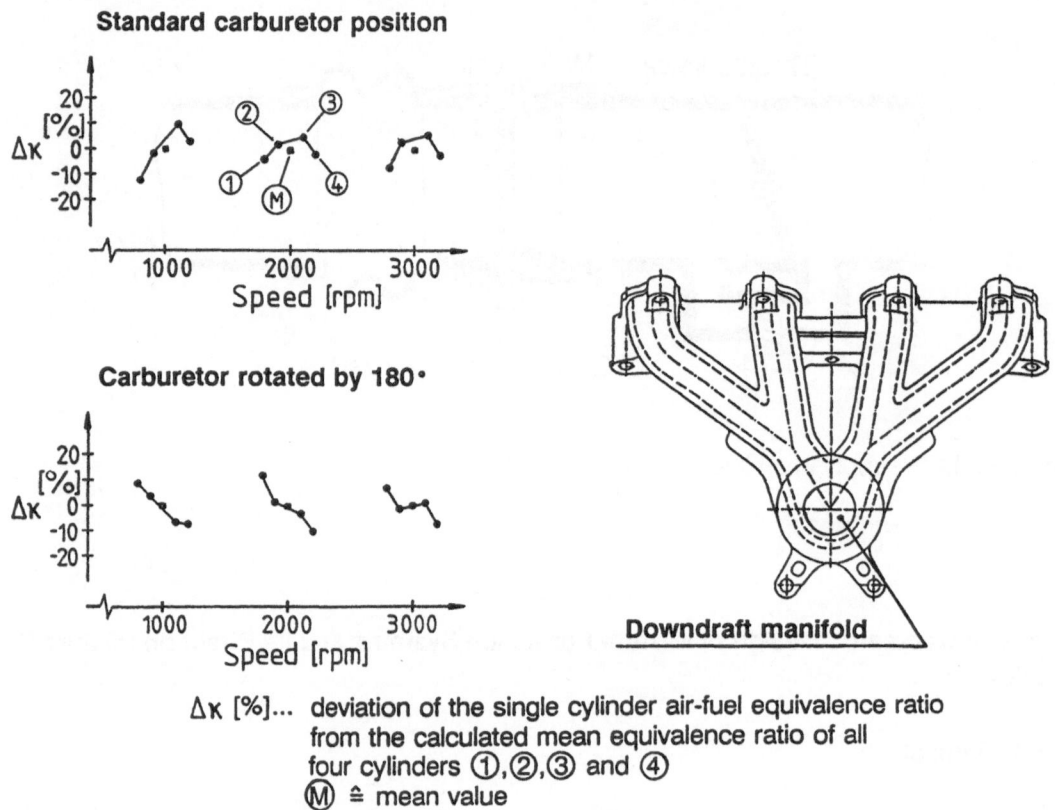

Δκ [%]... deviation of the single cylinder air-fuel equivalence ratio from the calculated mean equivalence ratio of all four cylinders ①,②,③ and ④
Ⓜ ≙ mean value

Fig. 4.99. Position of a single-flow downdraft carburetor and its effect on full load mixture distribution [4.59]

Below we will discuss the components of the mixture formation system which have the greatest influence on mixture distribution.

4.1.4.2. The Throttle Valve

The position of the throttle valve usually does not affect mixture distribution in the lower part load range since the fuel is effectively atomized while the throttle valve is still virtually closed.

The gradual opening of the throttle valve brings about a decline in the atomizing effect, and we find that - from an opening angle of about 30° - the throttle valve exerts a directional effect. Secondary atomization increasingly changes into deflection of the fuel droplets, resulting in a higher rate of wall deposits immediately downstream of the throttle valve [4.37]. Where the throttle valve axis runs in parallel instead of at a right angle to the symmetric manifold axis, the result, in addition to more wall deposits, is a heavy bias in the mixture distribution between the two manifold halves, depending on the throttle valve opening angle. The eddies generated at the back of the throttle valve will effectively prevent the asymmetric current from being homogenized even if some distance is provided for steadying the flow. Accordingly, the throttle valve axis must always be arranged at a right angle to the symmetric manifold axis (in parallel to the longitudinal engine axis). The same applies to the initial throttle valve, if provided.

Another factor to take into account with regard to throttle valves is that its axial play should be reduced as much as possible. Studies [4.35] have found that a play of 0.2 mm may already suffice to cause air-fuel ratio deviations ($\Delta\lambda$) of up to ±15% in the individual cylinders.

In spite of the negative effect that the throttle valve exerts on mixture distribution in the higher part load range, it is not advisable to replace the central throttle valve by individual throttle valves located at the manifold outlet. As indicated in Fig. 4.100, such an arrangement eliminates secondary atomization at the manifold inlet and causes an unacceptable deterioration in the overall part load mixture distribution plus a considerable increase in the quantity of wall film deposited in the manifold (high absolute manifold pressure!). The findings given in **Fig. 4.100** have since been confirmed by other manufacturers of manifolds.

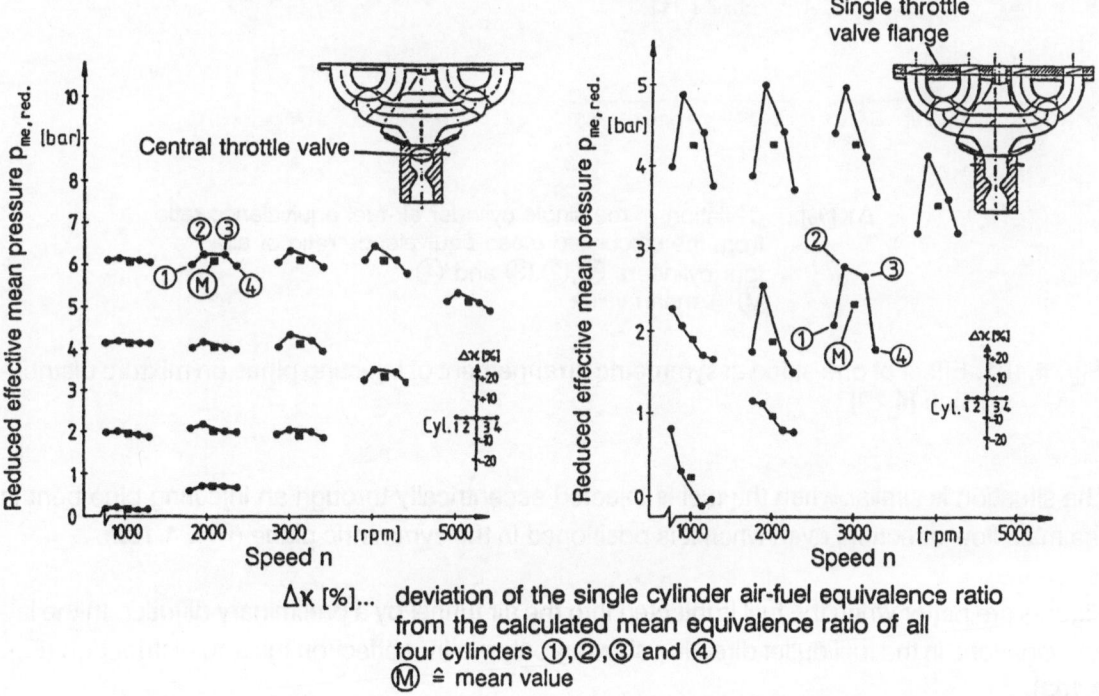

$\Delta\kappa$ [%]... deviation of the single cylinder air-fuel equivalence ratio
from the calculated mean equivalence ratio of all
four cylinders ①,②,③ and ④
Ⓜ ≙ mean value

Fig. 4.100. Comparison of part load mixture distributions occurring at different throttle valve locations in the same heated horizontal draft manifold [4.59]

4.1.4.3. The Mixing Chamber

The shape and design of the mixing chamber are of crucial importance for the quality of mixture distribution. Even the most minor asymmetric deviation may already cause major maldistribution biased towards the left or right half of the manifold.

Asymmetric feeding of the fuel into the air funnel (**Fig. 4.101**) may result in deviations of up to 40% in the single-cylinder equivalence ratios $\Delta\kappa$ between the left and right halves of the manifold, which cannot be corrected by any design measures in the manifold itself.

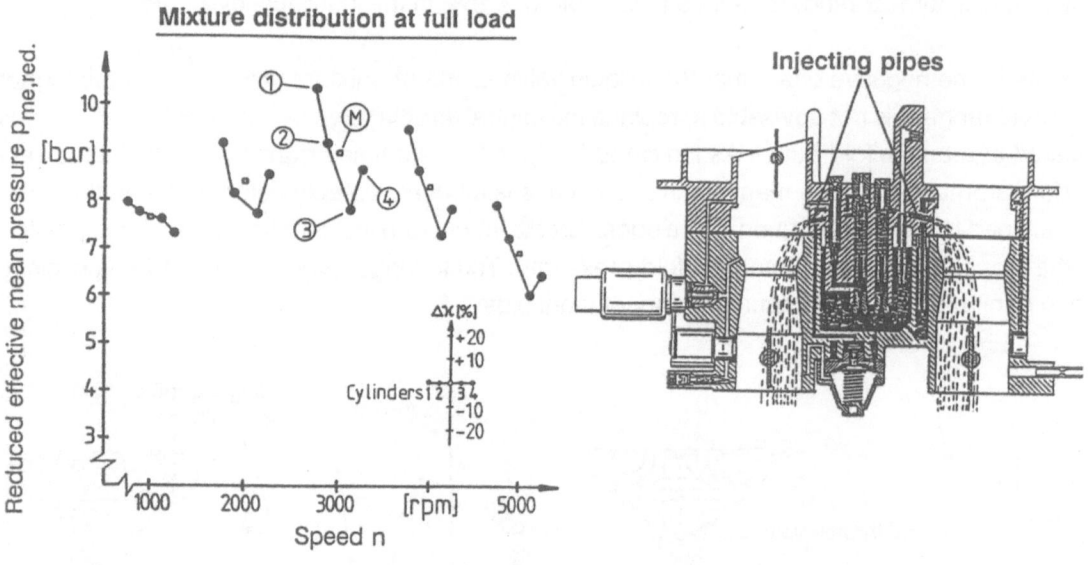

$\Delta\kappa$ [%]... deviation of the single cylinder air-fuel equivalence ratio
from the calculated mean equivalence ratio of all
four cylinders ①,②,③ and ④
Ⓜ $\hat{=}$ mean value

Fig. 4.101. Effect of one-sided assymmetric arrangement of injecting pipes on mixture distribution [4.29]

The situation is similar when the fuel is injected eccentrically through an injecting pipe bent in the main flow direction, even when it is positioned in the symmetric plane (**Fig. 4.102**).

Figures are better when the fuel is injected into the air funnel by a preliminary diffuser. In the latter, variations in the fuel outlet direction showed only a minor effect on mixture distribution (**Fig. 4.103**).

But even with this design, the slightest off-center arrangement of the preliminary diffuser in the air funnel will result in significantly biased fuel distribution in the manifold (**Fig. 4.104**).

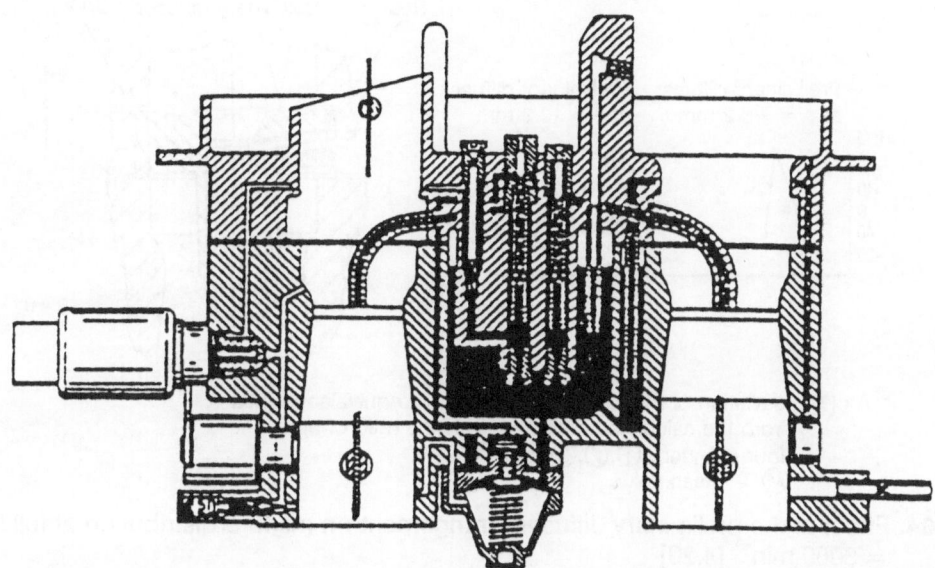

Fig. 4.102. Compound carburetor with injecting pipes injecting fuel in the main flow direction [4.29]

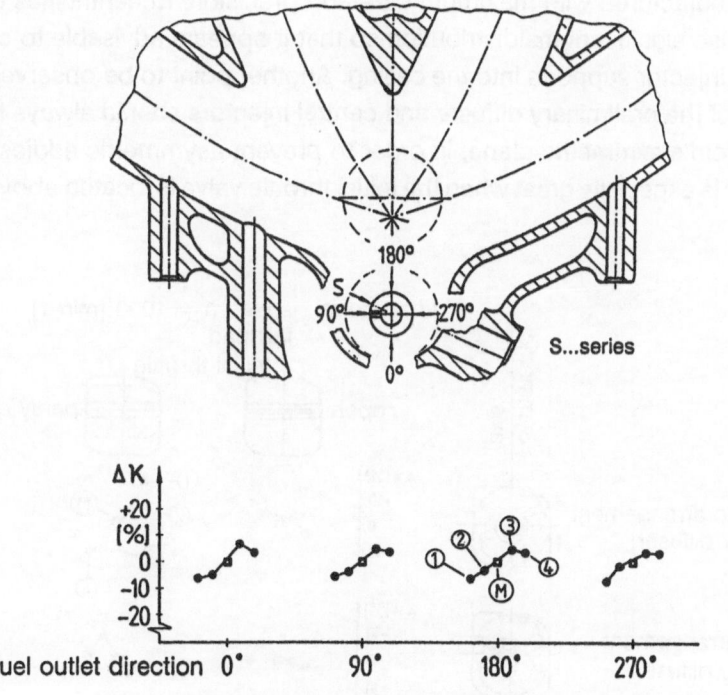

ΔK [%]... deviation of the single cylinder air-fuel equivalence ratio
from the calculated mean equivalence ratio of all
four cylinders ①,②,③ and ④
Ⓜ ≙ mean value

Fig. 4.103. Effect of the fuel outlet direction in the preliminary diffuser on mixture distribution, part load point n = 3000 min-1, p_{me} = 6 bars (first stage only opened) [4.59]

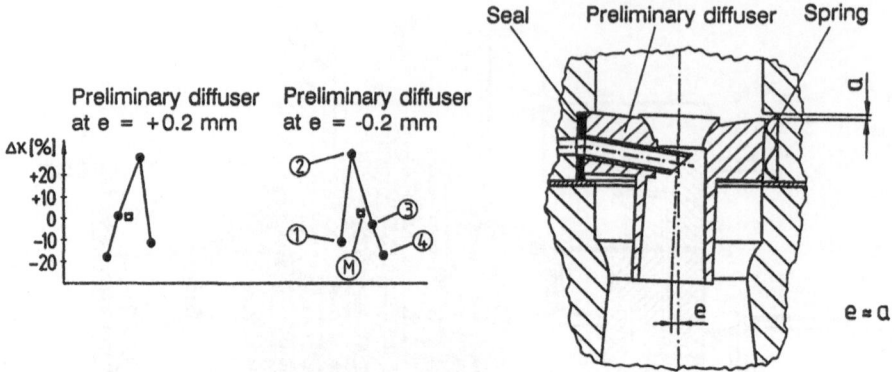

$\Delta\kappa$ [%]...deviation of the single cylinder air-fuel equivalence ratio
from the calculated mean equivalence ratio of all
four cylinders ①,②,③ and ④
Ⓜ ≙ mean value

Fig. 4.104. Effect of the preliminary diffuser arrangement on mixture distribution at full load, n = 3000 min⁻¹ [4.29]

Such dependencies between the geometric design of the mixing chamber and mixture distribution make it obvious that the mixing chamber in a single-point mixture formation system should be designed and manufactured with maximum care and precision. Eccentricities of 0.1 to 0.2 mm may already cause significant maldistribution so that it appears advisable to cast the preliminary diffuser and injector supports into the casing. Another point to be observed is that the supports (brackets) of the preliminary diffuser and central injectors should always be arranged in line with the manifold's symmetric plane, in order to prevent asymmetric eddies from being generated, a risk that is especially great when the initial throttle valve is located above and partly closed (**Fig. 4.105**).

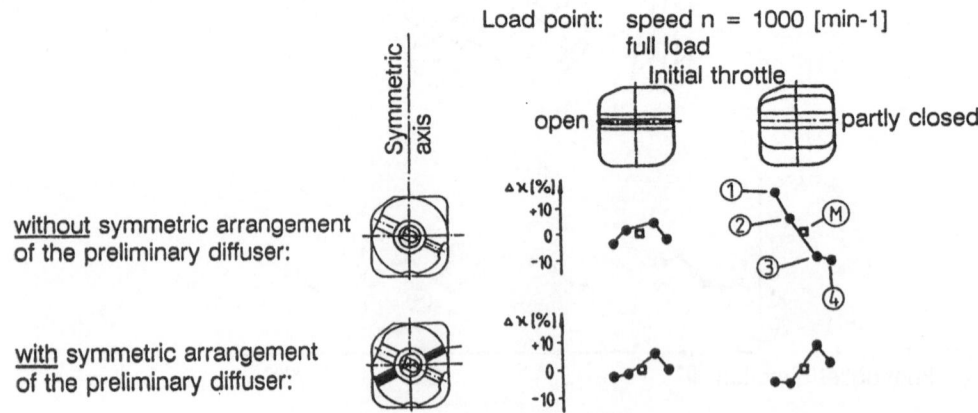

$\Delta\kappa$ [%]...deviation of the single cylinder air-fuel equivalence ratio
from the calculated mean equivalence ratio of all
four cylinders ①,②,③ and ④
Ⓜ ≙ mean value

Fig. 4.105. Effect of symmetric positioning of the preliminary diffuser support on mixture distribution, with the initial throttle open and partly closed [4.29]

The ridge design of the injector support in central fuel injection units has a similar - albeit slightly weaker - effect on mixture distribution. When the ridge is designed rectangularly (**Fig. 4.106**), its unfavorable flow characteristics cause a vacuum to be generated downstream of the ridge at high air throughputs. The fuel-air mixture is deflected in this direction which may cause unequal distribution of the fuel between the internal and external pairs of cylinders.

In single-point injection units, mixture distribution is easily affected by a flawed rotation symmetry below the injector. A slight slant (±0.5%) of the air guide fitted to the injector support may already cause considerable variations in fuel distribution (**Fig. 4.107**), a situation that shows a clear analogy to fuel maldistributions caused by the eccentric arrangement of the preliminary diffuser.

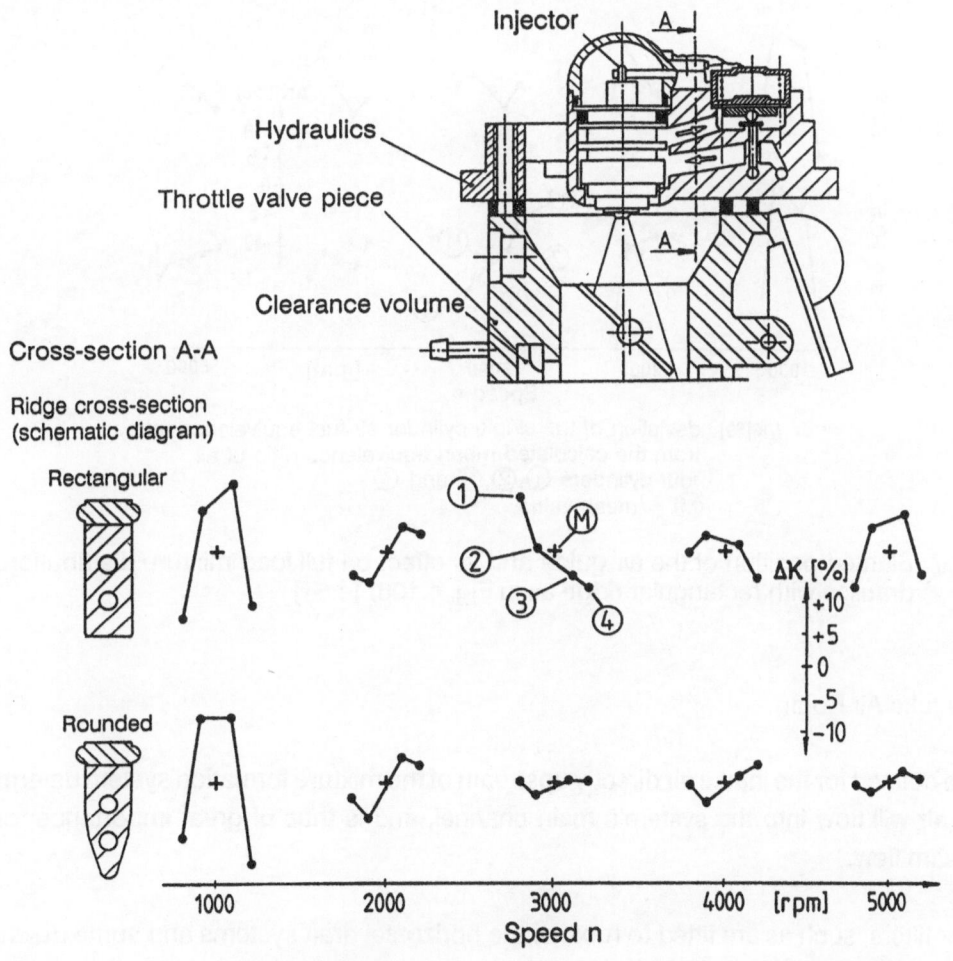

ΔK [%]...deviation of the single cylinder air-fuel equivalence ratio
from the calculated mean equivalence ratio of all
four cylinders ①,②,③ and ④
Ⓜ ≙ mean value

Fig. 4.106. Rectangular and rounded ridge in single-point injection units and its effect on mixture distribution at full load (λ ~ 0.85) [4.63]

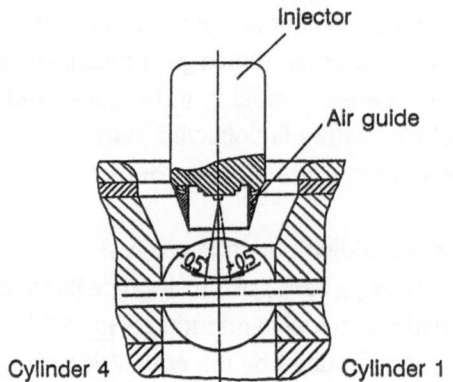

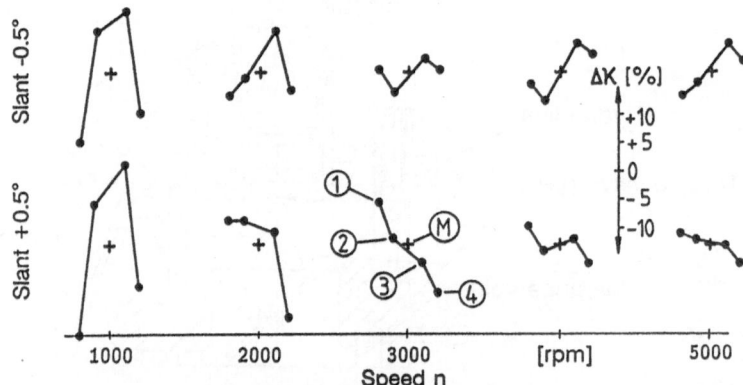

ΔK[%]...deviation of the single cylinder air-fuel equivalence ratio
from the calculated mean equivalence ratio of all
four cylinders ①,②,③ and ④
Ⓜ = mean value

Fig. 4.107. Slanted position of the air guide and its effect on full load mixture distribution (hydraulics with rectangular ridge as in Fig. 4.106) [4.64]

4.1.4.4. Intake Air Route

The route defined for the intake air directly upstream of the mixture formation system determines how the air will flow into the system's main channel, and is thus of great importance for the downstream flow.

Lateral air filters, such as are fitted to most of the horizontal draft systems and some downdraft systems, should be designed to ensure that the air is channeled in line with the symmetric manifold plane. Abrupt deflections directly upstream of the mixture formation system should be avoided in order to prevent secondary currents from forming. Air baffle plates may be used to direct the air flow.

Where the air filter is placed directly on the mixture formation system, mixture distribution may depend very much on the flow conditions of the intake air in the air filter body, in particular with single-flow systems (large diameter of the main channel). As indicated in **Fig. 4.108**, baffle plates may once again prove to be a remedy.

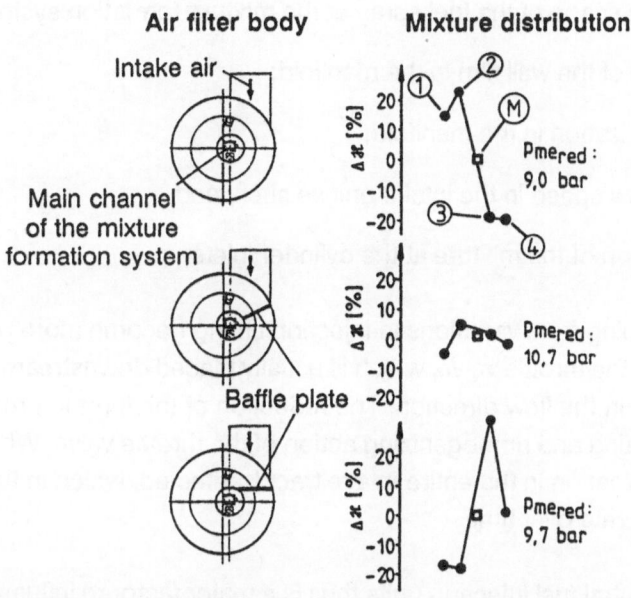

$\Delta\kappa$ [%]...deviation of the single cylinder air-fuel equivalence ratio
from the calculated mean equivalence ratio of all
four cylinders ①,②,③ and ④
Ⓜ ≙ mean value

Fig. 4.108. The effect of the intake air route in the air filter body on mixture distribution (central fuel injection system) at 4000 min⁻¹/full load [4.30]

4.1.4.5. Injection Timing and Its Effect on Mixture Distribution

In single-cylinder injection units, injection timing primarily affects the quality of the mixture entering the cylinder and of the fuel in the manifold wall film. In central fuel injection units on the other hand, injection timing is important for fuel distribution to the engine cylinders [4.68, 4.69, 4.70 and 4.71].

As established by studies of mixture distribution in carburetors, flow variations within the mixing chamber definitely affect fuel distribution in the manifold, particularly during full load operation. The flow conditions in the mixture formation system and manifold are, however, subject to drastic change during intake, so that injection timing obviously plays a key role with regard to mixture distribution in intermittent single-point fuel metering. Variations in injection timing during intake affect the following factors:

- Air mass flow rate during injection, and with it

- Flow-velocity dependent atomization quality of the injection system;

- Shape of the flow in the mixing chamber, and with it

- Geometric shape of the fuel spray at the mixture formation system outlet;

- Formation of the wall film in the manifold;

- Fuel vaporization in the manifold;

- Mixture flow speed in the intake unit as such, and

- Composition of the mixture at the cylinder inlets.

Such changes stemming from variations in injection timing become more pronounced with the degree of opening of the throttle valve, which is usually placed downstream of the fuel metering system when viewed in the flow direction. The reduction of the throttle cross-section activates the secondary atomizing and homogenizing action of the throttle valve. When the throttle valve closes entirely, air pulsation in the entire intake tract is affected, which in turn offsets the variations in air mass flow rate over time.

Injection timing in central fuel injection units thus is a major factor to influence mixture distribution and wall film formation in the manifold, especially when the engine runs in the upper part load and full load mode.

There are some other points to be taken into account with regard to the influence of injection timing on mixture distribution and the consequent requirements for central intermittent fuel injection units:

Mixture distribution at full load strongly depends on the air mass flow rate, as is shown in **Fig. 4.109**. Here, a ram pipe was used to shift the phases of air mass flow rate over time in the mixture formation system at the same load point. The mixture distributions found for the different injection times typically reflect the phase shift in the air mass flow rate. In summary we may note:

- Injection frequency equal to the ignition frequency or a multiple of it, and identical injection timing for all cylinders produce uniform fuel distribution, in particular during full load operation.

- The fuel should be injected at high air mass flow rates to ensure good atomization and fast flow. The start of injection should thus be shifted as a function of the speed, at least during full load operation.

- Optimum injection timing for each load point must be determined by tests in combination with the relevant manifold.

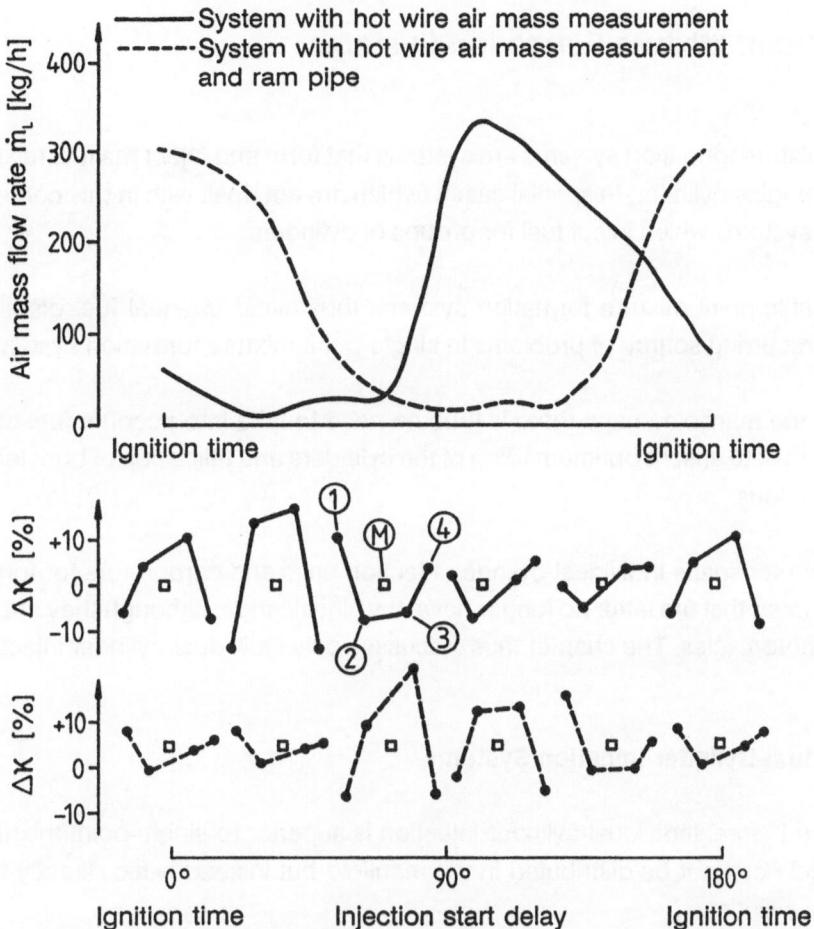

ΔK[%]...deviation of the single cylinder air-fuel equivalence ratio
from the calculated mean equivalence ratio of all
four cylinders ①,②,③ and ④
Ⓜ = mean value

Fig. 4.109. Change in the air mass flow rate at the mixture formation system inlet and its effect
on full load mixture distribution in an engine with four cylinders in line and a speed
of 2000 min-1 [4.68]

4.2. Multi-Point Mixture Formation Systems

Multi-point mixture formation systems are systems that form and inject the fuel mixture individually for each engine cylinder. In special cases (which are not dealt with in this context) they may also include systems which inject fuel for groups of cylinders.

In general, multi-point mixture formation systems thus avoid unequal fuel distribution in the manifold - a recurring source of problems in single-point mixture formation systems.

In designing the manifold shape there is thus no need to take into account fuel distribution so that it can be built to ensure optimum filling of the cylinders and utilization of booster effects from manifold vibrations.

While this chapter spans individual-cylinder injection units and carburetors for formal reasons, it should be noted that the latter no longer have any significance, although they are on occasion still used in motorcycles. The chapter thus discusses only individual-cylinder injection systems.

4.2.1. Individual-Cylinder Injection Systems

As mentioned before, individual-cylinder injection is superior to single-point mixture formation in that the fuel need not be distributed in the manifold but instead is fed directly to the appropriate engine cylinder.

Accordingly, it is possible to design the manifold for optimum cylinder filling. This individual type of injection is the classical type of injection system.

In World War II, aircraft engines were usually fitted with injection systems because their performance with regard to bearing stability and icing was superior to that of carburetors. After the war, injection systems were gradually accepted in automotive engines as well.

4.2.1.1. A History of Fuel Injection Systems

Car engines were first equipped with high-pressure direct injection systems which injected the fuel into the cylinder during the intake stroke at a pressure of 50 to 100 bars.

In the two-cycle engine, the fuel must be injected directly into the cylinder to avoid flushing losses. This is not necessary in the four-cycle engine, a fact which soon led to the development of indirect mechanical manifold injection systems.

Manifold injection systems offered several benefits:

- Simple and inexpensive arrangement of pumps and nozzles, since pressure requirements are low and the nozzles are not exposed to combustion chamber temperatures;

- The oil film in the cylinder is not washed off by the injected fuel;

- Simple installation of the injection nozzles in the engine.

Today, variations of manifold injection systems are standard features in cars while R&D work is being performed with regard to direct injection methods.

After some pilot tests involving direct cylinder injection in the aftermath of World War II, mechanical manifold injection was soon introduced into serial manufacturing of high-power cars. This was an intermittent type of mechanical injection (**Fig. 4.110**) where each engine cylinder was provided with an injection pump element. The injection pump itself was an in-line reciprocating pump, a variation of the diesel injection pump shown in **Fig. 4.111**.

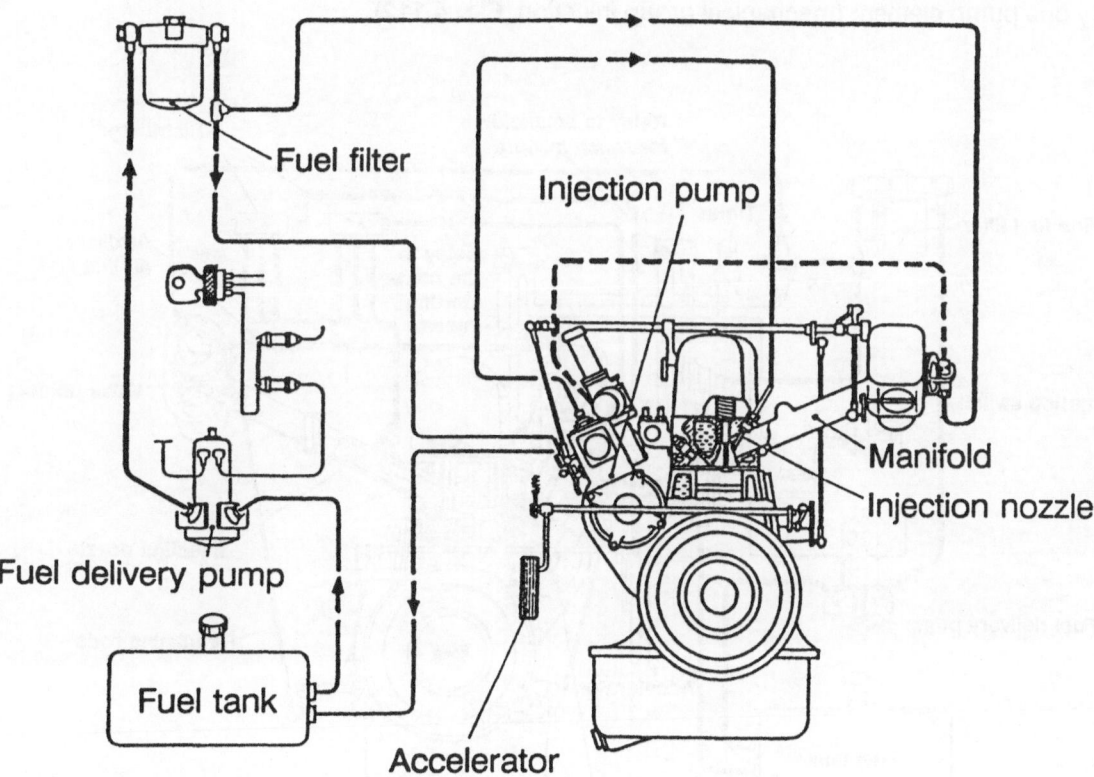

Fig. 4.110. Intermittent mechanical Bosch manifold injection system for Daimler-Benz

Injection line

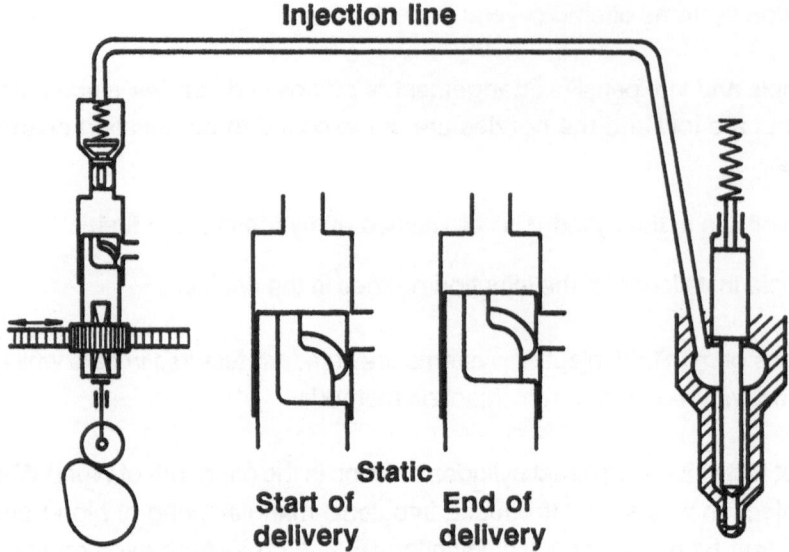

Static
Start of   **End of**
delivery   **delivery**

Fig. 4.111. Bosch in-line diesel injection pump [4.43]

For cost-saving reasons, designers occasionally provided for several cylinders being supplied by one pump element (mechanical group injection, **Fig. 4.112**).

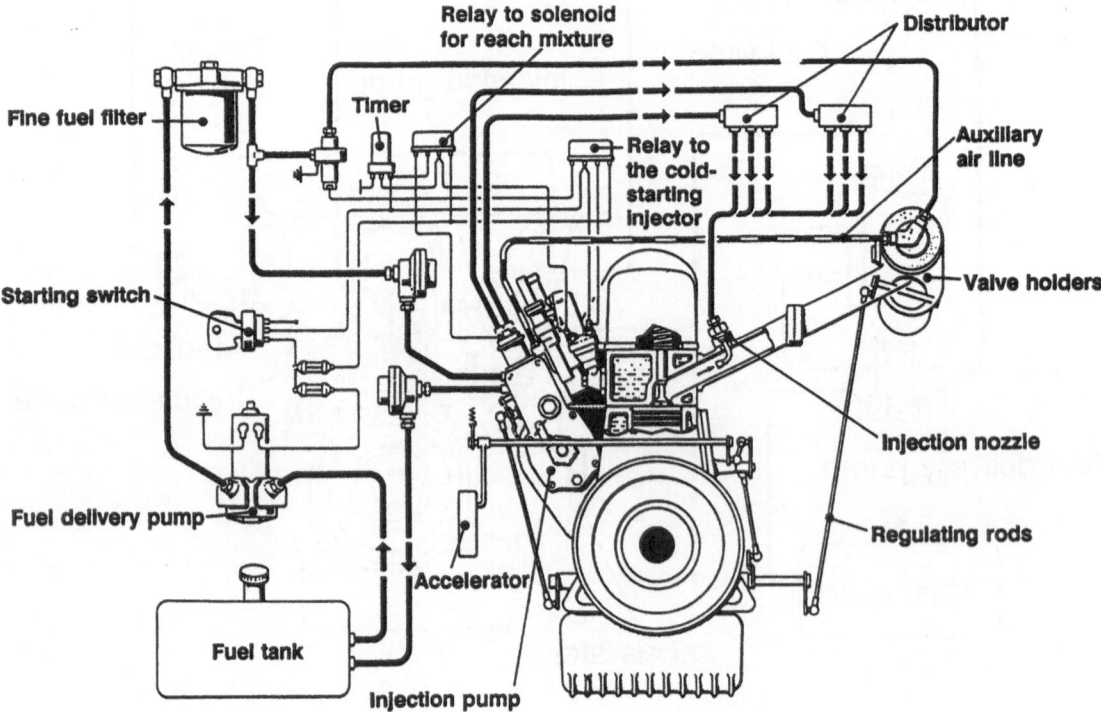

Fig. 4.112. Mechanical Bosch manifold group injection system for Daimler-Benz

The mixture was metered through a cam which was rotated or moved in line with the throttle valve position and speed (**Fig. 4.113**) and whose contours were sensed by a probe. The injection quantity per cycle was proportionate to the rise of the cam.

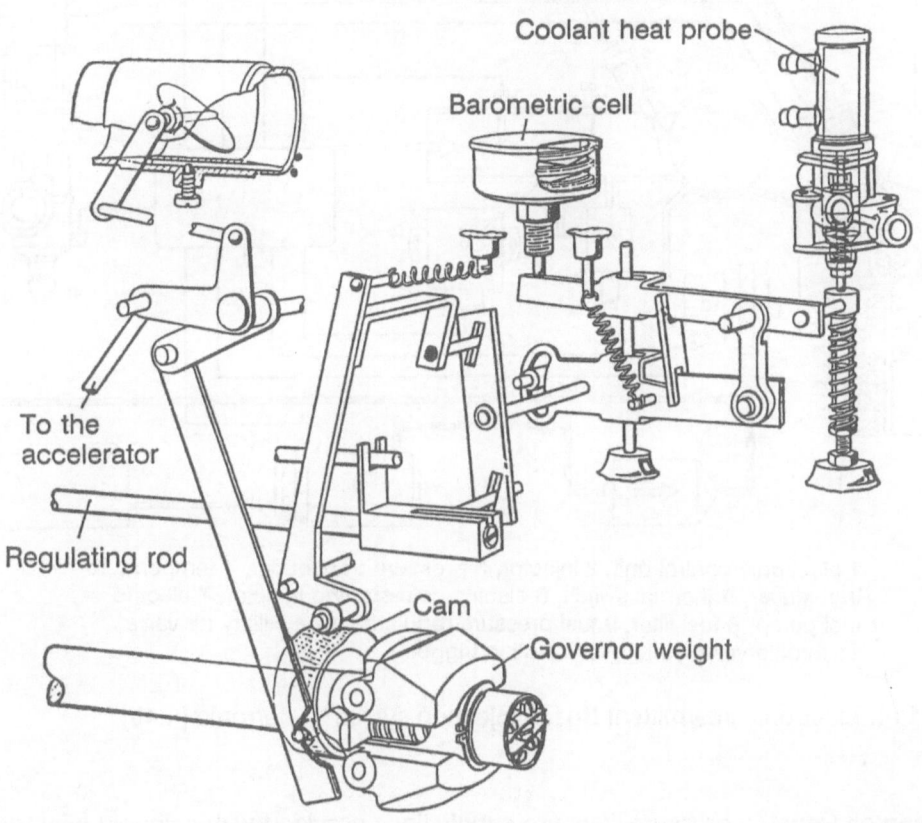

Coolant heat probe

Barometric cell

To the accelerator

Regulating rod

Cam

Governor weight

Fig. 4.113. Mechanical Bosch injection system for Daimler- Benz

It was corrected as a function of the actual atmospheric pressure and engine temperature.

As this type of mechanical injection system proved to be very expensive, an intermittent injection system was developed, the so-called Bosch D-Jetronic (where D stands for pressure-controlled).

The system, based on licenses from Bendix, was used by Volkswagen for its serial production in the U.S., to cater to the new and stricter exhaust emission laws, as the system promised to observe the limits better than would be the case with carburetors.

A schematic diagram of the D-Jetronic system is given in **Fig. 4.114**.

A roller cell pump delivers the fuel from the fuel tank through the fuel filter into the pressure line where a pressure regulator (overflow regulator) keeps the fuel at an overpressure of 2 bars. Excess fuel flows back into the tank.

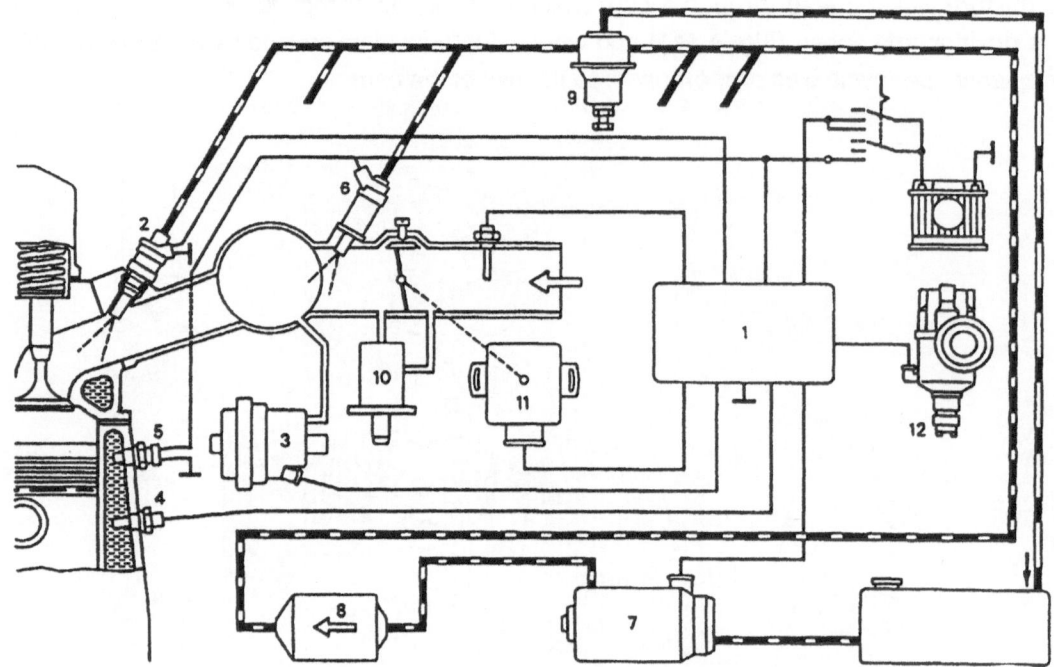

1 electronic control unit, 2 injector, 3 pressure transducer, 4 temperature transducer, 5 thermal switch, 6 electric cold-starting injector, 7 electric fuel pump, 8 fuel filter, 9 fuel pressure regulator, 10 auxiliary air valve, 11 throttle valve switch, 12 injection trigger

Fig. 4.114. Electronic intermittent Bosch injection system D-Jetronic [4.45]

Branching off from the pressure lines are supply lines provided with solenoid injectors at their end points which inject the fuel near the engine inlet valves.

When the magnetic coil is excited, it lifts the jet needle some 0.15 mm from its seat.

Each engine cylinder is allocated an injector, actuated once for each cam rotation. To make the arrangement less complicated, four-cylinder engines have two injectors each switched in parallel (three injectors each in six-cylinder engines). The fuel pressure at the injectors is a constant 2 bars. The fuel dosage to the cylinders for each cycle is controlled by the opening period of the solenoid valves. The electronic control unit (ECU) supplies the requisite control signals whose length depends on the absolute manifold pressure, engine speed, and other correction values.

The values that affect the length of the injection signal as a function of the injection quantity are entered in the form of mechanical-electrical values. The following transducers are used:

- A manifold pressure transducer containing evacuated barometric cells which shift the coil inductor and thus change its inductivity. The inductivity is used as a timer element in a monostable circuit in the ECU;

- A throttle valve switch to transmit switching pulses to the ECU at idle position and when the throttle valve moves, to produce an enrichment of the mixture during transition. Overrun fuel cutoff occurs when the throttle valve is closed and the speed exceeds 1200 rpm;

- Temperature transducers in the form of temperature-dependent resistors which are fitted at appropriate places at the engine wrapped in a protective cover and which extend the pulse length when the engine is not running at operating temperatures;

- A pulse trigger in the ignition distributor, designed as two additional, maintenance-free breaker contacts which are staggered by 180 degrees;

- An auxiliary air valve in the form of a slide valve actuated by a bimetal spring or thermostatic expansion element which supplies the engine with the additional air to get the engine properly running during warm-up;

- A cold-start valve to inject additional fuel into the intake distributor when the engine is started at low temperatures;

- The electronic control unit then converts the engine values into electric pulses whose length depends primarily on the manifold pressure and engine temperature. The ECU consists basically of a timer (monostable circuit) and one amplifier for each valve group. The timer is connected consecutively to the individual amplifiers.

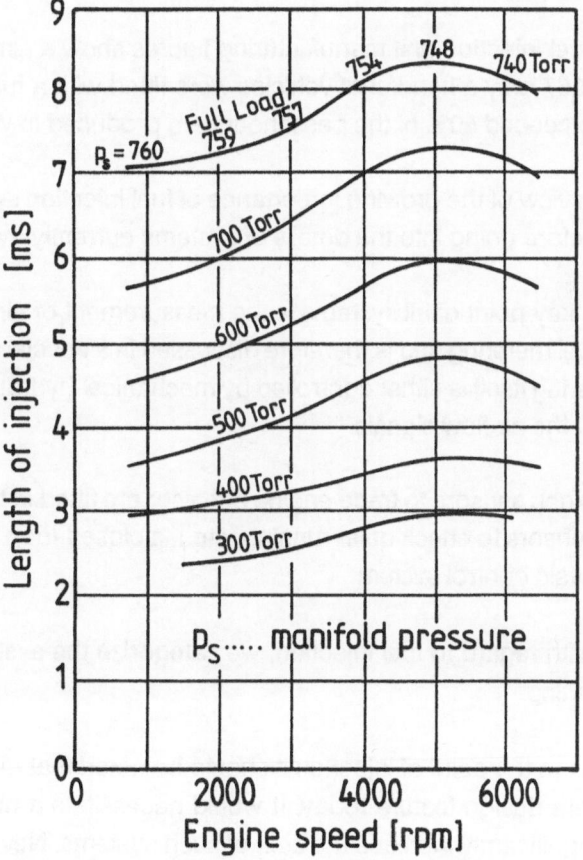

Fig. 4.115 illustrates the function of injection length and speed for several manifold pressures.

The introduction of the Bosch D-Jetronic system in 1967 initiated the widespread use of fuel injection as well as intense R&D activities in the field [4.39].

At its introduction the D-Jetronic was considered a complex control sys-

Fig. 4.115. Characteristic engine lines for the Bosch D- Jetronic

tem, but today's systems have long since surpassed it with regard to adaptability to engine operation parameters. A look at the number of data to be determined and optimized proves the point [4.39]:

Bosch D-Jetronic 1967:	about 18 alignment points for characteristic lines and constant values
Bosch LH-Jetronic 1980:	about 85 alignment points
Bosch L 3.2-Jetronic 1984:	about 200 data
Bosch Motronic ML 4.1 1986:	about 2500 data
Bosch Motronic M 2.5 1988:	about 3400 data

Future combined mixture formation and ignition systems will probably have to process 10,000 data, linked with other functions.

Fuel injection unit manufacturing figures show a similarly explosive development: Where up to 1967 only a handful of vehicles were fitted with a fuel injection system, their number has since exceeded 50% of the passenger cars produced in Western countries today.

In view of the growing importance of fuel injection systems, we will discuss some basic designs before going into the details of systems currently available.

A key point of all systems is the measurement of air in the engine cylinder fill. It is the basis for fuel metering and is therefore discussed in a special section on hot-wire anemometers. Fuel metering itself is either controlled by mechanical-hydraulical means or by the electronic conversion of the air flow signals.

When sensors to trace engine behavior are fitted in the vehicle (lambda probes, knock sensors, sensors to check quiet running, etc.), a closed-loop control circuit can be superimposed on the basic control system.

With regard to fuel injection, we categorize the systems by their injection point, pressure, and timing.

As to the point of injection it should be noted that direct injection into the cylinder is a relatively rare design feature today. It would necessitate a high level of injection pressure which would significantly increase the cost of such systems. Nevertheless it is a field of intense research activities. Developers hope to achieve a type of stratification that permits part load operation more or less without throttling, higher knock-free compression rates, and leaner mixtures. The goal is to utilize the advantages of the Otto and Diesel process in a combined system.

As to injection timing we have several options (**Fig. 4.116**):

- Simultaneous injection for all cylinders or groups of cylinders (simultaneous or group injection);

- Injection as a function of the inlet valve opening time (sequential injection), e.g. injection when the inlet valve is opening);

- Continuous injection, i.e. over the entire operating cycle;

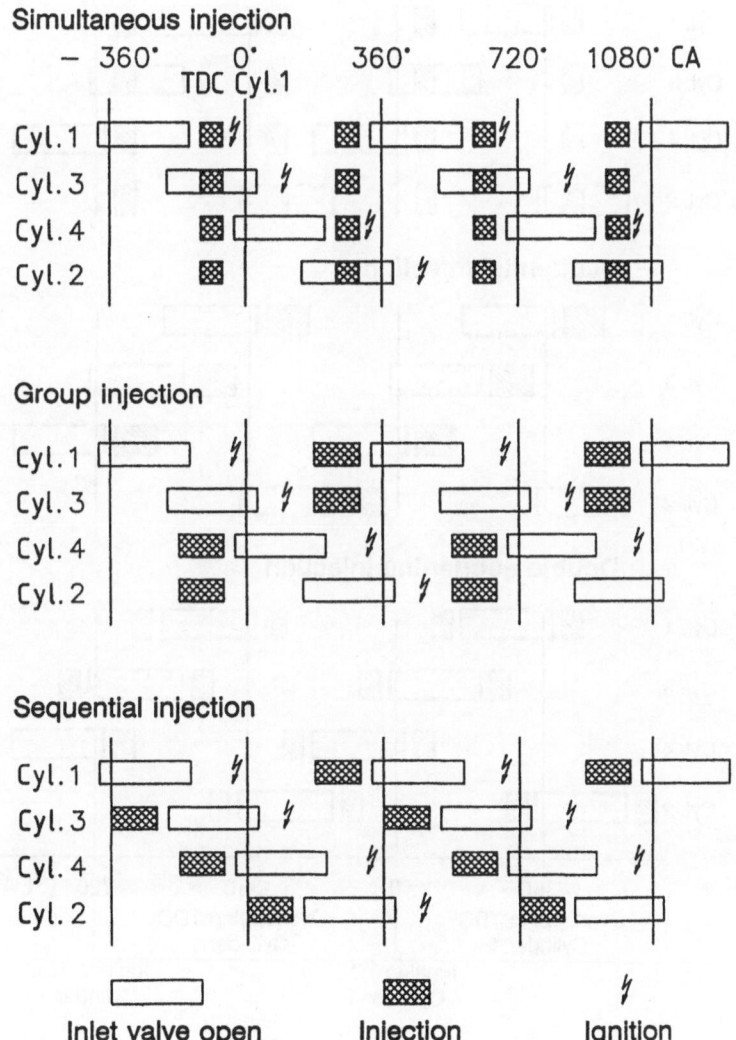

Fig. 4.116. Simultaneous, group, and sequential injection [4.39]

• Injection as a complex procedure in the form of multiple sequential injection.

Methods to improve engine operating performance by multiple sequential injection have been studied in depth by [4.53] and [4.73]. **Fig. 4.117** illustrates the types of injection examined for a four-cylinder engine. In a simultaneous system we generally have two injections per cycle, triggered by the cylinder ignition time, where mixture formation conditions vary for each individual cylinder. In the sequential system with one injection for each cycle we get identical mixture formation conditions for all cylinders so that injection timing can be better adjusted to engine requirements.

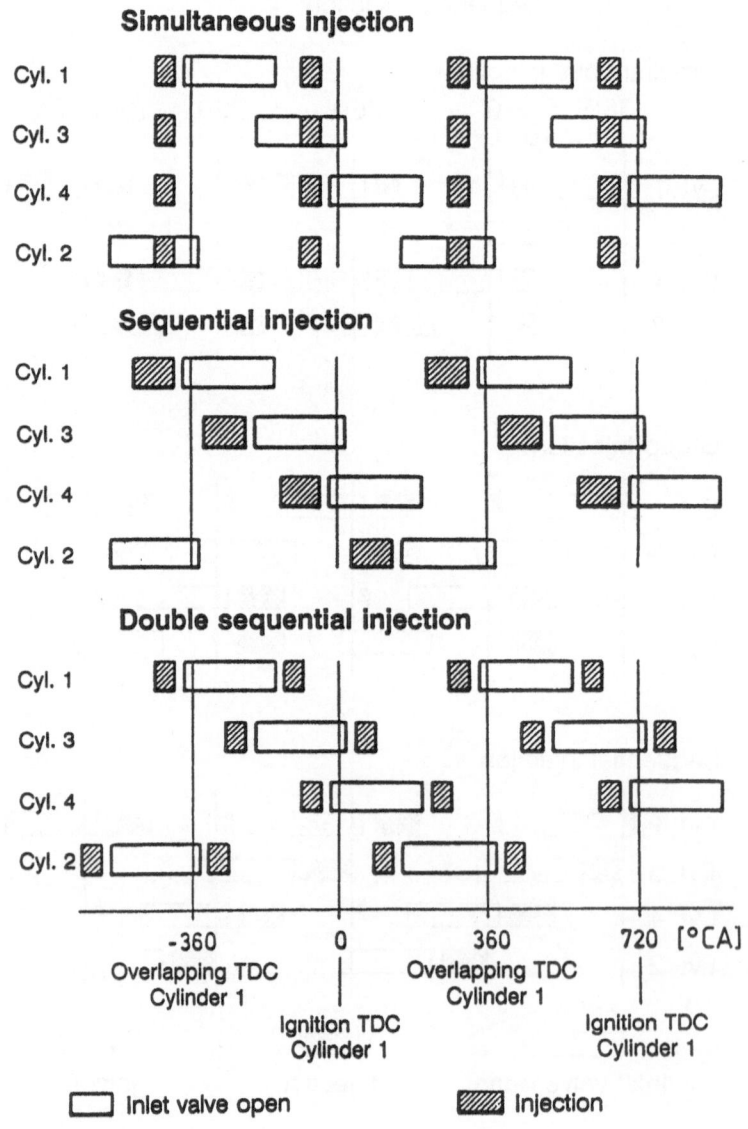

Fig. 4.117. Injection timing in simultaneous, sequential, and double sequential injection [4.53]

Figs. 4.118 and **4.119** demonstrate the effect of simultaneous and sequential injection on part load behavior of a lean-mix engine when the injector is positioned near the inlet valve. Engine behavior is clearly different depending on the injection time during intake. An injection time relatively late in the intake cycle translates into increased combustion stability, here described as the coefficient of variation of the subscripted specific work "V_{Pi}", and reduced fuel consumption on the one hand, but higher nitrogen oxide emission on the other.

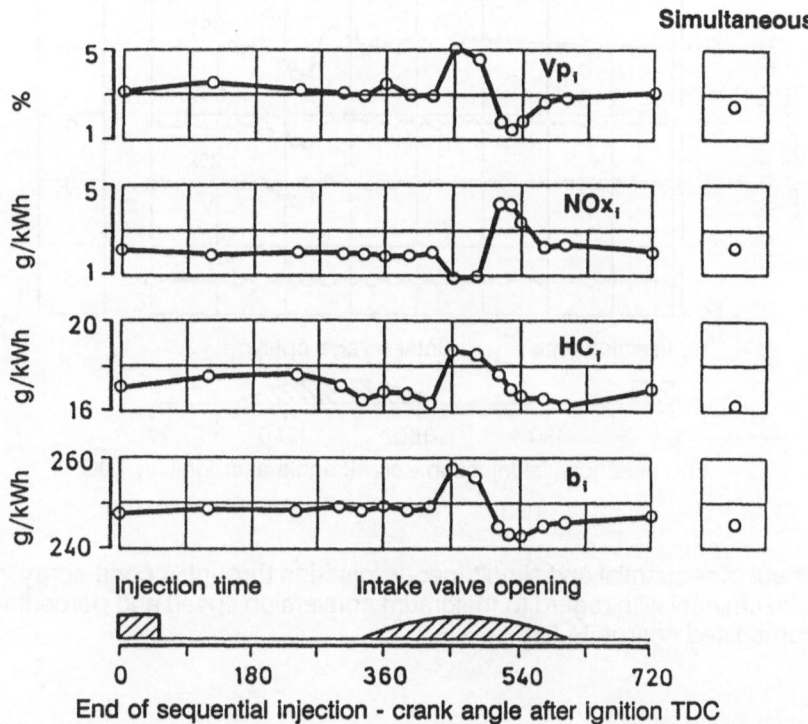

AVL research engine

Engine speed: 1550 [rpm]; Lambda = 1.45
Indicated mean pressure: 4.5 bars
Ignition time: crank angle of 30° before TDC
Injector: cone-spray valve in the intake channel

Fig. 4.118. Effect of sequential and simultaneous injection through a cone-spray valve in the intake channel on the coefficient of variation of the subscripted specific work V_{Pi}, the subscripted specific NO_{xi} and HC_i emissions and on the subscripted specific fuel consumption b_i [4.53]

From Figure 4.119 it can be seen that late intake-synchronous injection reduces both the ignition delay and combustion time, while early intake-synchronous injection results in unfavorable engine behavior, i.e. lower combustion stability and longer combustion time. It should be noted, however, that the results found for this particular engine cannot be generalized to cover all engines, as influences such as charge movement or the position of the injector may have some impact.

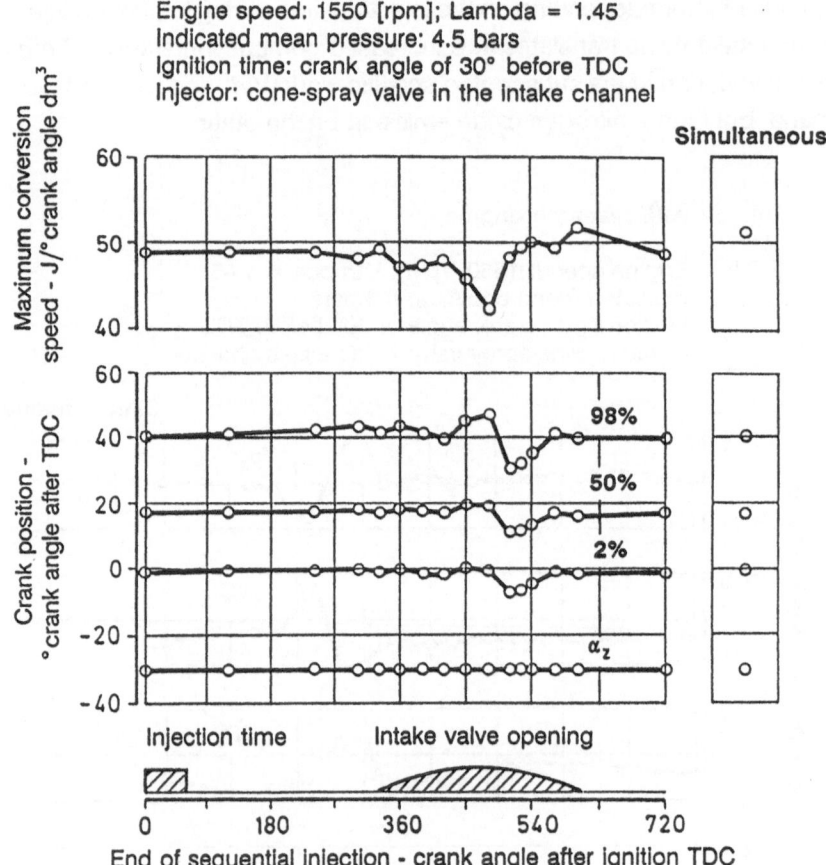

Fig. 4.119. Effect of sequential and simultaneous injection through a cone-spray valve in the intake channel with regard to maximum conversion speed and percentage rate of the combusted charge [4.53]

In summary it may be said that sequential intake-synchronous injection may improve ignition delay, combustion stability, lean limit, fuel consumption, and HC emissions, depending on the engine's load point. On the other hand we must expect higher NO_x emissions caused by stratification. Double sequential injection may slightly improve engine behavior because the injection quantity is fed in an initial quantity, which displays optimum formation characteristics, and a second quantity which improves stratification and thereby reduces HC emission rates.

Fig. 4.120 shows the injection time with regard to double sequential injection as a function of the fuel quantity relationships between first and second injection. From Fig. **4.121** we find that values for combustion stability, NO_x emission, and fuel consumption are better when the first injection comprises 50-75 percent of the fuel quantity.

AVL research engine

Indicated mean pressure: 2.5 bars; engine speed: 1780 [rpm]
Lambda = 1.3; inlet valve opening: crank angle of 330-600°
after ignition TDC
Injector: air-shrouded cone-spray valve in the intake
channel
Ignition time: 1st injection: max. advanced
2nd injection: into the open inlet valve
to optimize HC-emissions

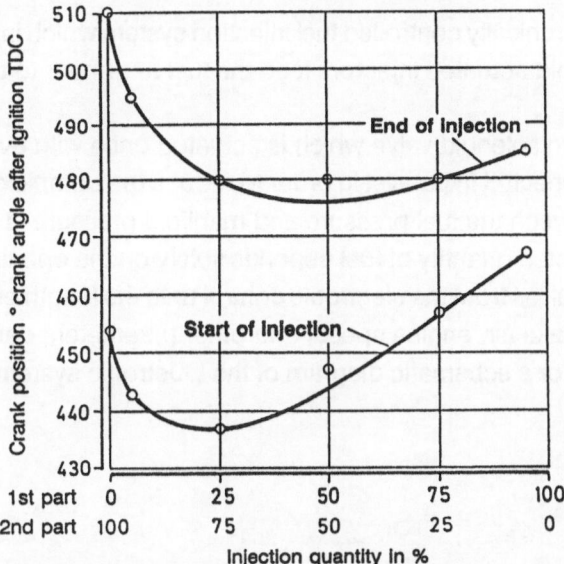

Fig. 4.120. Optimum injection time with regard to HC emission for the second injection in the double sequential injection system at different quantity relationships between first and second injection [4.53]

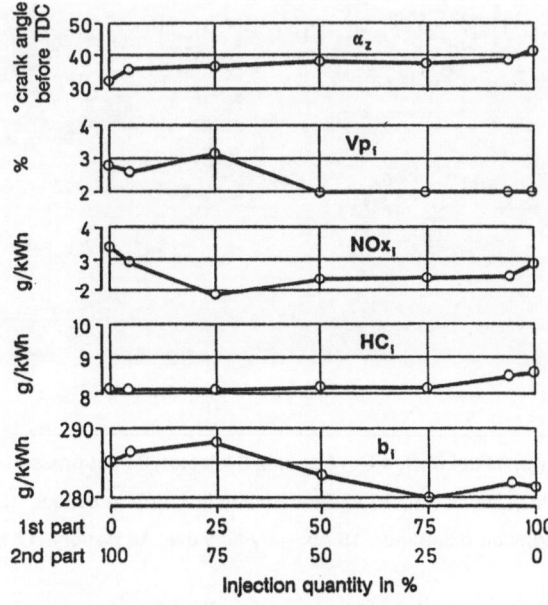

Fig. 4.121. Effect of the quantity relationships between first and second injection on the engine parameters; engine data as in Fig. 4.120 [4.53]

4.2.2. Electronic Intermittent Injection Systems

4.2.2.1. Bosch L-Jetronic

Basic function:

The L-Jetronic is an electronically controlled fuel injection system which injects fuel intermittently into the manifold. Solenoid-actuated injectors feed the fuel to or near to the engine inlet valves.

Each cylinder has its own solenoid valve which is actuated once with every crankshaft revolution. All injectors are connected in parallel in order to reduce the complexity of the circuitry. The differential pressure between the fuel pressure and manifold pressure is held at a constant 2.5 or 3 bars so that the injected quantity of fuel depends solely on the opening times of the valves. They are controlled by pulses from the electronic control unit; the length of these pulses depending on the quantity of intake air, engine speed, and other parameters detected by sensors and processed in the ECU. For a schematic diagram of the L-Jetronic system see **Fig. 4.122**.

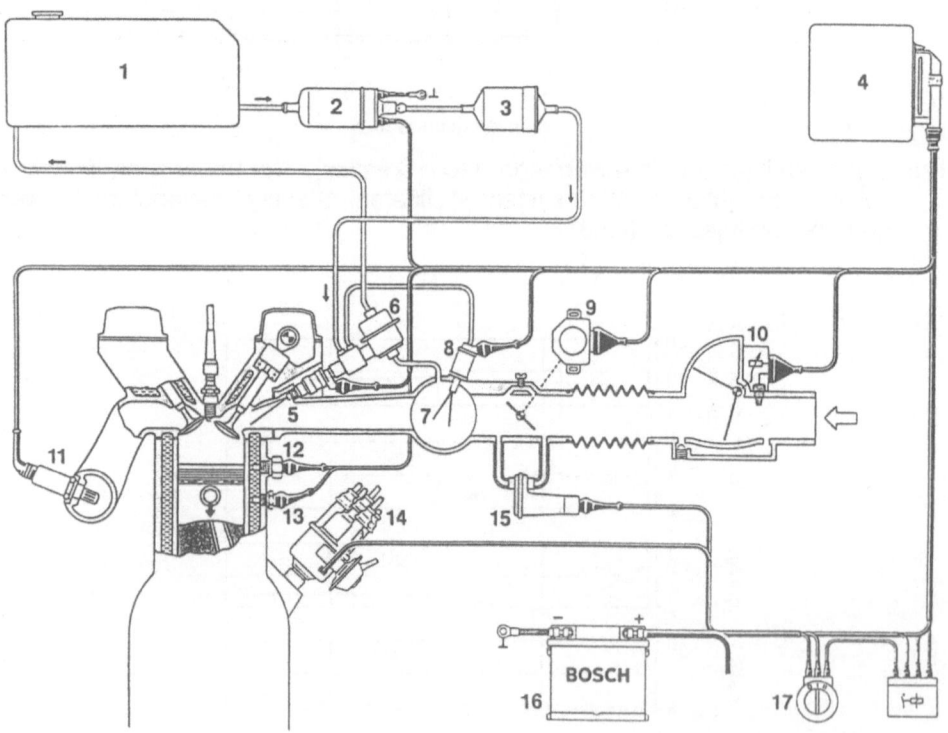

1 Fuel tank, 2 Electric fuel pump, 3 Fuel filter, 4 Control unit, 5 Injector, 6 Fuel-pressure regulator, 7 Intake manifold, 8 Electric start valve, 9 Throttle valve switch, 10 Air-flow meter, 11 Lambda sensor, 12 Thermo-time switch, 13 Engine-temperature sensor, 14 Ignition distributor, 15 Auxiliary-air valve, 16 Battery, 17 Ignition/starting switch

Fig. 4.122. Diagram of a lambda closed-loop controlled Bosch L-Jetronic system [4.22]

Fuel supply:

As illustrated in Fig. 4.122, electric fuel pump 2 supplies fuel from fuel tank 1 through fuel filter 3 to fuel rail/fuel pressure regulator 6. The fuel rail branches off into lines to the injectors 5. More fuel is drawn from the tank than can be consumed by the engine so that the excess fuel returns through the pressure regulator to the tank. The system is constantly flushed so that the fuel is always kept cool, air locks are avoided and hot-start behavior is improved. The fuel pump is of the roller-cell type, driven by a permanently excited electric motor. The rotor plate mounted eccentrically in the pump housing has metal rollers in notches around its circumference which are pressed against the pump housing by the centrifugal force and which act as a rotating seal. The fuel is carried in the hollows which form between the rollers. The pumping action is achieved by the rollers which, after closing the inlet bore, force the fuel trapped inside around with them until it can escape through the outlet bore (**Fig. 4.123**).

1 Intake side
2 Rotor plate
3 Roller
4 Roller race plate
5 Delivery side

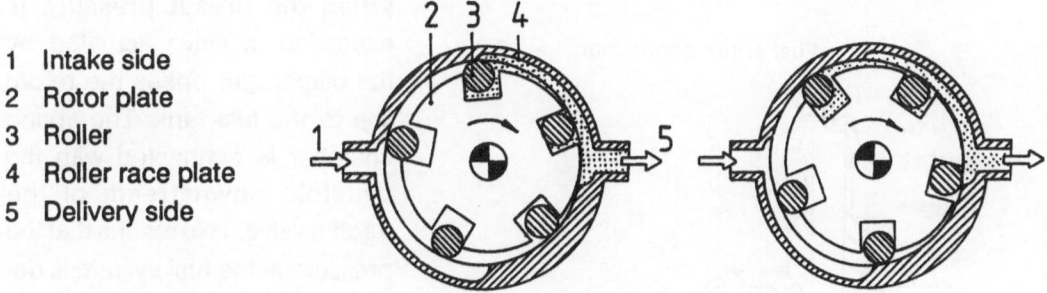

Fig. 4.123. Pumping action in a Bosch roller-cell pump [4.22]P

The fuel flows through the motor (**Fig. 4.124**) without the risk of explosion because the motor pump housing does not hold any ignitable mixture.

1 Intake side
2 Pressure relief valve
3 Roller-cell pump
4 Motor armature
5 Check valve
6 Delivery side

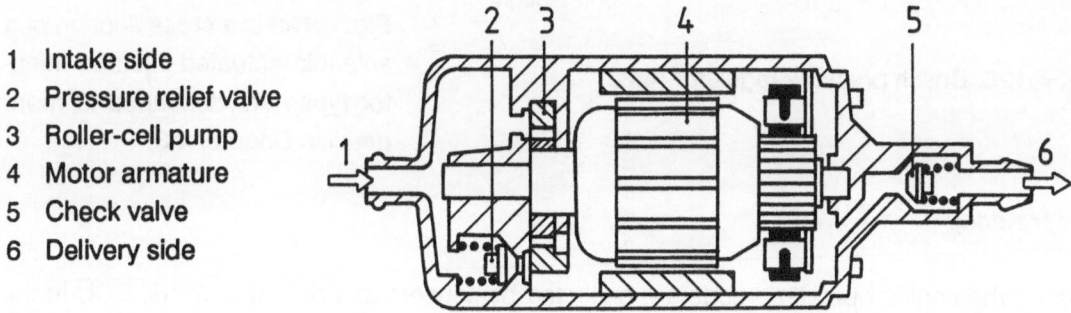

Fig. 4.124. Bosch electric fuel pump [4.22]

The electric fuel pump starts when the ignition and starting switch is actuated and continues to operate for as long as the engine runs.

The fuel filter (3 in Fig. 4.122) contains a paper element with an average pore size of 10 μm, which is backed up by a fluff strainer.

The fuel rail supplies all injectors with an equal quantity of fuel and ensures uniform fuel pressure at all injectors. This is possible by its storage function which provides for a larger volume in the rail compared to the quantity of fuel injected during each working cycle.

The pressure regulator keeps the fuel pressure at a constant 2.5 or 3 bars, depending on the system. It is a diaphragm-controlled overflow pressure regulator fitted at the end of the fuel rail, consisting of a metal casing, divided by a flanged diaphragm into a spring chamber to house the pretensioned helical spring which loads the diaphragm, and a fuel chamber (**Fig. 4.125**).

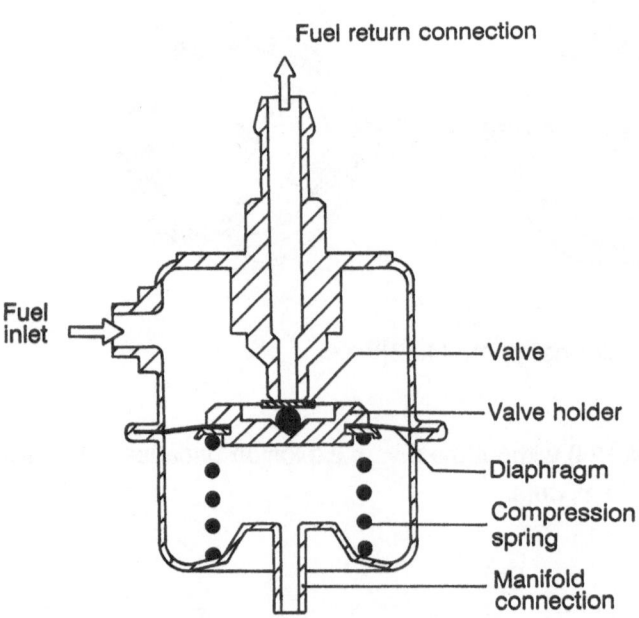

Fuel return connection

Fuel inlet

Valve

Valve holder

Diaphragm

Compression spring

Manifold connection

Fig. 4.125. Bosch pressure regulator [4.22]

When the preset pressure is exceeded, a valve actuated by the diaphragm opens the return line to the fuel tank. The spring chamber is connected with the manifold downstream of the throttle valve. This means that the pressure in the fuel system is dependent on the absolute manifold pressure, while the pressure drop across the injector does not change.

The solenoid-controlled injectors inject the fuel onto the engine inlet valves.

Fig. 4.126 is a cross-section of a solenoid-actuated injector. Injector types are described in more detail in Chapter 4.3.

Fuel Metering

Data on the engine operating mode are collected by sensors and delivered to the ECU in the form of electric signals. The key variables (air mass drawn in by the engine and engine speed) are used to derive the air quantity for each cycle which serves as a direct measure of the engine load condition.

The key variables must be supplemented by other variables to adapt the quantity of injected fuel to particular engine requirements such as cold start, warm-up, idle, part and full load, acceleration and deceleration behavior, and maximum engine speed limitation.

The information on speed and injection time is delivered to the L-Jetronic ECU by the contact breaker in the ignition distributor in breaker-triggered ignition systems, or, in contactless ignition systems, by terminal 1 of the ignition coil.

The sensor flap in the air flow sensor measures the air quantity drawn in by the engine. It is based on the measurement of the impact pressure exerted by the inducted air and the force acting on the sensor flap and counteracting the restoring force of a spring.

Fig. 4.127 provides a schematic diagram of an air flow sensor incorporated in the intake system, while Fig. 4.128 shows the sensor separately. For more details cf. Chapter 4.4 on hot-wire anemometers.

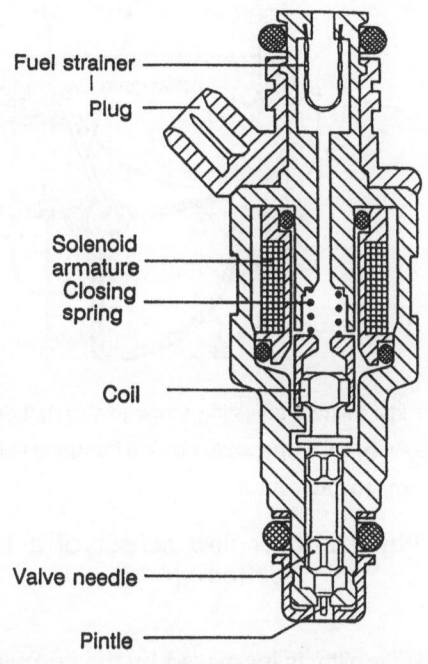

Fig. 4.126. Bosch solenoid injector [4.22]

1	Throttle valve	4	ECU
2	Air flow sensor-	5	Air flow sensor signal to the ECU
3	Intake air	6	Air filter
-	Temperature signal to the ECU	m_L	Inducted air quantity
		α	Deflection angle

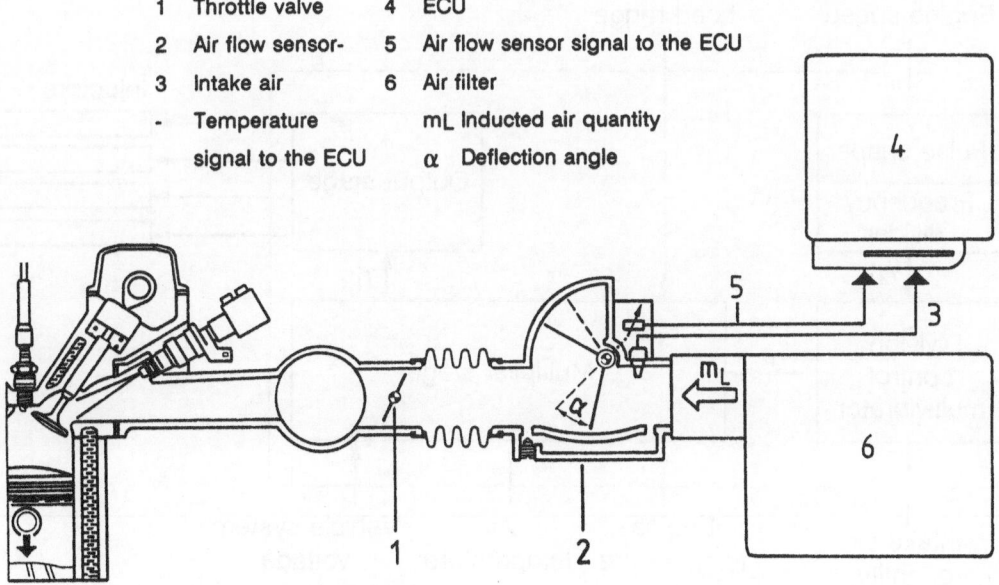

Fig. 4.127. Air flow sensor in the intake system of a Bosch L-Jetronic [4.22]

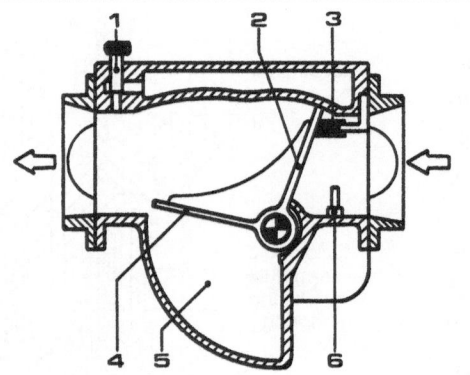

1 Idle mixture adjusting screw (bypass), 2 Sensor flap,
3 Stop, 4 Compensation flap, 5 Damping volume, 6 Air
temperature sensor

Fig. 4.128. Air flow sensor of a Bosch L-
Jetronic [4.22]

Electronic control unit (ECU):

As already mentioned, the ECU evaluates
the data delivered by the sensors on the
operating mode of the engine. Based on
these data it generates control pulses for
fuel metering through the injectors. The
quantity to be injected is determined by the
length of time that the injectors are open.

The ECU is placed in a splash-proof sheet-
metal casing, fitted so that it is not affected
by the heat radiating from the engine. Its
electronic components are arranged on PC
boards. By using integrated circuits and hybrid modules it has been possible to reduce
the number of components to a minimum.
Reliability is increased by the combination of functional groups into integrated circuits.

Figs. 4.129 (a block diagram of the ECU) and **4.130** (complete schematic pulse diagram of the
L-Jetronic for four cylinder engines) show how engine operating data are processed to produce
an injection pulse.

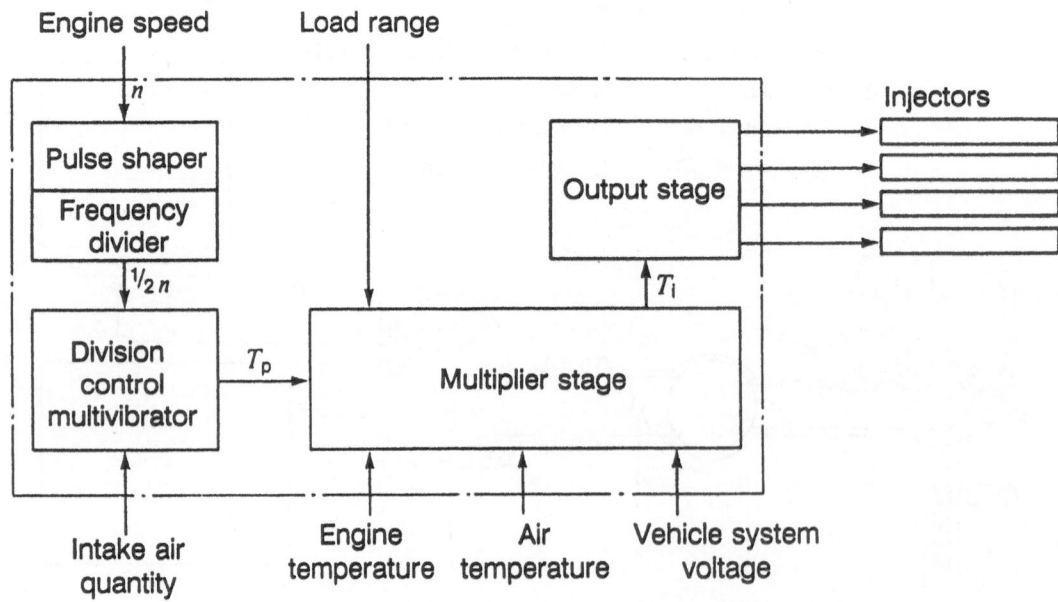

T_i injection pulses, corrected, T_p Basic injection time, n Engine speed

Fig. 4.129. Block diagram of the ECU in a Bosch L-Jetronic system [4.22]

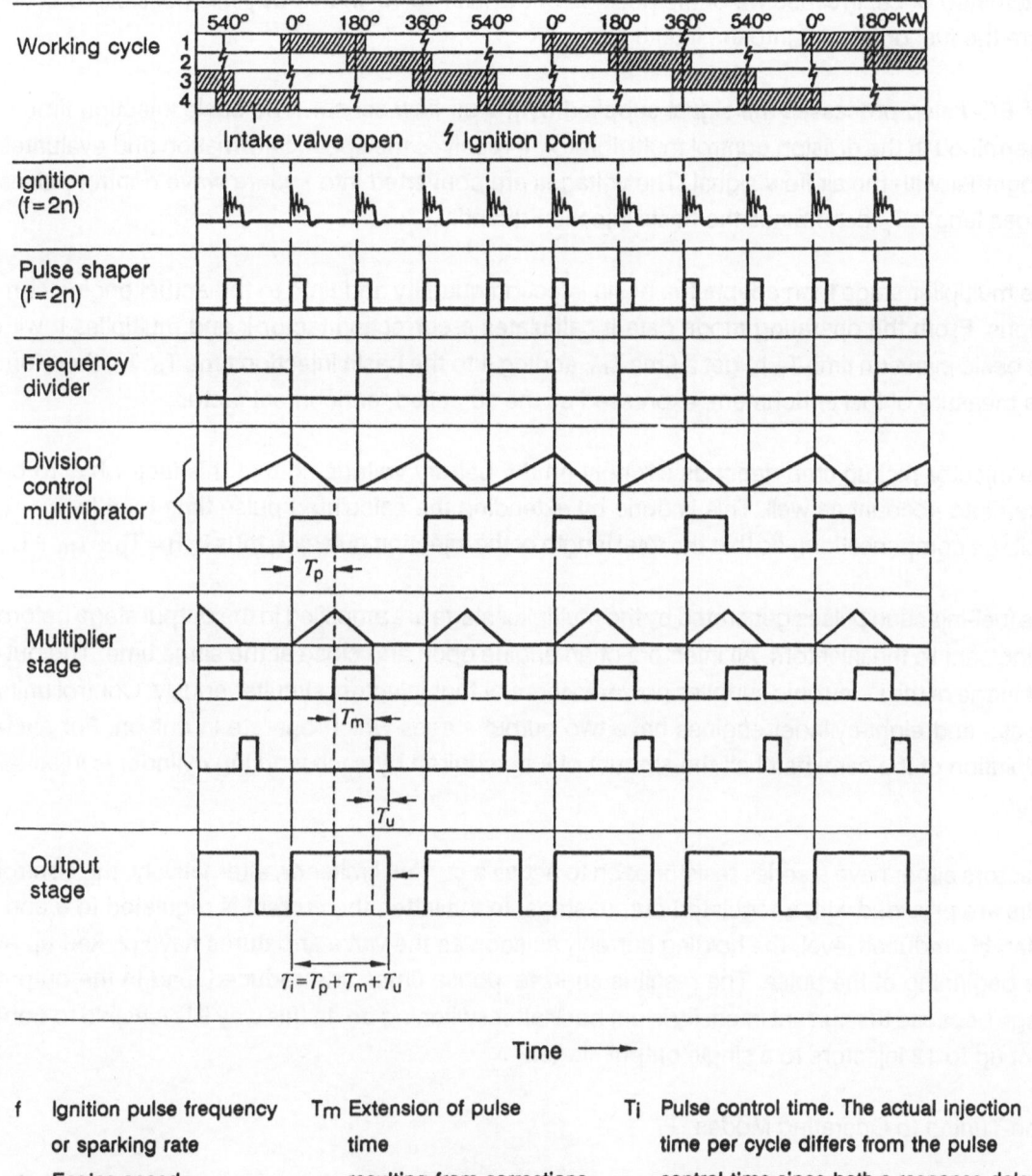

Fig. 4.130. Complete schematic pulse timing diagram of the L-Jetronic for four-cylinder engines [4.22]

The timing frequency of the injection pulses is determined on the basis of the engine speed. The pulses delivered by the ignition system are processed by the pulse shaper, which generates square-wave pulses from the signal and feeds them to a frequency divider. The frequency divider divides the pulse frequency given by the ignition sequence in such a way that two pulses occur for each working cycle regardless of the number of cylinders. The start of the pulse also serves as the start of injection by the injectors. For each revolution of the crankshaft, each injector thus is actuated once, irrespective of the position of the inlet valve. In this way it is possible to either store the fuel or inject it into the intake air.

The ECU also processes the signal supplied by the air flow sensor. The basic injection time is determined in the division control multivibrator. It receives the speed information and evaluates it together with the air flow signal. The voltages are converted into square-wave control pulses whose length T_p determines the basic injection quantity.

The multiplier stage then adapts the basic injection quantity and time to the actual engine conditions. From the operating mode data it calculates a correction factor k and multiplies it with the basic injection time T_p to get a time T_m, adding it to the basic injection time T_p. T_m therefore is a measure of fuel enrichment, expressed as the so-called "enrichment factor."

The injector pickup time depends strongly on the battery voltage so that this factor has to be taken into account as well. This is done by extending the calculated pulse time by a factor T_u (voltage compensation), so that the total length of the injection pulses T_i thus is $T_i = T_p + T_m + T_u$.

The fuel-injection pulses generated by the multiplier stage are amplified in the output stage before being sent to the injectors. All injectors of an engine open and close at the same time. The output stage of the L-Jetronic supplies power to three or four injectors simultaneously. Control units for six- and eight-cylinder engines have two output stages which operate in unison. For each revolution of the camshaft half the amount of fuel required by each working cylinder is injected twice.

Injectors either have a series resistor each to act as a current limiter or, alternatively, the control units are provided with a regulated output stage. In the latter, the current is regulated to a considerably reduced level, the holding current, as soon as the valve armatures have picked up at the beginning of the pulse. The result is short response times and reduced load in the output stage because the current intensity is cut back after switching on. In this way it is feasible to connect up to 12 injectors to a single output stage.

Fine-Tuning to Operating Modes

Cold-start enrichment:

When the engine is cold-started, the combustion mixture must be enriched to compensate for fuel condensation losses and less than optimum mixture preparation. This is done by either extending the injector injection times or by adding a special cold-start valve to the manifold which operates for a period limited by a thermo-timer.

The cold-start valve shown in **Fig. 4.131** has its winding inside the valve. In neutral position a helical spring presses the movable solenoid armature against a seal, thereby shutting off the valve. When a current is passed through the solenoid, the armature, lifted from the valve seat, allows fuel to flow along the sides of the armature to a nozzle where it is swirled to improve atomization.

The thermo-timer (**Fig. 4.132**) limits the injection time of the cold-start valve depending on the temperature of the engine. It consists of an electrically heated bimetal strip which opens or closes a contact depending on its temperature. Controlled through the ignition and start switch, it is fitted in a position representative of the engine temperature.

Post-start and warm-up enrichment:

An engine that has been cold-started needs substantially more fuel during its post-start phase since some of the fuel condenses on the still cold cylinder walls and fuel preparation has not yet been optimized. For example, at a temperature of -20 °C, enrichment is two or three times that of when the engine runs at normal operating temperature, so that 30 to 60 percent more fuel must be injected for some 30 seconds during the post-start phase.

Fig. 4.133 shows a typical enrichment curve in terms of time for a starting temperature of 20 °C.

An engine temperature sensor (**Fig. 4.134**) signals the engine temperature to the ECU. It is mounted in the engine block in air-cooled engines, or projecting into the coolant in water-cooled engines. It has a resistor to fit the appropriate temperature.

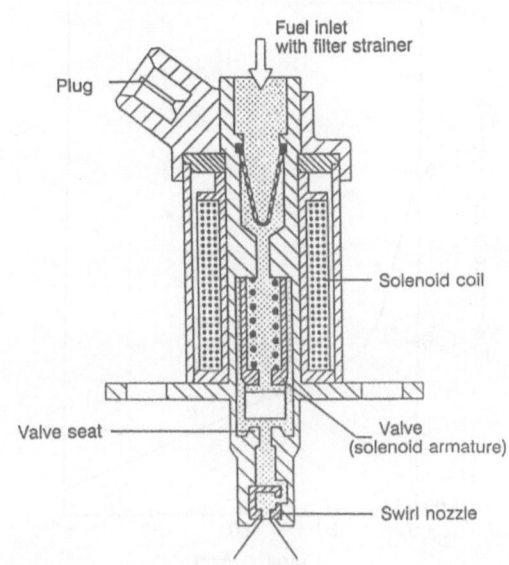

Fig. 4.131. Cold-start valve in a Bosch injection system [4.22]

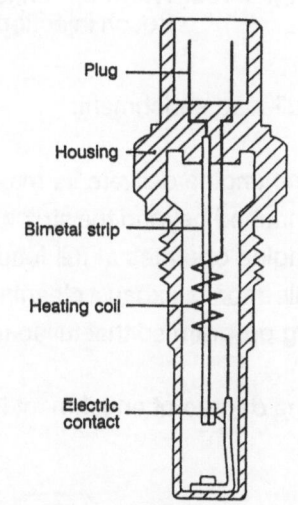

Fig. 4.132. Thermo-timer in a Bosch injection system [4.22]

Acceleration enrichment:

During acceleration, the L-Jetronic meters additional fuel to the engine to prevent momentary leaning of the air-fuel mixture when the throttle is opened abruptly. In this case the sensor flap, opened suddenly, overswings past the open throttle valve point, with the result that more fuel is metered to the engine.

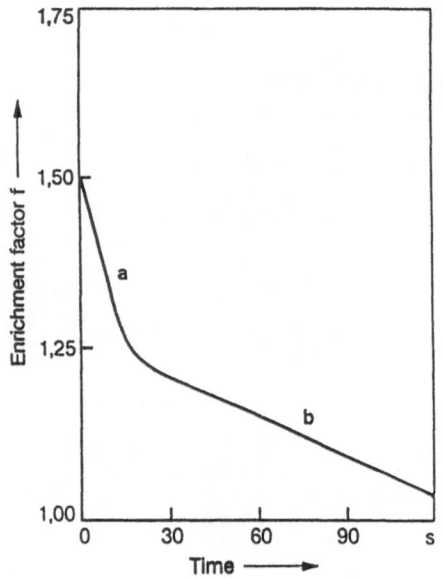

a) proportion mainly dependent on time
b) proportion mainly dependent on engine temperature

Fig. 4.133. Warm-up enrichment curve in a Bosch injection system [4.22]

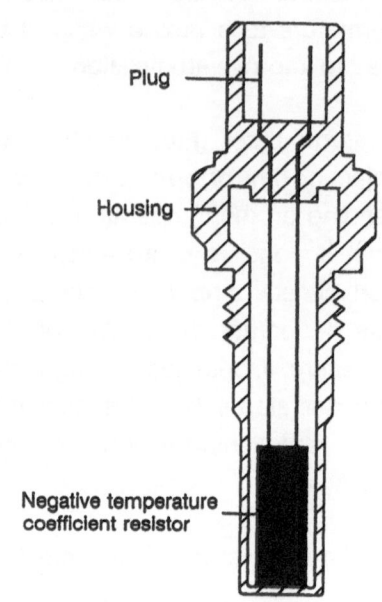

Fig. 4.134. Bosch engine temperature sensor [4.22]

Full-load enrichment:

The engine delivers its maximum output when supplied with a combustion fuel that has been enriched beyond the stoichiometric ratio. Accordingly, the mixture is usually enriched when the engine operates at full load. In a lambda-closed-loop-controlled three-way catalytic converter this impairs exhaust cleaning. But the full-load range is normally not included in the exhaust testing program so that full-load enrichment is permissible in spite of its negative ecological effect.

The degree of enrichment is programmed in the ECU specifically for each engine. Information on the load condition is supplied by the throttle valve switch which also signals the "idle" position.

Fig. 4.135 is a diagram of a throttle valve switch mounted to the throttle valve body. It is actuated by the throttle valve shaft which holds the throttle valve. A contoured switching guide closes the "idle" and "full load" contacts at the respective ends of the switch travel.

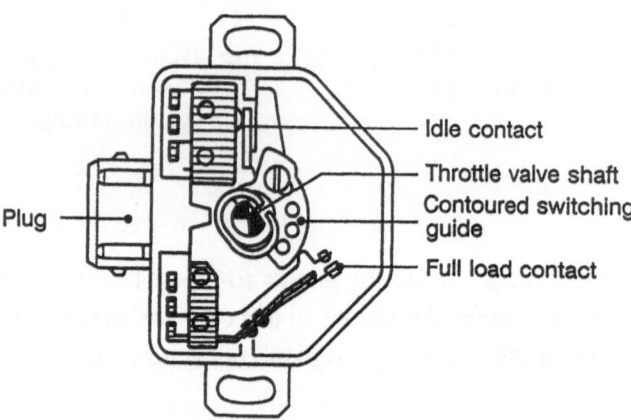

Fig. 4.135. Throttle valve switch in a Bosch injection system [4.22]

Controlling the idle speed:

The air flow sensor contains an adjustable bypass to allow a small quantity of air to bypass the sensor flap. A basic setting of the air-fuel ratio can be made by means of an idle mixture adjusting screw located in the bypass.

In order to achieve smooth idle running even when the engine is still cold, the idle speed may be increased, which incidentally leads to a more rapid warm-up of the engine. This is done by an auxiliary air device (**Fig. 4.136**) which allows more air to pass through in parallel with the throttle valve in order to overcome increased friction resistance in the cold engine.

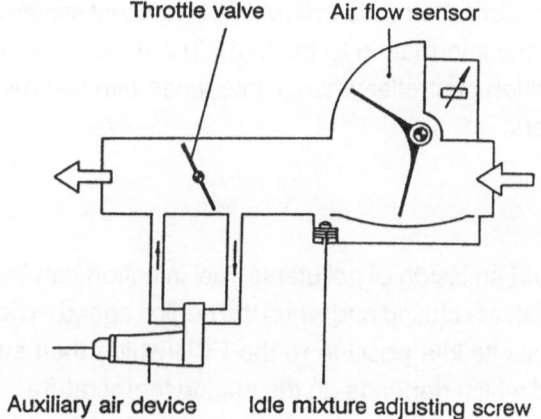

Fig. 4.136. Idle speed control in the Bosch L-Jetronic [4.22]

The auxiliary air device (**Fig. 4.137**) may be designed to incorporate an electrically heated bimetal strip which controls the cross-section of the opening as a function of the temperature.

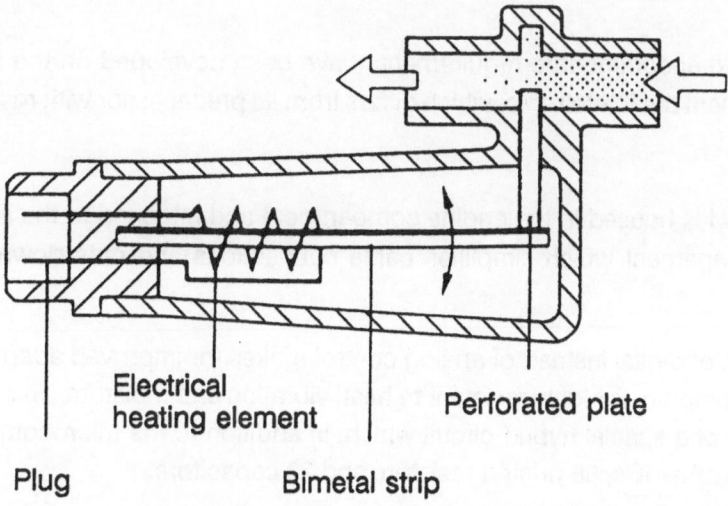

Fig. 4.137. Electrically heated auxiliary air device in the Bosch injection system [4.22]

Adaptation to the air temperature:

The air mass necessary for combustion depends on the temperature of the intake air. A temperature sensor is thus fitted in the intake duct of the air flow sensor to register this effect. The sensor passes the information on to the ECU which then controls the amount of fuel to be injected accordingly.

Lambda closed-loop control:

To ensure optimum operation of three-way catalytic converters (converters which oxidize carbon dioxide and hydrocarbons that have not been burned off and at the same time reduce nitrogen oxides), it is necessary to keep to a very narrow range of lambda = 0.98 to 1.02. This is possible only by closed-loop control. A lambda sensor in the exhaust system measures the partial oxygen pressure, sending the information to the ECU. There the signal is compared with a setpoint and sent to a two-position controller where it interfaces with fuel metering by changing the opening time of the injectors.

Overrun fuel cutoff:

To reduce fuel consumption and emission of pollutants, fuel injection can be suspended during overrun, i.e. when the throttle valve is closed and when the engine speed exceeds the idle speed. The throttle valve switch signals its idle position to the ECU which then suppresses injection pulses from a speed threshold which depends on the engine temperature.

Engine speed limitation:

When the maximum permissible engine speed is reached, a limiting system suppresses the injection signals, thus blocking the supply of fuel to the injectors.

Developments of the Bosch L-Jetronic

Specific systems to accommodate requirements have been developed on the basis of the L-Jetronic, among them the L3-Jetronic, which differs from its predecessor with respect to the following details:

- The ECU is housed in the engine compartment and attached to the air flow sensor, an arrangement which simplifies cable connections and cuts down on assembly costs.

- The use of digital instead of analog control makes for improved adaptation. Its ECU microcomputer is highly resistant to heat, vibration and moisture. This is achieved by the use of a special hybrid circuit which, in addition to the microcomputer, comprises five other ICs, 88 printed resistors and 23 capacitors.

- A "limp-home" function enables the motorist to drive the vehicle to the nearest workshop if the microcomputer fails. Input signals are checked for their plausibility, i.e. implausible input signals are ignored and replaced by default values from the ECU.

4.2.2.2. Bosch LH-Jetronic

The LH-Jetronic (**Fig. 4.138**) is closely related to the L- Jetronic; the main difference lies in the hot-wire anemometer used by the LH-Jetronic to measure the air mass. The intake air stream is routed past a heated wire (hot wire) which is part of an electrical bridge circuit. The current flowing through it keeps it at a temperature which is always above the intake air temperature by a constant amount. The heating current required is a measure of the air mass inducted by the engine. It is converted into a voltage signal. For further details cf. Chapter 4.4 on hot-wire anemometers. A rotary idle actuator controls the idle speed by supplying the engine with more or less air.

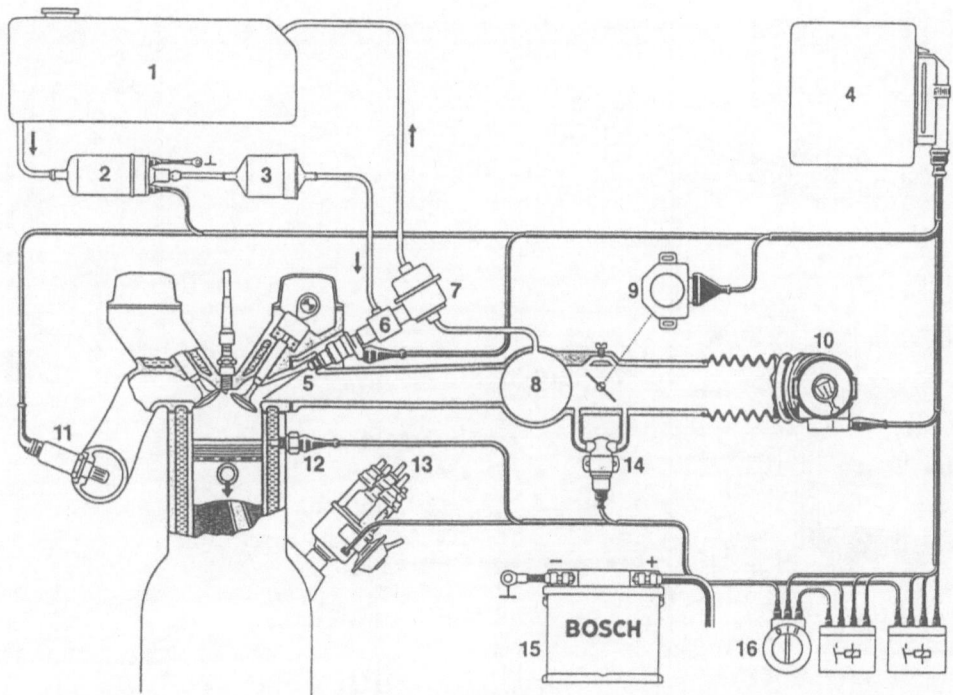

1 Fuel tank, 2 Electric fuel pump, 3 Fuel filter, 4 ECU, 5 Injector, 6 Fuel rail, 7 Pressure regulator, 8 Intake manifold, 9 Throttle valve switch, 10 Hot-wire air mass meter, 11 Lambda sensor, 12 Engine temperature sensor, 13 Ignition distributor, 14 Rotary idle actuator, 15 Battery, 16 Ignition and starting switch

Fig. 4.138. Schematic diagram of the Bosch LH-Jetronic [4.22]

4.2.3. Continuous Mechanical/Electronic Injection Systems

4.2.3.1. Bosch K-Jetronic

The K-Jetronic is a mechanically and hydraulically controlled driveless fuel injection system developed by Bosch. It meters the fuel as a function of the intake air quantity and injects it continuously onto the engine inlet valves.

The K-Jetronic was originally designed as a purely mechanical injection system. In the course of its development it has been fitted with auxiliary electronic functions such as lambda closed-loop control. The result was the KE-Jetronic discussed in a later section, where the basic mechanical system is supplemented by an electronic control unit.

The K-Jetronic consists of three functional elements: fuel supply, air mass metering, and fuel metering. **Fig. 4.139** provides a diagram of a lambda closed-loop controlled K- Jetronic.

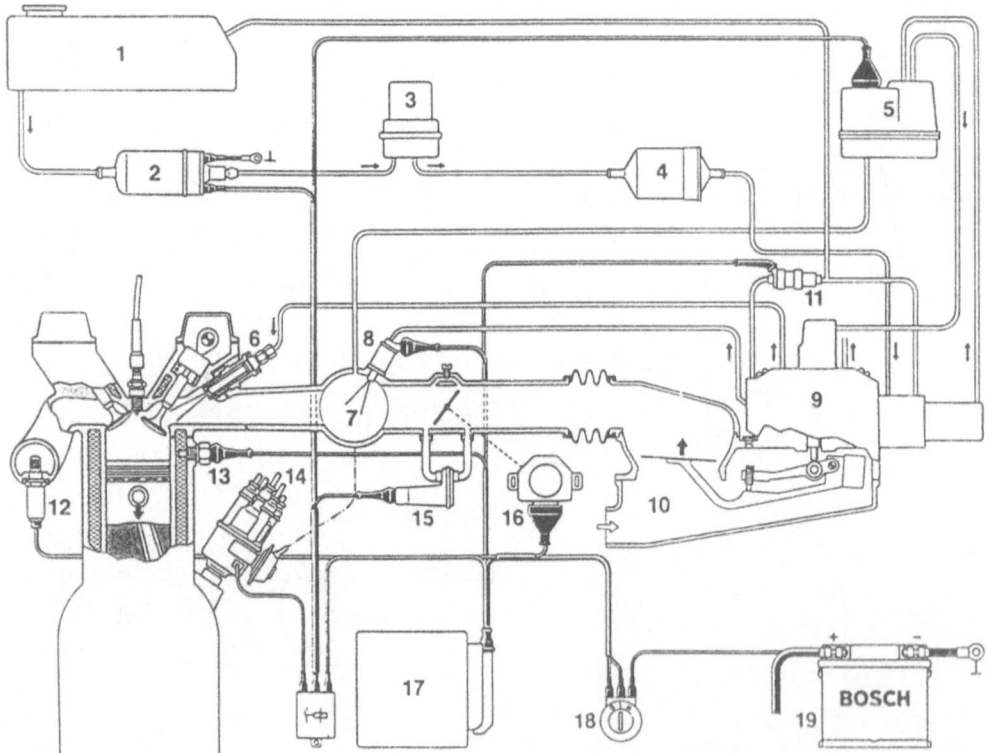

1 Fuel tank, 2 Electric fuel pump, 3 Fuel accumulator, 4 Fuel filter, 5 Warm-up regulator, 6 Injector, 7 Intake manifold, 8 Cold-start valve, 9 Fuel distributor, 10 Air flow sensor, 11 Frequency valve, 12 Lambda sensor, 13 Thermo-timer, 14 Ignition distributor, 15 Auxiliary air, 16 Throttle valve switch, 17 ECU, 18 Ignition and starting switch, 19 Battery

Fig. 4.139. Diagram of the K-Jetronic system with lambda closed-loop control [4.22]

The fuel supply system comprises the following elements:

- Electric fuel pump;

- Fuel accumulator;

- Fine filter;

- Primary pressure regulator, and

- Injectors.

An electrically driven roller-cell pump feeds the fuel from the fuel tank at a pressure of over 5 bars to a fuel accumulator and through a filter to the fuel distributor. From there the fuel flows to the injectors which inject it continuously into the intake ports of the engine. The primary pressure regulator maintains the delivery pressure in the system at a constant level and reroutes the excess fuel back to the fuel tank.

For a detailed description of the electric fuel pump see the section on the L-Jetronic system.

The fuel accumulator (**Fig. 4.140**) maintains the pressure in the fuel system for a certain time after the engine has been switched off in order to facilitate restarting, particularly when the engine is still hot. The special design of its housing dampens the sound of the fuel pump. A diaphragm divides the interior of the fuel accumulator into two chambers, one of which serves as an accumulator for the fuel while the other represents the compensation volume and is connected to the atmosphere or fuel tank by means of a vent. During operation, the accumulator chamber is filled with fuel, pushing the diaphragm against the force of the spring until it reaches the stop in the spring chamber. The diaphragm remains in this position, which corresponds to the maximum accumulator volume, for as long as the engine is running.

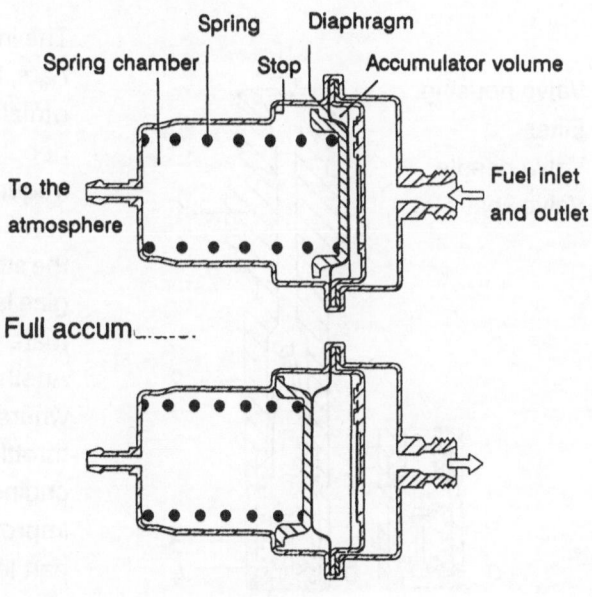

Fig. 4.140. Fuel accumulator in the Bosch K-Jetronic [4.22]

The fuel filter is similar to the one described for the L- Jetronic system.

The primary pressure regulator (**Fig. 4.141**) maintains the pressure in the fuel system at a constant level.

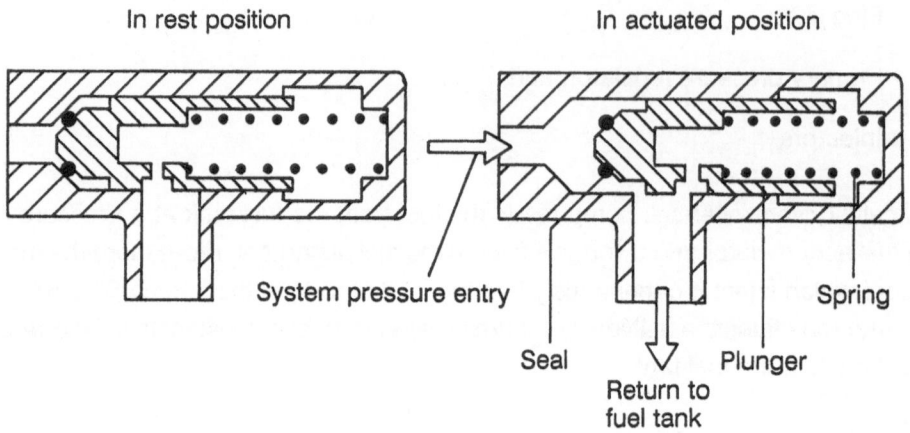

Fig. 4.141. Primary pressure regulator in the fuel distributor of the Bosch K-Jetronic [4.22]

1   Valve housing
2   Filter
3   Valve needle
4   Valve seat

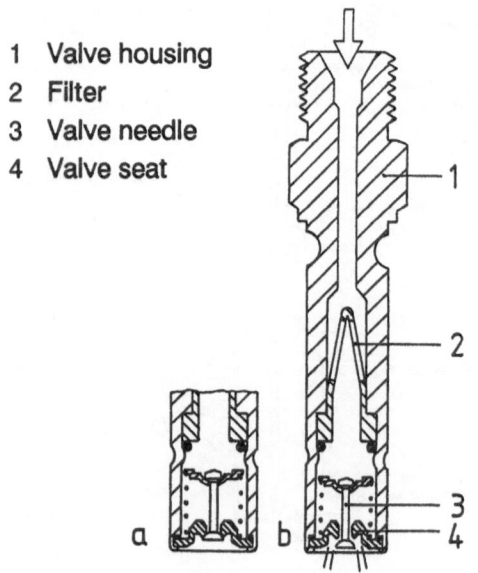

a) In rest position
b) In actuated position

Fig. 4.142. Injector in the Bosch K-Jetronic [4.22]

The injectors open at a given pressure, e.g. 3.5 bars, injecting the fuel into the manifold and atomizing it through oscillation of the valve needle. By oscillating ("chattering") at high frequency, the valve needle (**Fig. 4.142**) ensures excellent atomization of the fuel even with the smallest of injection quantities. When the engine is switched off, the injectors are closed, actuated by the drop in the fuel delivery pressure. Another alternative are air-shrouded injectors where, using the pressure drop across the throttle valve, a portion of the air drawn in by the engine is passed through the injectors, which improves atomization especially for idle and part load operation.

Air flow sensor:

The air flow sensor (**Fig. 4.143**) operates by the suspended-body principle. For more details see Chapter 4.4.

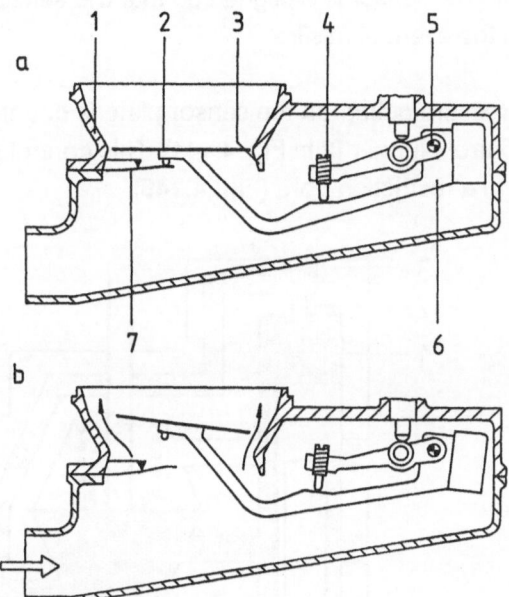

1 Air funnel
2 Sensor plate
3 Relief cross-section
4 Idle mixture adjusting screw
5 Pivot
6 Lever
7 Leaf spring

a) Sensor plate is in its
   zero position
b) Sensor plate is in its
   operating position

Fig. 4.143. Updraft air flow sensor in the Bosch K-Jetronic [4.22]

The intake air quantity serves as the main actuating variable for determining the basic injection quantity. Since the air is measured before it actually reaches the engine cylinders, an acceleration enrichment effect occurs when the air throughput is changed.

The air flow sensor consists of an air funnel in which a sensor plate is free to pivot. The air flowing through the funnel deflects the sensor plate by a given amount from its zero position. This movement is transmitted by a lever system to a control plunger which determines the basic fuel quantity (**Fig. 4.144**).

1 Intake air
2 Control pressure
3 Fuel inlet
4 Metered quantity of fuel
5 Control plunger
6 Barrel with metering slits
7 Fuel distributor

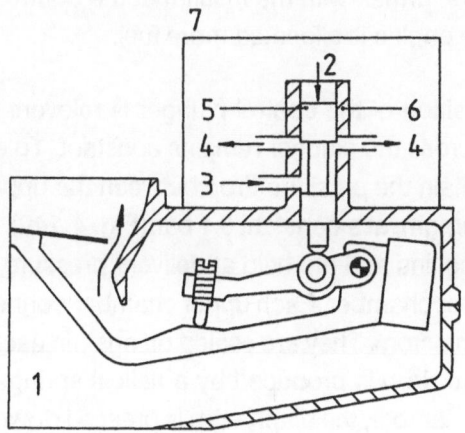

Fig. 4.144. Lever system with control plunger and barrel in the fuel distributor of a Bosch K-Jetronic [4.22]

The air flow sensor is designed so that the sensor plate can swing back in the opposite direction in the event of misfire.

The pressure acting on the sensor plate is counteracted by the pressure on the upper side of the control plunger (2 in Fig. 4.144). This control pressure is tapped from the primary pressure through a restriction bore (**Fig. 4.145**).

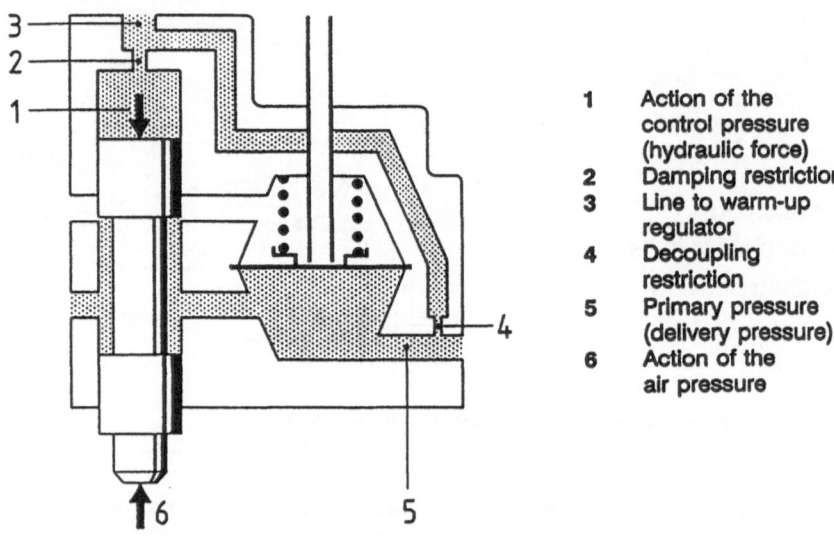

1   Action of the
    control pressure
    (hydraulic force)
2   Damping restriction
3   Line to warm-up
    regulator
4   Decoupling
    restriction
5   Primary pressure
    (delivery pressure)
6   Action of the
    air pressure

Fig. 4.145. Primary pressure and control pressure in the Bosch K-Jetronic [4.22]

In the primary pressure circuit a connection line joins the fuel distributor and the warm-up regulator (control pressure regulator). The control pressure is about 0.5 bar at cold start, and is raised to some 3.7 bars by the warm-up regulator as the engine warms up. The control pressure influences the fuel metering. When it is low, the air drawn in by the engine can deflect the sensor plate further, with the result that the control plunger opens the metering restrictions further and the engine is allocated more fuel.

The position of the control plunger is relevant for the fuel throughput only when the pressure drop across the plunger remains constant. To ensure this, differential pressure valves are used to maintain the pressure drop between the upper and lower chambers, which are separated by a diaphragm, at a constant 0.1 bar (**Fig. 4.146**). The lower chambers of all valves are connected by a ring line and are held at delivery pressure (primary pressure). The valve seat is located in the upper chamber. Each upper chamber connects to a metering slit and the corresponding line to the injectors. They are sealed off against each other. The diaphragms are spring-loaded. The pressure drop is produced by a helical spring: when a large basic fuel quantity flows into the upper chamber, the diaphragm is pressed downwards and opens the outlet cross-section of the differential pressure valve until the desired pressure drop once again prevails. A drop in the fuel quantity reduces the valve cross-section, owing to the equilibrium of forces at the diaphragm, until the pressure drop of 0.1 bar is again present. An equilibrium of forces thus prevails at the

1 Fuel intake
  (primary pressure)
2 Upper chamber of the
  differential pressure valve
3 Line to the fuel injector
  (injection pressure)
4 Control plunger
5 Control edge
  and metering slit
6 Valve spring
7 Valve diaphragm
8 Lower chamber of the
  differential pressure valve

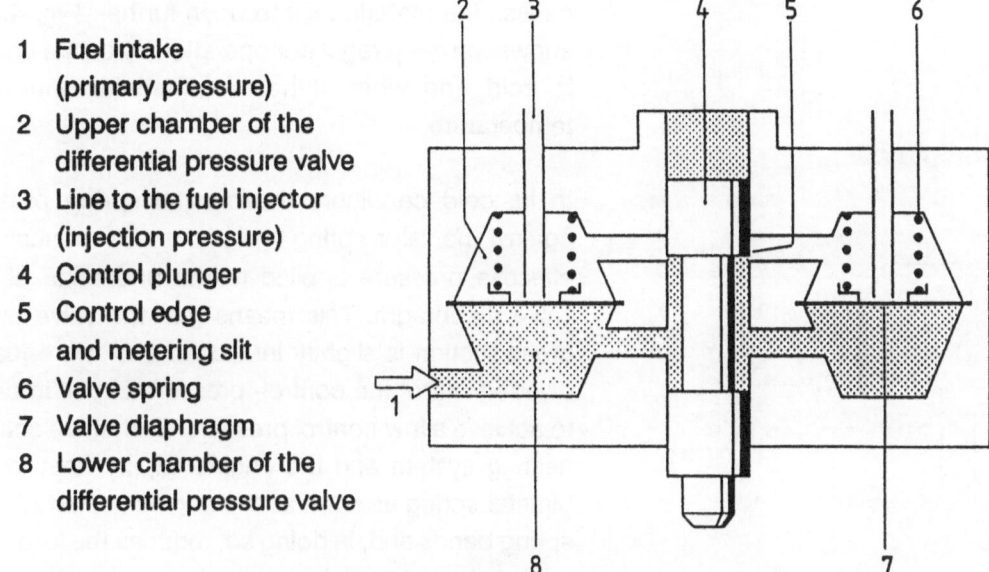

Fig. 4.146. Fuel distributor with differential pressure valves in the Bosch K-Jetronic [4.22]

diaphragm which is maintained for every basic fuel quantity by controlling the valve cross-section.

Fuel metering:

Basic adaptation of the air-fuel mixture is done by the shape of the air funnel. If the funnel has a regular conical shape, the result is a mixture with an approximately constant air-fuel ratio across the entire sensor plate range of travel. If the mixture is to be leaner or richer for specific load ranges, it is necessary to design the air funnel so that it becomes wider in stages. **Fig. 4.147** shows cone shapes of the air flow sensor for different operating modes.

1 For maximum power
2 For part load
3 For idle

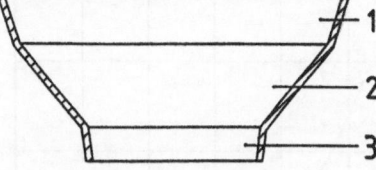

Fig. 4.147. Adaptations of the funnel shape in an air flow sensor in the Bosch K-Jetronic [4.22]

Cold-start enrichment is executed through a special cold- start valve which injects fuel into the intake manifold for a period limited by the thermo-timer. The cold-start valve and thermo-timer are similar to their counterparts in the L- Jetronic. Warm-up enrichment is controlled by the warm-up regulator. When the engine is cold, the warm-up regulator reduces the control pressure to a

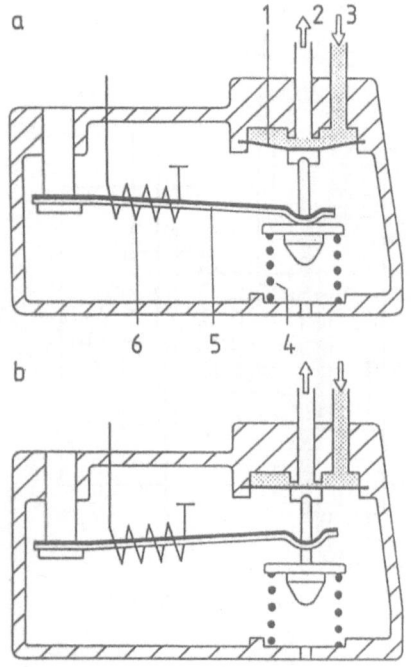

a) Cold engine
b) Engine at operating temperature

1 Valve diaphragm, 2 Return, 3 Control pressure (from the mixture control unit), 4 Valve spring, 5 Bimetal spring, 6 Electric heating

Fig. 4.148. Warm-up regulator in the Bosch K-Jetronic [4.22]

degree depending on the engine temperature, which causes the metering slit to open further. **Fig. 4.148** shows warm-up regulator operation when the engine is cold and when it has reached its operating temperature.

In its cold condition, the bimetal spring presses against the valve spring and as a result reduces the effective pressure applied to the underside of the valve diaphragm. This means that the valve outlet cross-section is slightly increased and more fuel is diverted out of the control- pressure circuit in order to achieve a low control pressure. Both the electrical heating system and the engine supply heat to the bimetal spring as soon as the engine is started. The spring bends and, in doing so, reduces the force opposing the valve spring which in turn increases the valve spring force applied to the flat-seat valve. This reduces the outlet cross-section and raises the pressure in the control pressure circuit. Warm-up enrichment is completed when the bimetal spring has lifted fully from the valve spring. With this, the control pressure is solely controlled by the valve spring and maintained at its normal level. The control pressure is about 0.5 bar at cold start and some 3.7 bars with the engine at operating temperature.

Fig. 4.149 illustrates characteristic warm-up regulator curves typical for various operating temperatures.

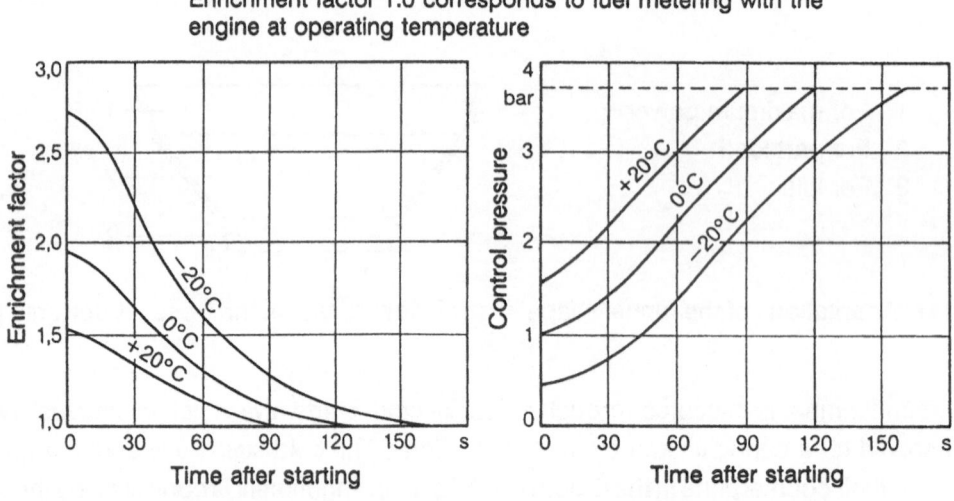

Enrichment factor 1.0 corresponds to fuel metering with the engine at operating temperature

Fig. 4.149. Warm-up regulator characteristics at various operating temperatures [4.22]

During warm-up the engine has to overcome increased friction which requires a larger mixture quantity during idling. The additional air is supplied by an auxiliary air device which is similar to the one described for the L-Jetronic.

Extra enrichment in addition to the mixture adaptation resulting from the air funnel shape for full load operation can be achieved by a specially designed warm-up regulator which regulates the control pressure as a function of the engine load.

When the engine is accelerated, i.e. the throttle valve is opened abruptly, this causes the sensor plate to "overswing," resulting in more fuel being metered to the engine (acceleration enrichment) and ensures good transition behavior.

Supplementary functions:

To achieve overrun fuel cutoff, it is possible to arrange a bypass around the sensor plate which is opened during overrun. The sensor plate then reverts to its zero position and blocks fuel metering. To adapt the injected fuel quantity to an air-fuel ratio of $\lambda = 1$, the pressure in the lower chambers of the fuel distributor is varied.

4.2.3.2. Bosch KE-Jetronic

The Bosch KE-Jetronic system is an enhanced version of the K-Jetronic. The basic system includes an electronic control unit to increase flexibility and add supplementary functions. The ECU receives signals from a number of additional sensors and uses them for controlling and if necessary adapting mixture formation.

Fuel supply is basically similar to the K-Jetronic system, except that while the K-Jetronic had its control pressure regulated by the warm-up regulator, the KE-Jetronic uses hydraulic counterpressure on the control plunger which is equal to the primary pressure. The control pressure must be maintained with maximum accuracy since variations would immediately affect the fuel-air ratio.

Fig. **4.150** provides a cross-section through the primary pressure regulator. Fuel enters on the left. The return fuel connection from the fuel distributor

1 Return line from
   fuel distributor
2 To the fuel tank
3 Adjustment screw
4 Counterspring
5 Seal
6 Inlet
7 Valve plate
8 Diaphragm
9 Control spring
10 Valve body

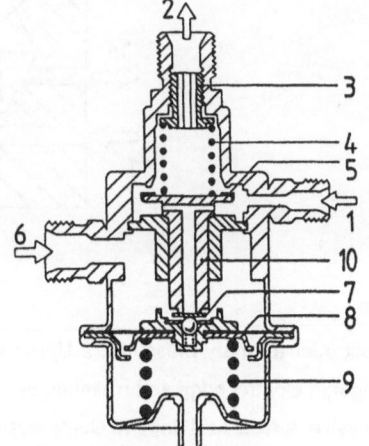

Fig. 4.150. Cross-section of the primary pressure regulator in the Bosch KE-Jetronic [4.22]

is located on the right. The return line to the tank is connected at the top. As soon as the fuel pump generates pressure at engine start, the control diaphragm of the pressure regulator moves downwards. The pressure of the counterspring forces the valve body to follow the diaphragm until, after a short travel, it encounters a stop and the pressure control function starts. The fuel returning from the fuel distributor, comprising the fuel flowing through the pressure actuator and the control-plunger leakage, can now flow back through the open valve seat to the tank together with the excess fuel.

When the engine is switched off, the electric fuel pump stops as well. The pressure in the fuel supply system drops and the valve plate moves back to its seat, pushing the valve body upwards against the force of the counterspring until the seal closes the return line to the fuel tank.

The pressure in the fuel supply system then sinks rapidly to the level of the closing pressure so that the injectors close tightly. The system pressure consequently rises again to a value determined by the fuel accumulator.

In the fuel distributor (**Fig. 4.151**), a damping throttle serves to suppress vibrations that could occur as a result of sensor-plate forces. When the engine is switched off, the control plunger sinks until it comes to rest against an axial seal ring due to the force of the compression spring and the residual primary pressure acting on it. This measure serves to prevent pressure loss due to leakage past the control plunger and emptying of the fuel accumulator. The fuel accumulator must remain full because it has to maintain the primary pressure above the fuel vapor pressure when the engine is switched off.

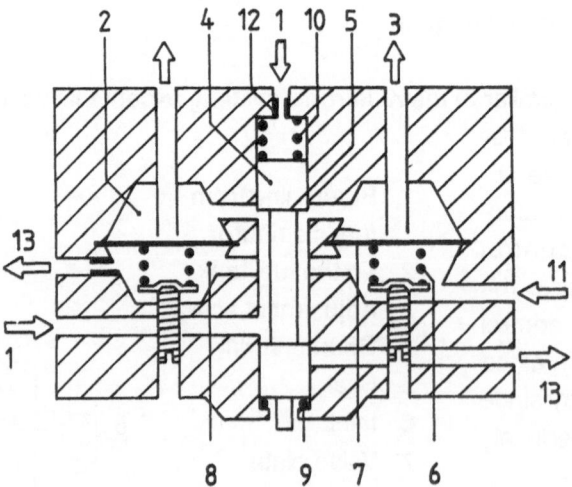

1 Fuel inlet (primary pressure), 2 Upper chamber of differential pressure valve, 3 Line to the injector, 4 Control plunger, 5 Control edge and metering slit, 6 Valve spring, 7 Valve diaphragm, 8 Lower chamber of differential pressure valve, 9 Axial seal ring, 10 Compression spring, 11 Fuel from the electro-hydraulic pressure actuator, 12 Throttling restriction, 13 Return line

Fig. 4.151. Fuel distributor with differential pressure valves in the Bosch KE-Jetronic [4.22]

Electronic control unit:

Operating data over and above the information coming from the intake air quantity are registered by sensors and signalled to the ECU which uses them to generate a control signal to the electro-hydraulic pressure actuator. The sensors and their characteristic values are:

- Throttle valve switch: full load, idle;

- Ignition triggering system: speed;

- Ignition and starting switch: start;

- Engine temperature sensor: engine temperature;

- Barometric cell: air pressure;

- Lambda sensor: air fuel mixture composition.

Fig. 4.152 is a block diagram of the KE-Jetronic's ECU in the analog version. ECUs with a more extensive range of functions are designed as digital versions.

Electro-hydraulic pressure actuator:

Depending on the operating mode of the engine and the resulting current signal received from the ECU, the electro-hydraulic pressure actuator varies the pressure in the lower chambers of the differential pressure valves. This changes the amount of fuel delivered to the injectors.

Fig. 4.153 illustrates the electro-hydraulic pressure actuator mounted on the fuel distributor. It is a differential pressure controller that operates according to the nozzle/baffle plate principle with a current-controlled pressure drop. In a housing of non-magnetic material, an armature is suspended from a frictionless suspension element between two double magnetic poles.

Operation of the electro-hydraulic pressure actuator is explained in **Fig. 4.154**: the magnetic fluxes of the permanent magnet

The correcting signals from the individual blocks are combined in an adder stage and sent to the electro-hydraulic pressure actuator.

VK   full load correction
SAS overrun fuel cutoff
BA   acceleration enrichment
NA   post-start enrichment
SA   voltage increase for starting
WA   warm-up enrichment
SU   adder stage
ES   output stage

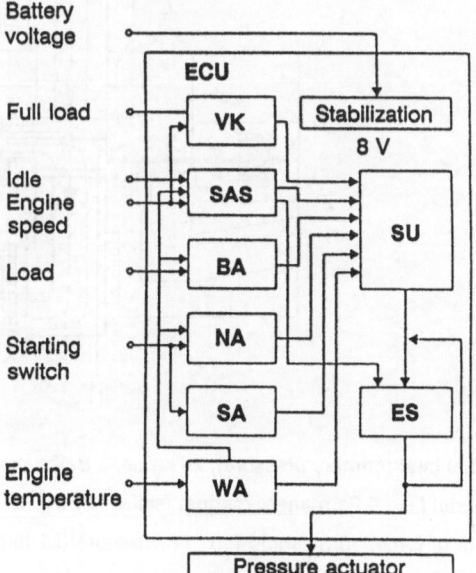

Fig. 4.152. Block diagram of the Bosch KE-Jetronic ECU, analog version [4.22]

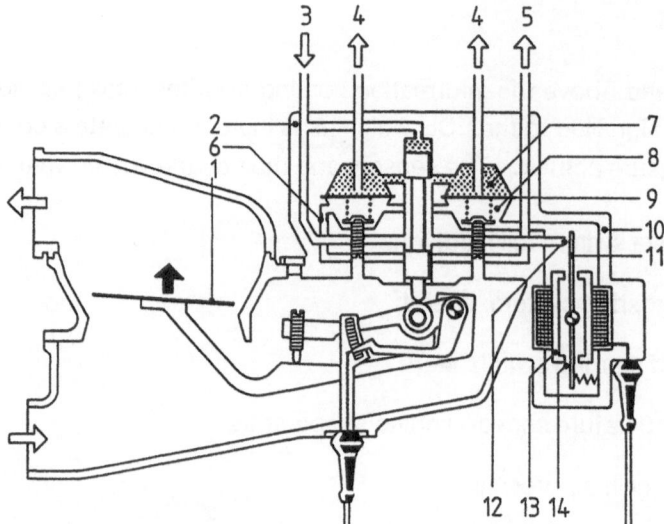

1 Sensor plate, 2 Fuel distributor, 3 Fuel inlet (primary pressure), 4 Fuel to the injectors, 5 Fuel return to the pressure regulator, 6 Fixed restriction, 7 Upper chamber, 8 Lower chamber, 9 Diaphragm, 10 Pressure actuator, 11 Baffle plate, 12 Nozzle, magnetic pole, 14 Air gap

Fig. 4.153. Electro-hydraulic pressure actuator fitted to the fuel distributor in the Bosch KE-Jetronic [4.22]

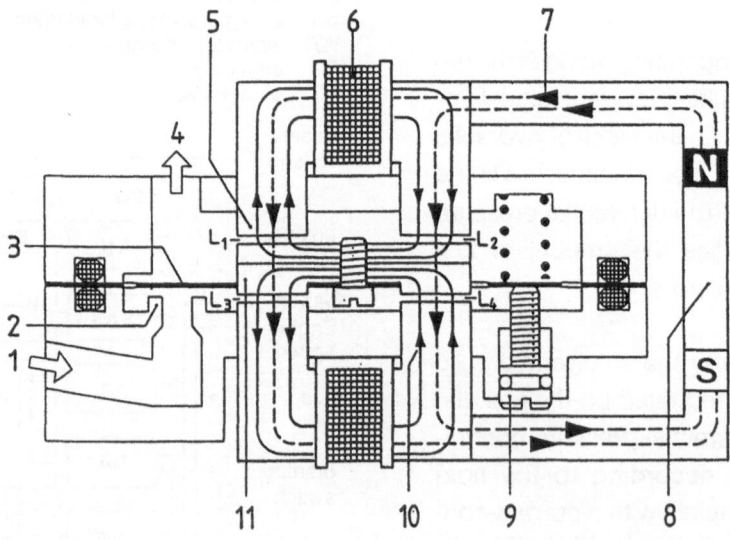

1 Fuel inlet (primary pressure), 2 Nozzle, 3 Baffle plate, 4 Fuel outlet, 5 Magnetic pole, 6 Magnet coil, 7 Permanent magnet flux, 8 Permanent magnet (shifted by 90° to fit into the drawing plane), 9 Adjustment screw for basic movement of force, 10 Solenoid flux, 11 Armature (L1 to L4 = air gaps)

Fig. 4.154. Section through the electro-hydraulic pressure actuator in the Bosch KE-Jetronic [4.22]

(broken line) and that of the solenoid (solid line) are superimposed upon each other in the magnetic poles and their air gaps.

The fuel jet entering through the nozzle attempts to bend the baffle plate away against the magnetic and mechanical forces. The differential pressure between the inlet and outlet, determined by a fixed restriction, is proportional to the current. The pressure drop at the nozzle, varied by the pressure actuator current, results in a variable pressure in the lower chamber which in turn affects the difference between upper chamber pressure and primary pressure (i.e. at the metering slits). It is thus a means to vary the fuel quantity delivered to the injectors.

Enrichment for cold-start, post-start, and warm-up is basically similar to the respective functions in the K- Jetronic. A difference is in acceleration enrichment where additional fuel is metered to the engine when it is still cold. As a result of the change in the load signal (in terms of time), the ECU recognizes that the engine is accelerated and triggers acceleration enrichment as a function of the temperature. It is triggered by a needle-shaped enrichment pulse with a length of about 1 second.

A potentiometer registers the sensor plate in the air flow sensor and is thus responsible for acceleration enrichment. It is of the film type on a ceramic base (**Fig. 4.155**).

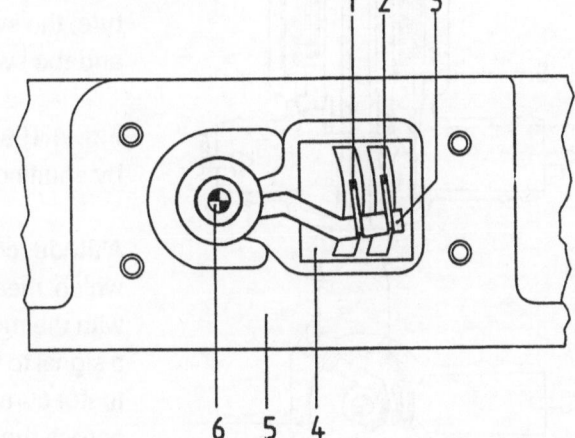

1 Pickoff brush
2 Main brush
3 Slide lever
4 Potentiometer plate
  (shifted out of the focal plane)
5 Air flow sensor housing
6 Sensor plate shaft

Fig. 4.155. Potentiometer to determine the sensor plate position in a Bosch KE-Jetronic [4.22]

Closed-loop idle speed control (**Fig. 4.156**) is by means of a rotary idle actuator which supplies the engine with more or less intake air through a bypass around the throttle valve. For a detailed view of the rotary idle actuator see **Fig. 4.157**.

Overrun fuel cutoff and engine speed limitation are controlled by an electro-hydraulic pressure actuator. In the fuel distributor, the differential pressure valves are closed by the springs in their

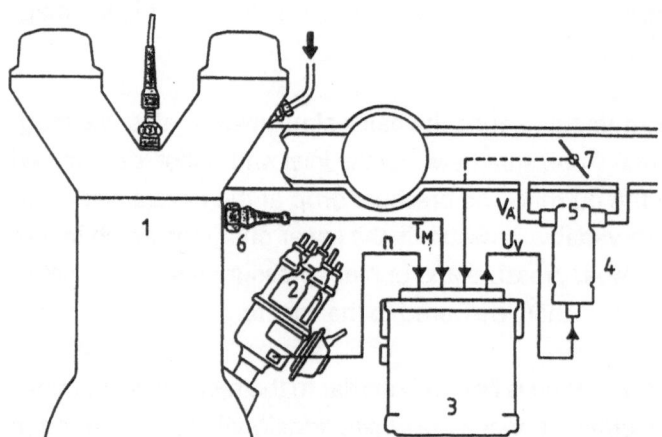

1 Control system: engine
2 Control variable:engine speed n
3 Controller: control unit
   (supplies control voltage U_v)
4 Actuator: rotary idle actuator
5 Actuating variable: bypass cross-
   section (intake air quantity V_A)
6 Auxilliary control variable:
   engine temperature T_M
7 Auxilliary control variable:
   throttle valve end position
   ($\alpha_{DK} = 0$)

Fig. 4.156. Control loop for the closed-loop idle speed control in a Bosch KE-Jetronic [4.22]

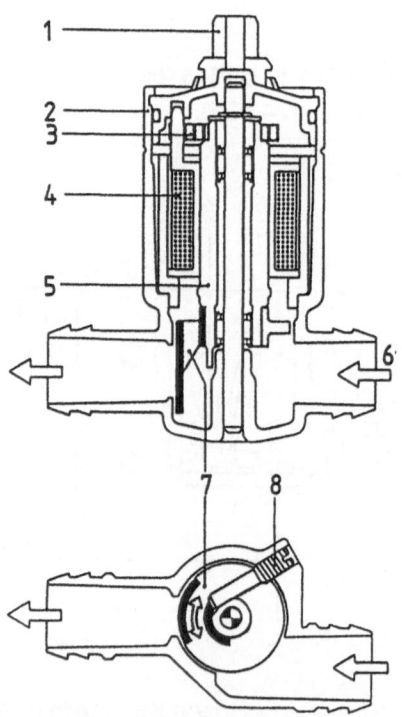

1 Plug, 2 Housing, 3 Return spring, 4 Coil, 5 Rotating armature, 6 Air passage as bypass around the throttle plate, 7 Rotating slide, 8 Adjustable stop

Fig. 4.157. Rotary idle actuator (single-coil rotary actuator) in the Bosch injection system [4.22]

lower chambers when the pressure drop in the actuator approaches zero, thereby interrupting the fuel flow to the injectors.

Fig. 4.158 indicates the minimum speed for overrun fuel cutoff as a function of the coolant temperature, the switching threshold for overrun fuel cutoff and the switching threshold for restoring injection.

Fig. 4.159 shows how the engine speed is limited by shutting off the fuel supply.

Altitude compensation is provided by a sensor which measures the air pressure. In accordance with the momentary air pressure, the sensor sends a signal to the ECU which changes the pressure actuator current, which in turn alters the pressure drop across the metering slits and with it the fuel quantity.

In engines provided with a lambda closed-loop control feature, the lambda sensor signal is processed in the ECU, and fuel metering is then controlled through the pressure actuator.

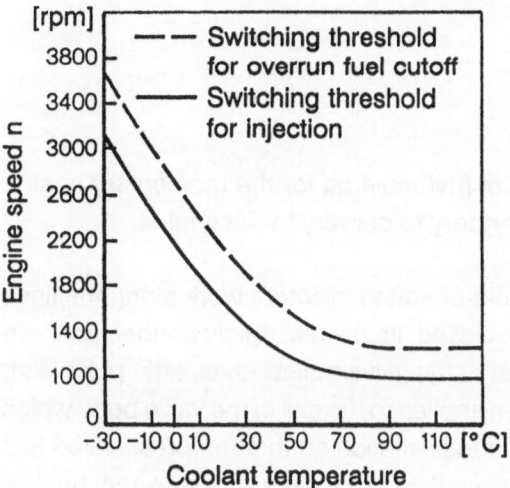

Fig. 4.158. Minimum engine speed for over-run fuel cutoff as a function of the coolant temperature in the Bosch KE- Jetronic [4.22]

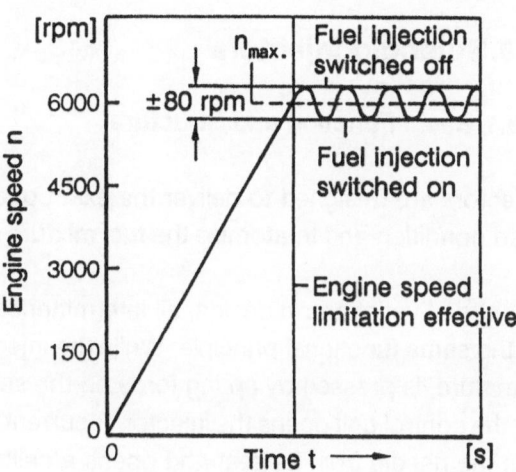

Fig. 4.159. Limiting the maximum engine speed n_{max} by stopping the fuel delivery to the injectors in the Bosch KE- Jetronic [4.22]

4.3. Solenoid Injectors

4.3.1. Basic Function and Structure

Injectors are designed to deliver the exact quantity of fuel required for the momentary engine load condition and to atomize the fuel mixture preparatory to delivery to the engine.

In spite of variations in design, all intermittent solenoid-operated injectors work along the lines of the same functional principle: While the injector is closed, its needle, rigidly connected to an armature, is pressed by spring force on the seal seat at the valve outlet. An electric pulse sent by the control unit opens the injector. A current is then applied to the coil in the valve body which lifts the needle from its seat and opens a calibrated cross-section so that the pressurized fuel can flow through the valve opening. Needle travel is restricted to 0.1-0.2 mm by a mechanical stop (**Fig. 4.160**).

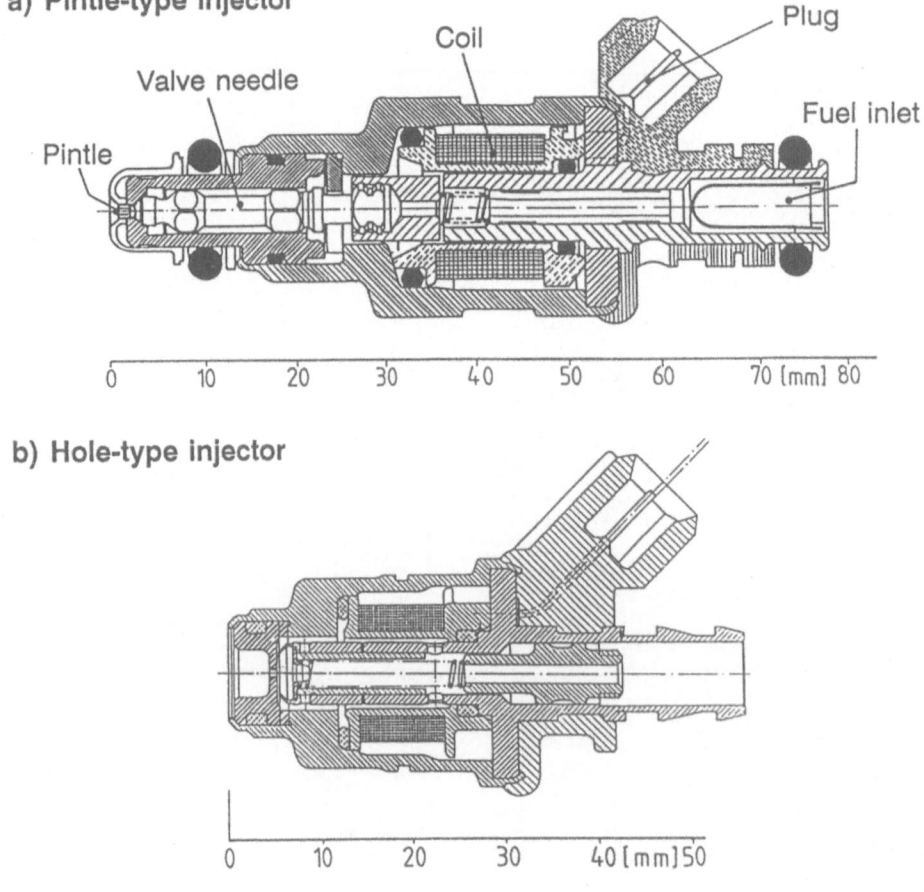

Fig. 4.160. Solenoid-operated injectors [4.48]

Opening of the valve takes some 0.5 to 1 ms, depending on its design: the time required to build up a magnetic field to overcome the closing force of the pretensioned spring, the hydrostatic force at the valve seat and the mass acceleration force. The response time, i.e. the delay from the start of the electric injection pulse to the moment that the needle starts to open, greatly depends on the needle design and on the coil and circuit arrangement. When the mass to be accelerated is kept small, response times will be short. In Bosch EV 1.1 injectors, the valve needle, which is connected rigidly to the armature, has a mass of 4.05 g, a figure that could be reduced in later versions (down to 2.7 g in the Bosch EV 1.4). Designers are striving to develop light and simple-shaped needles that have a small mass (**Fig. 4.161**).

Injectors differ primarily in the design of their metering cross-section and the shape of the valve needle. **Fig. 4.162** shows schematic views of the orifices in a pintle-type injector, a single-hole injector, and a multi-hole plate injector, together with the respective sprays generated by them.

Fig. 4.162 further includes a design that uses a front attachment to divide the spray of a single-hole-type injector into two branches and ensure that the fuel is injected accurately into the two intake ports of a four- valve engine.

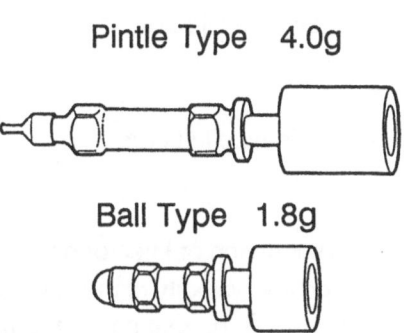

Pintle Type   4.0g

Ball Type   1.8g

Fig. 4.161. Valve needle designs [4.53]

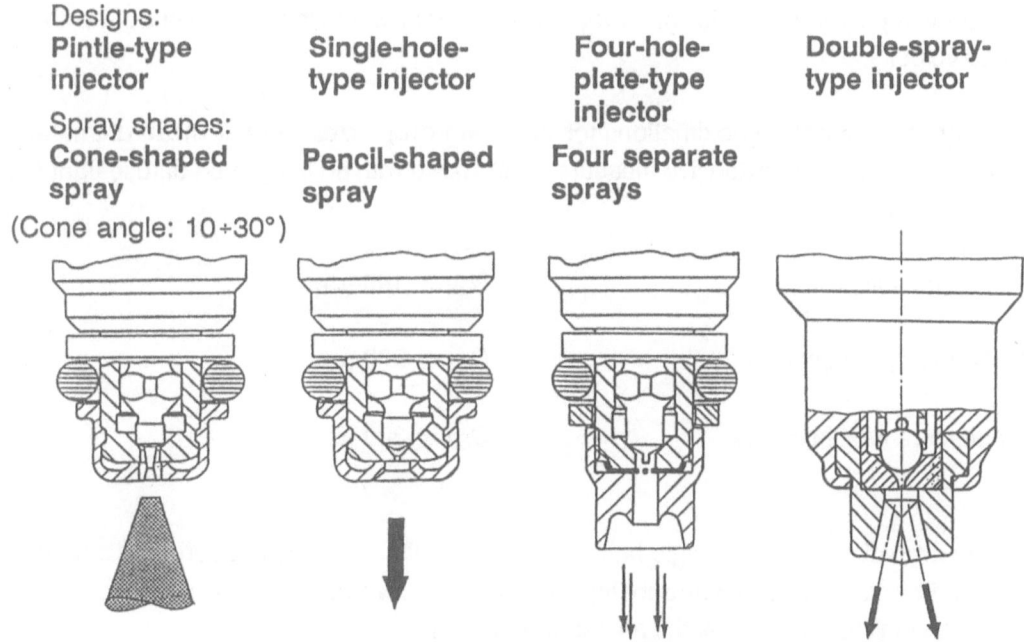

Designs: **Pintle-type injector**	**Single-hole-type injector**	**Four-hole-plate-type injector**	**Double-spray-type injector**
Spray shapes: **Cone-shaped spray** (Cone angle: 10÷30°)	**Pencil-shaped spray**	**Four separate sprays**	

Fig. 4.162. Typical designs of injectors, distinguished by the shape of their spray [4.47 and 4.55]

4.3.2. A Comparison of Injector Designs

4.3.2.1. General

The most common injector designs and spray types are:

Design:	Spray type:
Pintle-type injector	Cone-shaped spray (cone angle: 10-30°)
Multi-hole-type injector	Single spray (made up of several sprays) or double spray (for four-valve engines)
Single-hole-type injector	Pencil-shaped spray

A direct comparison of injector designs helps clarify the differences between them with regard to the shape and velocity of their spray, the range of drop sizes and response times. **Fig. 4.163** shows the sprays generated by the most commonly used injectors. The photographs, taken during intermittent operation, indicate the time sequence of an injection process. They were taken by stroboscopic flash shutter release with a delay T_v of 2, 4, and 6 ms. The injection pulse had a length T_i of 4.5 ms.

The mean velocity of the drops comprised in the spray can be identified as the spray velocity. The velocity depends on the design of the metering bore which is decisive in converting pressure into velocity.

In a similar way we get approximations for the mean drop sizes (Sauter mean diameter D_{32}). The figures are based on extensive measurements made with an integrated diffuse-light system (cf. Chapter 3.4).

Pintle-type injectors:	$D_{32} > 100 \ \mu m$;
Multi-hole-type injectors:	$D_{32} \approx 200 \ \mu m$;
Single-hole-type injectors:	$D_{32} \approx 300 \ \mu m$.

The mean drop diameter for a single-hole-type injector with a pencil jet is derived from the mean values for the many droplets surrounding the compact core of the spray, which latter is interpreted as a single large drop by the measuring system.

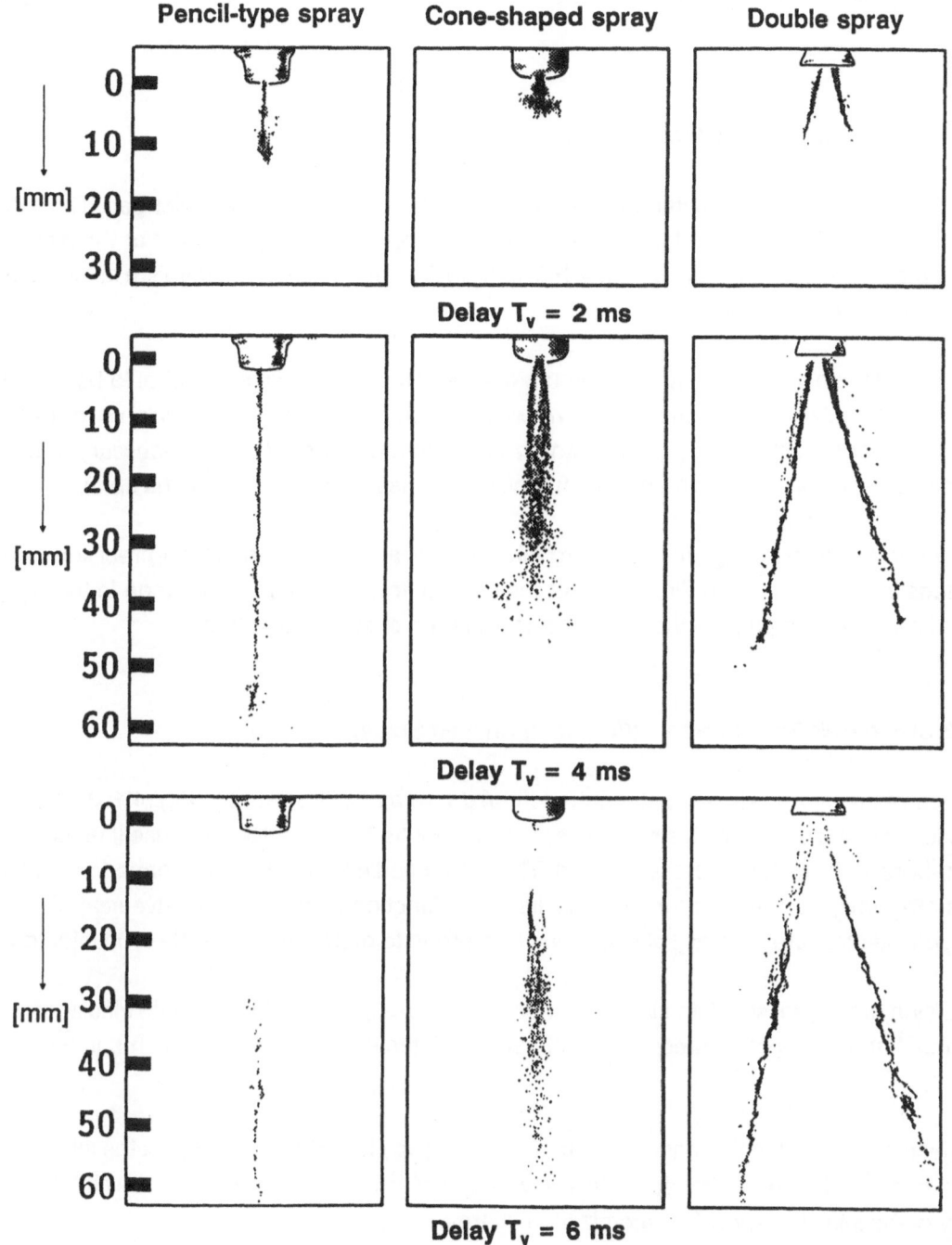

Fig. 4.163. A comparison of spray types generated by solenoid-operated injectors

A method to improve atomization of intermittent injectors is to shroud their orifice with air. A disadvantage is that in contrast to continously injecting systems, air-shrouded intermittent injectors can utilize only a small part of the atomization air supplied by the manifold vacuum because of the short injection periods at low engine loads and the long injection intervals at low engine speeds.

4.3.2.2. Pintle-Type Injectors (For Cone-Shaped Spray)

Pintle-type injectors have a metering cross-section that is shaped as an annular gap. The valve needle forms a conical pintle at its tip (Fig. 4.162). The ejected fuel is deflected at the pintle to form a spray enveloping the pintle (Fig. 4.163, center). The atomization mechanism active in this spray is described in Section 2.5.1.1.

The shape of the pintle determines the circumference of the spray. The spray should be as wide as possible to achieve fine atomization. The wider it is the more surface is available for the intake air to become active and produce atomization. On the other hand it is necessary to keep manifold wall wetting to a minimum, which requires a small circumference of the spray.

Manufacture of the metering cross-section, valve needle, and pintle is thus high-precision work. Variations in the annular gap directly affect the throughflow. The slightest change in the pintle design has a major impact on the spray shape and droplet size distribution.

4.3.2.3. Single-Hole-Type Injectors (For String-Shaped Spray)

This type of injector produces a straight and narrow spray as illustrated in Fig. 4.163 (left). Its orifice is designed along much simpler lines than the pintle-type injector, because it uses a calibrated bore as its metering cross- section. The valve needle is simplified or shaped as a plate (Fig. 4.162) which is sufficient for it to fulfill its single function of sealing the valve seat. Single-hole-type injectors are normally cheaper and less prone to deposits than pintle-type injectors.

Conversion of fuel pressure into kinetic energy is better owing to the higher nozzle efficiency η_D. At 20 $m.s^{-1}$ and an injection pressure of 3.0 bars, the injector has a relatively high spray velocity.

The injector is located in the manifold so that the fuel spray is injected directly on the inlet valve head, where it is diffused on impact. In this way it is possible to effectively convert the high kinetic energy of the injection spray into atomization energy.

4.3.2.4. Multi-Hole-Type Injectors

The orifice in a multi-hole-type injector is designed as a metal plate that contains several metering bores and is held in place by a metal sleeve threaded to the orifice.

This design allows variations both in the number of metering bores and in the location of the bore axes vis-à-vis the valve axis which affect the shape of the spray and its atomization behavior.

When used in four-valve engines, hole-type injectors can be designed to divide the injection spray into two branches, with the result that fuel supply to the twin injectors of an engine cylinder will be symmetrical. The spray is divided by metering bores suitably positioned to follow the branching manifold in multi-hole-plate-type injectors or, alternatively, by a front attachment placed at the orifice in single-hole-type injectors (Fig. 4.162, right). The spray shape of a double-spray-type injector is illustrated in Fig. 4.163 (right).

4.3.3. Key Parameters of Solenoid-Operated Injectors

Major criteria to evaluate an injector are the shape of its spray, its circumference and velocity, its droplet size distribution and, not least, its dynamic behavior.

Depending on the engine operating conditions, the fuel quantity to be injected is determined by the period that the injector orifice is open. By varying the opening time, it must be possible to achieve a fuel throughput range of some 1:10 without changing the ratio of injection frequency to engine speed [4.47].

The flow behavior is described by the factors of steady fuel throughput and dynamic flow volume. The steady fuel throughput is the fuel volume flow generated at a given differential pressure when the injector is open continuously. For the linear relationship between pulse length and injected fuel quantity see **Fig. 4.164**.

$$q = T_i \cdot V \qquad\qquad (4.27)$$

The following theoretical relationship applies:

where:

q	[mm^3]	..... injected fuel volume per cycle
T_i	[ms]	..... pulse length
V	[mm^3/ms]	..... steady flow volume.

Actually, the steady curve is shifted towards higher pulse lengths due to the delay from pulse start and the time that the valve needle is lifted from its seat (the so-called pickup time) to valve closing.

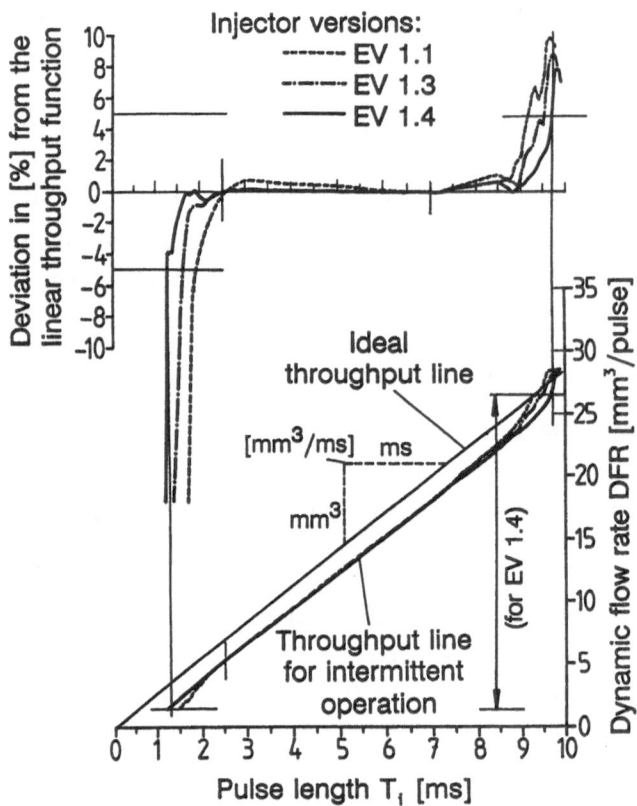

Fig. 4.164. Relationship between pulse length and injected fuel quantity (idealized and actual behavior) [4.47]

Current-controlled injectors (low impedance; **Fig. 4.165**, bottom) have shorter pickup times than voltage-controlled ones (high impedance; Fig. 4.165, top). From the needle travel curve it can be seen that the needle rebounds during opening and closing. This also generates vibrations which reduces fuel throughput. Consequently, we find a significant difference between the dynamic and steady curves in the opening and closing phases.

Injector linearity is the period of injection where the deviation of the actual fuel throughput from its linear throughput curve is still within a certain margin of tolerance. Minimum injection length is limited by the opening and closing times of the injector and the resultant throughput scattering in the non-linear area, while maximum injection length is limited by the low cycle lengths at high engine speeds.

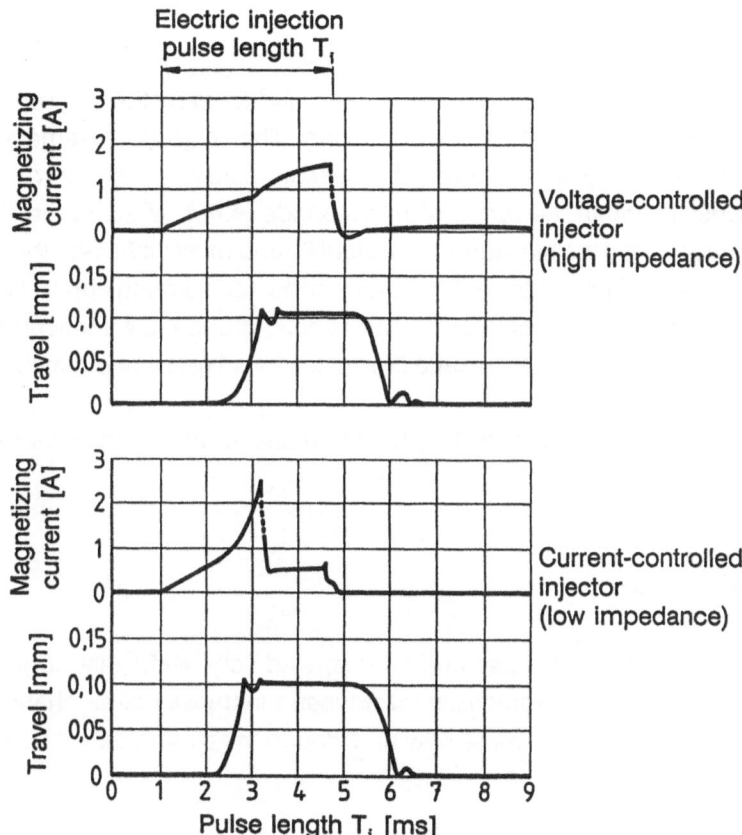

Fig. 4.165. Pickup times in current-controlled and voltage-controlled injectors [4.49]

4.4. Air Mass Flow Measurement

As mentioned earlier, determination of the air mass inducted by the engine cylinder is the most important prerequisite for optimum fuel metering. The major requirements to be met by a measurement system used are coverage of a measurement range of approx. 65:1 with 3% error and a time constant within the range of milliseconds [4.46]. Measurement systems used, of course, also have to ensure a high long-term stability and must withstand the temperature range common in engine design (-40 to +130°C) as well as accelerations up to 80g. Another important characteristic is immunity against electromagnetic interference. Measurement systems also have to be capable of handling the intake flow pulsations typical of engines.

Only few of the numerous theoretically possible measurement principles have found general use.

4.4.1. Air Flow Sensors

Devices of this type are widely used and have proved very useful, although they have the disadvantage of producing a characteristic error when the density of air changes. In accordance with Bernoulli's equation, the air mass flow m_L through the air flow sensor can be expressed as:

where:

$$m_L = \alpha \cdot A \cdot (\rho_L \cdot \Delta p)^{0,5} \qquad\qquad (4.28)$$

$\alpha \cdot A$	$[m^2]$	.....the area of the smallest flow cross section
α	$[-]$	.....the flow index
ρ_L	$[kg.m^{-3}]$	.....the air density before constriction of the stream
Δp	$[N.m^{-2}]$	.....the pressure loss at the sensor plate
m_L	$[kg.s^{-1}]$	.....the air mass flow rate

Keeping the antagonistic force to the dynamic pressure constant is crucial, since only in this case there will be a direct relationship between air flow rate and deflection of the sensor plate.

As described earlier, a sensor plate deflection is used for direct control of the gasoline quantity in the case of the Bosch K-Jetronic system and is transformed, via a potentiometer, to a voltage signal in the case of the Bosch L-Jetronic system.

If the air density changes, an error corresponding to the root of the air density change is produced in accordance with the above equation. Therefore, in many cases an air temperature sensor and in specific cases an atmospheric pressure sensor are additionally used.

Based on the quadratic relationship between pressure loss and air velocity according to Bernoulli, the mean values of these two parameters generally are not equal in pulsating flows as is the case in internal combustion engines. Therefore, all measurement systems of this type produce a so-called "pulsation error", which increases with increasing pulsation - especially in the full-load range, since under full load, contrary to part-load conditions, these pulsations are not dampened due to the throttle valve being fully opened. There is evidence [4.46] that pulsations always lead to excess readings of the flowmeter. These excess readings will be the smaller the greater the pressure loss in the induction system and the greater the inducted volume and the pulsation frequency (engine speed).

4.4.2. Vortex Flowmeters

Flowmeters of this type have found a comparatively limited use. Their operation is based on the Kármán vortex principle, see **Fig. 4.166**. A special vortex generator placed in the intake air stream creates vortices whose frequencies are a measure of the volume flow. These frequencies are determined by emitting ultrasonic waves in transverse direction to the intake air stream - the propagation speed of these waves, which is influenced by the vortices, is then measured and evaluated by an ultrasonic receiver.

4.4.3. Thermal Sensors

Presently, sensors of this type are the most suitable devices for measurement of the intake air flow.

The theoretical principle of thermal sensors is King's law, which expresses a semi-empirical relationship between the required heating current and the air mass flow rate. In a simplefied form, this relationship can be expressed as [4.46]:

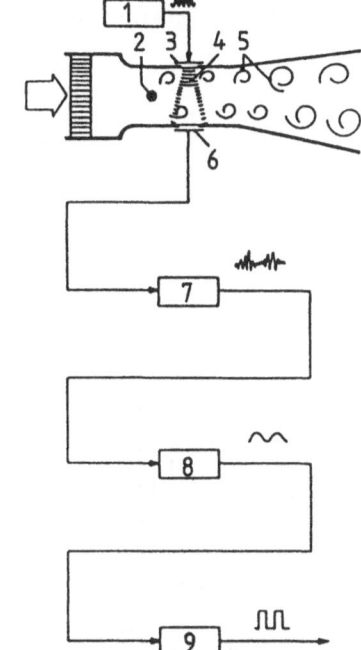

1 Oscillator, 2 Vortex generator, 3 Transmitter, 4 Ultrasonic waves, 5 Vortices, 6 Receiver, 7 Amplifier, 8 Filter, 9 Pulse shaper

Fig. 4.166. Kármán Vortex flowmeter [4.21]

$$I_H \approx K_1 + K_2 \cdot (m_L)^{1/4} \tag{4.29}$$

I_H ..... the sensor heating current required for maintaining constant temperature

K_1, K_2 .....   are constants

m_L .....   air mass flow rate.

This relationship applies - irrespective of the geometrical shape of the flow obstacle - for both the hot-wire and the hot-film air mass flowmeters.

Design variants of thermal sensors are the hot-wire and the hot-film air mass flowmeters.

4.4.3.1. Hot-Wire Air Mass Flowmeter

A wire is suspended in the intake air stream and kept at a constant temperature - see **Fig. 4.167**. This heated wire is part of an electric bridge circuit whose output voltage is controlled to zero. This yields the above-shown relationship between air flow and heating current.

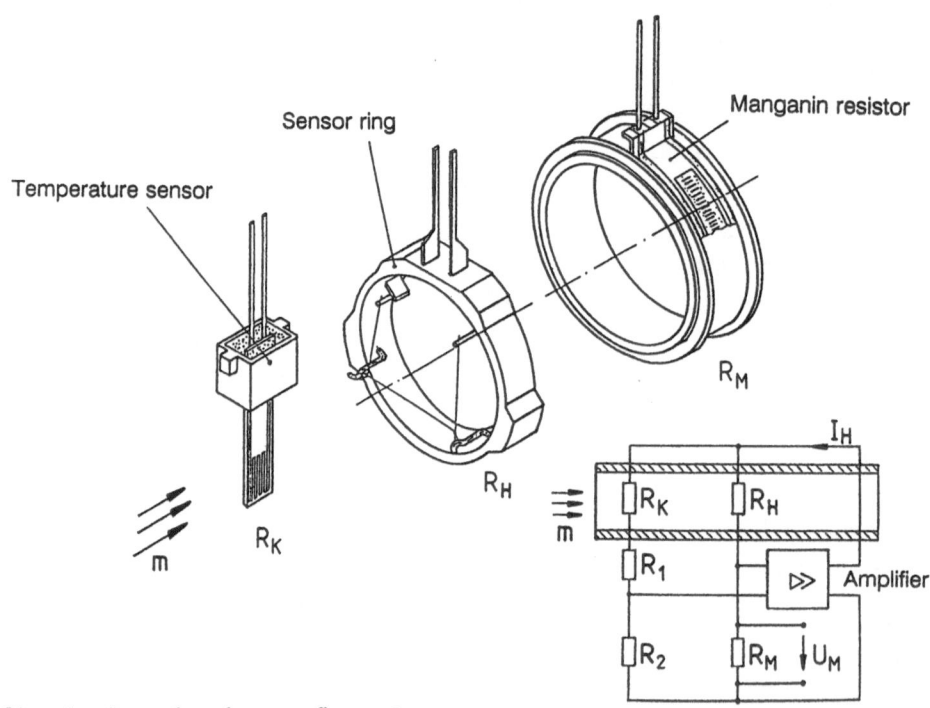

U_M	Signal voltage for air mass flow rate
I_H	Heating current
m	Intake air mass flow per unit time
R_H	Hot-wire resistor
R_K	Temperature compensation sensor
R_1, R_2	High-impedance resistor
R_M	Precision resistor

Fig. 4.167. Components of a hot-wire air mass flowmeter [4.22 and 4.46]

Fig. 4.168 shows the design of a hot-wire air mass flowmeter. The heated wire, a platinum wire 70 μm in diameter, is suspended in the inner measurement channel; the control circuit and the clean-burn circuit are part of a hybrid circuit incorporated in the outer channel. Owing to the small wire mass, cooling and control take place so quickly that limit frequencies in the kilohertz range are achieved.

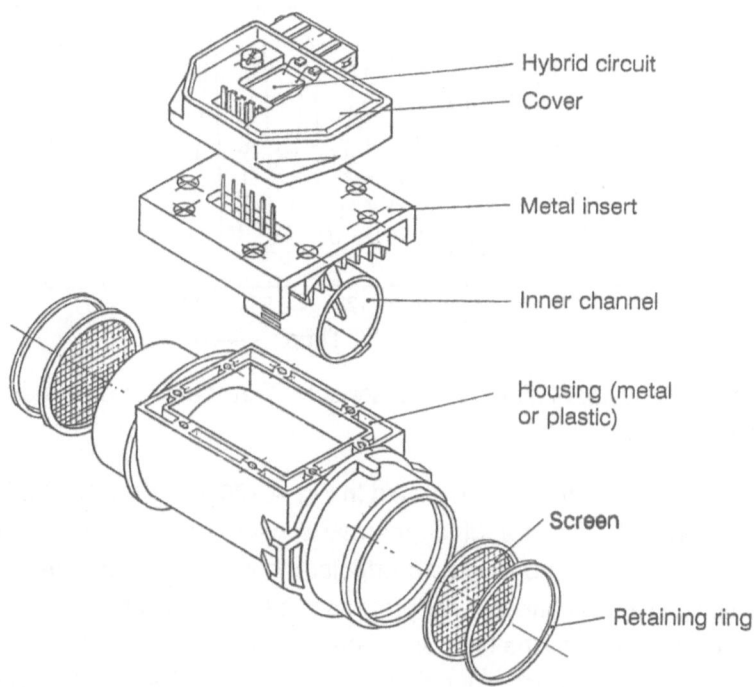

Fig. 4.168. Design of a hot-wire air mass flowmeter [4.46]

In pulsating flows, mean value errors could basically be avoided with this arrangement. This also applies to return flow pulsations - **Fig. 4.169** and **4.170**. However, this is not yet general practice.

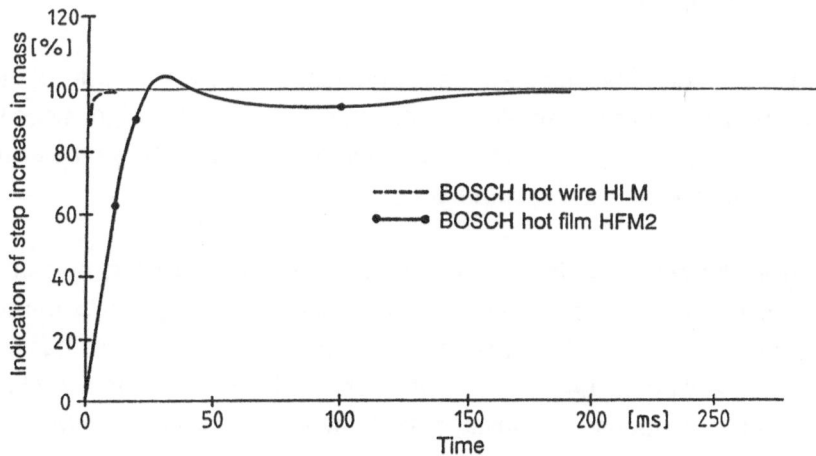

Fig. 4.169. Temporal trace of the indicated mass flow when the instrument is excited by a step change in flow rate from 10 kg/h to 310 kg/h [4.46].

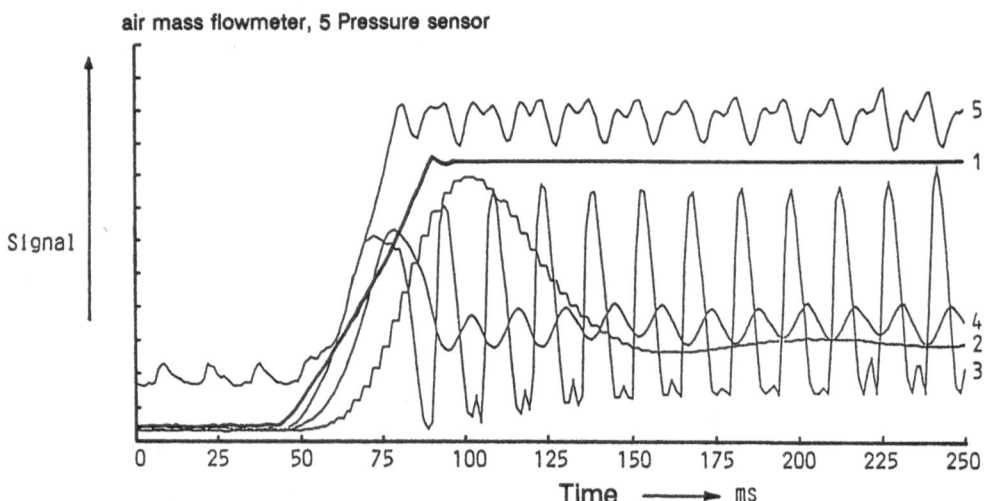

1 Throttle valve angle, 2 Air flow sensor (L-Jetronic), 3 Hot wire air-mass flowmeter, 4 Hot film air mass flowmeter, 5 Pressure sensor

Fig. 4.170. Dynamic behavior of air mass flow sensors. Throttle valve opens at n = 2000 1/min [4.46]

The extent of return flow pulsation, as illustrated in Fig. 4.170, is shown by the inverse peaks at the lower turning points of the curve. Since thermal sensors do not recognize the direction of flow, return flows are indicated positively. In principle, it would be possible to measure these peaks electronically and to evaluate them appropriately after linearization. However, this has not yet been done in practice due to the tedious procedure involved, so that, at present, pulsation errors still have to be expected when using thermal sensors. The main problem of flow measurement by hot-wire flowmeters is their susceptibility to soiling on account of the dust being transported in the intake air stream. Dirt accumulations of this type are eliminated by a short-term increase of the heating current after switching off the engine to achieve a hot-wire temperature between 1000°C and 1050°C, which causes the accumulated dirt to be burnt off.

4.4.3.2. Hot-Film Air Mass Flowmeter

The main disadvantage of the hot-wire air mass flowmeter, that is the sensitivity of the wire, does not exist in the hot-film air mass flowmeter. Moreover, it is simpler in design, since major single components can be integrated on a ceramic base plate. **Fig. 4.171** shows the arrangement of a hot-film air mass flowmeter in a measurement channel.

Fig. 4.172 shows the sensor with mounted resistors: heating resistor, sensor resistor, air temperature sensor resistor, trimming resistor. Fig. 4.172 additionally shows the bridge circuit of the hot-film air mass flowmeter. The response behavior of thermal sensors basically depends on the sensor mass and on the heat losses at the sensor attachment. The hot-wire sensor naturally exhibits the best response. However, if the heating resistor and sensor resistor are designed appropriately, satisfactory response behavior can be achieved also in the case of the hot-film sensor.

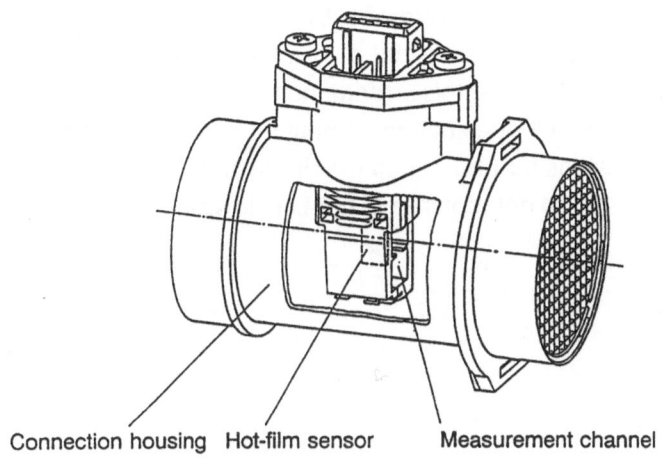

Connection housing Hot-film sensor   Measurement channel

Fig. 4.171. Incorporation of a Bosch hot-film air mass flowmeter in a measurement channel

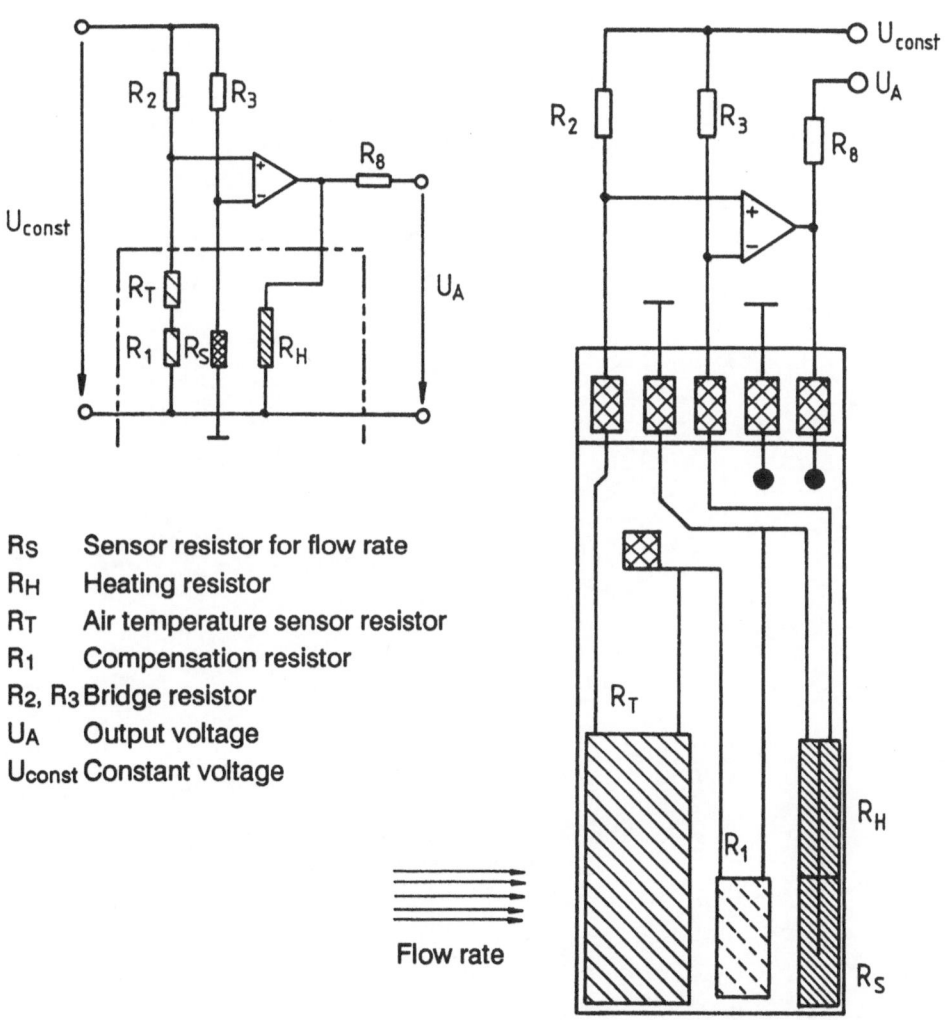

R_S Sensor resistor for flow rate
R_H Heating resistor
R_T Air temperature sensor resistor
R_1 Compensation resistor
R_2, R_3 Bridge resistor
U_A Output voltage
U_{const} Constant voltage

Fig. 4.172. Circuit and layout of a Bosch HFM2 hot-film air mass flowmeter [4.46]

In hot-film air mass flowmeters, the problem of soiling is solved by placing those sensor-resistor and heating-resistor zones which are decisive for heat transfer (see Fig. 4.172) on the downstream side of the sensor unit. By this arrangement, the inevitable dirt accumulation on the upstream edge of the sensor unit does not affect heat transfer and thus the characteristic sensor curve. Therefore, provisions for burning off the dirt (which would not be capable of providing sufficient heating power) are not required in hot-film flowmeters.

4.5. Combined Mixture Formation/Ignition/Engine Management Systems

A common electronic control system covering several functions such as ignition and fuel injection at once can prove practical. It cuts down on electronic components, and improves the compactness of the engine design. The Bosch Motronic is a typical representative of such systems. It is characterized by the high number of three-dimensional maps, which can be programmed at discretion for many subfunctions. The heart of the Motronic is its electronic control unit incorporating a digital microcomputer. The microcomputers available today make it possible to link fuel injection and ignition by a single control process, performed by one microcomputer, working with one power supply in a single housing for the ECU. The sensors for fuel injection and ignition can be used jointly.

The system described in **Fig. 4.173** and **4.174** combines an L- or LH-Jetronic with an electronically controlled ignition system. Its ECU uses a microcomputer to process the high number of input functions. Engine speed is measured directly at the crankshaft. The ignition time is calculated in the ECU, which leaves the ignition distributor with the sole task of distributing the high voltage. In addition to processing the signals from the flap-type air flow sensor or hot-wire anemometer to determine the engine load condition, the Bosch M2 unit performs closed-loop idle speed control, tank ventilation control, exhaust gas recirculation, sequential injection, knock control, and self-testing. Functions such as changing pressure control, manifold selection, or ignition distribution are in the process of being integrated.

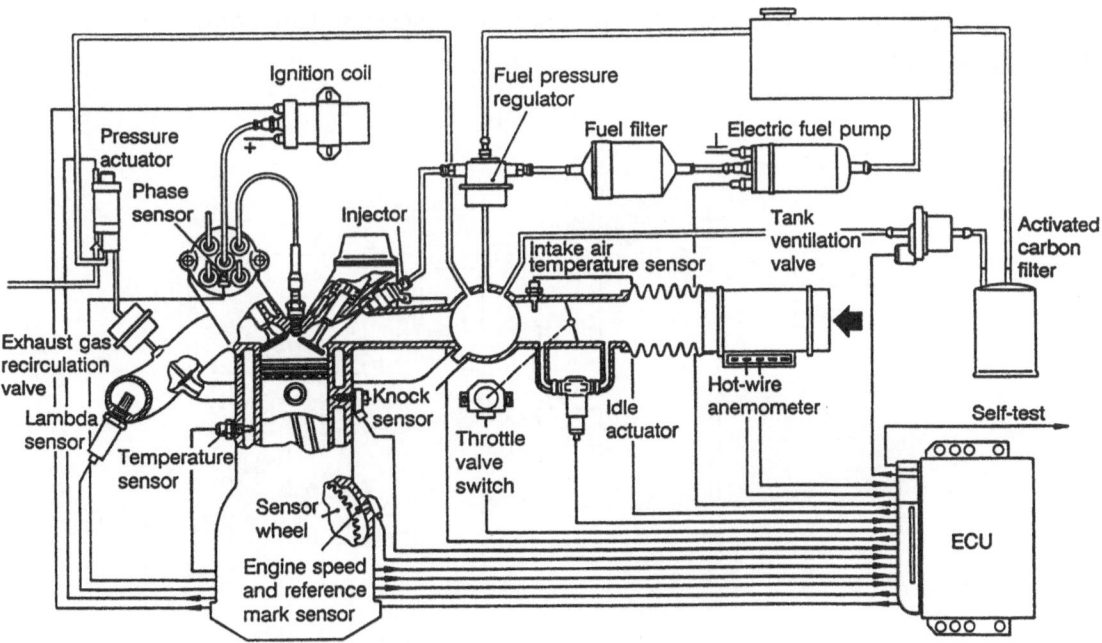

Fig. 4.173. Bosch Motronic M2 [4.39]

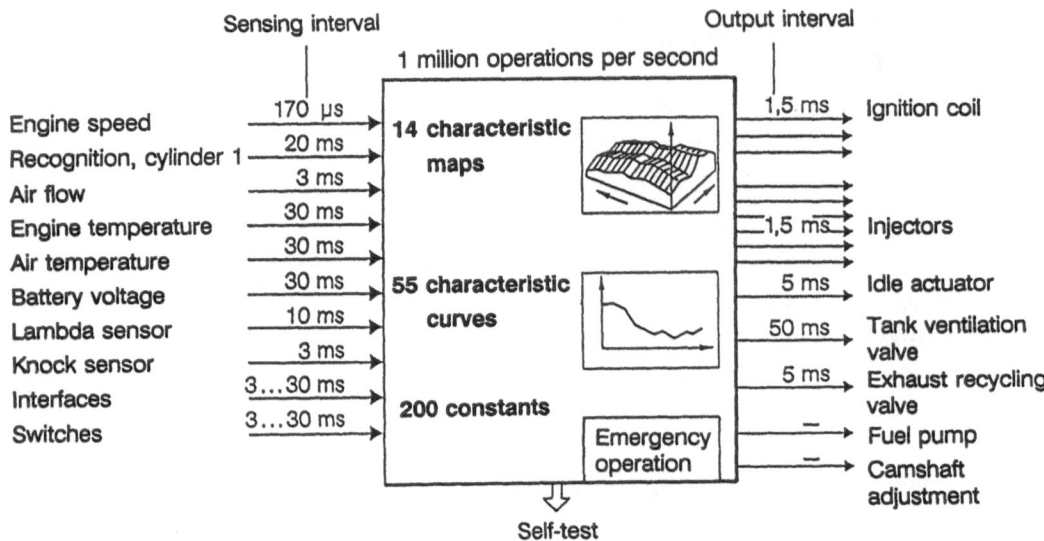

Fig. 4.174. Computer processes in engine control systems [4.39]

The BMW 12-cylinder engine [4.50] provides another comprehensive solution for a combined mixture formation/ignition system that also controls several other drive management functions such as tank ventilation, engine torque adaptation to check vehicle stability, and gear control. This design is described below as an exemplary mixture formation system since it integrates practically all the developments currently introduced in the field.

The design combines five electronic systems that communicate and are harmonized with each other (**Fig. 4.175**):

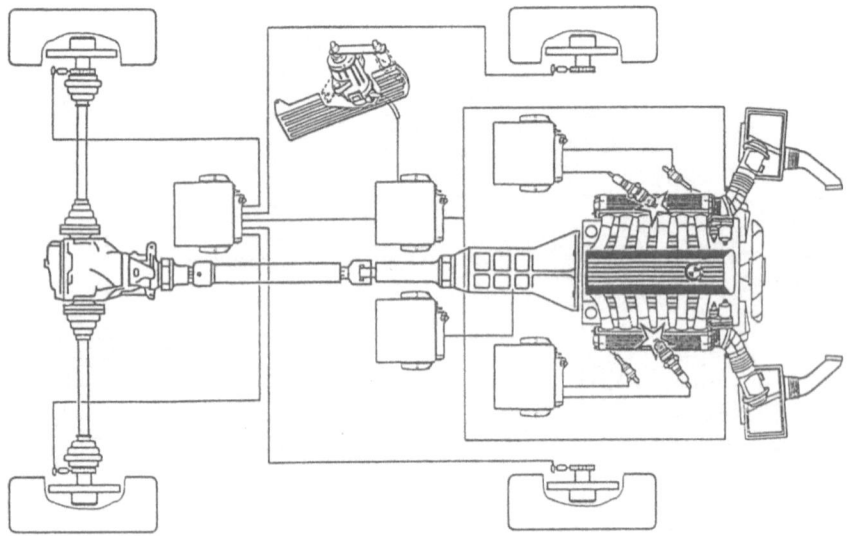

Fig. 4.175. Drive management in the BMW 750 iL [4.50]

- Two ECUs for one digital engine electronic system (DME) each for a cylinder bank of the V-12 engine,

- Double electronic closed-loop engine management controller (EML),

- Electronic gear controller (AEGS),

- Anti-blocking system with stability check (ASC) and closed-loop overrun mode control.

The drive management system continually receives and processes some 70 input data from sensors and electronic interfaces. On this basis it actuates the two electrically driven throttle valves, controls the injectors and ignition timing and adjusts the gear position and engine torque whenever necessary.

In the DME the two cylinder banks of the V-12 engine have separate closed-loop controls to improve metering accuracy for air and fuel in the range of low air throughputs. In addition to basic adaptation using the characteristic map, the DME has the following main functions:

With regard to the combustion mixture:

- Acceleration enrichment;

- Soft start after overrun cutoff;

- Cold start;

- Lambda closed-loop control

- Tank ventilation;

- Injection timing as a function of the characteristic map (sequential group injection).

With regard to ignition:

- Adaptation of the injection timing to the driving behavior;

- Correction of the injection timing depending on the engine temperature;

- Knock suppression.

Other functions (i.a.):

- Lambda sensor heating depending on the characteristic map;

- Closed-loop engine torque control to check vehicle stability;

- Fuel consumption signal (measuring the injection pulse length);

- Diagnostic and emergency functions.

Operation of the EML is described in **Fig. 4.176**:

Valuators signal the driver's action to the EML control unit which calculates and implements the throttle valve

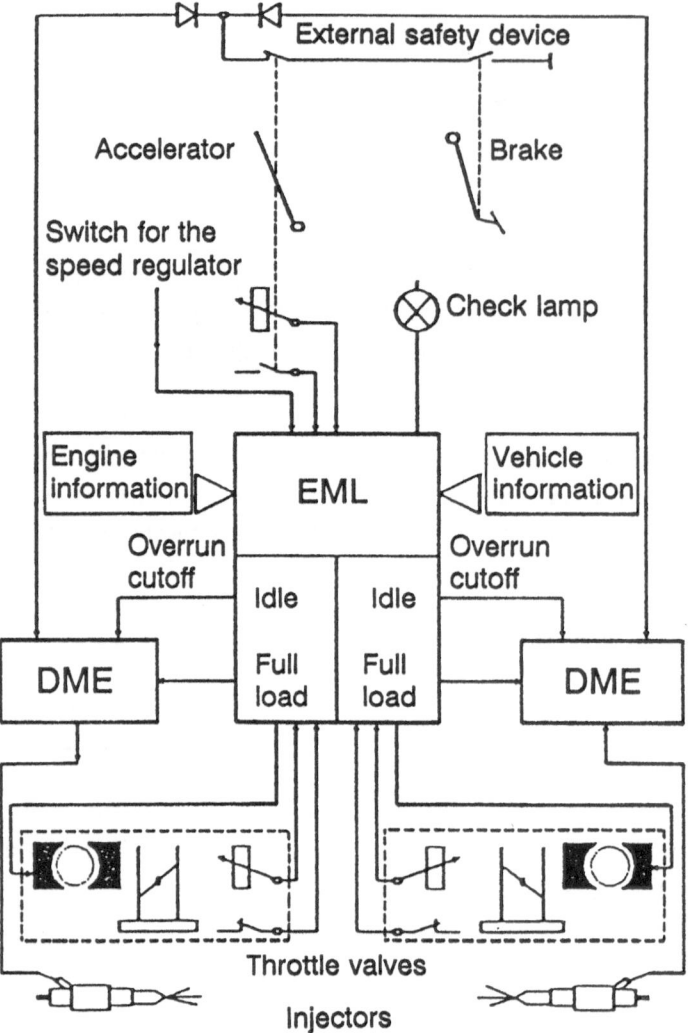

Fig. 4.176. Functional diagram of the EML [4.50]

angles, taking into account momentary operating conditions. The throttle valve position can be adapted to account for the following factors:

- Optimum start regardless of the engine temperature;

- Automatic synchronization of the two cylinder lines;

- Simple speed control and limitation;

- Prevention of excessive slipping.

4.6. Mixture Formation Requirements of Multivalve Engines

4.6.1. Differences Between Two- and Multivalve Engines with Regard to Mixture Formation

The primary consideration for fitting one or more inlet valves for each cylinder is the fact that both the geometric and the effective cross-section of the inlets gets larger with each valve. **Fig. 4.177** [4.51] compares the effective inlet cross-sections of AUDI five-cylinder engines with a cubic capacity of 2.2 l as found in flow measurements using two, four, and five valves, where the travel/bore ratio was always 86.4 mm to 81 mm. The hatched sections limiting the curves on their right sides indicate the maximum valve travels.

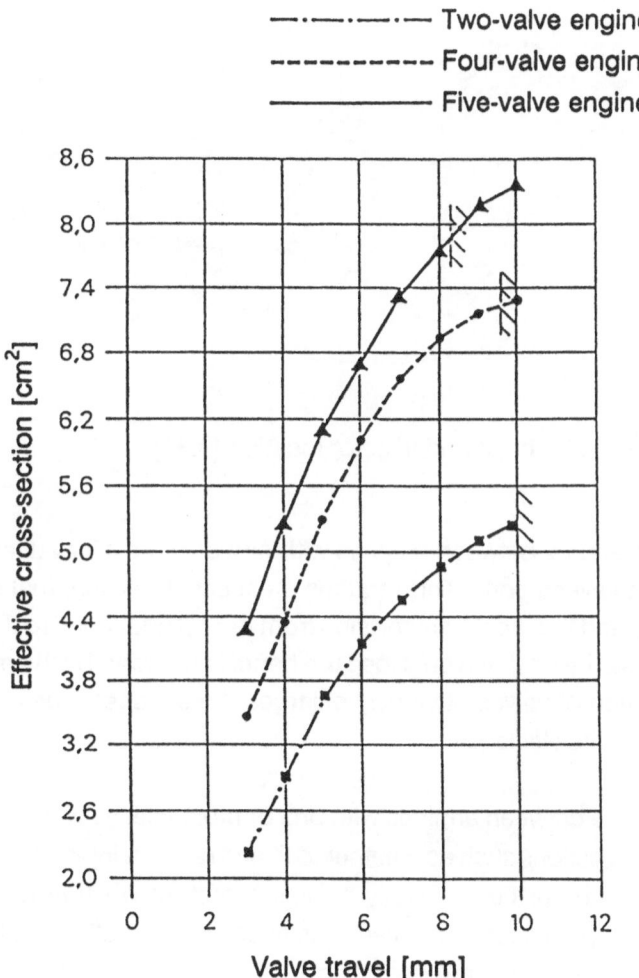

Fig. 4.177. A comparison of effective inlet cross-sections [4.51]

Fig. 4.178 illustrates the design of a five-valve cylinder head. The larger inlet cross-section of multi-valve engines makes for a lower flow loss. Such engines furthermore have smaller valve masses to move, which can be translated into a higher engine speed and improved overall engine performance.

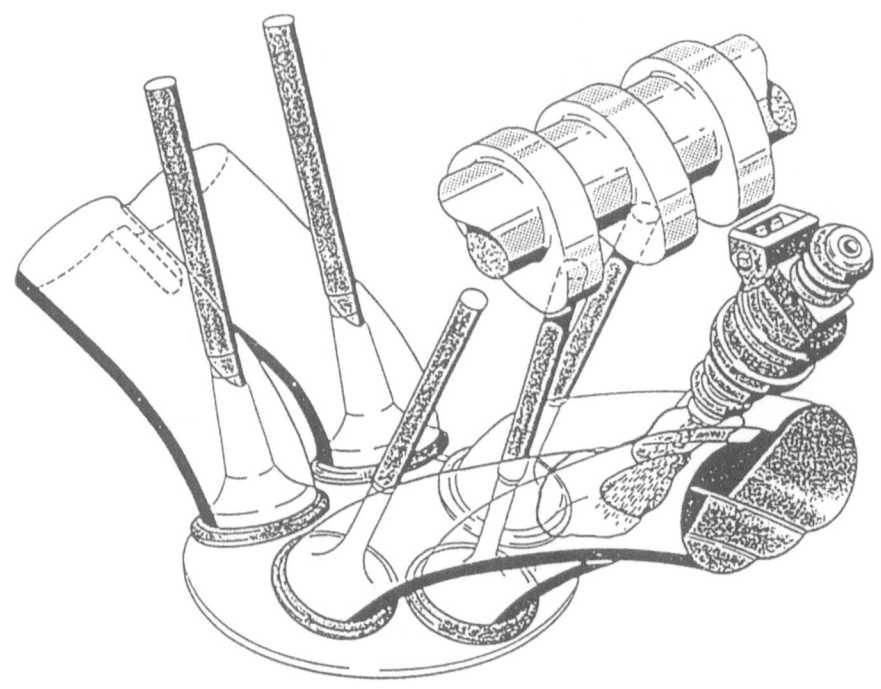

Fig. 4.178. Five-valve cylinder head and injector location [4.51]

A second important difference between engines with one or more inlet valves is the division of the intake manifold into several ports. This requires great care to ensure that good flow characteristics are maintained and that heat transmission from the cylinder head to the intake air is kept to a minimum. Otherwise the design would lose the benefits afforded by the multi-valve system. Nevertheless, the addition of valves obviously enlarges the surface/volume ratio of the intake system and thus reduces its efficiency.

As a third major difference between engines with one or more inlet valves, designers find themselves confronted with questions such as whether to fit one or more injectors, whether, if a single injector is used, it should be placed centrally or near a port, or whether to prefer a single- or multi-spray injector. If several injectors are used it is necessary to optimize injection by defining the ranges when one or all injectors are operating.

4.6.2. Effects on Mixture Formation and Optimization

The higher air and fuel throughputs made possible by the larger inlet cross-section in multi-valve engines allow higher engine speed and better engine performance. The idle speed must remain basically unchanged, and fuel consumption during idling is not much affected by the number of valves used, so that the deciding factor is the enlargement of the effective throughput for fuel and air. The following figures show the difference in effective fuel throughput for two- and five-valve cylinders in an AUDI five-cylinder engine [4.51]:

Engine:	Two-valve engine	Five-valve engine
Performance:	121 kW	230 kW
Rated speed:	6300 min^{-1}	7000 min^{-1}
Idle speed:	700 min^{-1}	700 min^{-1}
Effective fuel throughput:	1:58	1:105
Effective fuel throughput per work cycle	1:6.5	1:11.1

The effective fuel throughput increases from 1:6.5 to 1:11.1 for each cycle. The injection period is 17.1 ms at rated performance and an engine speed of 7000 min^{-1}, and 1.54 ms in idle.

In this engine the idle injection period has been reduced so much that reproducibility may be affected. Possible solutions to this problem would be to provide two injectors for each cylinder or to fit a two-stage pressure regulator in the fuel injection system. Alternatively, manufacturers of injectors are also working on extending the effective throughput.

Such an extension would also pose more demanding requirements to air metering. Problem areas could include throttling of the intake air or inadequate resolution of signals. Such circumstances must be taken into account appropriately and may have to be compensated for by more sophisticated engineering.

A large inlet port cross-section in multi-valve engines impairs mixture atomization at low engine speeds, particularly in the idle mode, due to the low flow velocity, which may add to the problems with regard to fuel metering.

Another factor to be taken into account is the larger surface of the intake system. This increases the deposition of fuel at the walls and consequently affects engine response during transient operation and when the engine is still cold.

Injection timing is of particular importance in multi-valve engines. Long storage times may have a positive effect in idle and steady-state operation but not in the transient mode. **Fig. 4.179** supplies injection timing ranges that have proved suitable in the AUDI five-valve turbo engine: injection into the opening inlet valve during low engine speed and part load, and injection into the open inlet valve in the high load range. For high throughputs we come into the range where the inlet valve is permanently open. At high engine speeds and low load, fuel storage was found to improve vehicle drivability.

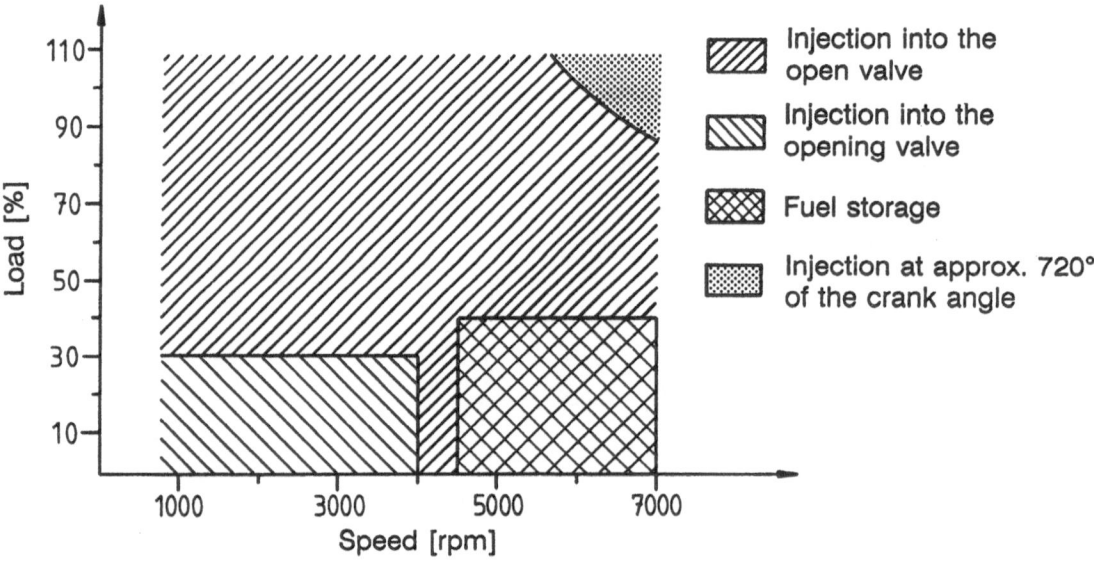

Fig. 4.179. Injection timing ranges [4.51]

Studies performed in Japan [4.52] also stress the importance of proper injection timing. When injection is timed at low inflow velocities, atomization is affected and the combustion mixture becomes heterogeneous, with the result that it is incompletely burned and emits a high level of carbon monoxide, and that consumption is increased while the mean pressure is reduced.

[4.52] found that the push-back motion of the mixture from the cylinder into the manifold was strengthened in multi- valve engines because of their larger inlet port cross- sections which may cause an imbalance in mixture distribution between cylinders. To reduce that "push-back" into the manifold it is advisable to inject at a very early stage in the inflow cycle.

Fig. 4.180 plots the effect of injection timing on mixture distribution in an engine as defined in **Table 4.2.**

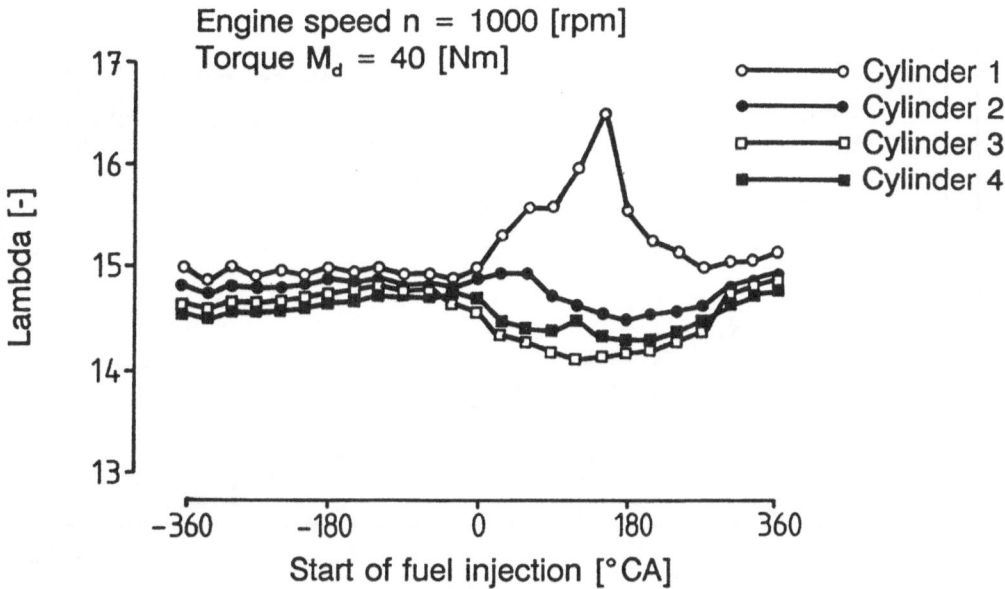

Fig. 4.180. Effect of injection timing on mixture distribution in a four-valve engine as defined in Table 4.2 [4.52]

	Four-valve engine	Two-valve engine
Bore stroke	81(mm) x 77(mm)	81(mm) x 77(mm)
Displacement	1587(cc)	1587(cc)
Compression ratio	9,4	9,0
Inlet valve diameter	30,5(mm) x 2	36(mm)
Exhaust valve diameter	25,5(mm) x 2	31(mm)
Valve timing	$\dfrac{9 \mid 9}{51 \mid 51}$	$\dfrac{18 \mid 12}{46 \mid 52}$
Valve lift	7,2(mm)	9,25(mm)
Combustion chamber shape	Pent roof	Wedge

Table 4.2. Engine specifications for Fig. 4.180 [4.52]

Apart from proper injection timing, designers of multi-valve engines must take into account the increased fuel deposits on the walls by providing for careful and flexible acceleration enrichment.

Fig. 4.181 shows the qualitative sequence of acceleration enrichment found to be optimal by [4.51].

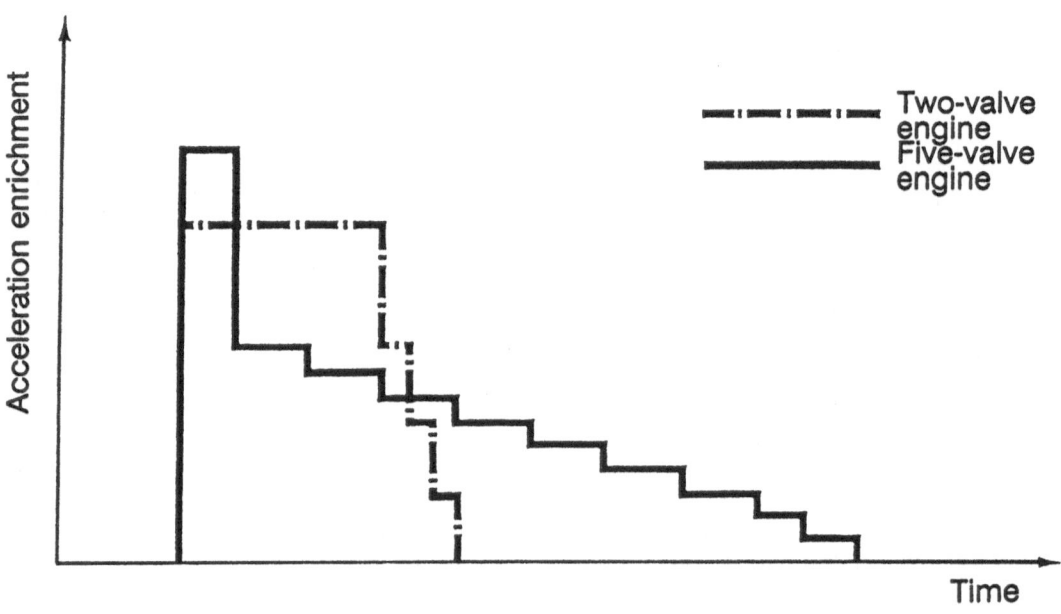

Fig. 4.181. Qualitative sequence of acceleration enrichment [4.51]

Acceleration enrichment is a key factor in multi-valve engines for another reason as well: the change in the flow when the throttle valve opens to acute throttle valve angles is greater than in the two-valve engine.

Fig. 4.182 illustrates the relationship between air throughput and throttle valve angle in an engine as described in Table 4.2.

In general, engine response can be influenced more strongly in multi-valve engines than in two-valve engines through the medium of suitably adapted acceleration enrichment.

From this it follows that injection timing and adaptation of acceleration enrichment are much more important in multi- valve engines than in two-valve engines. It thus appears advisable to use high-quality, flexible, sequential injection systems which can be adapted to the specific engine requirements.

The question of whether two injectors would be preferable over a single two-spray injector in four-valve engines cannot yet be answered conclusively but appears to be dependent on the relevant engine circumstances. Nevertheless, both the two-injector version and the two-spray version can be expected to produce better results than a central single-spray injector, as has been confirmed by [4.51]. According to [4.52], engine response is better when the injector is arranged centrally than when it is located in either of the two ports.

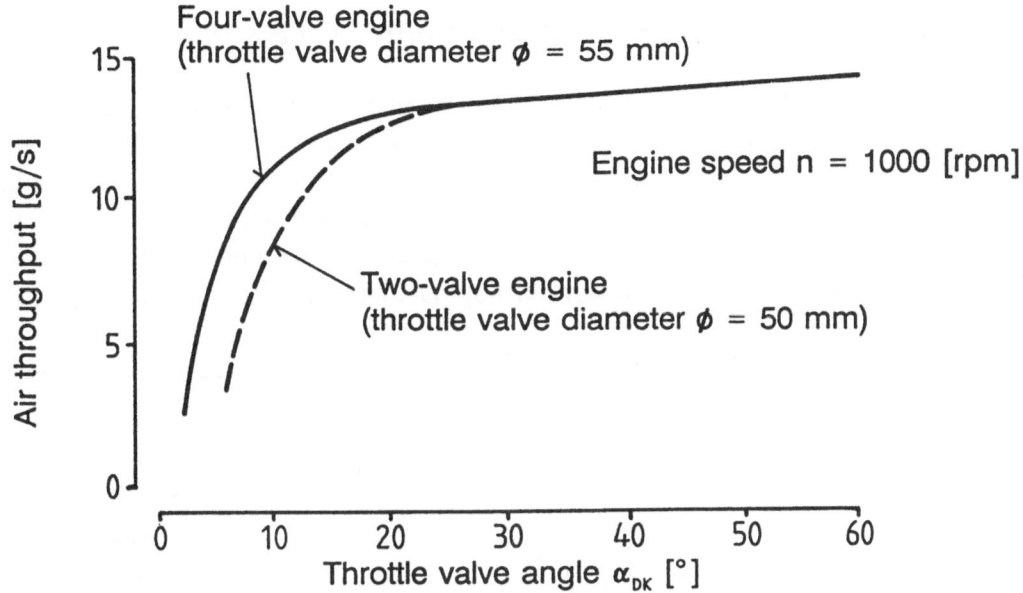

Fig. 4.182. Relationship between air throughput and throttle valve angle in an engine as described in Table 4.2 [4.52]

Tests are currently being carried out to determine whether three- spray injectors would provide for balanced distribution of the fuel over the three inlet ports in a five-valve engine. As seen by [4.51] the main problem appears to be that the three sprays could attract each other and thus impair atomization.

In summary it can be said that the more complex design requirements for multi-valve engines as compared to two- valve engines will necessitate the use of superior injection systems if developers want to exploit the full potential of such engines. Injection systems should be of the electronically controlled sequential type and should allow for incorporating numerous additional control functions, in particular with regard to dynamic features.

Injectors in multi-valve engines must be designed to ensure quick actuation and excellent atomization.

The inclusion of injection systems into multi-valve engines is much more cost-intensive owing to the very high number of parameters to be taken into account.

4.7.A Comparative Evaluation of Mixture Formation Systems

Basically, we distinguish between single-point and multi- point mixture formation systems. The single-point category comprises carburetors and central fuel injection units, while the multi-point systems consist of multiple-carburetor systems and individual-cylinder injection systems. Multiple- carburetor systems are rarely used today because they are difficult to synchronize, take up a large construction volume, and are expensive to build. Consequently, we have restricted our discussion of multi-point systems to individual-cylinder injection systems.

Fig. 4.183 outlines the "family tree" of mixture formation systems that traces the categories discussed before and subdivides the systems even further.

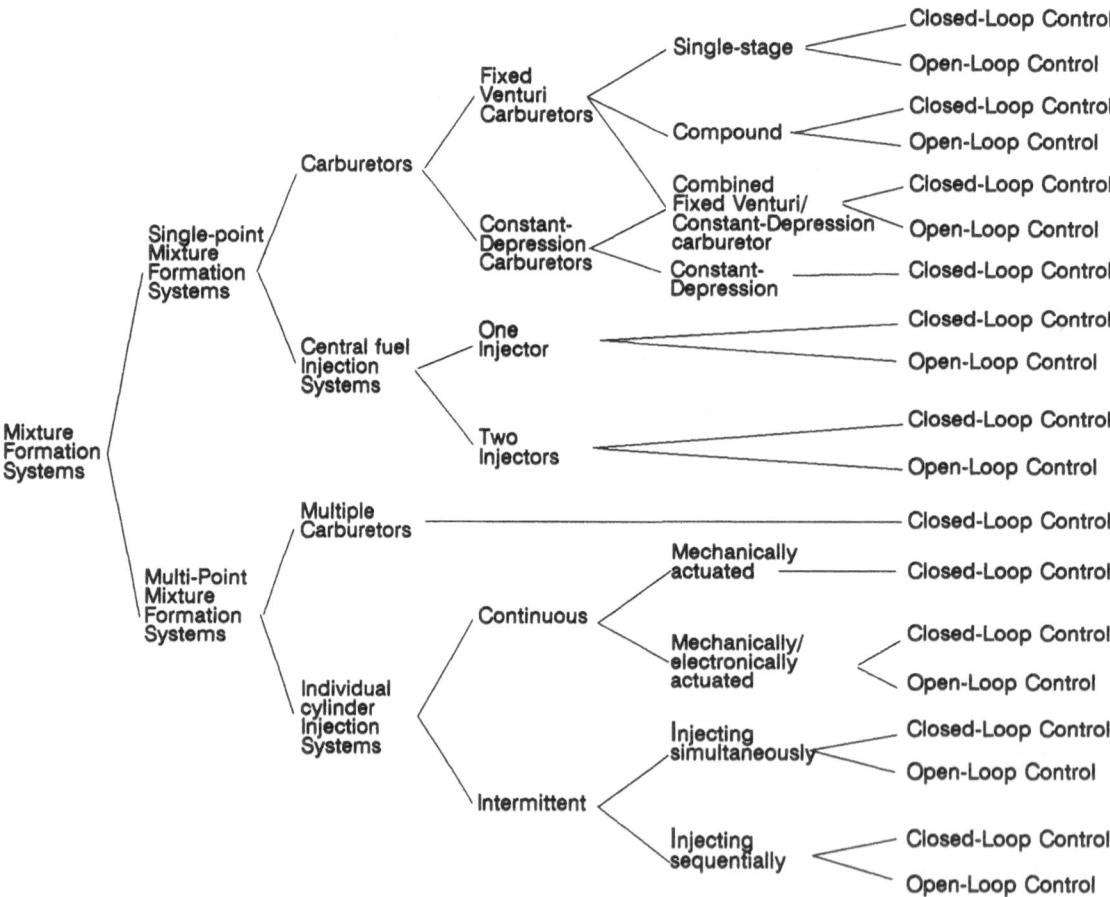

Fig. 4.183. "Family tree" of mixture formation systems

Experts have argued heatedly about the pros and cons of the different systems, and the discussion has frequently been affected by business interests in addition to the purely technical viewpoints. **Table 4.3** attempts to arrange the advantages and disadvantages in some form of evaluation. We find that the simple carburetor comes out on top with regard to costs but requires

Evaluation parameter	Single-point mixture formation systems				Multi-point mixture formation systems			
	Carburetors			Central-fuel inject. systems	Continuous		Intermittent electr. actuation	
	Simple-carburetor	Compound carburetor	Electronic carburetor		Mechanically actuated	Mech./electronically actuated	Injecting simultaneously	Injecting sequentially
Expenditure on work and cost	low	medium	medium	medium	high	very high	hight	very high
Maximum output Maximum torque Response behavior	low	medium	low/medium	medium	high	high	high	very high
Consumption (steady-state, transient mode, cold/warm)	medium	medium	low	low	very low	very low	very low	very low
Optimum exhaust cleaning (closed-loop control possible)	unsuitable	unsuitable	suitable	suitable	unsuitable	suitable	suitable	suitable
Expenditure for development, incl. manifold system	high	very high	very high	high	medium	high	medium	very high

Table 4.3. Advantages and disadvantages of mixture formation systems

considerable know-how of the complex flow conditions in its development and adaptation, which translates into significant R&D efforts. Where its performance is adequate for a particular engine it is the most simple and robust option available to engineers.

When requirements are more stringent and when closed-loop control is a must, the best solutions are central fuel injection systems and electronic carburetors. The two systems are very similar in their performance and cost expenditure. Central fuel injection systems are simple in their structure and easy to harmonize. On the other hand they require a mixture-distributing manifold when used in single-point mixture formation systems that is difficult to install and dismantle and where engineers encounter the intricate problem of balancing the fuel film distribution over the walls. As a result, the manifold cannot be designed to achieve optimum filling and maximum performance and torque.

For maximum requirements with regard to performance, torque, driving behavior, comfort, and exhaust cleaning it is necessary to opt for multi-point mixture formation systems.

Systems that inject continuously or simultaneously into all cylinders will not achieve the fine-tuning possible with sequential, i.e. timed injection. Nevertheless, the pros are increasingly counterweighed by the high expenditure necessary for R&D.

The future for high-performance multi-point mixture formation systems will be comprehensive electronic circuits and sequential injection timing with open- and closed-loop control.

Seen on a medium time scale, (closed-loop controlled) central fuel injection is a transitional solution while carburetors will definitely cease to be employed in automotive engines, at least in Western Europa and the U.S.

Fig. 4.184 projects the trend in mixture formation systems for passenger cars.

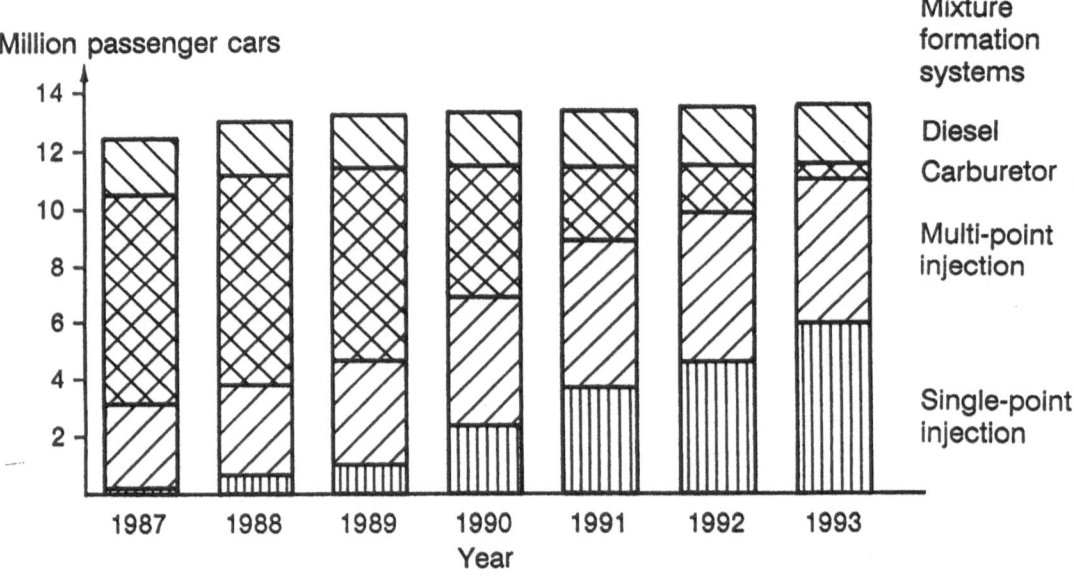

Fig. 4.184. The spread of injection systems in Europe [4.72]

5. Intake Manifold Design

5.1. Intake Manifolds for Single-Point Mixture Formation Systems

5.1.1. Intake Manifold Requirements

As modern engines are expected to feature good response characteristics, maximum perform-ance, i.e., high engine torque, and low fuel consumption while keeping pollutant emissions at a minimum, intake manifolds have to meet the following, clearly defined, main requirements:

- Fast mixture delivery;

- Low pressure loss;

- Uniform mixture distribution;

- Low fuel-film accumulation.

Moreover, the intake manifold in its capacity as the connecting element between the mixture for-mation system and the cylinder head is subject to substantial mechanical, thermal, and chemi-cal wear.

The design engineer now has to carefully adjust intake manifold geometry so that these require-ments, which partly even contradict each other, can be fulfilled as best possible. Due to the com-plexity of the problem, compromises as well as fine tuning during engine testing will be inevi-table.

Below follows a description of the most important factors influencing volumetric efficiency, mix-ture distribution, and wall-film formation in induction systems with single-point mixture forma-tion. It not only describes relationships but also gives practical hints as how to improve intake manifold design as well as simple formulae for approximate intake manifold dimensioning.

5.1.2. Design Principles

Some design principles apply to all intake manifolds with single-point mixture formation inde-pendent of a particular induction system, or engine design, or of the chosen intake manifold type. These general guidelines can be grouped under three requirements that have to be met by the intake-manifold geometry:

- Spatial symmetry;

- Temperature symmetry;

- Time symmetry.

The first requirement seems to be a perfectly logical fundamental precondition for the uniform distribution of the air/fuel mixture from the mixture formation system to the individual cylinders. As shown in **Fig. 5.1**, intake manifold design must always be symmetrical in relation to a plane through the mixture formation system center and vertical to the engine axis. The mixture formation system is always deployed in the middle between the outer and inner inlet duct pairs. As almost all flow disturbances caused in the mixture formation system (see Chapter 4.1).are a very sensitive matter, asymmetrical intake manifold designs cannot be recommended as a solution to equalize asymmetrical flows that may have developed in the mixture formation system. While such flow disturbances might be successfully compensated in one mixture formation system, they could easily be amplified when using another mixture formation system, often simply because of manufacturing tolerances.

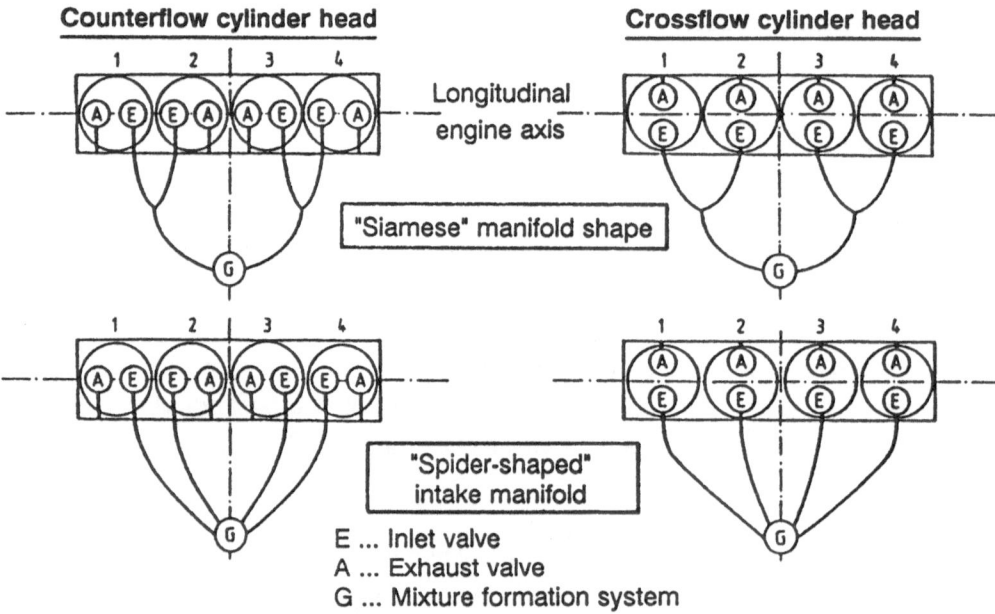

Fig. 5.1. Various basic manifold designs according to cylinder head types [5.1]

The demand for temperature symmetry seems just as logical as the one for spatial/dimensional symmetry. While the latter eliminates the danger of an asymmetrical temperature distribution inside the intake manifold that could result from an asymmetric design of the manifold heating system (using engine coolant or exhaust gas), an asymmetrical temperature distribution inside the intake manifold can well be feasible by intake manifold positioning in the engine compartment. Depending on motoring conditions, the individual runners facing longitudinally deployed engines are exposed to considerably higher variations in temperature (head wind, cooling fan), which leads to differing fuel vaporization in the front and rear intake manifold halves [5.2 and 5.3]. While the influence the temperature distribution exerts on the fuel mixture distribution largely depends on intake manifold design and, as a whole, seems to be quite negligible, the effects of temperature distribution during transient engine operation (accelerating, decelerating) are no longer negligible regardless of the chosen type of intake manifold.

Thus, with longitudinally deployed engines, the demand for temperature symmetry should be met by adequately shielding the intake manifold against head and cooling fan winds.

With transverse engines, the problem of unilateral intake manifold cooling rarely arises. It seems more sensible to deploy the induction system at the rear of the engine in order to enable faster intake manifold warmup.

Contrary to what has been said so far about intake manifold design requirements, the third demand, i.e. time symmetry ("induction symmetry") has been attributed varying importance in the relevant literature.

"Induction or intake symmetry" means equal intake intervals between the cylinders drawing fuel from one common intake manifold section. Time asymmetry, i.e., varying intake intervals within one pipe section, can be the result of two cylinders inducting from a common pipe with the induction strokes of these cylinders following immediately upon each other according to a set induction cycle. Thus temporal intake symmetry is possible only with a Siamese intake manifold shape (Fig. 5.1.).

During the suction break the fuel film inside an individual pipe section vaporizes off into the cylinder inducting first after the suction break thus affecting mixture distribution. This influence is the greater

- the later the header branches off to the individual cylinders;

- the more fuel film has accumulated in the header ;

- the more fuel film evaporates in the header, and

- the lower the speed, i.e., the longer the intervals during which the fuel film can evaporate between the individual induction strokes.

Thus, in a four-cylinder engine with a Siamese intake manifold geometry (Fig. 5.1) with the common induction cycle of 1-3-4-2 (i.e., cylinders 2 and 3 induct from the two common headers before cylinders 1 and 4) cylinders 2 and 3 would be fed more fuel vapor as a result of the preceding longer suction break.

A reversal of the induction cycle to 1-2-4-3 would also reverse the above-described beneficial effect, as has been proven in numerous empirical studies - partly by using differing camshafts, [5.3, 5.4, 5.5, 5.6, 5.7, 5.8, 5.9, 5.10].

The influence of the induction cycle can partially be mitigated by installing a compensation line between the two header ends (where they branch off to the individual cylinders) on the left and right manifold halves so that each cylinder inducts from its own as well as from the adjacent header [5.3, 5.5, 5.11, 5.12]. Consequently, the suction breaks in the headers will no longer vary in time.

The only alternative would be to use another intake manifold geometry closer to the spider-type intake manifold (Fig.5.1.).

5.1.3. Basic Intake Manifold Geometry

The basic geometry of the intake manifold is defined by the type of induction system as well as by the type of manifold chosen.

5.1.3.1. Type of Induction System

Analogous to the above breakdown of the various types of mixture formation systems we distinguish between horizontal, transverse-draft and downdraft intake manifolds. In practice, transverse-draft intake manifolds differ from horizontal intake manifolds only in their inclined fitting position and can thus be classified as horizontal intake manifolds in the wider sense of the word.

Which induction system should best be chosen will, above all, depend on the design of the engine (**Fig.5.2**), on how much space there is in the engine compartment, and in the mixture formation system.

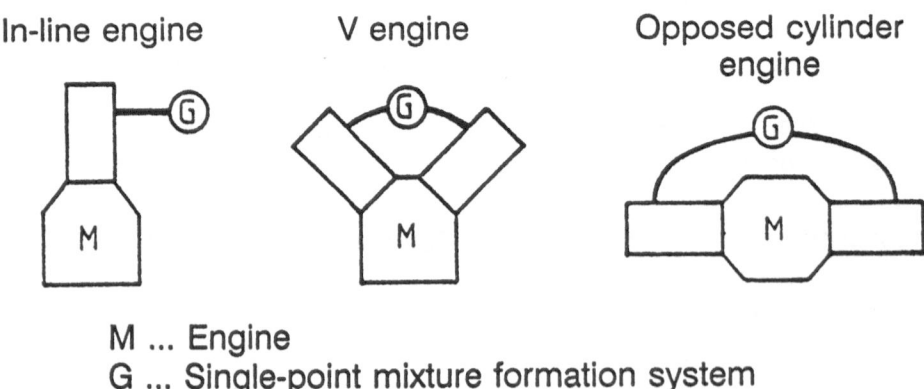

In-line engine   V engine   Opposed cylinder engine

M ... Engine
G ... Single-point mixture formation system

Fig. 5.2. Conventional reciprocating-piston engine designs

From the mixture formation point of view, in the case of in-line engines, horizontal induction systems should be given preference over downdraft induction systems, as horizontal induction systems enable highly flow-favorable designs. From the system point of view, however, in down-draft induction systems the multiphase mixture has to pass through quite a sharp, i.e., rectangular deflection immediately after entering the intake manifold (**Fig. 5.3**) which results in a significantly higher total-pressure loss as well as in significantly increased manifold floor wall wetting.

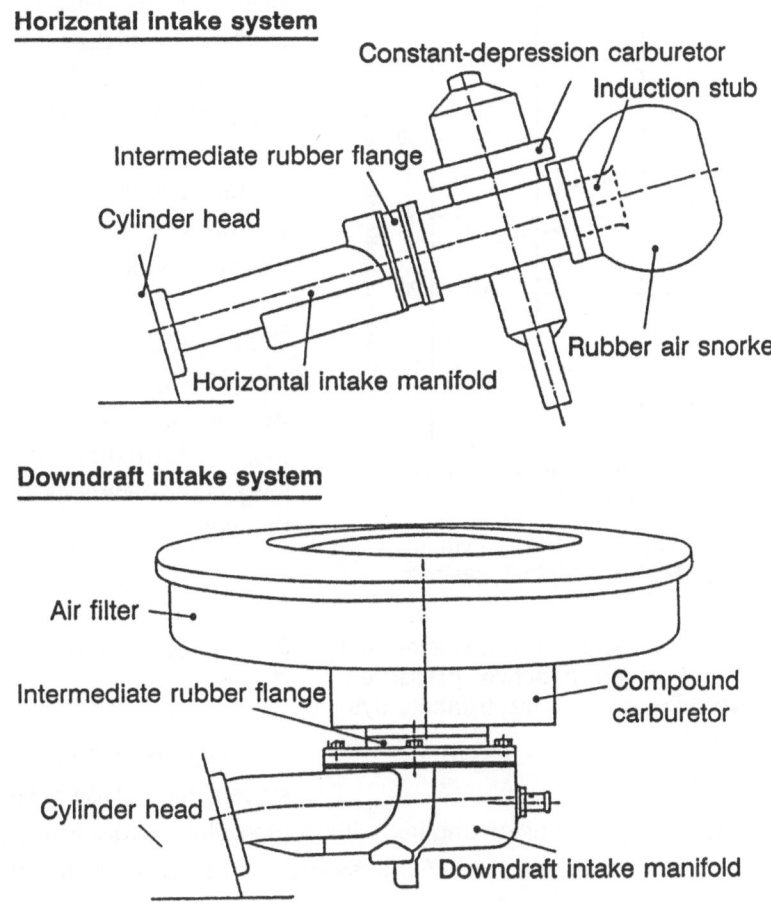

Horizontal intake system

Constant-depression carburetor

Induction stub

Intermediate rubber flange

Cylinder head

Rubber air snorkel

Horizontal intake manifold

Downdraft intake system

Air filter

Intermediate rubber flange

Compound carburetor

Cylinder head

Downdraft intake manifold

Fig. 5.3. Side view of various induction systems of a 15° inclined in-line engine [5.1]

Fig. 5.4 compares the mean effective pressures determined under the same external wide-open throttle conditions on the same four-cylinder in-line engine using, however, various induction systems. In spite of identical runner diameters, identical lambda in the full-load range, optimum ignition timing and approximately similarly favorable mixture distribution (scatter < 4%) the best values that can be obtained in downdraft induction systems at speeds of more than 1500 rpm are significantly lower than the values obtained in horizontal induction systems; on an average they differ by 0.5 bar. However, in the last few years there has been an obvious trend towards downdraft induction systems which seems to contradict the above-mentioned results. In reality this is due to the more compact design of these systems as well as to the fact that the majority of the mixture formation systems on the market use downdraft induction systems when it comes to meeting the most recent requirements (e.g. lambda closed-loop control).

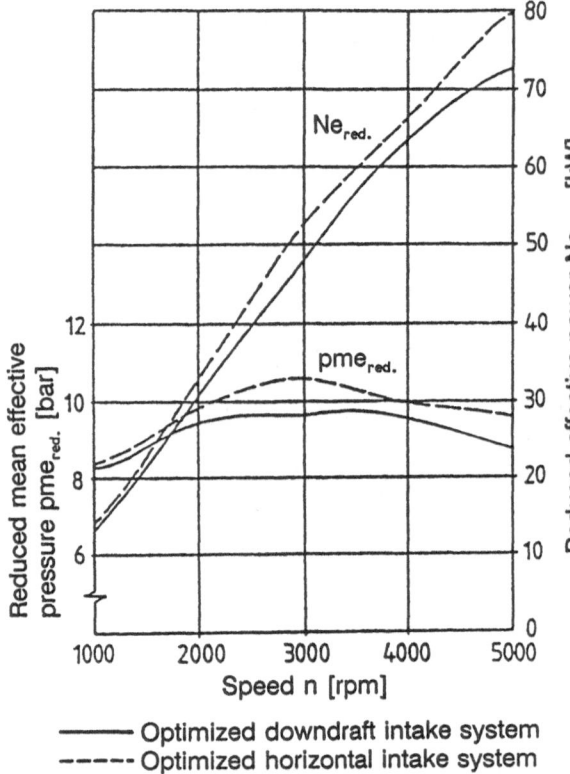

——— Optimized downdraft intake system
- - - - - Optimized horizontal intake system

Fig. 5.4. Comparison of carbureted in-line engine outputs and mean effective pressures, using two different, optimized intake systems [5.1]

Things are different in the V and opposed-cylinder engines where practically only downdraft induction systems are feasible. However, due to the fact that the cylinder arrangement of a V engine offers significant advantages for single-point mixture formation, and due to the fact that a V engine is shorter in length than an in-line engine with the same number of cylinders, V engines feature very short intake distances and are thus expected to render good transient behavior.

The above described preconditions are significantly less favorable in opposed-cylinder engines where single-point mixture formation has to encounter long intake distances as can be seen in Fig. 5.2.

5.1.3.2. Type of Intake Manifold Geometry

Independent of the type of induction system, almost all intake manifolds presently known and used in spark-ignition engines with single-point mixture formation go back to two different intake manifold shapes, i.e., "Siamese manifold geometry" and the "spider-type manifold geometry" (Fig.5.1.).

The Siamese manifold geometry is characterized by common pipe sections (headers) from which two or more individual cylinders induct, depending on the number of cylinders and on engine design. The intake passage of a four-cylinder engine with a Siamese manifold as shown in Fig. 5.1. features two successive branchings with the first branch immediately downstream of the mixture formation system and the second immediately upstream of the cylinder head intake ports. In Siamese horizontal intake manifolds the mixture is both times deflected at the same level (horizontally or slightly inclined) while in downdraft intake manifolds it is first deflected vertically and than horizontally.

Theoretically, six- or eight-cylinder engines with only one mixture formation system can also feature three successive branches; at present, however, almost all multi-cylinder engine designs have two or more mixture formation systems (e.g. two individual carburetors or one duplex carburetor) as a result of which the intake manifold is broken down into two separated three- or four-cylinder intake manifolds.

The advantages of a Siamese intake manifold geometry are: - equal intake passages to the individual cylinders, which, at least in theory, enable a gasdynamic intake manifold adjustment (in practice, due to the carburetor manifolds which are designed as short as possible with regard to transient engine operation, almost no gasdynamic boosting effect is feasible); - high mean mixture flow velocities in common pipe sections (two instead of one induction stroke per two crankshaft revolutions); - less wettable surface as well as lower overall intake manifold weights. As final mixture distribution to the individual cylinders takes place at a very late point in time as compared to spider-type manifolds, another advantage of the Siamese manifold is that flow disturbances (e.g. dissymmetries) are largely eliminated on their way through the header, which is interspersed with deflections, and thus reach the second point of distribution only in a very reduced form. Logically, the effectiveness of mixture distribution correctors in the distributor (which will be described later) will also be reduced.

The main disadvantages of the Siamese manifold geometry lie in the already mentioned influence the induction cycle exerts on the mixture distribution and, in the case of in-line engines, in the unilateral pipe bends immediately before the fuel is finally distributed to the individual cylinders. Due to secondary flows developing in the manifolds, the fuel film deposited at the first point of distribution flows in helical lines to the second point of distribution. Which cylinder will finally induct these amounts of fuel film depends not only on intake manifold geometry (pipe diameter, length, and bend) but also on the air flow rate, on the absolute amounts of fuel film as well as on the mixture flow inlet conditions, i.e., on the throttle valve angle and can hardly be predicted for the whole engine characteristic map. Actually, in experiments with Siamese manifolds it has been proven by [5.13] that, depending on the airflow rate and bending radius, the wall film is sometimes inducted by the left and sometimes by the right cylinder of each header. Since, with regard to transient engine operation, it cannot be recommended to install longer straight pipe sections with low turbulence upstream of each branching as suggested in [5.14], the only efficient way to minimize the filmlike fuel fraction is intense intake manifold heating.

As opposed to "Siamese intake manifolds" so-called "spider-type manifolds" feature only one point of distribution directly at the intake manifold inlet where the mixture is fed into the cylinders via separate pipe sections (Fig. 5.1.). Thus the fuel/air mixture coming from the mixture formation system is distributed from one central point which means that, on the one hand, mixture distribution can be simply and efficiently corrected at a later stage while, on the other hand, the incoming mixture flow will be more sensitive to dissymmetries. As all engine cylinders draw from one common distributor with identical suction breaks before each induction stroke, the vaporized fuel film is distributed evenly to all cylinders and the induction cycle can no longer influence mixture distribution.

Moreover, as in spider-type manifolds the mixture is generally deflected at smaller angles, the overall pressure loss as well as wall wetting are less substantial.

However, the above-mentioned advantages of the spider-type manifold are offset by numerous disadvantages. The fact that each cylinder is allocated a runner not only reduces the construction weight and the wettable surface but also the mean mixture transport velocities in the intake manifold, since in each runner there is only one induction stroke accomplished per two crankshaft revolutions.

The main disadvantage of the spider-type manifold, however, lies in the varying horizontal deflection angles when using it as a horizontal intake manifold.

While in the part-load range hardly any significant lambda differences will occur in the individual cylinders because of the intense "secondary fuel atomization" at the largely closed throttle valve gap (**Fig. 5.5**), **Fig.5.6.** clearly indicates that cylinders 2 and 3 are over-enriched in the full-load range.

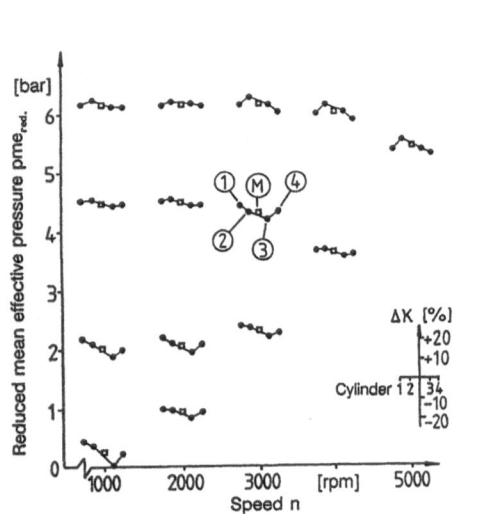

$\triangle$K [%]...deviation of the single cylinder air-fuel equivalence ratio from the calculated mean equivalence ratio of all four cylinders ①,②,③ and ④
Ⓜ ≙ mean value

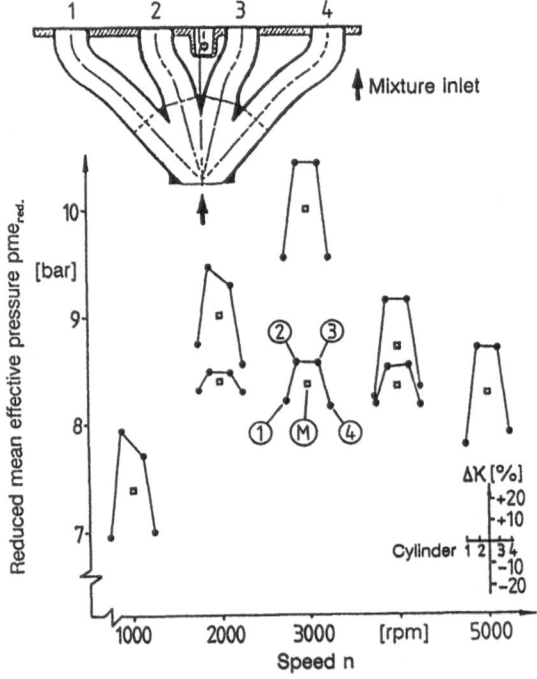

Fig. 5.5. Mixture distribution during part-load operation of a non-optimized horizontal spider-type manifold [5.1]

Fig. 5.6. Mixture distribution during full-load operation of a non-optimized horizontal spider-type manifold [5.1]

Computing the arithmetical means of the individual cylinder lambda λ_{ges} will give a correct engine lambda λ_i only if the fuel mass fed into the individual cylinders is identical. If, however, the engine lambda λ_{ges} is calculated by computing the arithmetical mean of the reciprocal value of the individual cylinder lambda $1/\lambda_i$:

$$\frac{1}{i} \; \Sigma_i \frac{1}{\lambda_i} = \frac{1}{\lambda_{ges}} \qquad (5.1)$$

λ_i [-]   ..... Individual-cylinder lambda
λ_{ges} [-]   ..... lambda of the entire engine

the result will only be correct with the intake air evenly distributed to the individual cylinders. As in carbureted engines insufficient mixture formation is mainly caused by irregular fuel distribution, the engine lambda is calculated according to (5.1) as recommended in [5.8].

The reciprocal value of the lambda $\kappa = 1/\lambda$ is called the "air-fuel equivalence ratio". The relative deviation of a cylinders air-fuel equivalence ratio from the air-fuel equivalence ratio of the entire engine $\delta\kappa_i$ corresponds to the percentage by which the fuel mass fed into the cylinder deviates from the fuel mass the cylinder should receive under uniform fuel distribution conditions. Positive $\delta\kappa_i$ values indicate that in relation to the entire engine the cylinder "i" is fed an over-enriched mixture, while negative $\delta\kappa_i$ values are characteristic of an overly lean mixture in the cylinder "i".

The relative deviation $\delta\kappa_i$ of the air-fuel equivalence ratio κ_i of the i-th individual cylinder from the calculated mean air-fuel equivalence ratio κ_m of all cylinders characterizes the mixture distribution quality:

$$\kappa_m = \frac{1}{\lambda_m} = \frac{1}{4} \; \Sigma_i \frac{1}{\lambda_i} = \frac{1}{4} \Sigma_i \left(\frac{m_K}{m_L}\right)_i \qquad (5.2)$$

$$i = 1, 2, 3, 4$$

κ_m [-]   ..... Mean air-fuel equivalence ratio
λ_m [-]   ..... Mean lambda
m_K [kg]   ..... Fuel mass
m_L [kg]   ..... Air mass

The reason for this is that an increasing throttle valve opening angle reduces throttle valve atomization and thus leads to the formation of increasingly larger droplets which can no longer be deflected to cylinders 1 and 4 after having entered the intake manifold. As will be described later on, a uniform fuel distribution can be obtained only by additional integral uniform mixture quantities in the point of distribution area.

5.1.4. Intake Manifold Heating

Numerous comprehensive and systematic theoretical, as well as empirical, studies of how heating influences mixture distribution and fuel film formation are mentioned in the relevant literature [5.2, 5.4, 5.6, 5.7, 5.13, 5.33, 5.34, 5.35, 5.36, 5.37, 5.38, 5.39, 5.40, 5.41, 5.42, 5.43, 5.44, 5.45, 5.46, 5.47, 5.48, 5.49, 5.50, 5.51, 5.52, 5.53, 5.54, 5.55, 5.56, 5.57, 5.58, 5.59, 5.60]. The authors of these studies unanimously conclude that they prefer to preheat the mixture by heating the intake manifold wall especially in the area of the point of distribution immediately downstream of the mixture formation system ("hot spot") rather than using any other heating system, since this type of heating mainly concentrates on the unwanted film-like fuel fraction while leaving the remaining vapor and droplet fraction of the mixture flow almost unheated. Thus the power loss as well as an increased knocking tendency that are to be expected with mixture preheating will be kept within reasonable bounds. Moreover, vaporization of the heavier fuel fractions of the wall film can only be guaranteed with intake manifold wall heating. **Figs. 5.7** and **5.8** show the mixture distribution measured in a standard coolant-heated six-cylinder downdraft intake manifold at coolant temperatures of 10°C and 85-90°C (operating temperature).

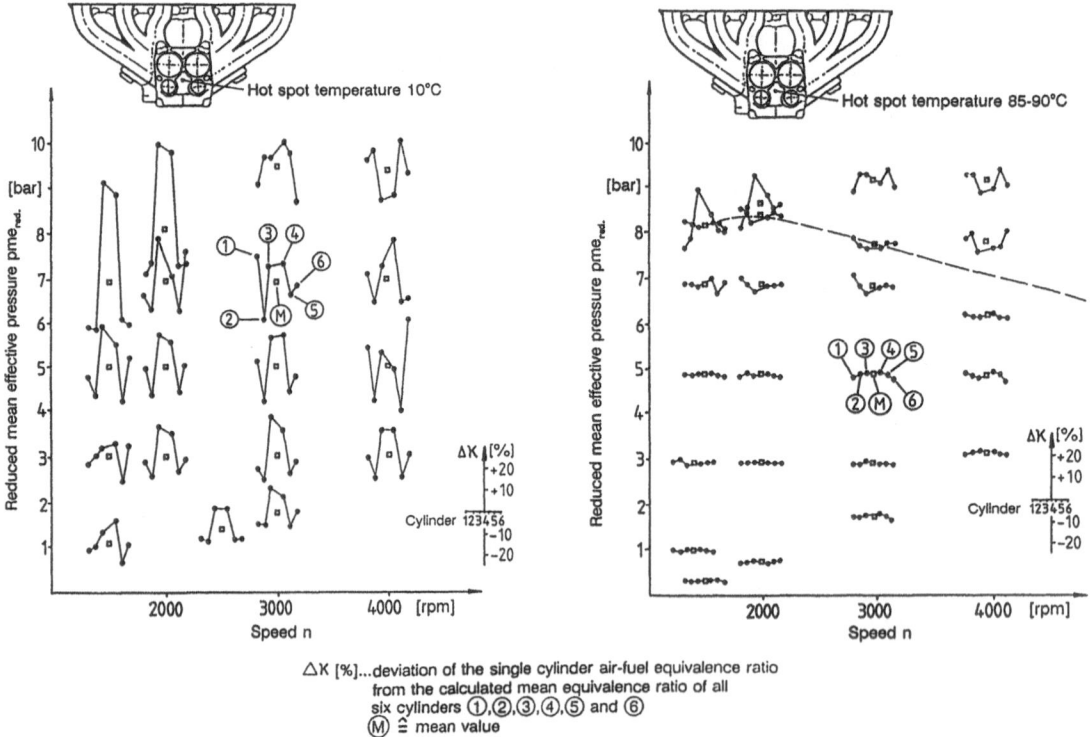

ΔK [%]...deviation of the single cylinder air-fuel equivalence ratio
from the calculated mean equivalence ratio of all
six cylinders ①,②,③,④,⑤ and ⑥
Ⓜ ≙ mean value

Fig. 5.7. Mixture distribution in the cold intake manifold (hot-spot temperature 10°C) [5.33]

Fig. 5.8. Mixture distribution in the manifold at operating temperature (hot-spot temperature 85-90°C) [5.33]

The less favorable mixture distribution in the full-load range coincides with the response of the second carburetor stage, which is characterized by a broken line in Fig. 5.8. At the operating points on this line, the throttle valves of the first carburetor stage open approximately 50° while the throttle valves of the second carburetor stage remain shut for the time being.

The improvement in mixture distribution with increasing hot-spot temperatures is as obvious as the resulting increase of the WOT mean effective pressure curve. As shown in **Fig. 5.9**, a further increase in heating can improve mixture distribution further.

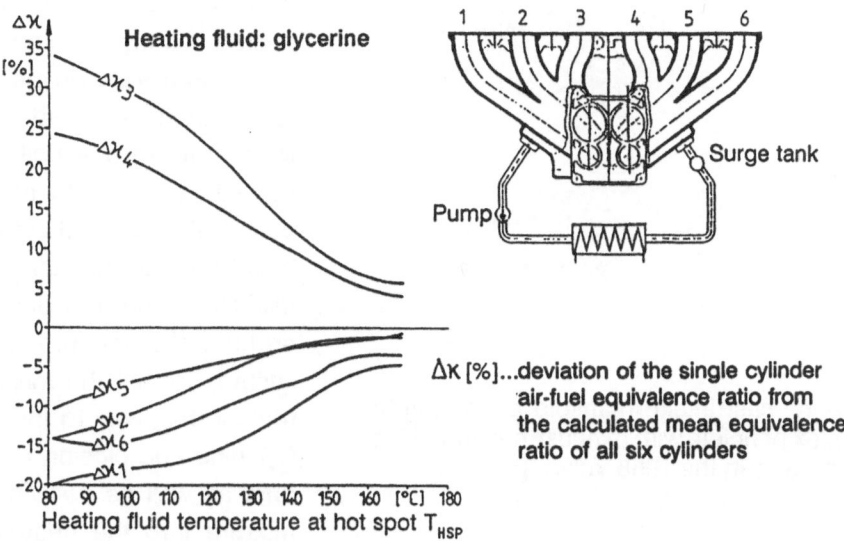

Fig. 5.9. Effects of the hot-spot temperature on mixture distribution at n = 2000 min^{-1} at wide open throttle conditions, [5.33]

When comparing intake manifold wall heating to preheating of the air, the latter improves mixture distribution only slightly (**Fig. 5.10**).

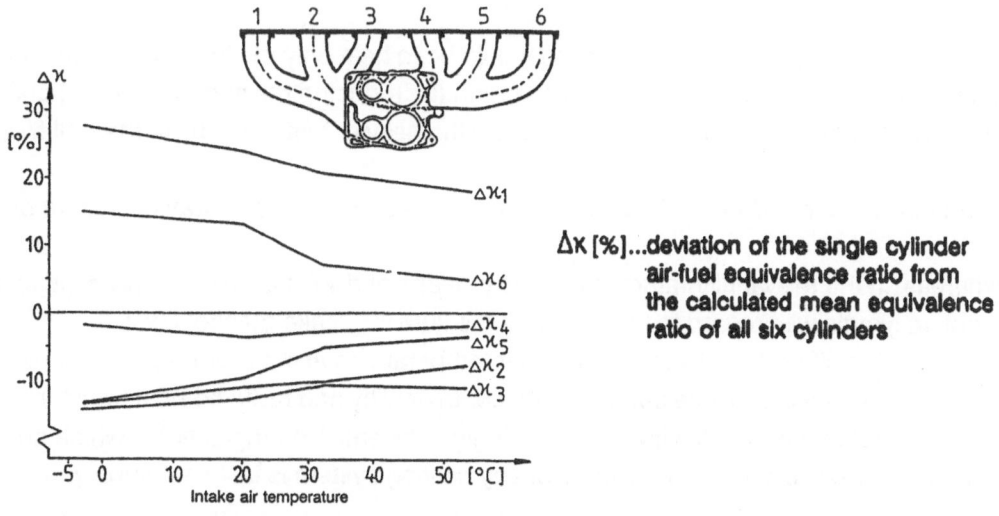

Fig. 5.10. Effects of the intake air temperature on mixture distribution at n = 2000 min^{-1} [5.33] at wide open throttle conditions, see Ref. [5.33]

Moreover, as air heating has an adverse effect on volumetric efficiency and knocking behavior, this type of preheating is useful only in the part-load range. However, when providing a suitable control arrangement which limits the intake air temperature in the part-load range to about 45°C, air heating can be a relatively simple means to improve engine warmup behavior [5.47]

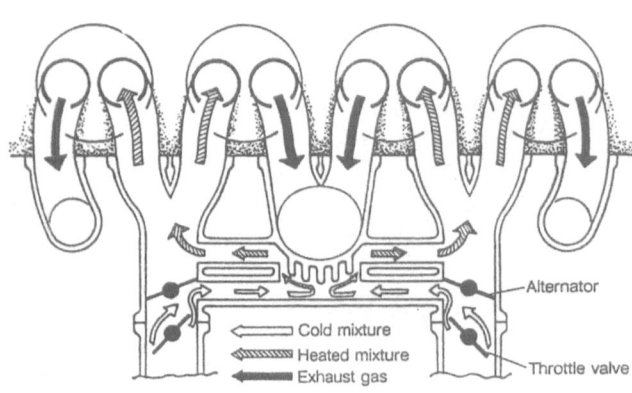

Fig. 5.11. Integral cast intake manifold according to duplex principle with exhaust-heating of the mixture as in the 1968 Volvo, [5.58]

With the so-called duplex principle, **Fig. 5.11**, not only the intake air but also all the air-fuel mixture is pre-heated. During warmup, and with some models also in the operating temperature part-load range, the total mixture flow is deflected from the intake manifold and heated outside the manifold before it is re-circulated into the manifold. In the full-load range and during acceleration, the mixture is not de-flected, so that the mixture takes the shortest way to the manifold. Opening and closing of the main duct as well as deflection of the mixture into the heated bypass lines is accomplished by a second throttle valve installed in the intermediate flange duct.

However, as with air heating, the disadvantage of the duplex system lies in the adverse influence it exerts on the volumetric efficiency and knock characteristics and, additionally, in an increased pressure loss. Since with single-point mixture formation, the main problems regarding mixture distribution and wall film formation arise mainly in the full-load range, the "duplex" principle offers no satisfactory solution for an improved mixture distribution.

Another possibility of mixture preheating would be to heat only the fuel entering the mixture for-mation system. This would improve atomization (by reducing the surface tension) and intensify fuel vaporization while, however, not reducing the heavier fuel fractions in the wall film [5.38].

It can thus be concluded that intake manifold wall heating is the best way of mixture preheating.

Whereas, in the relevant literature, a relatively clear standpoint is taken on preheating, the case is not so simple with regard to the optimum medium for intake manifold wall heating. While ex-haust heating offers the advantage of higher temperatures and a more rapid availability of heat, its disadvantages are a more cumbersome controllability and reduced heat transfer when com-pared to coolant heating. In view of increasingly strict emission regulations, which are being in-stituted world-wide, the utmost priority of any heating system is fast heat supply.

In practice, the question of which heating fluid should actually be used for intake manifold heating will mainly depend on the construction and design of the given engine (type of cooling, cylinder head, exhaust routing, etc.).

In the case of counterflow cylinder heads, where the inlet and exhaust ports are located on the same side of the cylinder head as can be seen in Fig.5.1., preference should be given to exhaust heating since optimum exhaust-gas routing and emission control are easy to effect and guarantee quick and trouble-free intake manifold warmup. Exhaust gas intake manifold heating is, however, more difficult in the case of cross-flow cylinder heads (i.e, in almost all V and opposed-cylinder engines) since long distances must be bridged between inlet and exhaust manifolds. With a valve arrangement as favorable as shown in the left half of Fig.5.1., (counterflow cylinder head), the exhaust gas can be drawn from the exhaust ducts, i.e., through bores on the intake manifold side in the cylinder head and then fed into the intake manifold. In most cases, however, long pipes will be inevitable to recirculate the exhaust gas into the exhaust system. Moreover, exhaust gas temperature control is difficult with this design.

Feeding the air-fuel mixture into the exhaust gas, as described in connection with the "duplex" system, poses even greater problems because of the long distances and the inherent danger of fuel film formation. As shown in **Figs. 5.12** and **5.13**, this principle has nevertheless been realized in some English cars.

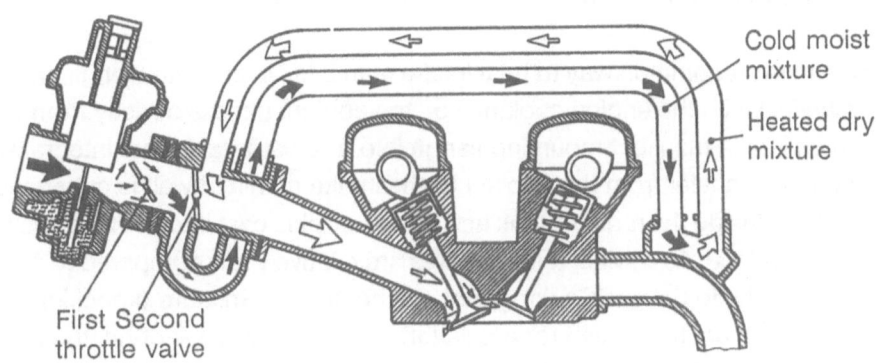

Fig. 5.12. Duplex induction system with exhaust-heated mixture as in the 1968 Lotus-Elan [5.58]

Another possibility of transferring heat from the exhaust gas to the fresh mixture, which is still at a trial stage, is to use so-called heat pipes. The lower end of these mostly vertically mounted and fluid-filled pipes protrudes into the exhaust system whereby the fluid at the manifold floor is heated and vaporized a few seconds after starting. The rising vapor then condenses at the cold upper pipe end in the intermediate flange situated between the mixture formation system and the intake manifold or on the intake manifold hot spot, where it releases the nascent condensa-

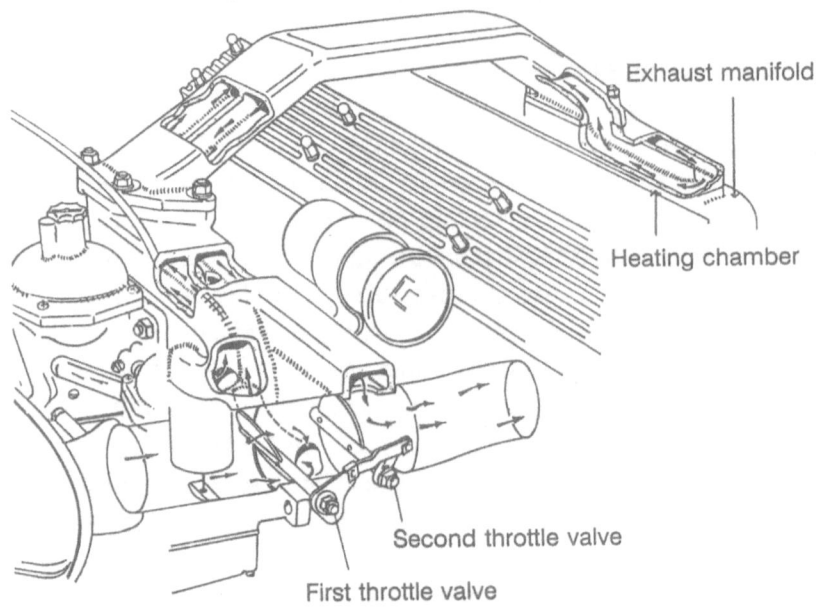

Fig. 5.13. Duplex induction system with exhaust-heated mixture as in the 1968 Jaguar [5.58]

tion heat and flows back to the manifold floor. The result is a continuous heat transfer from the hot exhaust gases to the cool fresh mixture [5.61, 5.62, 5.63, 5.64].

The simplest and most common way to heat intake manifolds in crossflow engines, however, is to heat the manifold with the engine coolant, i.e., the coolant passes directly from the cylinder head through an intake manifold mounting flange into a radiator tank cast integral with the intake manifold. Heat transfer from the coolant to the intake manifold wall is enhanced by small flow cross sections inside the radiator tank and by fins or pins cast into the wall and thereby increasing the wall surface. As coolant temperatures are relatively low compared to exhaust heating, and consequently do not pose a danger of overheating the mixture (knocking, loss of performance) there is generally no need for regulation in the case of coolant-heated manifolds. As mentioned above, the main problem with coolant-heated manifolds is that it takes quite long until the heat is available. Since CO and HC emissions and total fuel consumption are, however, highest during the initial state of the emission tests (**Fig. 5.14**), coolant-heated intake manifolds are often additionally equipped with an electric heating system.

From the design point of view, electric auxiliary heating devices are classified into electrical preheaters (**Fig. 5.15**), which are mounted on the manifold floor (hot spot) where most of the wall film impacts, and honeycomb heaters, which, in most cases, are located in an intermediate flange directly beneath the mixture formation system (in the case of compound carburetors generally only beneath the first stage) and are thus penetrated by the fresh mixture flow (Fig. 5.3 and **Fig. 5.16**).

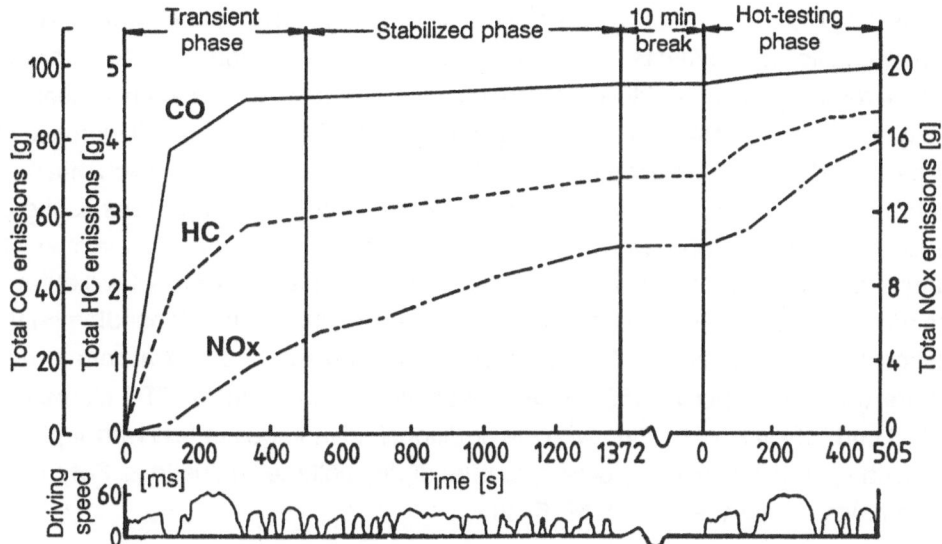

Fig. 5.14. Total emissions during the 1975 US FTP (US FTP 75) driving cycle as a function of time

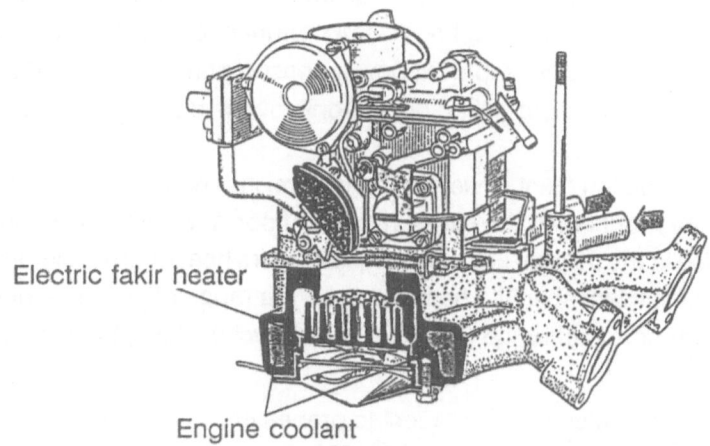

Fig. 5. 15. Coolant-heated intake manifold with an additional electric preheater ("fakir heater") [5.65]

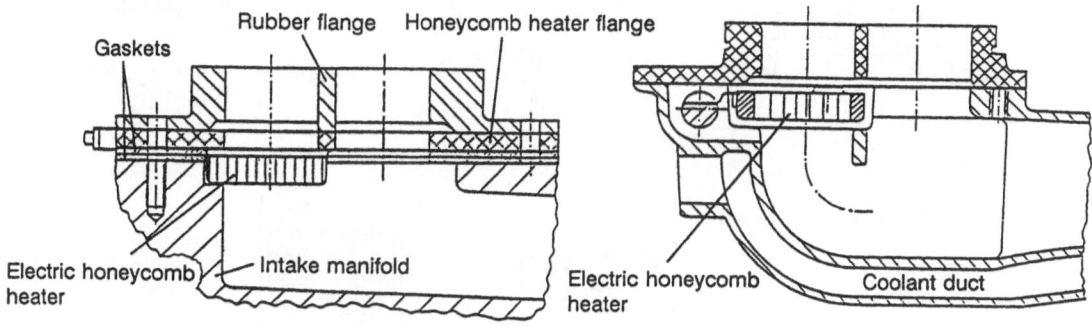

Fig. 5.16. Intake manifolds with an electric honeycomb heater mounted beneath first stage of compound carburetor [5.1]

The fakir heating elements as well as the honeycomb heaters consist of ceramic PTC (positive temperature coefficient) elements which are characterized by an approximately constant, low electric base resistance over a wide temperature range. When the PTC element is energized and heated beyond a certain temperature limit, its internal electric resistance will increase abruptly thereby inevitably limiting the electricity circulating in the heating element, which means that the honeycomb heater cannot be heated further. The internal resistance increases as the ceramic lattice structure changes from hexagonal to cubic. Since this effect can be reversed when the temperature falls below the above-mentioned limit, the PTC heating element will automatically draw the amount of electricity required to reach the temperature limit - depending on the cooling situation (mixture quantity and temperature). This and the fact that a certain maximum temperature (generally approx. 150°C) can never be exceeded, make PTC elements a very economical and safe way of heating with the additional advantages of fast warmup (approx. 4 sec.) and a comparative constancy of temperature during voltage fluctuations. For more details on PTC elements please see [5.66, 5.67, 5.68].

Fakir heaters as well as honeycomb heaters are thermally insulated against the intake manifold. Therefore full heating takes effect in a few seconds after turning the ignition key, i.e. during engine start-up. After the engine starts running, the electric auxiliary heating devices remain switched on until the coolant has reached a temperature of 60°C. Upon reaching that temperature the auxiliary heating devices (approx. power consumption: 200 - 400 W) switch off in order to save energy, and the manifold will be heated only by the engine coolant.

From the fuel film reduction point of view, fakir heaters, independent of their geometrical design (**Fig.5. 17**), that are mounted in the intake manifold floor have proven superior to honeycomb heaters suspended in the mixture flow, since fakir heaters heat only the film-like fuel fraction. On grounds of the insignificant absolute heater output, the range of influence of fakir heaters is restricted to the lower part-load range (e.g. ECE urban cycle) which, in most cases, corresponds

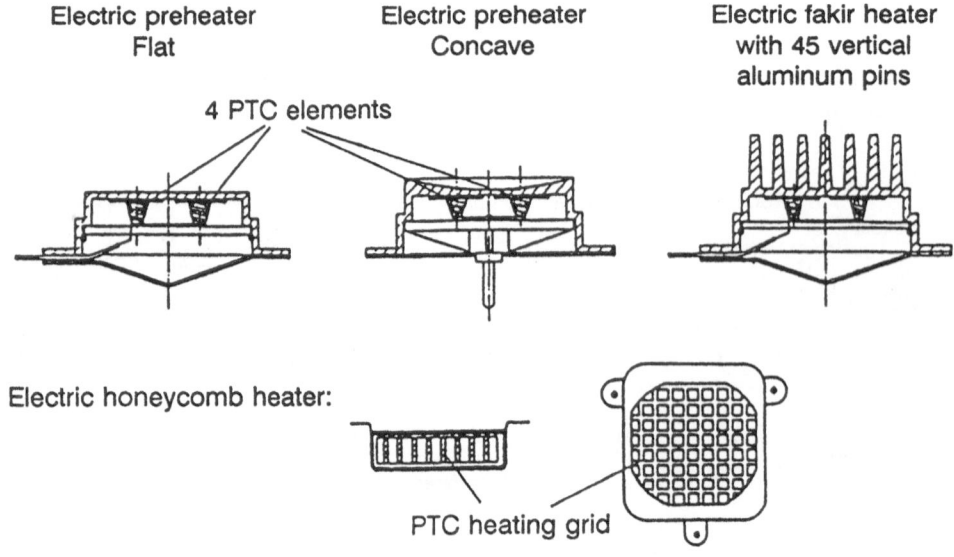

Fig. 5.17. Various PTC heaters [5.1]

to the practical requirements during warmup. After the fakir heaters switch off, i.e. at engine operating temperatures, since they are thermally insulated against the intake manifold, cool down rapidly and are thus a potential fuel condensation point. Here, honeycomb heaters are a better solution, which allow for the entire manifold floor being heated by engine coolant. The efficiency of honeycomb heaters can be slightly increased by placing them in a somewhat lower position so that they are passed only by larger droplets, which are not deflected together with the mixture. Moreover, especially in single-flow formation systems, the bypass forming upstream of the honeycomb heater leads to a reduced pressure loss, **Fig. 5.18**.

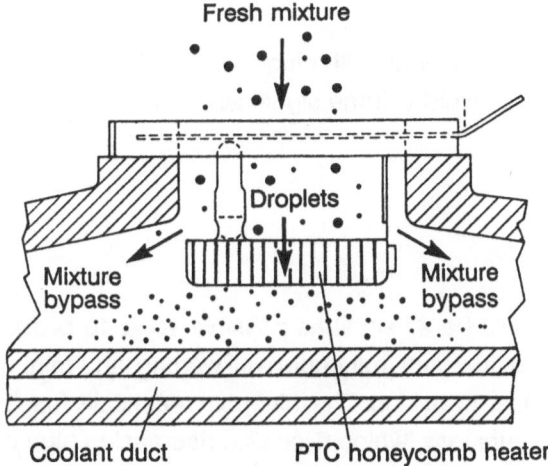

Fig. 5.18. Coolant-heated intake manifold with suspended PTC honeycomb heater [5.66]

Fig. 5.19 shows the fuel film fractions deposited on the intake manifold floor directly beneath the mixture formation system under various engine operating conditions. In one case, the

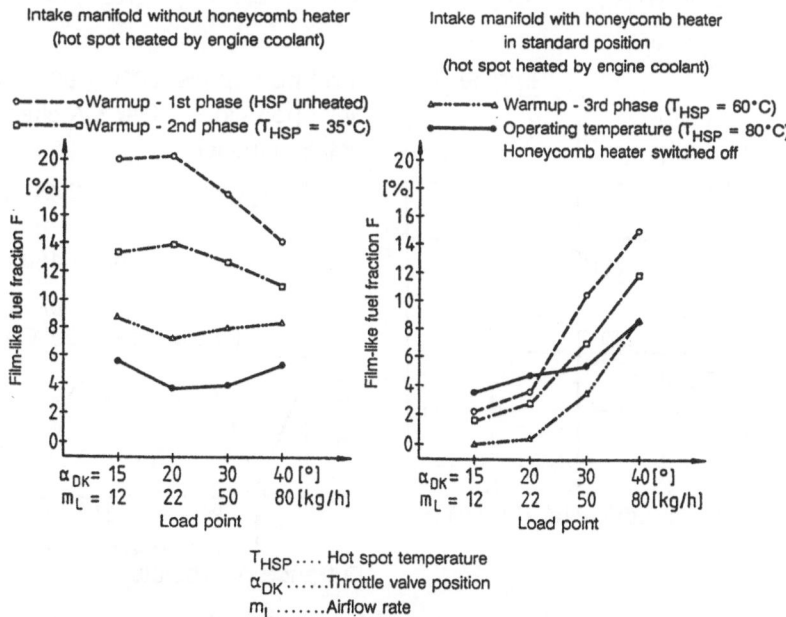

Fig. 5.19. Fuel film deposits on the intake manifold hot spot at various load and operating conditions with and without utilization of a honeycomb heater [5.69]

coolant-heated intake manifold was also equipped with an PTC honeycomb heater (with a heating capacity of approximately 200W) while, in the second case, no such heater was installed. The honeycomb heater clearly reduces wall film formation, especially near idle.

5.1.5. Intake Manifold Volume

The volume of an intake manifold is defined by the length and cross-section of its runners as well as by the volume of the distributor.

Transient intake system investigations with single-point mixture formation have shown [5.69, 5.70, 5.72] that the intake manifold volume significantly influences the response behavior and the exhaust gas quality.

5.1.5.1. Runner Cross-Section

Runner cross-sections should be as small as possible so as to provide for minimum mixture delivery times (gas, droplets, fuel film), minimum wall wetting, and minimum construction weight with maximum heat and phase transition. The minimum practically feasible cross-section will depend on the frictional pressure loss, which rises with decreasing pipe cross-sections and, beyond a certain level, leads to an unacceptably steep drop in volumetric efficiency in the full-load range.

Fig. 5.20 shows the fuel droplet flight times in a 250 mm long straight intake manifold whose diameter varies, computed for a constant full-load point. The beneficial effect of smaller runner cross-sections is obvious.

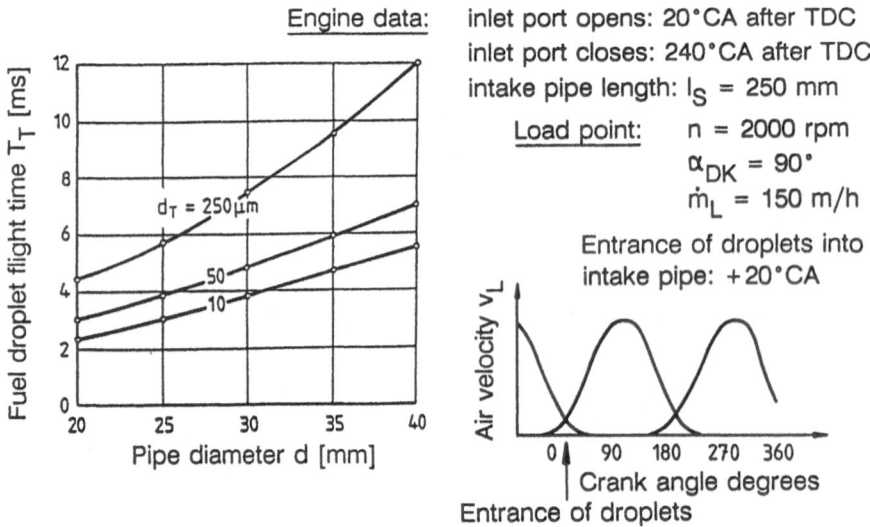

Fig. 5.20. Effects of the pipe diameter d and droplet diameter d_T on the droplet flight time T_T in a straight intake pipe [5.21]

Similarly, a reduction of the runner cross-section will lead to a drastic increase in fuel film flow rates as can be seen in the results computed for a steady-state pipe fuel flow as depicted in the left half of **Fig. 5.21**. Based on the continuity theorem, higher fuel-film flow rates reduce the fuel film thickness which, in view of the reduced wettable manifold area, leads to a reduction of the amount of fuel film accumulated in the intake manifold (right half of Fig. 5.21). As already described in Chapter 2.5.2, higher fuel film flow rates also increase heat transfer from the manifold wall to the fuel film and the phase transition taking place at the gas/fuel phase boundary.

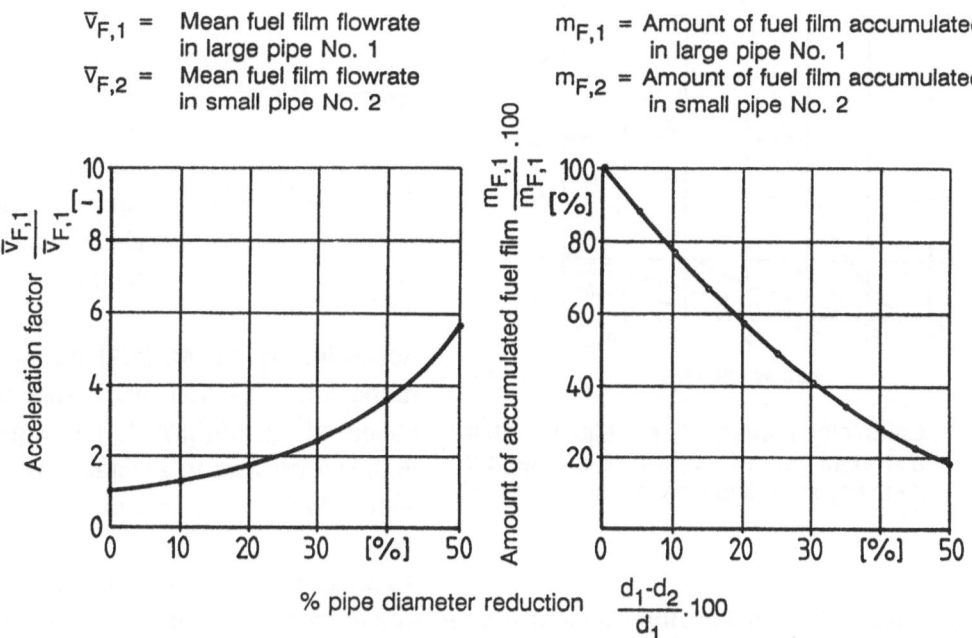

Fig. 5. 21. Influence of pipe diameter reductions on mean flow rates and accumulated quantities of a steady-state manifold fuel film flow [5.1]

In **Fig. 5.22** the frictional pressure loss, which rises significantly with decreasing pipe diameters, is shown for a straight pipe of constant length. The pressure loss curves are calculated according to Chawla [5.73], which is an appropriate method for the calculation of horizontal, turbulent two-phase pipe flows.

As can be seen in Fig. 5.22, the pressure loss rises steeply only when the diameter falls below a certain minimum depending on the maximum mixture flow rate. With a maximum air flow rate of 300 kg/h as in the present case the minimum diameter of the runner will be approximately 35 mm. The design engineer therefore has to strike a beneficial compromise between maximum mixture flow rates and minimum frictional pressure losses when determining the diameters of the runners.

Assumption: $T_L = T_K = 20°C$, $p_{atm.} = 750$ Torr, $\lambda = 1$
Referenced pipe roughness $k_s/d = 10^{-6} \div 10^{-2}$

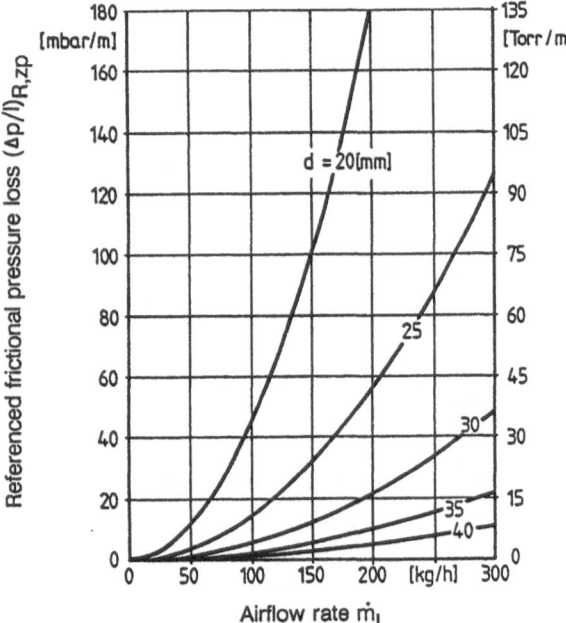

Fig. 5.22. Calculated friction loss per pipe length in a straight pipe with varying diameter d in a steady-state two-phase flow [5.1, 5.73]

As a general rule for the rough calculation of the minimum possible "optimum" runner cross-section a simple and practical relation can be applied which has been found by evaluating comprehensive intake manifold statistics:

$$A_{s, opt} = f_{SE} \cdot N_{e, max} \qquad (5.3)$$

$A_{s, opt}$ [mm²]   ..... Optimum runner cross-section

$N_{e, max}$ [kW]   ..... Maximum effective performance (engine rating)

f_{SE} [mm².kW⁻¹]   ..... Specific runner cross-section factor

According to Pachta [5.1] the specific runner cross-section factor was determined as 12 mm².kW⁻¹. The equation (5.3) is applicable to all intake manifolds with a circular cross-section.

As pressure losses are higher in runners with rectangular cross-sections than in circular ones, runner cross-sections must be designed larger for intake manifolds with rectangular cross-sections. For a single-phase turbulent pipe flow the required cross-section expansion factor "f" can be computed by substituting the respective hydraulic diameter d_h in the Blasius pressure loss equation [5.77]. On the assumption of equal pressure losses at equal air flow rates, the equation yields the cross-section expansion factor required for rectangular runners:

$$f = \frac{A_{rect.}}{A_{circ.}} = \varepsilon \cdot \left\{ \frac{(1 + \varepsilon)^{10}}{\varepsilon^{24} \cdot \pi^{5}} \right\}^{1/19} \qquad (5.4)$$

$A_{rect.}$ ..... Rectangular runner cross-section

$A_{circ.}$ ..... Circular runner cross-section

$\varepsilon = a/b$ ..... Width/length ratio of the rectangular cross-section $A = a \cdot b$

The values yielded by equation (5.4) are graphically depicted in **Fig. 5.23**. Equation (5.4) can be modified so as to compute length "a" of the rectangular cross-section as directly related to the respective circular cross-section instead of the cross-section expansion factor; see equation (5.5.) and/or right diagram of Fig. 5.23.

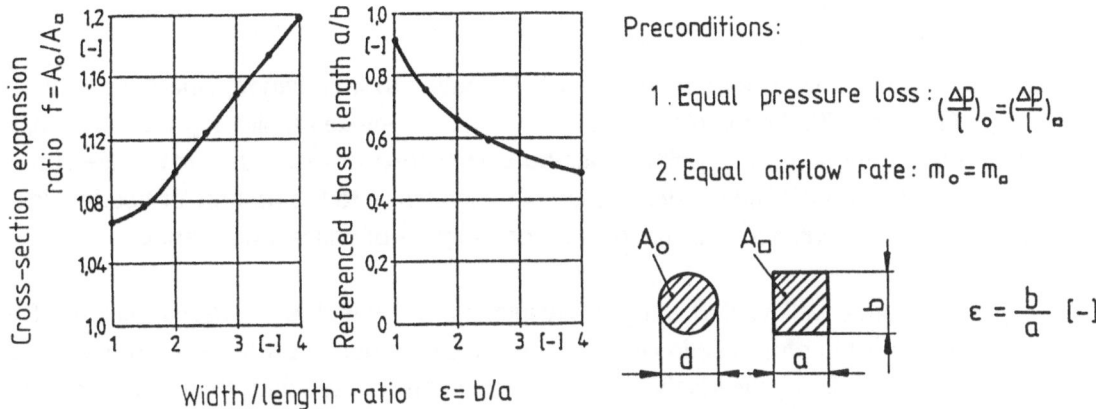

Fig. 5.23. Cross-section expansion required for the transition from a circular to a rectangular pipe cross-section with a single-phase turbulent gas flow ($2300 \leq Re \leq 10^5$) [5.1]

$$a = \frac{d}{2} \left\{ \frac{\pi^7 \cdot (1 + \varepsilon)^5}{\varepsilon^{12}} \right\}^{1/19}$$

(5.5)

a   [m]   .....Length of rectangular runner cross-section
d   [m]   .....Diameter of circular runner cross-section

5.1.5.2. Length of Runners

The length of runners for single-point mixture formation intake manifolds is determined by factors completely different from the ones determining the length of runners for multi-point injection intake manifolds. While in multi-point injection systems, the runners are designed as long as possible so that beneficial boosting effects in a certain speed range can be obtained by utilizing the developing intake manifold vibrations, in single-point injection pipes and carburetor manifolds the runners should always be as short as possible in order to keep fuel delivery times and fuel film deposits at a minimum.

With spider-type intake manifolds this can, in most cases, easily be achieved without adverse effects by installing almost straight connection pipes between the cylinder head intake ports and the central intake-manifold distributor. Siamese intake manifolds, however, often require longer common pipe sections in order to warrant symmetric inlet conditions at the 2nd branching. Moreover, with both spider-type and Siamese intake manifolds, care should be taken that the radii of the pipe knees do not fall below a given minimum so as to avoid unnecessarily high increases in pressure loss.

5.1.5.3. Distributor Volume

The area immediately downstream of the mixture formation system flange, where the runners (spider-type intake manifold geometry) or headers (Siamese intake manifold geometry) join, is referred to as the distribution area, point of distribution or simply as distributor. In this area great amounts of fuel film deposit and the distribution of fresh charge to the runners or headers is irreversibly determined, which means that distributor design is of critical importance.

In the distribution area, wall wetting is especially intense, which is due to various reasons: The sudden expansion of the cross-section leads to drastically reduced flow rates; moreover, in all types of downdraft intake manifolds the mixture is substantially deflected; the throttle valve installed immediately upstream of the distributor causes intense eddy current formation and turbulence; and, finally, almost the entire range of droplet sizes, including insufficiently atomized large droplets, are still present in this section.

Although designs which keep the distributor volume as small and clearance-volume free as possible (e.g., by rounding the edges) cannot prevent droplet deposits on the intake manifold floor, they can drastically reduce subsequent fuel accumulation and thus create important preconditions for stumble-free, low noxious emission transient engine operation.

5.1.6. Mixture Deflections and Branchings

Deflections and branchings are necessary to distribute the centrally prepared mixture to the individual cylinders.

In intake manifolds, the mixture can be deflected by pipe elbows as well as by knees. However, as, in knees, the burbling downstream of the inner edge of the knee causes further pressure losses additional to the direct pressure loss caused by the formation of secondary flows, pipe elbows are to be clearly preferred over knees so as to keep direct pressure losses as low as possible:

The pressure loss in single-phase steady-flow pipes can be characterized as follows (while neglecting the acceleration pressure loss) [5.1],

$$\Delta p_{ges} = \Delta p_R + \Delta p_E = \lambda_R \cdot \frac{l \cdot \rho \cdot v^2}{d \cdot 2} + \Sigma \cdot \zeta \frac{\rho \cdot v^2}{2} \qquad (5.6)$$

Δp_{ges} [Pa]   .....Total pressure loss

Δp_R [Pa]   .....Frictional pressure loss

Δp_E [Pa]   .....Pressure loss due to individual resistances (deflections, constrictions, expansions, etc.)

λ_R	[-]	..... Pipe friction coefficient
l	[m]	..... Pipe length
d	[m]	..... Pipe diameter
ζ	[-]	..... Drag coefficient
ρ	[kg.m^{-3}]	..... Density
v	[m.s^{-1}]	..... Velocity

with the last term in equation (5.6) indicating the sum total of all individual flow resistances with equal or different drag coefficients. While in straight pipes the pipe friction coefficient λ_R will lie between 0.01 and 0.07 [5.77], depending on the roughness of the pipes and the Reynolds number of the flow, the effective drag coefficients are considerably higher in deflected flows, especially in knee pipes (**Table 5.1**).

Pipe elbow

	ζ_{smooth}					ζ_{rough}
δ	15°	22,5°	45°	60°	90°	90°
R = d	0,03	0,045	0,14	0,19	0,21	0,51
R = 2d	0,03	0,045	0,09	0,12	0,14	0,30
R = 4d	0,03	0,045	0,08	0,10	0,11	0,23
R = 6d	0,03	0,045	0,075	0,09	0,09	0,18
R = 10d	0,03	0,045	0,07	0,07	0,11	0,20

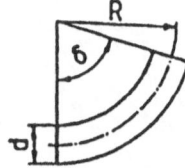

"smooth ".... technically smooth pipes
"rough " .... technically rough pipes

Sharp-edged knees

δ	10°	15°	22,5°	30°	45°	60°	90°	105°	120°
ζ_{smooth}	0,034	0,042	0,066	0,13	0,236	0,471	1,129	1,80	2,26
ζ_{rough}	0,044	0,062	0,154	0,165	0,32	0,684	1,265	2,00	2,54

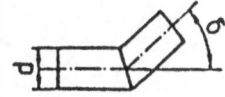

Table 5.1. Drag coefficients ζ [-] for pipe elbows and sharp-edged pipe knees with circular cross section [5.78]

Ref. [5.74] gives a good overview of the drag coefficients occurring in various pipe branching geometries.

As described in Chapter 2.6, the danger of droplet deposits on the intake manifold wall rises with increasing deflection angles. Moreover, developing secondary flows direct the fuel film into helical paths. Therefore the fuel-air mixture should be deflected as little as possible, particularly deflections immediately upstream of branchings should be avoided.

If deflection angles to the individual cylinders vary, as is often the case with horizontal spider-type intake manifolds, mixture distribution will in most cases only be possible by installing additional deflection devices.

5.1.7. Geometry of Pipe Cross Sections

The effects of pipe cross section geometry on the pressure loss has already been described. Due to their low circumference-to-area ratio, circular pipe cross sections cause the lowest frictional pressure loss. However, since the latter is relatively insignificant in comparison to the pressure losses caused by deflections, branchings, etc., it seems more sensible to chose the runner cross section from the point of view of what is best suited for an optimum distributor design. From the distributor design point of view, rectangular runner cross sections, especially rectangles standing on their narrow side, are more advantageous as they allow for both more compact distributor designs (**Fig. 5.24**, left) and smaller maximum deflection angles (Fig. 5.24, right).

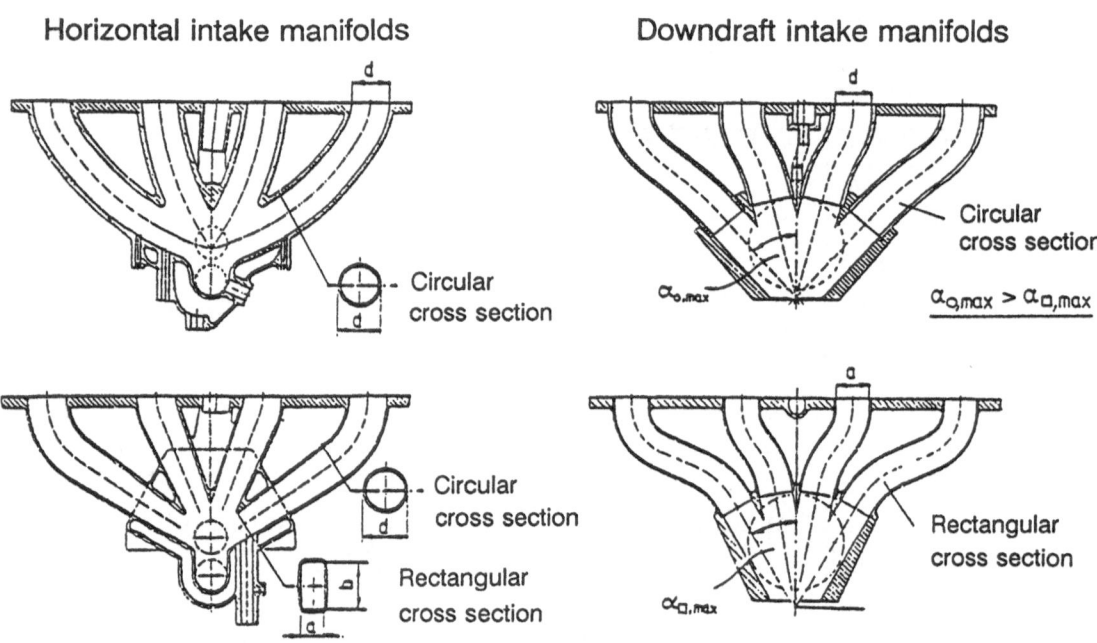

Fig. 5.24. Effects of pipe cross section geometry on distributor geometry [5.1]

Moreover, as opposed to circular cross sections, rectangular cross sections have no ribs on the manifold floor and ceiling, which considerably affect fuel film distribution in the intake manifold.

Also, fuel film vaporization is more intense in exhaust or coolant-heated runners, since the fuel flowing down the pipe walls can be distributed more evenly on the plain floor.

5.1.8. Distributor Geometry and Integral Distributor Parts

In most cases, distributor design is the most important factor influencing mixture distribution both in downdraft and in horizontal draft intake systems. Often minor geometrical modifications in this area suffice to change or completely redirect the mixture flow to another cylinder or pair of cylinders.

Basically, the distributor design should be as compact as possible for all types of intake manifolds. Clearance volumes, which, e.g., in the case of downdraft intake manifolds are preferably placed beneath the mixture distribution system (Fig. 5.16, left), should be avoided by adequate roundings as in the right half of Fig. 5.16, since they lead not only to increased pressure losses but also to more substantial fuel deposits in this area. Moreover, the entire distributor area should be intensely heated by either exhaust gas or engine coolant.

In addition to these general guidelines there are a number of geometric parameters which depend on the given basic intake manifold geometry and decisively influence mixture distribution.

With downdraft intake manifolds, the position of the inlet ports through which the mixture enters the intake manifold as well as the angle by which the distributor expands play an important role.

By changing the position of the inlet port(s) or the direction into which the throttle plate opens in the part-load range, the mixture flow can be completely redirected to another pair of cylinders. **Fig. 5.25** shows how the mixture distribution structure can be almost reversed by shifting a

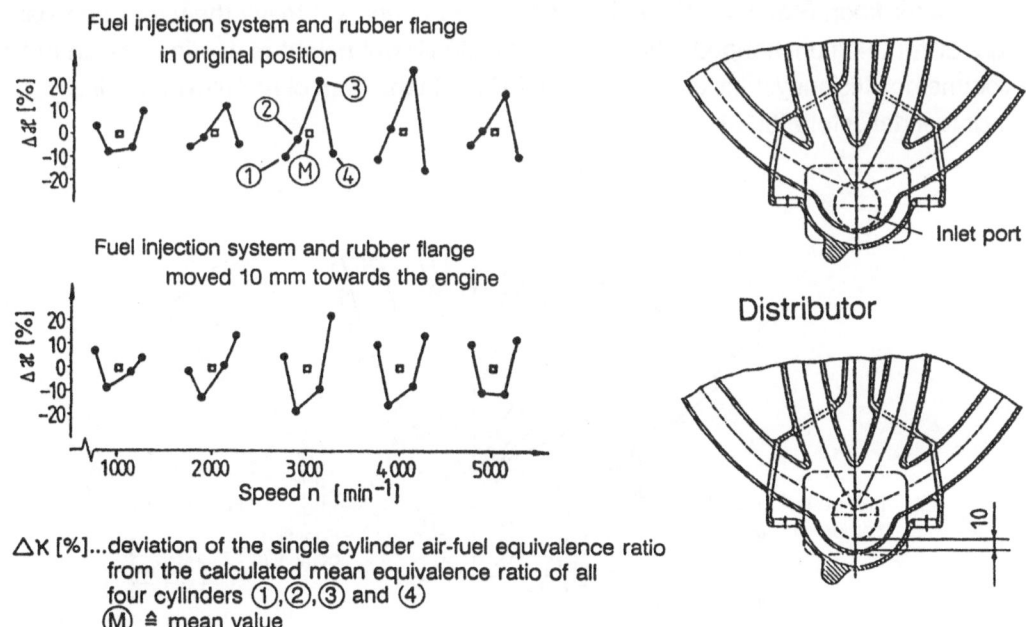

Fig. 5.25. Effects of inlet port position on mixture formation in the intake manifold [5.117]

single-flow central injection system just slightly closer towards the intake manifold center. The main reason of this redistribution effect seems to be that the inlet port protrudes from the rear distributor recess which deflects the fuel flow vertically entering the intake manifold towards cylinders 2 and 3. On the other hand, cylinders 1 and 4 can be leaned out by constricting the point of distribution (**Fig. 5.26**). This not only channels the flow to cylinders 2 and 3 but also disrupts the fuel film which, due to the dynamic pressure downstream of the inlet port(s), flows towards the outer cylinders.

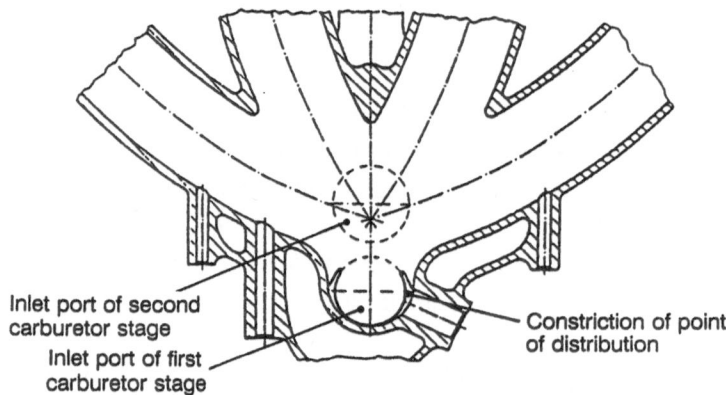

Fig. 5.26. Spider-type downdraft intake manifold with constricted point of distribution beneath first stage of compound carburetor [5.1]

Another measure to improve mixture distribution in downdraft manifolds, which is especially effective in the near full-load range, is to mount approximately 3 mm high fuel film baffle fins on the distributor floor, **Fig. 5.27**. Such fins can be added and yet leave the very good part-load mixture distribution unchanged, and due to the insignificant height of the fin, they do not affect the volumetric efficiency. Fins can be mounted either in the form of fuel film baffle fins [5.1, 5.80, 5.40] or fuel deflectors [5.75].

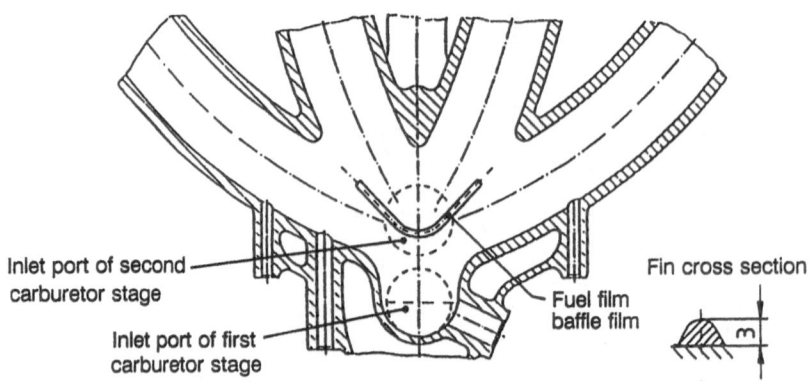

Fig. 5.27. Spider-type downdraft intake manifold with fuel film baffle fin at the hot spot [5.1]

In the case of horizontal intake manifolds where all the mixture entering the intake manifold is mainstreamed into the direction of cylinders 2 and 3, as can be seen in Fig. 5.6., 3 mm high fuel film baffle fins on the intake manifold floor are not sufficient. Due to unequal deflection angles in horizontal intake manifolds, maldistribution results not only from the fuel film but, to a large extent, also from the droplets floating in the mixture which means that, upon entering the intake manifold, the entire fuel flow must first be preliminarily deflected to cylinders 1 and 4. Here, vertical webs running from the manifold floor to the manifold ceiling have proven successful, see **Fig. 5.28**. Any fuel film deposited on the web breaks off at the back of the web at the latest, thus preventing the development of any uncontrollable fuel film flows as would be the case in Siamese manifolds. Moreover, the advantage of equal intake intervals in the distributor body is preserved.

Ref. [5.75] shows an interesting possibility of solving the problem of unequal deflection angles inherent to horizontal spider-type intake manifolds. Instead of the conventional in-line arrangement of the runner ports, the authors of [5.75] arrange them in a circle positioned concentrically and vertically to the inlet port, **(Fig. 5.29)** so that the mixture coming from the carburetor feeds the various cylinders according to the induction cycle. Grave disadvantages of this arrangement are the strong dependence of the mixture on the throttle position (preferential treatment of the upper or lower runner pair, depending on the throttle plate position during opening) and the fuel film flow on the manifold floor which, for gravity reasons, always goes to the lower runners.

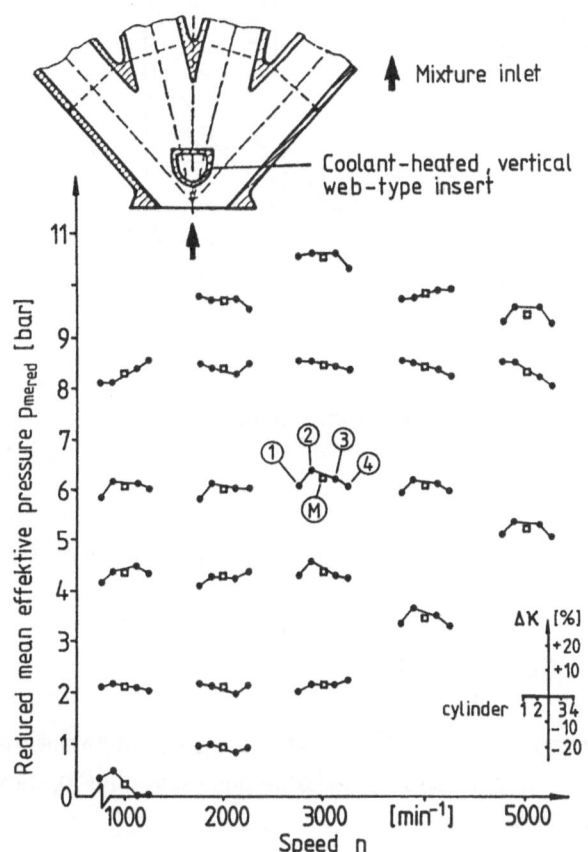

ΔK [%]...deviation of the single cylinder air-fuel equivalence ratio from the calculated mean equivalence ratio of all four cylinders ①,②,③ and ④
Ⓜ ≙ mean value

Fig. 5.28. Mixture distribution obtained in an optimized horizontal spider-type intake manifold with vertical web-type insert [5.71]

Additional integral parts for pressure loss reduction, e.g. baffle plates and guide blades (pure air stream) as used in air conditioning, should not be used in the case of intake manifolds with single-point mixture formation as they significantly influence mixture distribution.

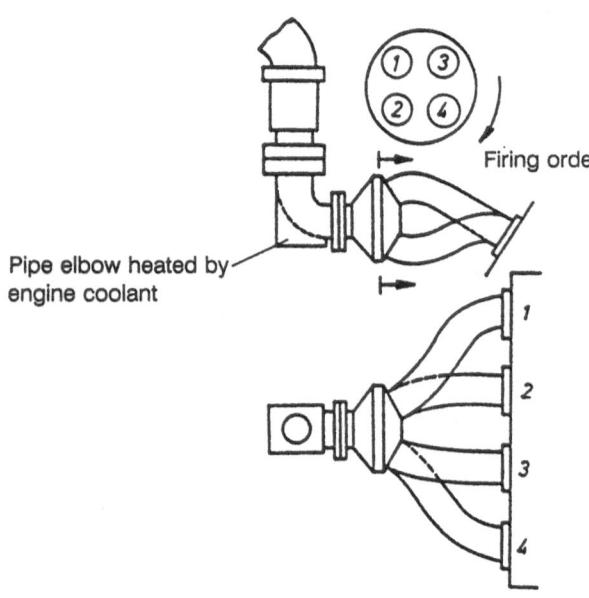

Firing order

Pipe elbow heated by engine coolant

Fig. 5.29. "Totally symmetric" spider-type horizontal intake manifold

Integral distributor parts for the improvement of fuel vaporization during warmup, however, are quite useful although they, too, must be adapted to the rest of the intake manifold geometry with regard to mixture distribution.

5.1.9. Inclination of the Intake Manifold

In view of the above-mentioned gravitational effect on fuel film transport, any lateral inclinations of the intake manifold, as may result from an inclined position typical of longitudinally deployed engines, must be avoided under any circumstances.

In contrast to this, a longitudinal inclination of the intake manifold in the flow direction will improve fuel film transport while reducing the danger of fuel accumulations and can thus be recommended.

5.1.10. Connection Bores

As practically no fuel transport takes place in intake manifold connection bores such as bores for vacuum extraction or exhaust gas recirculation, they influence mixture distribution only to a minor extent.

However, if these bores are unfavorably positioned, e.g., in the distributor floor area or in deadwater areas of the intake manifold, they can fill up with fuel at wide open throttle conditions. During overruns, this fuel then vaporizes in addition to the normal amount of fuel film and leads to an unwanted enrichment of the mixture. Therefore it is advisable to position bores, especially the large ones like the one for brake boosting, at the top of the intake manifold. A spot with as little wall wetting as possible should be determined by means of prior testing.

5.1.11. Intake Manifold Material

The material of carburetor manifolds should meet the following requirements:

* Castability;

- Good thermal conductivity;

- Low specific heat capacity;

- High mechanical strength;

- Chemical resistance against fuels, oils, and additives;

- Good resistance to thermal loads

- Good economy.

High mechanical strength is required as the intake manifold material is subject to great stress caused by engine vibrations. A low specific heat capacity guarantees that the cold intake manifold heats up fast during warmup whereas good thermal conductivity provides for high temperatures at the internal pipe walls.

The above requirements are largely met by numerous light metal alloys, especially aluminum alloys.

As far as the effects of intake-manifold coatings (e.g. teflon) on mixture distribution are concerned, there are no uniform results to be found in the relevant literature although numerous studies have been carried out in this field [5.33, 5.81, 5.82]. Coatings modify, e.g., the wettability of the pipe surfaces thus improving the quality of some intake manifolds while decreasing the quality of others. The effects of intake manifold coatings therefore depend on the specific manifold geometry.

5.1.12. Surface Roughness

The surface roughness has less effect on the frictional pressure loss in multiphase flows of intake manifolds with single-point mixture formation than in single-phase flows. Probably this is due to the fact that the phase boundary of the wave-like film flow is smoothed by the small abrupt elevations of the cast surface [5.77], which in some cases even leads to decreasing total pressure losses with increasing pipe roughness [5.77, 5.73].

Moreover, experiments [5.1] with intake manifolds of the same shape but of various roughness (grain sizes from 0.1 to 1 mm) indicated no influence of the roughness of the pipes on mixture distribution and fuel film formation, see **Fig. 5.30**.

Summing up it can thus be said that the surface roughness of the intake manifold plays only a minor role in intake manifold design.

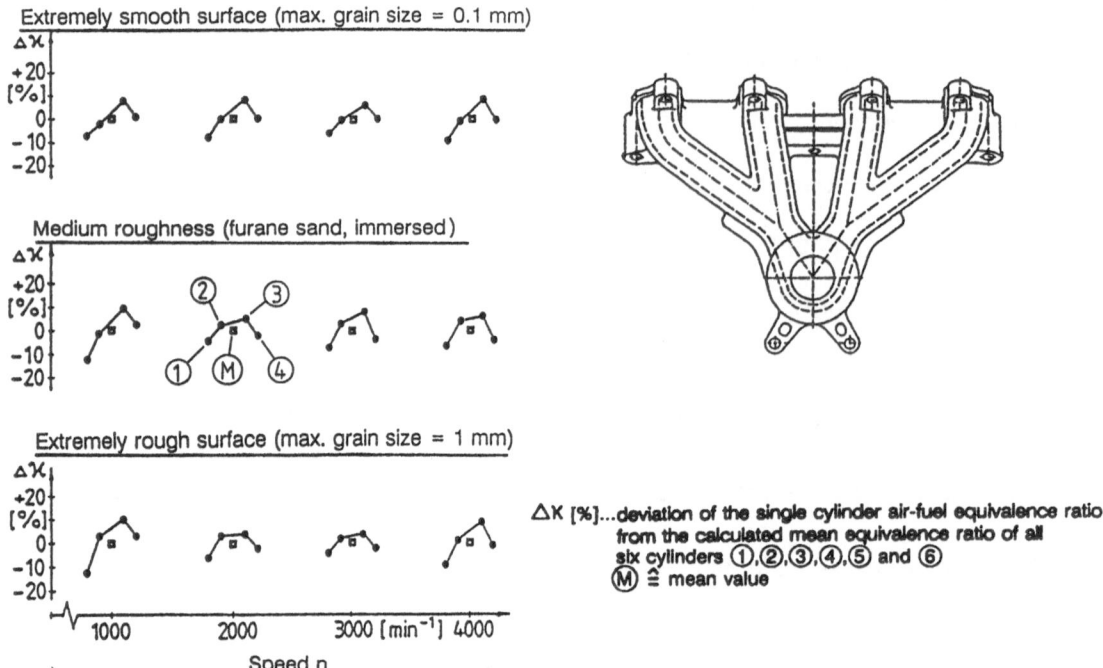

Fig. 5.30. Effects of intake manifold surface roughness on mixture distribution at wide open throttle conditions [5.1]

5.1.13. Intermediate Flange between Intake Manifold and Mixture Formation System

The main tasks of the intermediate flange between the intake manifold and the mixture formation system are:

- Thermal insulation of the mixture formation system against the exhaust or coolant-heated intake manifold in order to prevent vapor locks in the mixture formation system (float chamber, ducts, lines)

- Dampening of the vibrations the engine exerts on the mixture formation system, so as to prevent metering errors and premature wear of the moving parts.

Consequently, the material used for intermediate flanges will in most cases be a temperature resistant rubber compound which is vulcanization-reinforced by "floating" sheets.

In addition, mixture preparation and distribution can be improved by special components built into the intermediate flange.

According to their functions, such integral parts can be classified as follows:

- Integral parts for improved fuel vaporization such as the before-mentioned PTC honeycomb heaters, heatpipes, or duplex systems;

- Inserts that detach the fuel film deposited on the walls of the main mixture formation duct, such as deflector rings [5.75] or orifice plates (with internal teeth) (**Fig. 5.31**);

- Integral parts for mixture homogenizing (turbulent mixing, multiple deflections [5.84, 5.85], etc) which, however, always lead to an enormous pressure loss. An exception could be an exhaust-gas turbocharger installed between the mixture formation system and the intake manifold, whose main purpose would be to homogenize the mixture, in which case, however, it must be made sure that the turbocharger does not effect an unacceptable heatup of the fuel/air mixture.

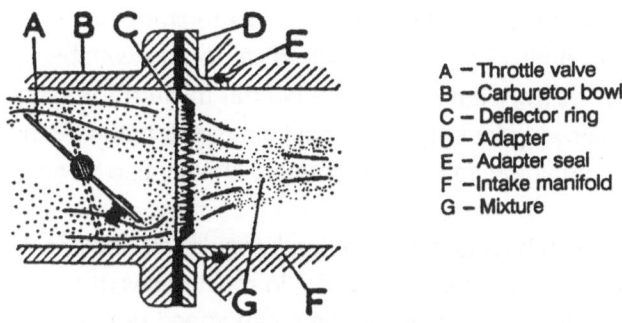

A – Throttle valve
B – Carburetor bowl
C – Deflector ring
D – Adapter
E – Adapter seal
F – Intake manifold
G – Mixture

Fig. 5.31. Deflector ring or fuel deflector in the carburetor flange for fuel film detachment [5.58]

- Inserts deflecting the mixture flow into the preferred direction or deflecting it by a certain angle just before it enters the intake manifold distributor. Thus, flaws in the manifold design as well as flow asymmetries developing in the mixture formation system can be compensated (**Fig. 5.32**). With these types of inserts, however, it must be kept in mind that even minor changes of their shape or placement can radically change mixture distribution in the intake manifold. Therefore precise manufacturing and a conscientious fine tuning on the engine test bench are an absolute must.

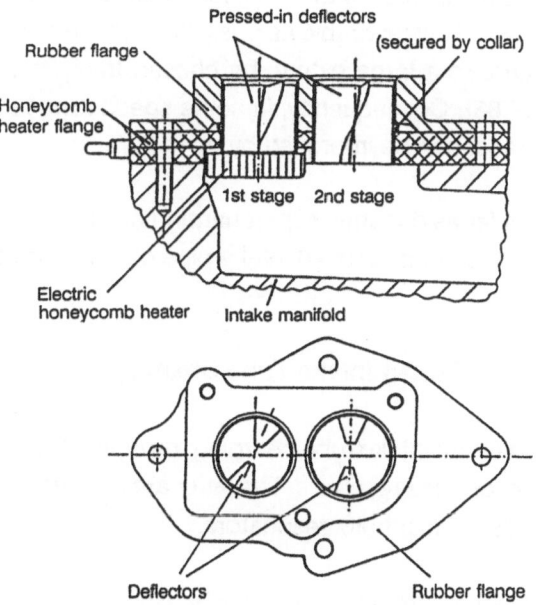

Fig. 5.32. Intermediate flange of a compound downdraft carburetor with integral deflectors [5.1]

5.2. Intake Manifold Design for Multi-Point Mixture Formation Systems

5.2.1. General

As described in Chapter 4.23, multi-point mixture formation systems can be single-barrel carburetors as well as multi-point injection systems. However, as single-barrel carburetion systems are almost extinct, this chapter only deals with intake manifold design for fuel-injected engines.

In relation to manifold geometry for engines with single-point mixture formation, the design engineer is quite restricted in his choice by a number of requirements [5.1]. Buildup and vaporization of fuel film deposits during transient operation call for the wettable surface to be as small as possible. Fuel droplet delivery times from the central mixture formation point to the cylinder should be short thus allowing for only short manifold pipe lengths. As high flow rates considerably improve mixture formation, manifold diameters should also be small.

The torque curves that can be reached by naturally aspirated engines with single-point mixture formation can thus be regarded as basis since, in most cases, they feature the lowest torque values. Their maximum torque often lies in the mean speed range around 3500 min^{-1}. Varying torque curves for low and high speed ranges respectively can be obtained by varying the valve timing.

The above problems in single-point mixture formation engines can be avoided by using multi-point injection with the injector near the intake valve. In that case, induction system design can be as flow-favorable as possible, and the geometrical dimensions of the entire induction system can, to a large extent, be chosen freely thus utilizing the gasdynamic supercharging effects [5.86]. Consequently, in some speed ranges the mean effective pressure, i.e. engine performance, can be improved by up to 20%.

As far as dynamic supercharging of internal combustion engines is concerned, ram pipe supercharging, and tuned intake pipe charging (**Fig. 5.33**), are commonly known and used.

5.2.2. Tuned-Intake Tube Charging

In tuned intake tube charging, groups of cylinders with equal ignition intervals are connected to resonance receivers. The latter are connected to a common receiver via tuned intake pipes and act as Helmholtz resonators.

Supercharging is effected when the fundamental harmonic of the excitation produced by the periodic induction cycles of the cylinders corresponds to the natural frequency of the system.

Ram pipe supercharging   Tuned-intake tube charging

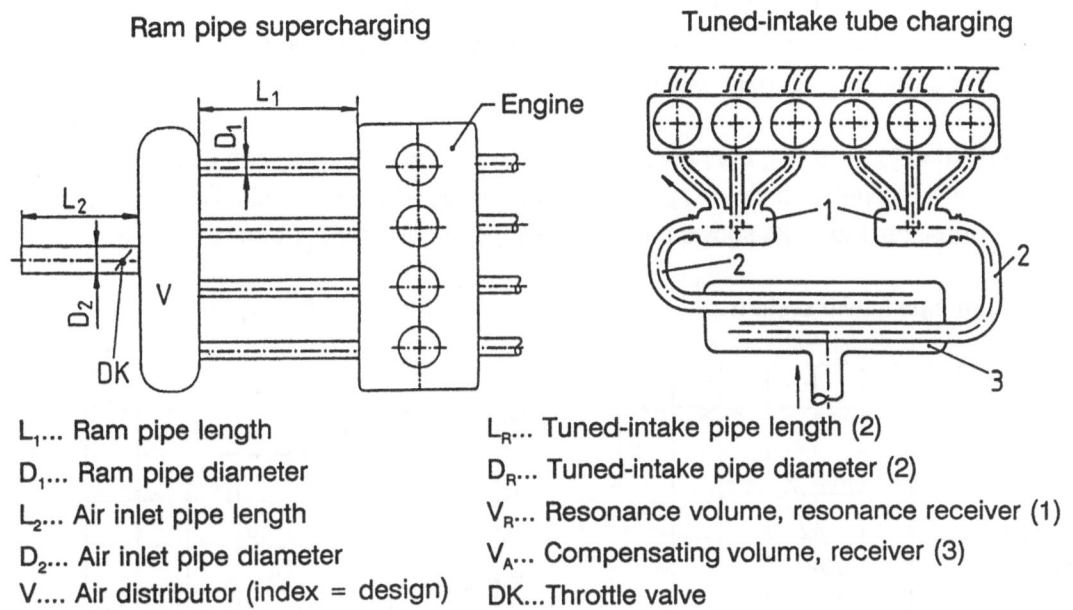

L_1... Ram pipe length L_R... Tuned-intake pipe length (2)
D_1... Ram pipe diameter D_R... Tuned-intake pipe diameter (2)
L_2... Air inlet pipe length V_R... Resonance volume, resonance receiver (1)
D_2... Air inlet pipe diameter V_A... Compensating volume, receiver (3)
V.... Air distributor (index = design)   DK...Throttle valve

Fig. 5.33. Induction systems for ram pipe supercharging and tuned-intake tube charging [5.86 and 5.87]

The critical thing here is the concurrence of three cylinders with induction phases staggered by 240°CA each, which do not overlap in time. The suction lines connecting the cylinders to the resonance receiver must be as short as possible.

The right-hand illustration of Fig. 5.33 shows a possible solution of tuned intake tube charging. The cylinders of a six-cylinder engine (firing order: 1-5-3-6-2-4) are broken down into two groups each of which is connected to the resonance receiver. According to the above precondition, the induction periods of the cylinders allocated one common resonance receiver do not overlap in time.

Tuned intake tube charging is characterized by the almost undampened vibration of the intake manifold and by a decreasing mass flow curve in the middle of the inlet phase which is due to the unfavorable compression ratio prevailing during this phase. The mass flow curve peaks at the beginning as well as at the end of the inlet phase.

The disadvantage of tuned intake tube charging in induction engines is that only a small increase in air consumption is reached just outside the resonant range, which means that only one marked torque peak occurs.

Tuned intake tube charging has become of practical importance due to the fact that in combination with exhaust-gas turbocharging (combined super/turbocharging) in automobile engines (mainly truck diesel engines) it can improve the low torque level of turbocharged engines in the lower speed range. For more detailed information please refer to the relevant literature [5.88,

5.89., 5.90]. In most cases, such a combination is chosen so that, due to the tuned-intake pipe system, in the 1100 min^{-1} to 1400 min^{-1} speed range an increase in air consumption can be reached in comparison to mere exhaust-gas turbocharging.

As described in [5.91, 5.92], the principle of combined supercharging can also be used for automotive Otto engines with a high specific power output, where it can significantly improve engine performance. **Fig. 5.34** shows the respective power and torque curves.

Ram pipe intake system   Tuned-intake pipe system

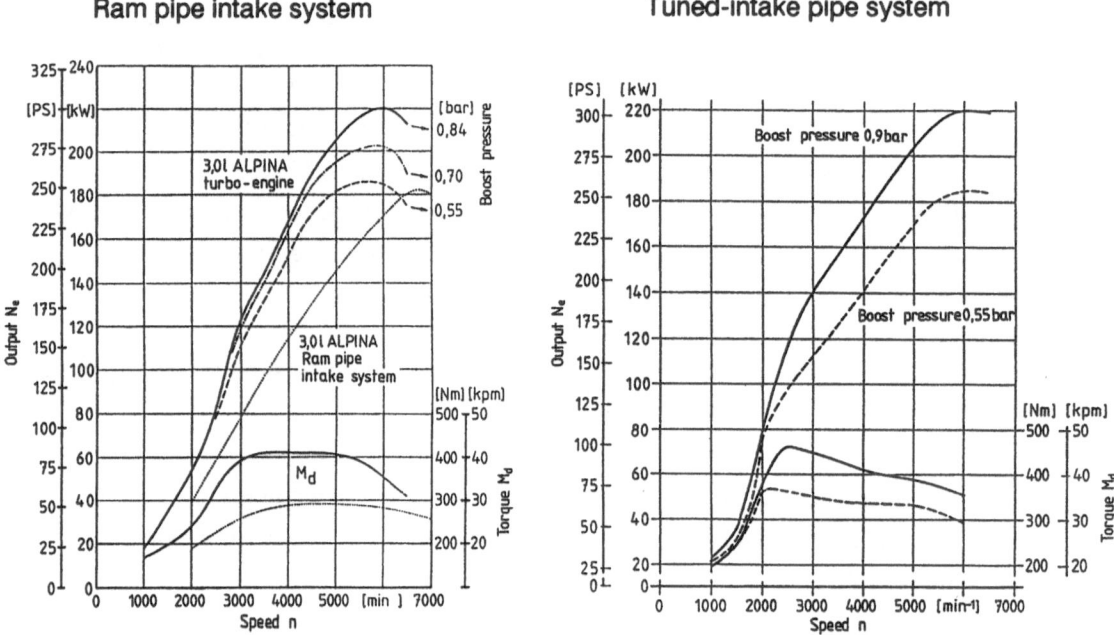

Fig. 5.34. Output and torque curves of the Alpine turbo-engine with ram pipe intake system or tuned-intake pipe system [5.91]

Ref. [5.93] shows that with high-speed passenger car aspirating Otto engines, too, tuned intake tube charging leads to excellent air consumption values (**Fig. 5.35.**).

Critical tuned intake tube charging design dimensions are:

- Tuned intake pipe length;

- Tuned intake pipe diameter;

- Resonance receiver volume;

- Resonance speed

- Displacement volume of the engine.

As already mentioned, the suction lines (connecting the cylinders to the resonance receiver) should be as short as possible with sufficiently large diameters.

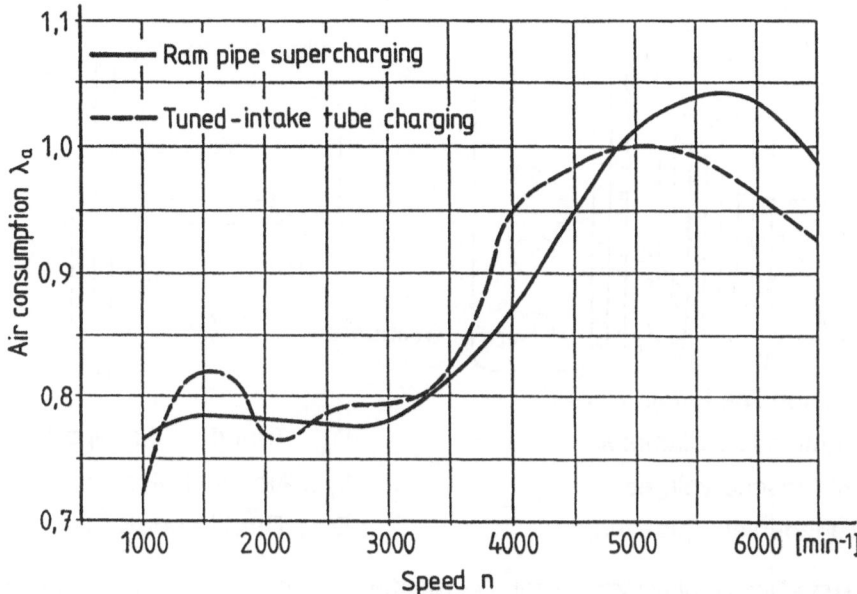

Fig. 5.35. Air consumption λ_a of the 911 Carrera engine with tuned-intake pipe system [5.93]

Increasing the length of the tuned intake pipe, decreasing its diameter, or increasing the resonance receiver volume results in lower resonance speeds.

The described variety of possible applications of tuned intake tube charging at the same time shows how largely the parameters can vary. The tuned intake pipe can have a length of 0.4 m to 1.4 m and a diameter of 40 mm to 80 mm, while the resonance receiver volume can vary between 0.15 l to 4 l.

In four-cylinder in-line engines whose induction phases are staggered by 180°CA, a number of the essential characteristics of the above-defined tuned intake tube charging cannot be produced.Overlapping of the induction cycles must be prevented by combining two by two cylinders whose induction phases are staggered by 360°CA. This means that cylinders 2 and 3 and cylinders 1 and 4 must be combined. The described principle requires short connection lines from the cylinders to the resonance receiver, and receiver volumes of approximately 0.2 dm^3, which however, contradicts the given geometric distances between cylinders 1 and 4 so that such an induction system can hardly be put into practice; see **Fig. 5.36**.

In an experimental arrangement, all 4 cylinders were connected to one resonance receiver, so as to keep the suction lines as short and the resonance receiver as small as possible; see Fig. 5.36, right. The geometrical dimensions thus corresponded to the principle of a resonance induction system described in [5.86]; the decisive difference to [5.86] is that in the experimental arrangement the induction cycles, and consequently the excitation phases, of the cylinders connected to the receiver overlapped in time.

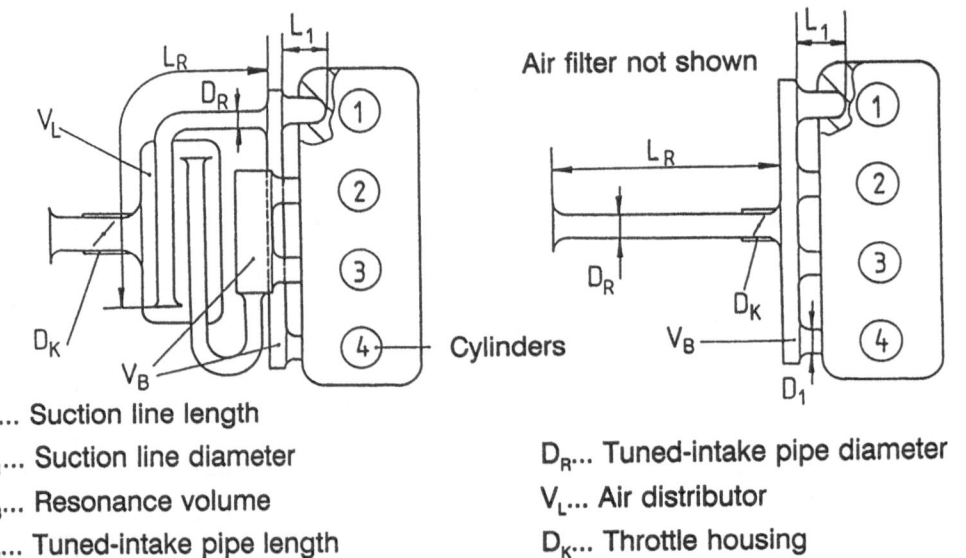

L₁... Suction line length

D₁... Suction line diameter

V_B... Resonance volume

L_R... Tuned-intake pipe length

D_R... Tuned-intake pipe diameter

V_L... Air distributor

D_K... Throttle housing

Fig. 5.36. Tuned intake pipe systems for a 4-cylinder inline-engine with one and two resonance receivers respectively [5.94]

Fig. 5.37 shows the mean effective pressure curves measured on various tuned intake pipe designs. Contrary to the systems described in [5.86], these designs feature a broad speed range with an increased torque with no marked torque peak, which is due to the fact that the induction cycles of the respective cylinders overlap.

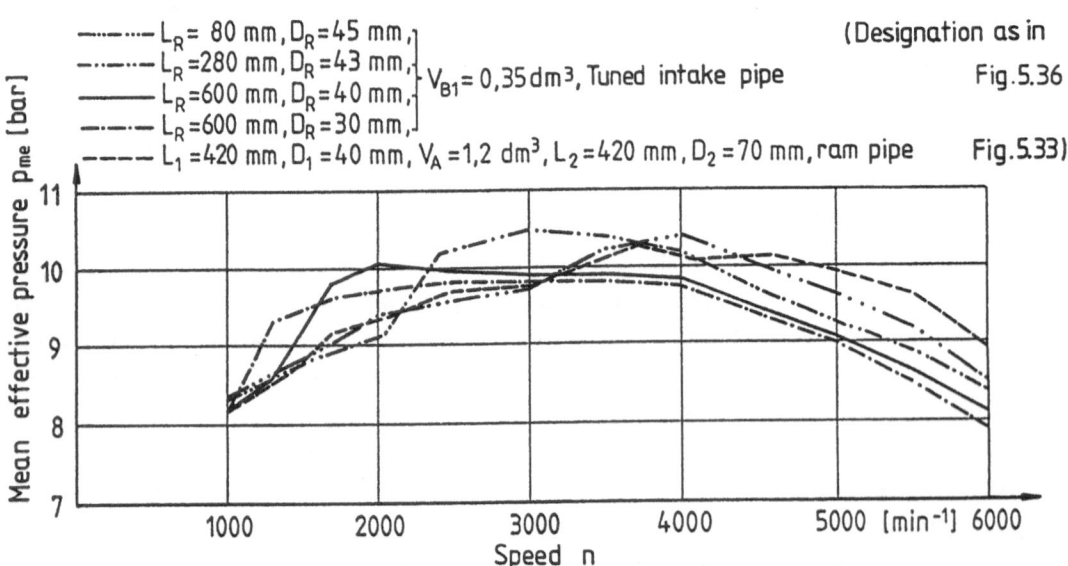

Fig. 5.37. Mean effective pressure versus the speed of various tuned intake pipe systems with one tuned-intake pipe for four cylinders as compared to a ram pipe system according to Fig. 5.33 [5.94]

What is remarkable is the torque curve of a tuned intake pipe with a length of 600 mm and a diameter of 40 mm. In the 1700 min^{-1} to 4000 min^{-1} speed range the mean effective pressure is almost constant at 9.9 bar. With a tuned intake pipe length of 80 mm, which in the present case is determined by the throttle housing, the mean effective pressure curve corresponds exactly to that obtained with a carburetor.

For comparison purposes, the mean effective pressure curve of a conventional ram pipe system was also plotted in Fig. 5.37.

5.2.3. Ram Pipe Supercharging

Here, each cylinder has its own intake pipe of a specific length, which is connected to an air distributor . Ram pipe supercharging is based on a physical principle different from that of tuned intake tube charging, i.e., a steep pressure drop at the intake valve at the beginning of the inlet phase. This vacuum wave runs through the intake manifold and is reflected inversely as a compression wave at the open pipe end, and, with correct timing of the phases, shortly before intake valve closing this compression wave leads to a supercharging effect resulting in high air consumption and consequently an optimum mean effective pressure. Via the air distributor the individual cylinders can influence each other, although this is not necessary to obtain the desired effect.

With ram pipe supercharging, the intake manifold pressure rises in the middle of the inlet phase due to the reflection of the pressure wave in the air distributor. Accordingly, the mass flow curve through the intake valve peaks in the middle of the inlet phase. Ideally, after intake valve closing, the air column continues to vibrate dampedly with the natural frequency of a unilaterally closed pipe.

Fig. 5.38 shows the mean effective pressure curves of various induction systems. With ram pipe supercharging, the speed range with an optimum mean effective pressure is relatively wide. However, with speeds exceeding the tuned speed range, there is a significant mean effective pressure loss.

Ram pipe supercharging should be used when a significant increase in volumetric efficiency, as compared to a directly aspirating engine, is desired over a wider speed range.

In References [5.94 and 5.108] the parameters were systematically varied. The left illustration of Fig. 5.33 shows the setup of such a ram pipe system. The parameter study investigated the effects of the ram pipe length, the air inlet pipe length, the air inlet pipe diameter, and the air distributor on the maximum mean effective pressure. The following figures show the speeds at which mean effective pressures reach a maximum or minimum, and also the extent of mean effective pressure variations within very narrow speed limits as well as the relative differences between the individual curves.

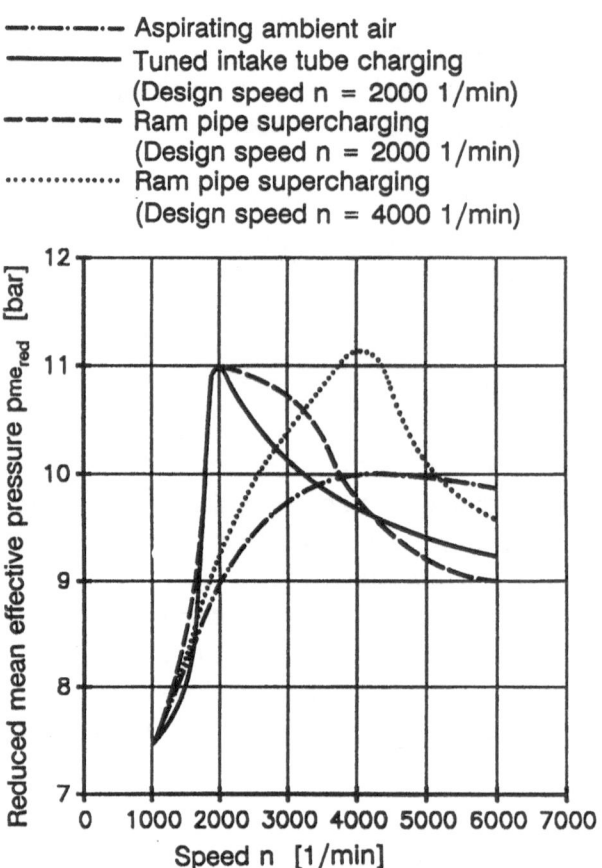

Fig. 5.38. Parameters of various induction systems

Fig. 5.39 shows the effect the ram pipe length L_1 has on the maximum mean effective pressure. Long ram pipes effect high torques at low speeds, but lead to a considerable loss of maximum power, i.e. at high speed. Short ram pipes guarantee high maximum power but low torque at low speeds. Attention must be paid to the fact that the maximum mean effective pressure can vary by up to 20%.

As can be seen in **Fig. 5.40**, small ram pipe diameters D_1 are advantageous at low speeds, while with large ram pipe diameters a high power output can be obtained at high speeds. Again, attention must be paid to the extent of the differences.

As to the air distributor, it was found that its volume has less effect on the mean effective pressure curve (**Fig. 5.41**) than its design, i.e. the geometrical arrangement of the air inlet and the ram pipes at the air distributor (**Figs. 5.42, 5.43,** and **5.44**).

Fig. 5.45 shows the effects of the air distributor design on the mean effective pressure. The index stated with the air distributor volume V indicates the air distributor design.

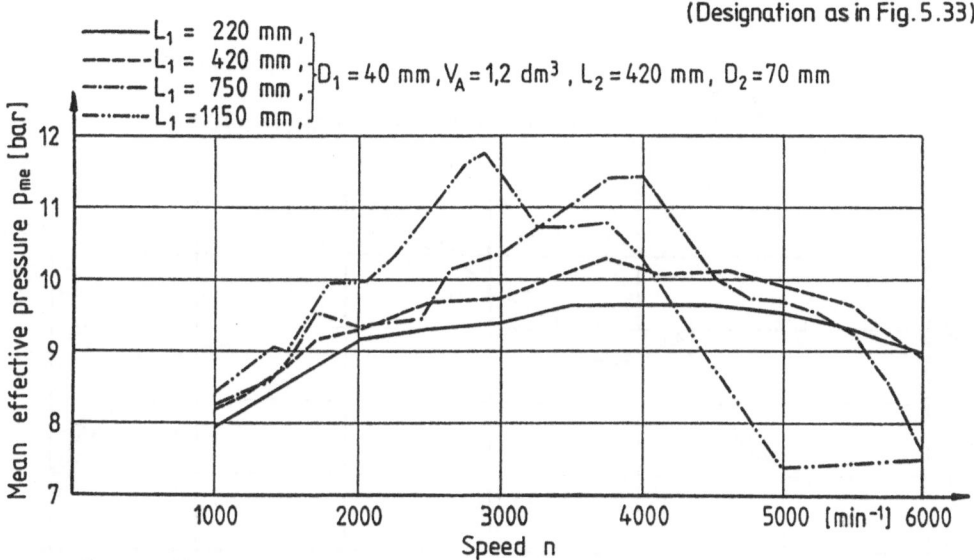

Fig. 5.39. Effect of the ram pipe length L₁ on the maximum mean effective pressure as a function of the speed [5.94]

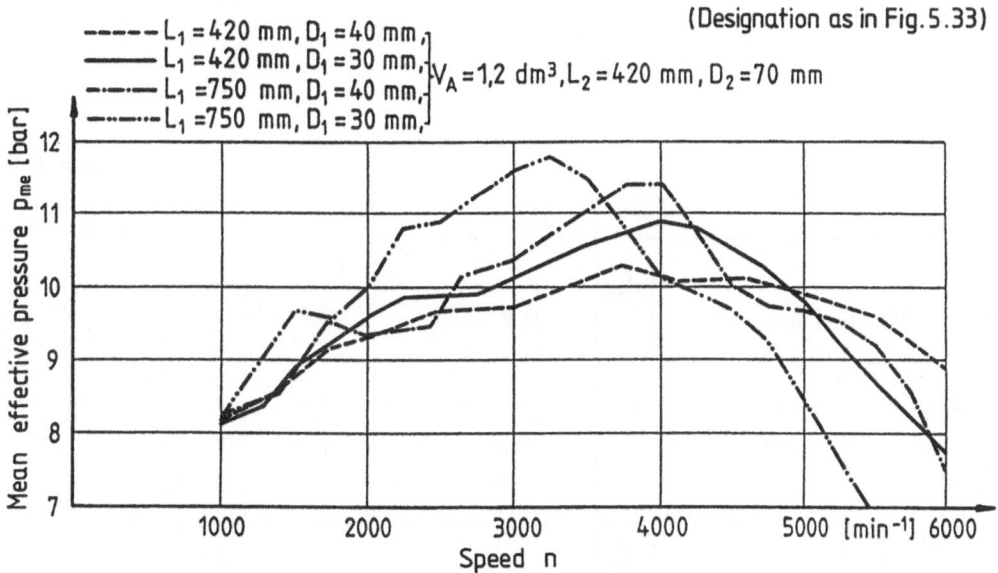

Fig. 5.40. Effects of the ram pipe diameter D₁ on the maximum mean effective pressure as a function of the speed for two different ram pipe lengths [5.94]

The length of the air inlet L_2 affects, above all, the torque curve in the lower speed range (**Fig. 5.46**). An air inlet pipe length of 600 mm is advantageous for a speed of 1600 min^{-1}, whereas a very short air inlet of only 80 mm length results in mean effective pressure gains at speeds of 2500 min^{-1}. At speeds above 4000 min^{-1}, the effects of present production air inlet pipe lengths with sufficiently large diameters on the maximum mean effective pressure are hardly measurable.

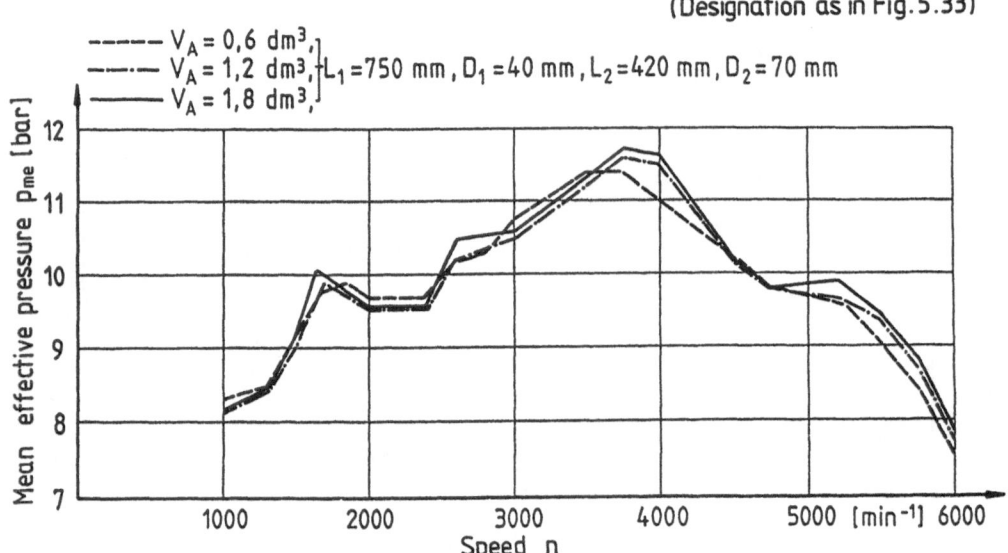

Fig. 5.41. Effects of air distributor volume V_A on the maximum mean effective pressure as a function of the speed

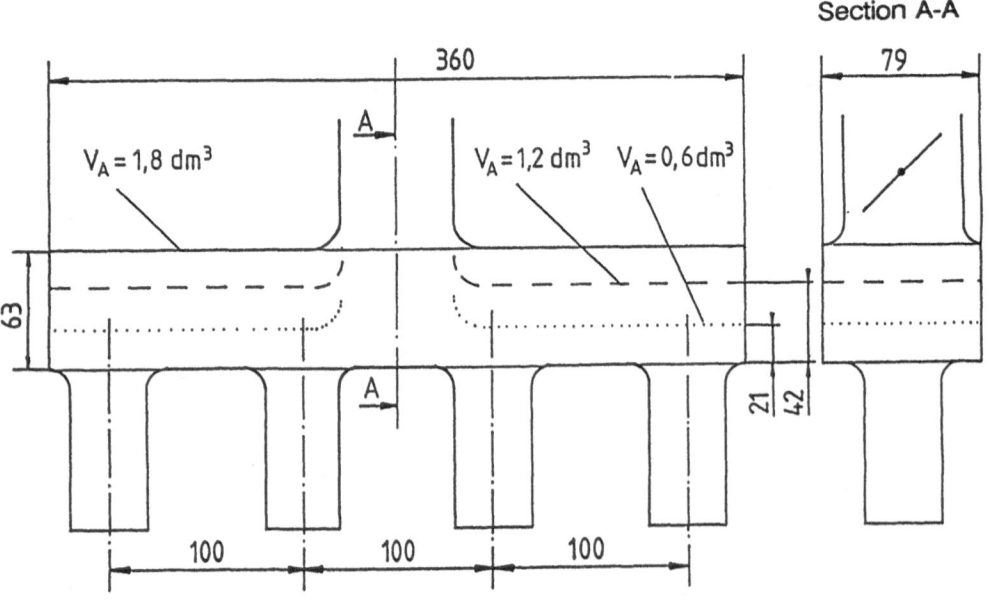

Fig. 5.42. Air distributor with parallel inlet and outlet pipes at one level, design A [5.94]

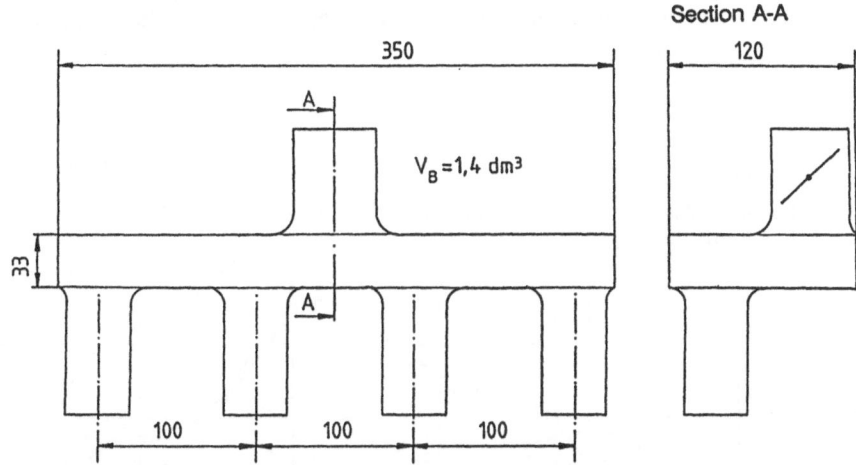

Fig. 5.43. Air distributor with parallel inlet and outlet pipes at staggered levels, design B [5.94]

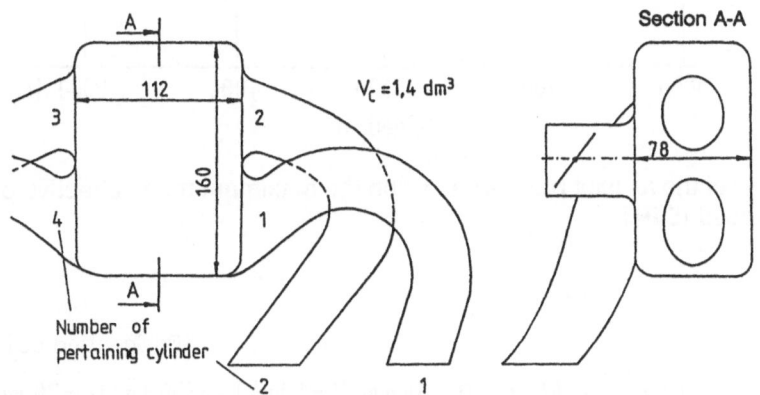

Fig. 5.44. Air distributor with inlet perpendicular to outlet, design C [5.94]

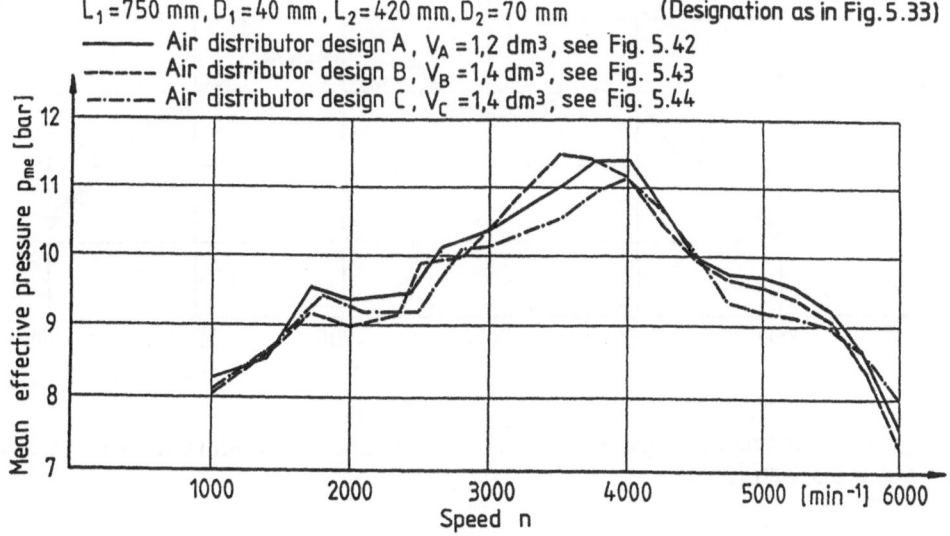

Fig. 5.45. Effects the air distributor design has on the maximum mean effective pressure versus the speed [5.94]

As shown in **Fig. 5.47**, the diameter of the air inlet D_2 affects, above all, the mean effective pressure peak in the lower speed range.

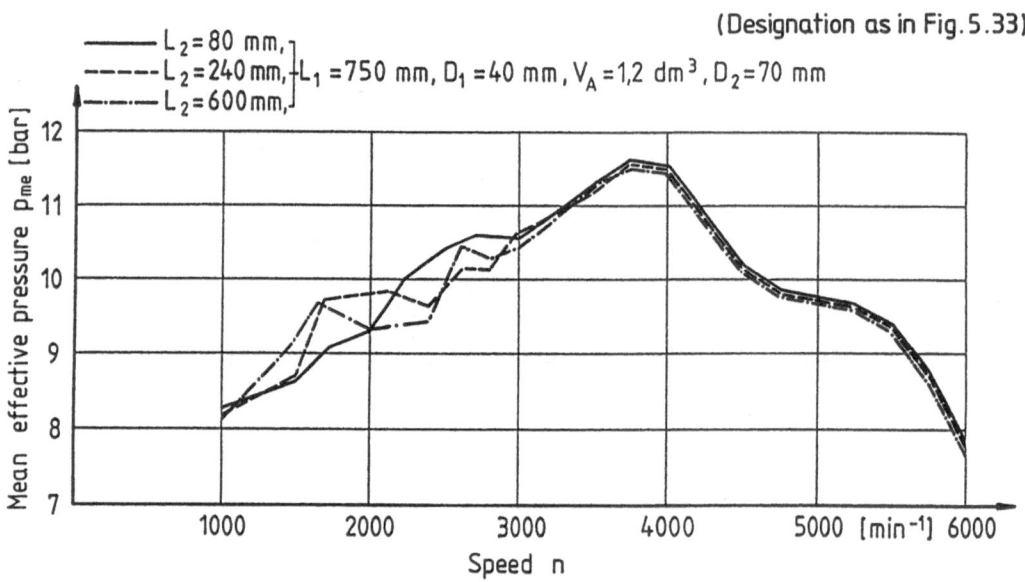

Fig. 5.46. Effects of the air inlet pipe length L_2 on the maximum mean effective pressure versus the speed [5.94]

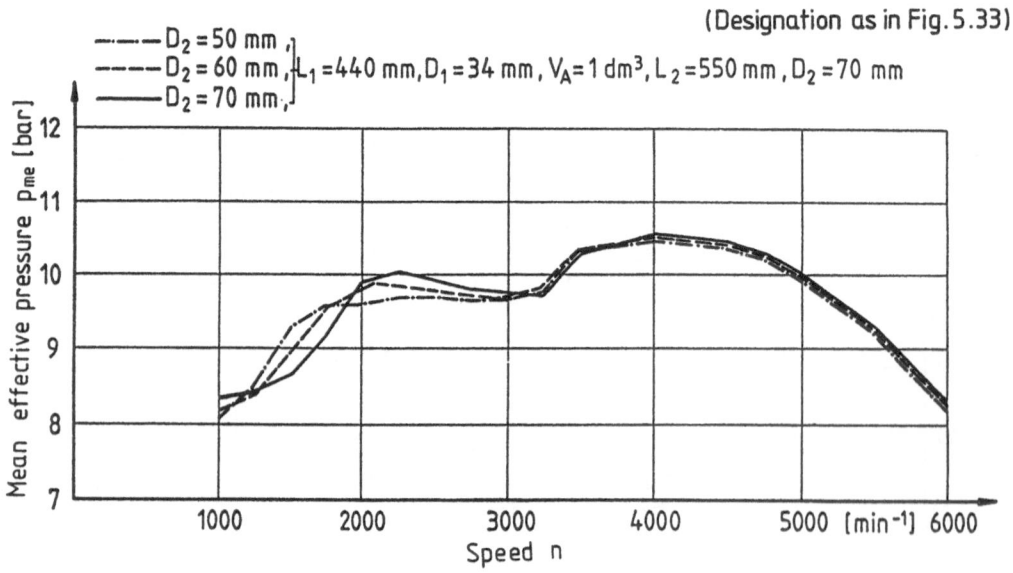

Fig. 5.47. Effects of the air inlet pipe diameter D_2 on the maximum mean effective pressure versus the speed [5.94]

Consequently, the ram pipe length L_1 and the ram pipe diameter D_1 are the main factors influencing the mean effective pressure curve providing the possibility of increasing the torque curves in the lower speed range by appropriately tuning the air inlet and the air distributor.

Fig. 5.48 compares the maximum mean effective pressure curve of an LH Jetronic induction system which has been tuned according to the maximum mean effective pressure setting for the mean speed range to the mean effective pressure curve of a standard LH Jetronic induction system. With the same maximum performance, the mean effective pressure can be increased by an average 5%.

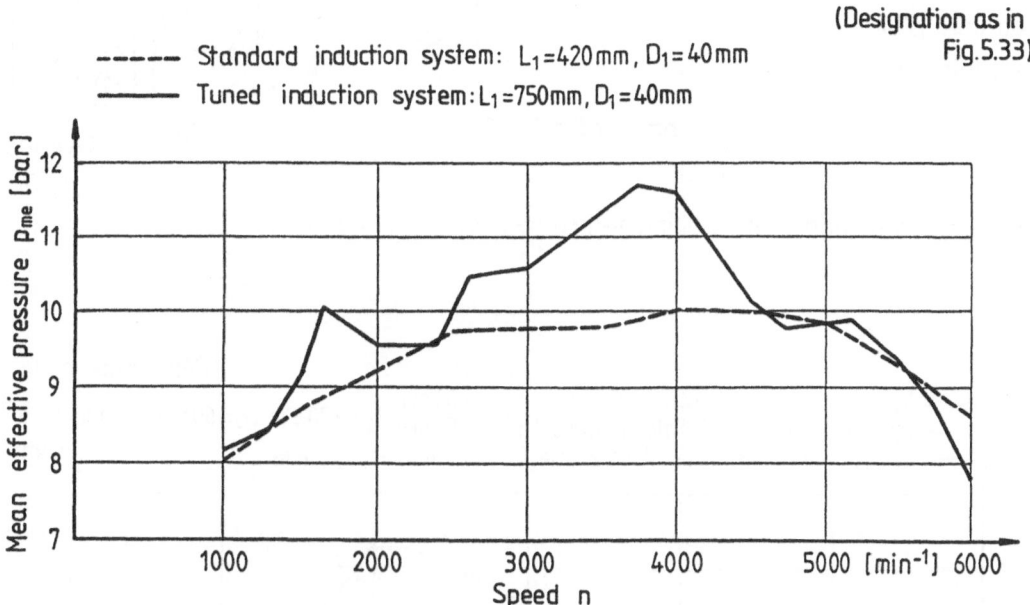

Fig. 5.48. Maximum mean effective pressure curve of a standard and of a specifically tuned induction system [5.105]

5.2.4. Unconventional Induction Systems Without Variable Dimensions

In Ref. [5.94] two intake manifold combinations without movable or adjustable elements were investigated as a preliminary stage of variable induction systems. **Fig. 5.49** shows a so-called "ignition sequence intake manifold", where two by two cylinders, whose induction order was staggered by 360°CA, were coupled via a tee. The question was, on the one hand, whether the residual vibration of one cylinder can favorably influence the induction process of the second cylinder and, on the other hand, whether a common ram pipe section could be of advantage.

The results shown in **Fig. 5.50** indicate a rather small torque gain in the lower speed range and a high torque loss at high speeds.

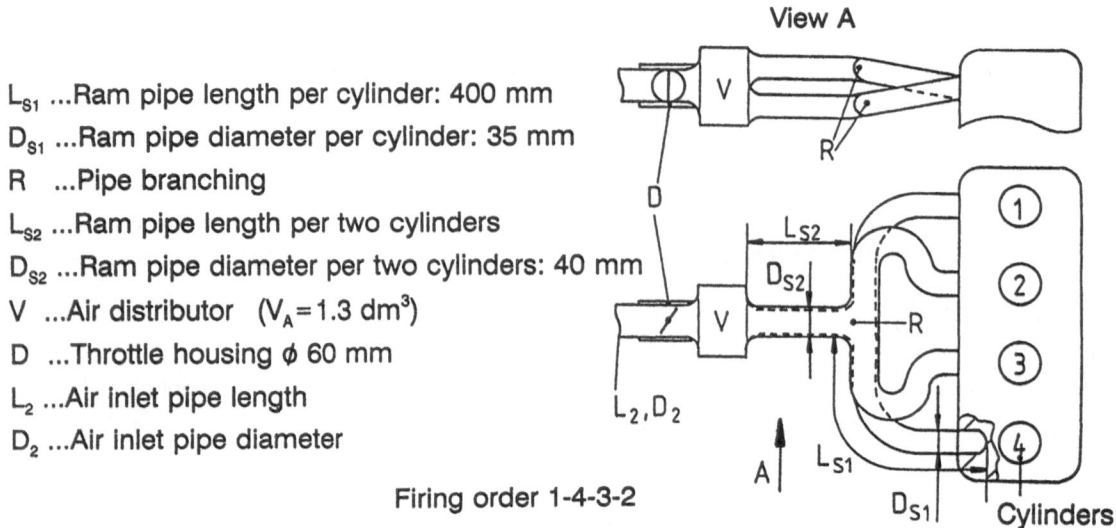

L_{S1} ...Ram pipe length per cylinder: 400 mm

D_{S1} ...Ram pipe diameter per cylinder: 35 mm

R   ...Pipe branching

L_{S2} ...Ram pipe length per two cylinders

D_{S2} ...Ram pipe diameter per two cylinders: 40 mm

V   ...Air distributor   ($V_A = 1.3$ dm^3)

D   ...Throttle housing ϕ 60 mm

L_2 ...Air inlet pipe length

D_2 ...Air inlet pipe diameter

Firing order 1-4-3-2

Fig. 5.49. Construction of an "ignition sequence intake manifold" [5.94]

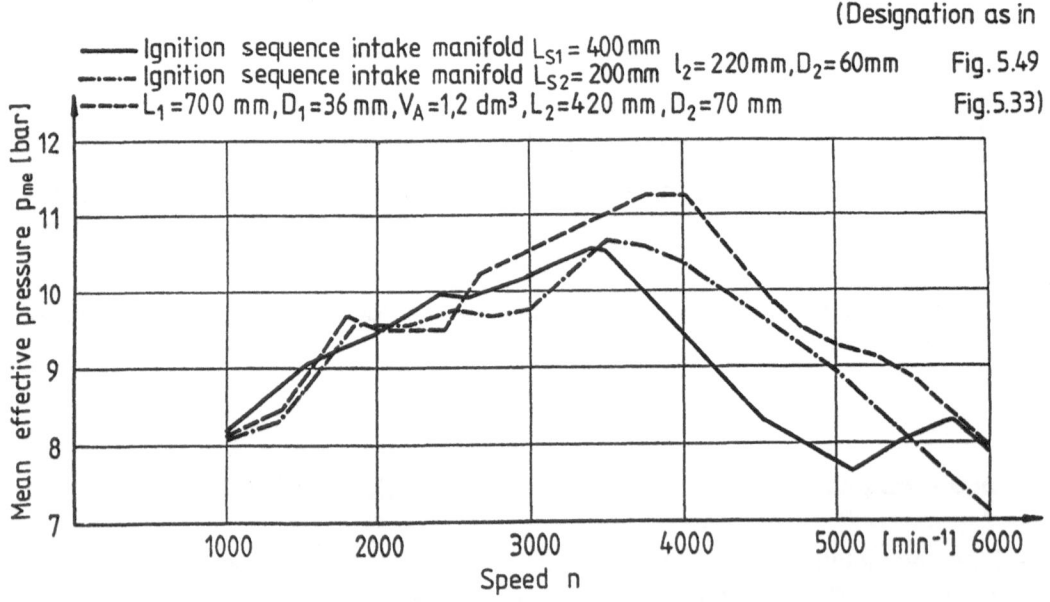

Fig. 5.50. Mean effective pressure versus the speed of two different ignition sequence intake manifolds according to Fig. 5.49. and a conventional ram pipe system according to Fig. 5.33 [5.94]

Another possibility to increase the torque was envisaged in a relatively long, U-shaped ram pipe which, due to a cross connection, contains a second ram pipe length (**Fig.5.51**).

L_{S1} = 750 mm long ram pipe

D_{S1} = 40 mm

R ...Pipe branching

L_{S2} = 400 mm short ram pipe

D_{S2} = 37 mm

V ...Air distributor (V_A = 0.6 dm³)

D ...Throttle housing ⌀60 mm

L_2...Air inlet pipe length

D_2...Air inlet pipe diameter

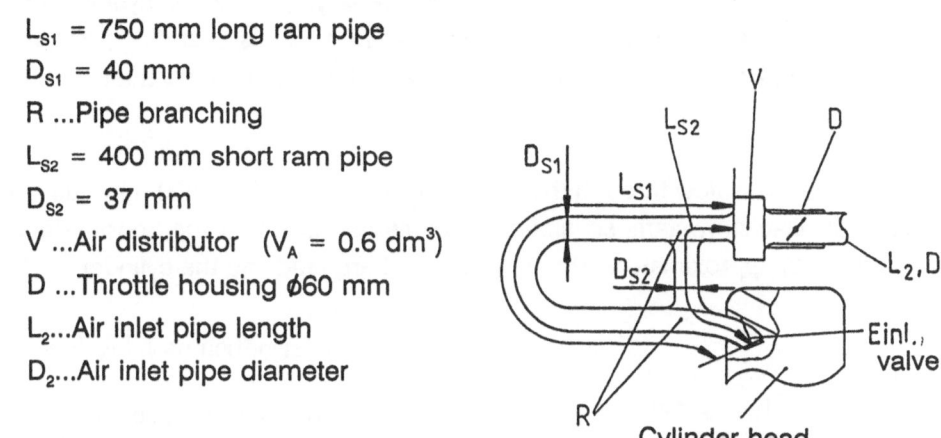

Fig. 5.51. Cross-sectional view of U-shaped ram pipe [5.94]

As can be seen in **Fig. 5.52**, the results were considerably less favorable than the ones obtained with the long ram pipe with no cross connection. Moreover, the expected increase in maximum performance could not be obtained either.

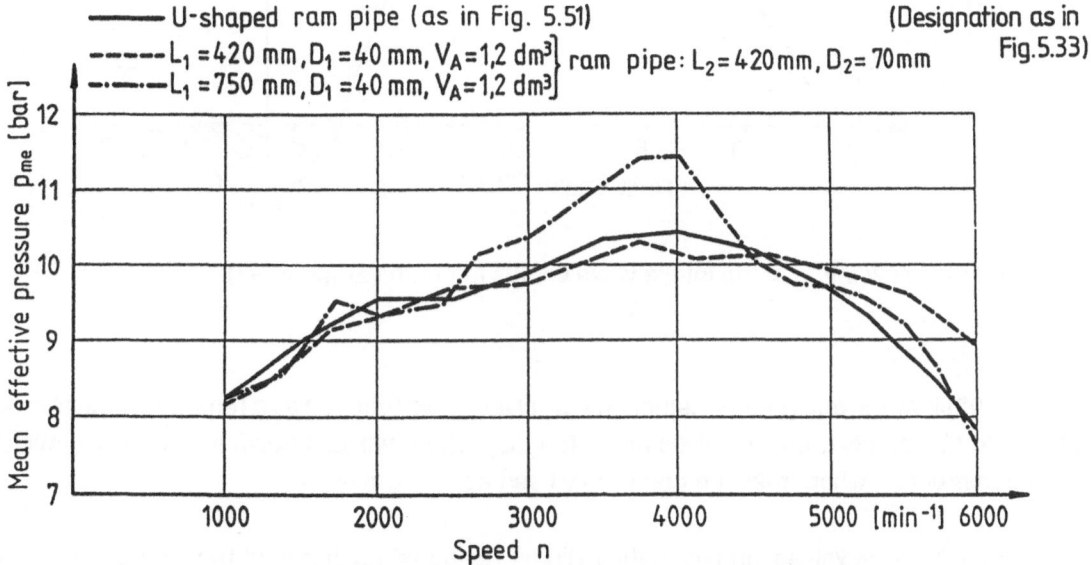

Fig. 5.52. Mean effective pressure versus the speed of the U-shaped ram pipe as in Fig. 5.51 and of two conventional ram pipe systems as in Fig. 5.33 [5.94]

From these results it can be concluded that in combined induction systems elements should be used which clearly determine the effective ram pipe length so that the given potential can be fully utilized and that there are no superpositions which would cause low pressure amplitudes or unfavorable phase displacements thereby no longer warranting an optimum air consumption.

Fig. 5.39. indicates that at speeds of 1000 min⁻¹, strongly varying induction system dimensions have extremely little influence on the mean effective pressure. At this operating point, the intake manifold vibration amplitudes are too insignificant to affect the backflow of the fresh charge already inducted by the cylinder.

Fig. 5.53 shows the mass flow through the intake valve at a speed of 1500 min⁻¹ as a function of the crank angle. Backflow starts several crank angle degrees after the bottom dead center whereby, in this case, approximately 10% of the fresh charge fed into the cylinder are lost .

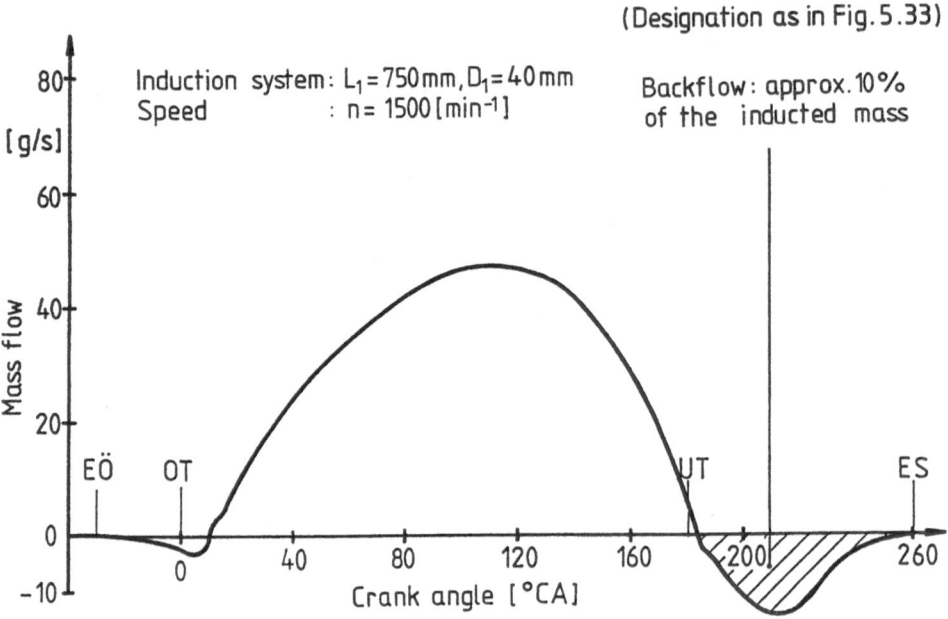

Fig. 5.53. Mass flow curve at the intake valve versus the crank angle [5.94]

In Ref. [5.105] it was investigated whether check valves could function as a remedy for the above problem. In this context, check valves mean the diaphragm valves known in two-stroke-cycle engine construction where they are commonly used as inlet governors.

Mounting such check valves directly at the cylinder head just upstream of the intake valve can prevent the fresh charge from flowing back out of the cylinder, while the negative effect of the clearance volume between the check and intake valves is negligible.

Fig. 5.54 shows the high torque gain at speeds below 3000 min⁻¹. The torque loss in the upper speed range is caused not only by pressure losses at the check valve but also by an unfavorable pressure curve phase displacement shortly before intake-valve closing and is thus a gasdynamics problem.

(Designation as in Fig. 5.33)

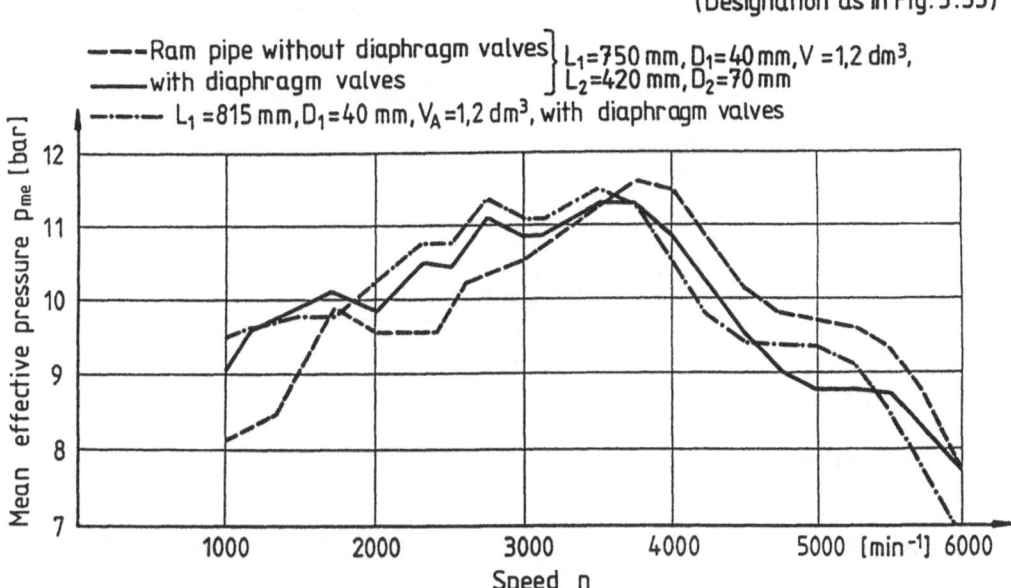

Fig. 5.54. Effects of check valves (diaphragm valves) on the maximum mean effective pressure [5.105]

Investigations were also carried out on the prevention of torque losses in the upper speed range when using check valves. The only promising measure seems to be the complete opening of the membrane tongues or tilting away of the diaphragm valves in the upper speed range.

5.2.5. Variable Induction Systems

As the processes in the ram pipe take place at sonic speed, with ram pipe supercharging each ram pipe length has a specific speed for optimum air consumption. Thus it seems logical to design variable induction systems for optimum utilization of the advantages of ram pipe supercharging in a very wide speed range.

In recent years, a lot of work has been accomplished in this direction [5.95, 5.96, 5.97, 5.98, 5.99, 5.100, 5.101].

From experimental findings and computations the author concludes that pipes that are as smooth as possible, i.e. that have no branchings and cross connections while featuring clear reflectance conditions, are the most favorable for complete utilization of the given potential.

Fig. 5.55 shows the construction of an induction system whose ram pipe length can be varied and which was developed and investigated at the Institute of Internal Combustion Engines and Automotive Engineering at the Technical University of Vienna. As can be seen from the sectional view, a relatively long ram pipe is routed through the flat air distributor. The ram pipe contains a sliding pipe piece which allows the ram pipe to open while in the air distributor. At low speeds,

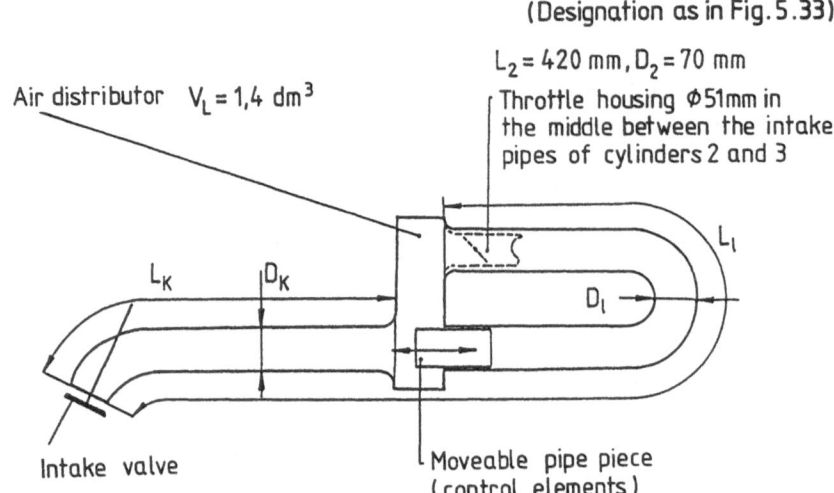

(Designation as in Fig. 5.33)

$L_2 = 420$ mm, $D_2 = 70$ mm

Throttle housing $\phi51$mm in the middle between the intake pipes of cylinders 2 and 3

Air distributor $V_L = 1,4$ dm^3

L_l

D_l

L_K

D_K

Intake valve

Moveable pipe piece (control elements)

Fig. 5.55. Cross-sectional view of a two-stage, variable-length induction system with one air distributor

this pipe piece is positioned so as to make the whole ram pipe length effective, while at high speeds the pipe piece is moved inside the intake pipe so that the mixture can be drawn from the air distributor via the short ram pipe length. The advantage of this design is that it does not increase the overall induction system volume.

In **Fig. 5.56** the torque curve obtained with the above system is compared to that of a standard induction system. As can be seen, the above-described system renders higher maximum performance while the mean effective pressure curve in the lower speed range is significantly increased.

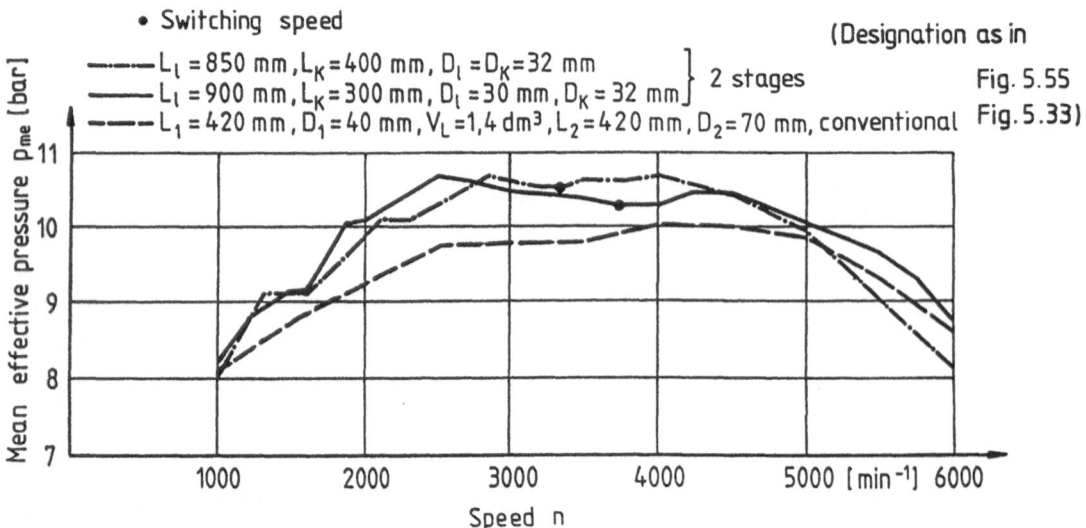

● Switching speed   (Designation as in

——— $L_l = 850$ mm, $L_K = 400$ mm, $D_l = D_K = 32$ mm ⎫
———— $L_l = 900$ mm, $L_K = 300$ mm, $D_l = 30$ mm, $D_K = 32$ mm ⎬ 2 stages   Fig. 5.55
– – – $L_1 = 420$ mm, $D_1 = 40$ mm, $V_L = 1,4$ dm^3, $L_2 = 420$ mm, $D_2 = 70$ mm, conventional ⎭   Fig. 5.33)

Mean effective pressure p_{me} [bar]

11

10

9

8

7

1000   2000   3000   4000   5000 [min^{-1}] 6000

Speed n

Fig. 5.56. The mean effective pressure versus the speed of a double stage switchable induction system as in Fig. 5.55 and versus the speed of a conventional ram pipe system as in Fig. 5.33 [5.94]

Another investigation was carried out on the arrangement depicted in **Fig. 5.57**. The point in question was whether an additional vessel connected to the air distributor via four ram pipe halves suffices to allow an optimum air consumption, while the control elements are open in the higher speed range.

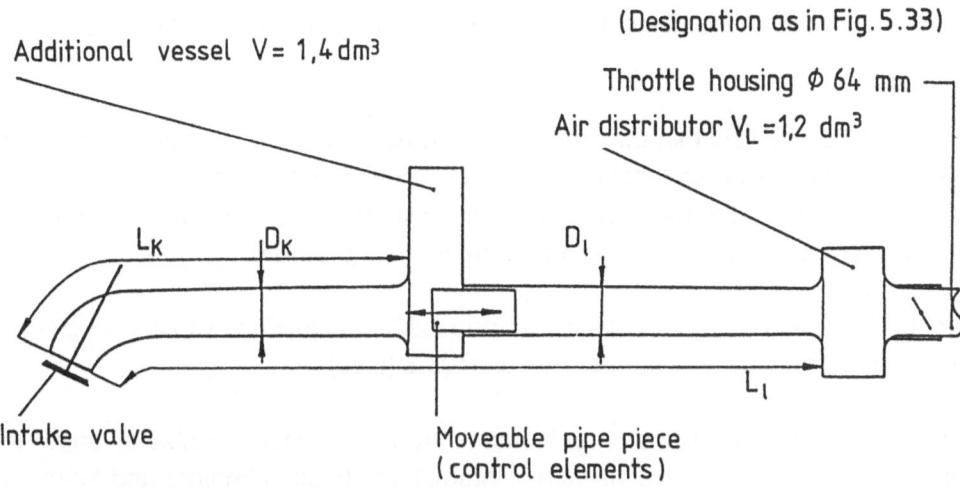

Fig. 5.57. Cross-sectional view of a double-stage variable-length induction system with an air distributor and an additional vessel [5.94]

As shown in **Fig. 5.58**, this was the case with an additional vessel volume of 1.4 dm^3. The mean effective pressures reached were even better than the ones reached with the previously described arrangement. As the ram pipes were straight with optimum sealing of the control elements, the depicted arrangement corresponded to a conventional ram pipe system as long as the control elements were closed.

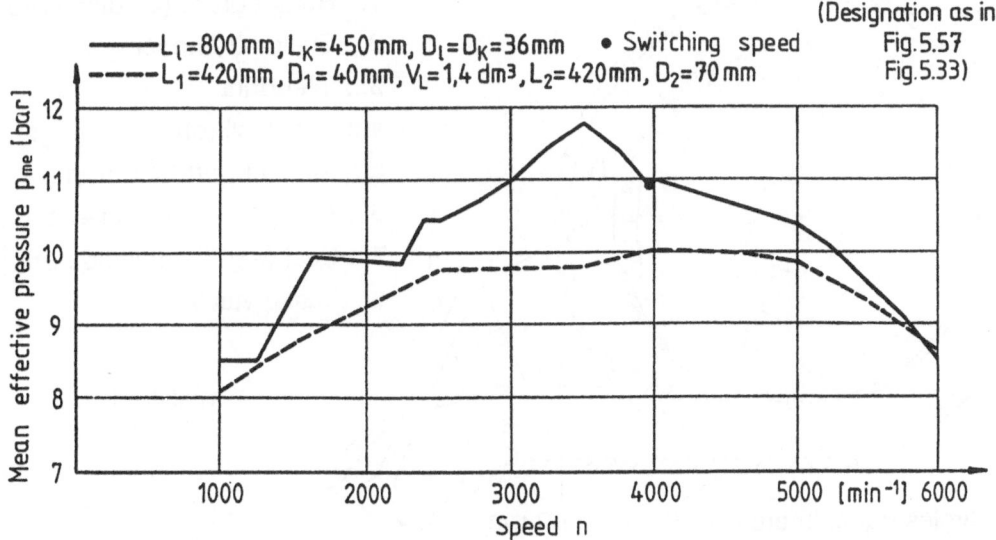

Fig. 5.58. The mean effective pressure versus the speed of a double stage switchable induction system as in Fig. 5.55 and versus the speed of a conventional ram pipe system as in Fig. 5.33 [5.94]

Consequently, this arrangement renders an optimum mean effective pressure curve in the lower speed range, too, and thus more favorable than in the previously investigated variable induction system.

The disadvantage of the described induction system with an additional vessel and air distributor lies in the fact that, in the present case, the induction system volume was increased by 1,2 dm^3.

Summing up it can be said that the double-stage variable-length induction system with an air distributor and additional vessel reached 2% more maximum performance than the standard induction systems, 18% more maximum mean effective pressure, and - when averaging the results for the entire speed range - 9% more mean effective pressure.

In one of their research papers Ford, too, introduced a double-stage variable-length induction system function along the above-described principles [5.96].

Fig. 5.59 shows the construction of an induction system infinitely variable in length that was developed and investigated at the Institute of Internal Combustion Engines and Automotive Engineering at the Technical University of Vienna. The length of this induction system can be varied infinitely over a wide range without affecting its space requirements, which means that each desired speed can continually be allocated the respective optimum ram pipe length. This allocation results in a maximum possible torque curve over a wide speed range and also prevents local unsteadinesses like peak torques or torque drops.

Similar induction systems have been patented in [5.97] and [5.98].

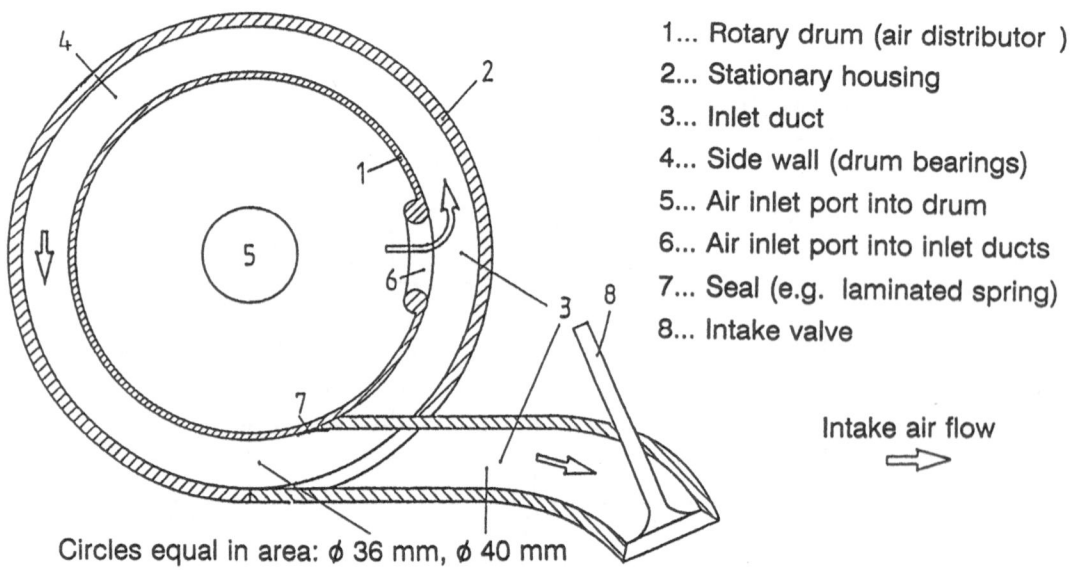

1... Rotary drum (air distributor)
2... Stationary housing
3... Inlet duct
4... Side wall (drum bearings)
5... Air inlet port into drum
6... Air inlet port into inlet ducts
7... Seal (e.g. laminated spring)
8... Intake valve

Intake air flow

Circles equal in area: ⌀ 36 mm, ⌀ 40 mm

Fig. 5.59. Cross-sectional view of an infinitely variable induction system [5.102]

The induction system shown in Fig. 5.59 consists of an inner rotary drum which runs in a bearing within an outer housing fixed to the engine. This stationary outer housing constitutes the outer wall of the rectangular inlet ducts, while the bearings between the housing and the drum function as side walls. The inner wall of the curved inlet ducts are made up by the hollow drum which, at the same time, functions as air distributor.

Through the port in the outer cap fixed to the housing the intake air enters the air distributor and flows through the ports in the movable drum into the outer inlet ducts. The length of the individual pipe through which the intake air passes can be infinitely varied by rotating the drum against the engine-fixed housing while the external dimensions of the entire intake manifold remain unchanged. Moreover, the air flow meter, or the air flow meter and the air filter, could be integrated into the drum so as to reduce the space requirements of the entire induction system.

The mean effective pressure curves shown in **Fig. 5.60** prove the correct functioning of the described arrangement and indicate the results obtained with 3 ram pipe lengths. In **Fig. 5.61** the mean effective pressure curve for a standard induction system is compared to the one of an infinitely variable induction system whose ram pipes get continually shorter with increasing speeds. In accordance with the chosen maximum length of the variable induction system, the mean effective pressure curve rises drastically until a speed of 3000 min^{-1} and drops again as the ram pipe supercharging effect decreases with rising speeds.

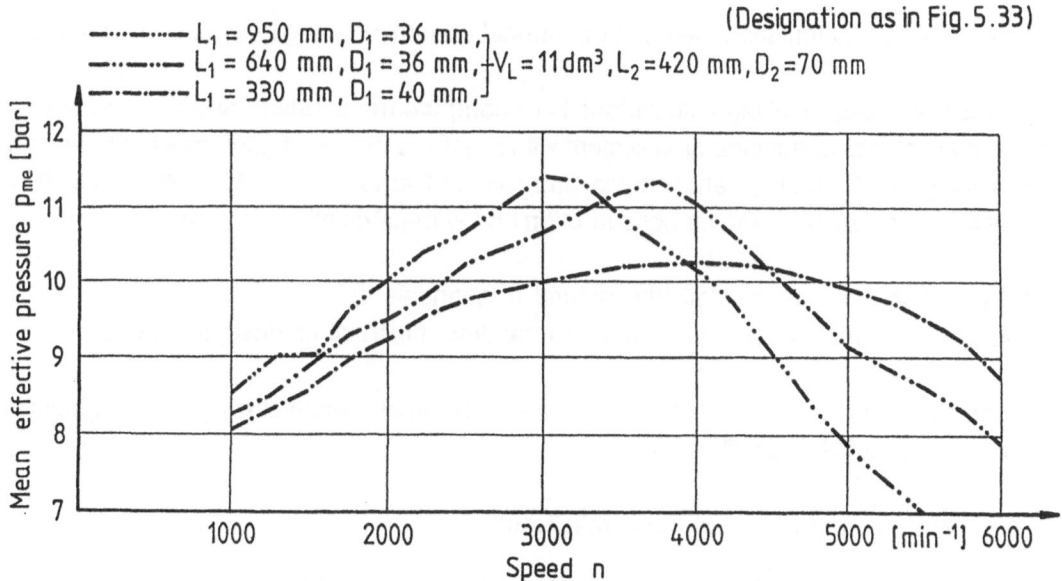

Fig. 5.60. Mean effective pressure versus the speed with three recorded lengths of an infinitely variable induction system as in Fig. 5.59 [5.102]

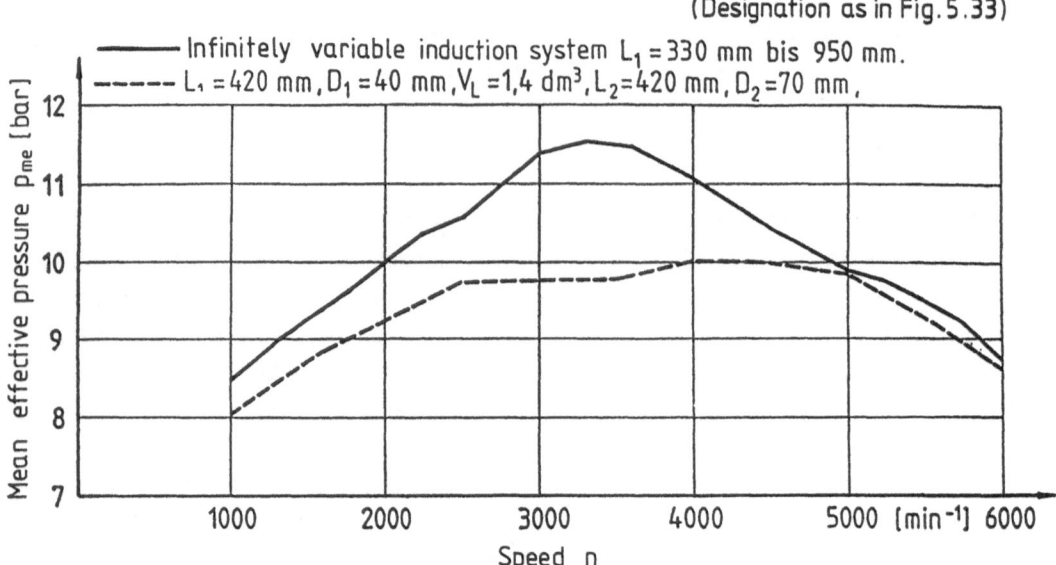

Fig. 5.61. Mean effective pressure versus the speed of an induction system with infinitely variable length and versus the speed of a conventional ram pipe system [5.102]

5.2.6. Computation of Intake Manifolds for Multi-Point Injection

There are numerous calculation methods for internal combustion engine intake manifolds.

The purpose of intake manifold calculations is to compute the pressure and mass flow vibrations in certain places of the induction system with regard to the given geometrical dimensions. When knowing the vibrations, status curves, pressure and temperature in the cylinder as well as the gas exchange variables, like air consumption and volumetric efficiency, can be computed.

Another purpose is to precalculate the natural frequencies of the induction system so that desirable or undesirable resonance speeds can be determined when designing the engine.

Apart from the trivial case of constant pressure in the intake manifold, manifold calculation methods can be roughly classified into:

- Quasi-steady filling and discharging method;

- Vibration observation, sometimes also called acoustic method;

- Computation according to the acoustics theory;

- Transient calculation method.

The calculations get more comprehensive in the order of the above list. The last two methods can, in fact, be carried out only by means of electronic data processing.

The above list gives only a rough overview. In practice, numerous modifications and combinations of the methods listed above are applied, all of which aim at obtaining the desired accuracy with the minimum possible calculation work.

Recently, the trend in intake manifold calculation has been to use vibration observation at the beginning in order to roughly determine the intake manifold dimensions for the desired natural frequencies of the system, and then use the transient calculation method with numerous variation calculations so as to optimize the gas exchange.

6. Special Mixture Formation Varieties

Striving for an optimum mixture formation, i.e., exact metering, fast delivery, uniform distribution as well as good mixture preparation at all engine operating conditions has led to a great variety of possible solutions, which, however, never went into mass production. In most cases this was due to overly high expenditures in relation to the advantages that could have been obtained. The following chapter thus touches only briefly on a few of these special types of mixture formation.

6.1. Special Single-Point Mixture Formation Varieties

As the distribution of the fuel mixture in the intake manifold is one of the major problems in single-point mixture formation, the objective has been a single-point mixture formation system that provides for high-quality mixture preparation over the entire engine map. This is of critical importance insofar as a highly atomized fuel-air mixture enhances uniform mixture distribution. If conventional single-point mixture formation systems can meet this requirement at low engine load, the atomization qualities obtained are partly very unsatisfactory because of low air velocities, especially during operation near the full load range and at low speeds

Fig. 6.1 shows a single-point injection system with an ultrasonic throttle that functions as follows:

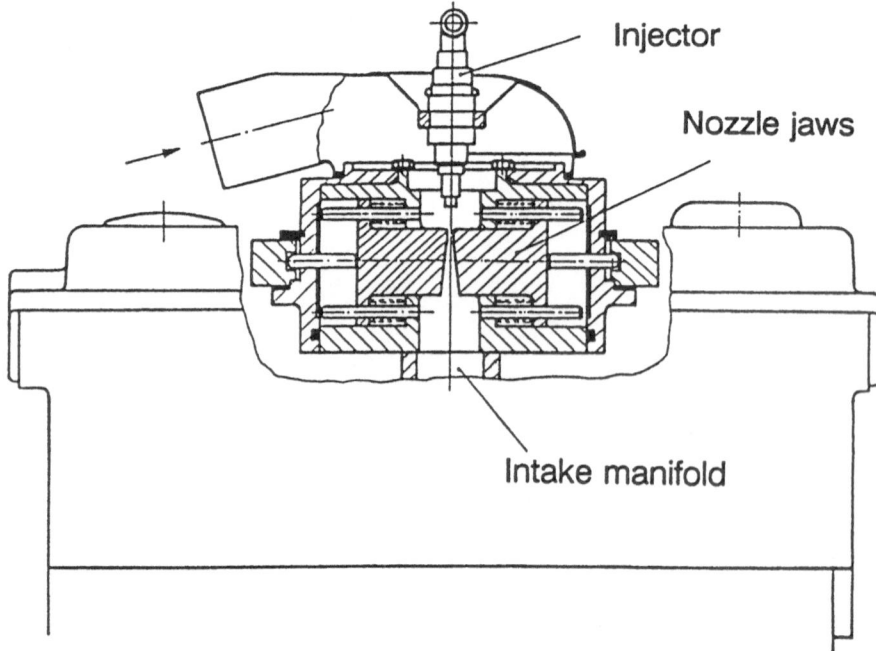

Fig. 6.1. Single-point injection with ultrasonic throttle [6.1]

While the fuel is delivered conventionally via the central injector, the air flow is tuned to the desired value by shifting the nozzle jaws. Depending on the pressure ratio of intake pipe and environment, very high air velocities can be reached in the narrowest gap between the nozzle jaws, and these velocities consequently effect good preparation of the injected fuel. While this arrangement clearly improves mixture distribution at low engine speeds in the foll-load range, the required mechanical expenditures still remain problematic.

Fig. 6.2 shows another single-point injection design. Here, the fuel is preliminarily prepared with part of the combustion air and delivered at the narrowest point of the Laval nozzle type air funnel. This system, too, rendered excellent mixture.distribution, especially during part load operation. The problem, however, was the dynamic behavior: during sudden accleration the nozzle body was drawn out of the air funnel to such an extent that the drop in the flowrate within the air funnel slowed down the fuel supply from the spiral ring groove.

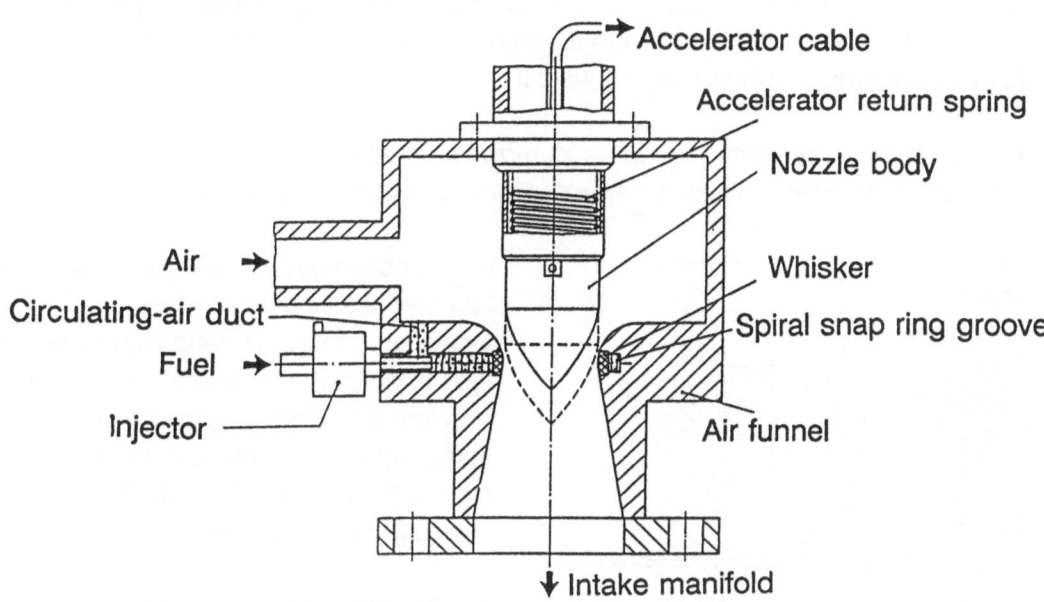

Fig. 6.2. Single-point injection with Laval nozzle throttle [6.1 and 6.5]

Another problem was that the nozzle body was either inducted into the intake manifold vacuum or jerked into the other extreme position by the actuating force.

Surely the above single-point injection problems can be technically solved. However, with all these systems it was observed that great expenditures are necessary to obtain good mixture preparation and consequently good mixture distribution to the individual cylinders. When comparing the costs of single-point injection to the costs of multi-point injection, it can be generally said that, especially with 4-cylinder engines, 3 additional multi-point injectors would cost less than the additional expenditures for the complicated throttle housing.

A combination suction carburetor/turbocharger also seemed feasible and was tested [6.1]. With such an arrangement the compressor stage with its additionally coolant-heated housing is responsible for the secondary atomization as well as for the preparation of the fuel film contained in the fuel-air mixture supplied by the carburetor.

The high-speed turbo-compressor impeller additionally atomizes and mixes the air with the fuel, thereby providing for good distribution of the compressed mixture to the individual cylinders of the internal combustion engine, while the favorable results are largely independent of the intake manifold design. From the construction point of view, however, such a system requires considerable expenditures.

With regard to the fact that fuel atomization is unfavorable while the throttle is open and at low air velocities, several experiments have been carried out with supersonic atomizers.

Actually, extremely fine fuel atomization is possible with relatively little power required from the oscillator, if the fuel can be metered onto the vibrating surface of the ultrasonic atomizer. For details of the atomizing process see Chapter 6.2.

Fig. 6.3 shows an experimental ultra-sonic atomizer arrangement in a downdraft carburetor. In the sixties the author also conducted similar experiments with constant-depression carburetors.

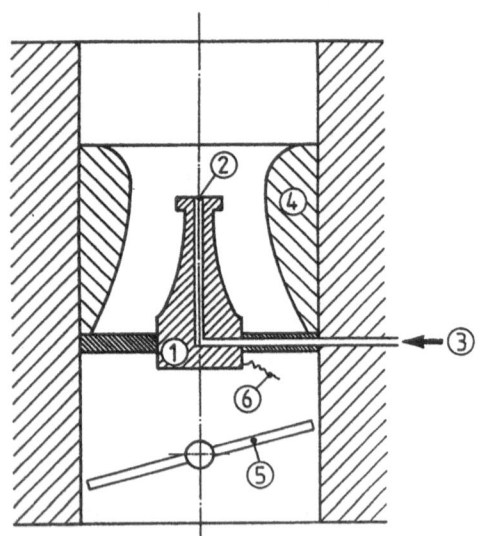

1   Ultrasonic oscillator
2   Oscillator head with fuel outlet
3   Fuel supply line
4   Air funnel
5   Throttle valve
6   Electric supply

Fig. 6.3. Ultrasonic atomizer inside carburetor

One problem has always been to meter the fuel onto the vibrating surface so as to provide appropriate atomization by an adequate fluid level on the oscillator, while the other problem has been to prevent the fine fuel spray from re-contacting the carburetor or manifold walls, which would lead to a renewed formation of large droplets.

In 1967 there was a US suggestion to use fluidic devices in the carburetor [6.3 and 6.4]. **Fig. 6.4** shows the basic principle: The proportional characteristics of the fluidic device installed in this type of carburetor direct the fuel pump flow into the carburetor for injection, i.e. proportional to the carburetor vacuum. Consequently, more fuel is injected into the carburetor when the vacuum is high, i.e. in the case of a high airflow rate in the carburetor, than with a low airflow rate; the remaining fuel flows back into the tank. There were quite a number of experimental versions of such a carburetor built.

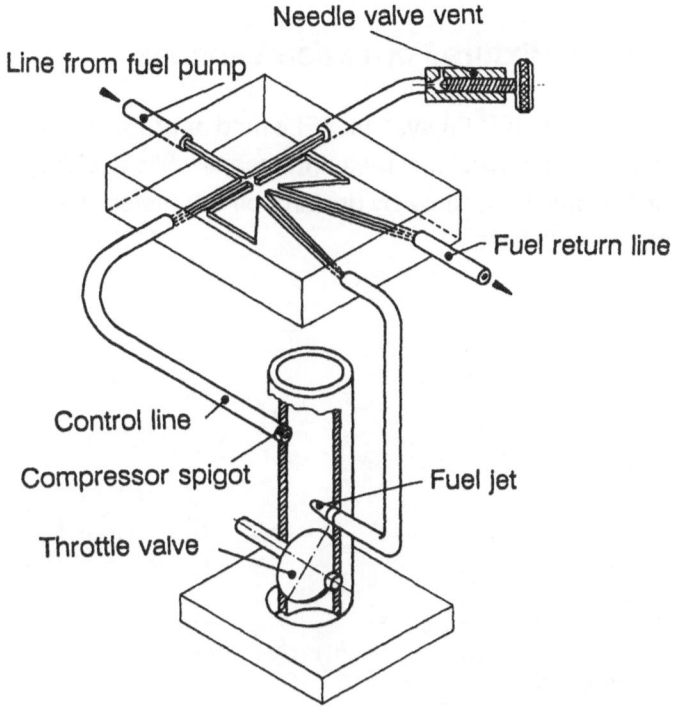

Fig. 6.4. Fluidic carburetor

6.2. Special Multi-Point Mixture Formation Varieties

Fig. 6.5 shows an air-fuel injection (AFI) system. AFI aimed at preserving the advantage of the carburetor as a simple and economical fuel metering means while avoiding the disadvantages of fuel film formation in the intake manifold by drawing off the metered fuel and delivering it to the cylinders via special, narrow lines.

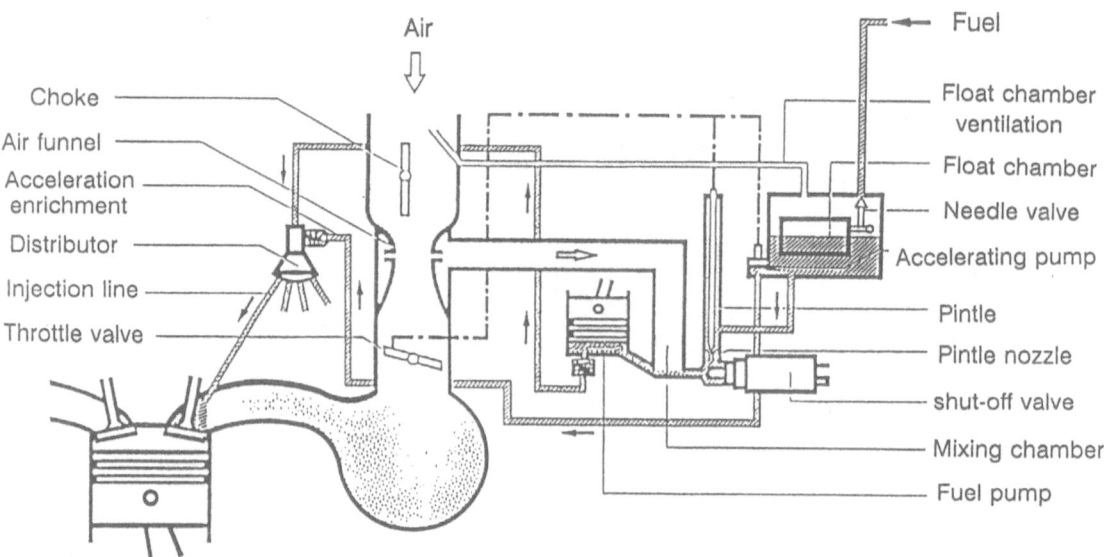

Fig. 6.5. Air-fuel injection (AFI) [6.1]

The left part of Fig. 6.5. shows an air routing similar to the one known from carbureted engines. The air enters the engine via the throttle valve. In a carbureted engine, the fuel would be drawn by the vacuum generated in the venturi, which in this case, however, is prevented by a fuel pump (as can be seen in the right part of Fig. 6.5.). This small pump sucks off the proportioned fuel together with a respective portion of combustion air. Via a corresponding number of injection lines the extremely rich gasoline/air mixture is delivered into the individual intake manifold branches to the cylinder. Thus carburetor technology helps to make multi-point injection feasible.

However, while good cold start, warmup, and hot start behaviors could be obtained with this system, transient engine operation was problematic with regard to uniform distribution of the rich fuel-air mixture in the distributor and also with regard to the additional expenditures necessary when using a lambda closed-loop control.

Fig. 6.6 shows another way of improving mixture preparation, i.e. a gasoline injection system where the gasoline metered from the solenoid valve is collected by a circulating air stream and then prepared in the heated mixture preparation section thereby producing an extremely well mixed, directed gasoline/air stream. By this method the mixture can be injected past the engine

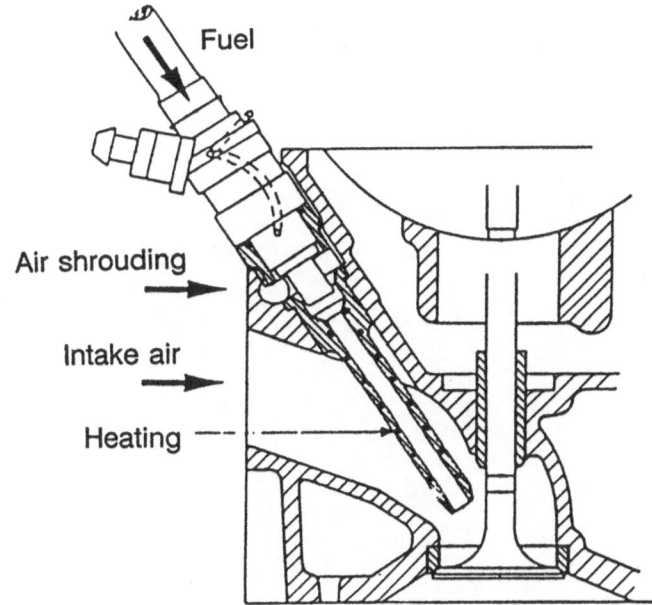

Fig. 6.6. Fuel injection with fuel preparation and fuel heating [6.1]

intake valve directly into the combustion chamber so as to drastically reduce wall wetting in the intake manifold, which is an advantage during transient engine operation and may also render charge stratification effects.

Another possibility for an optimum fuel preparation in multi-point injection as well as in single-point mixture formation systems is to use ultrasonic atomizers [6.2]. Fuel preparation takes place on a vertical atomizing surface where the liquid film is incited to form capillary waves. Consequently, the fuel must be properly delivered onto the atomizing surface where it is finely atomized.

When vibration reaches a certain level, the so-called basic amplitude that depends on the mass values of the fluid to be atomized as well as on the exciting frequency, vertical capillary waves form a checkered pattern on the fluid surface. The basic amplitude is predetermined by the oscillator energizing the tranformation. When the basic amplitude is reached, i.e. when capillary waves are forming, the amplitude continues to rise. Finally, the wave crests break and are flung off the fluid surface. The rise of the wave amplitude depends, above all, on the mass values of the fluid but also on other, largely unknown, stabilizing effects.

Fig. 6.7 shows an ultrasonic atomizer arrangement. The atomizers are flange-mounted in the lower intake pipe bends and connected to the header by hoses. Due to the pressure difference between the header and the intake pipes, the atomizers are fed a carrier air stream from behind. This air stream serves to stabilize the fine fuel mist that can be affected by minute air movements and to thoroughly mix it with most of the intake air.

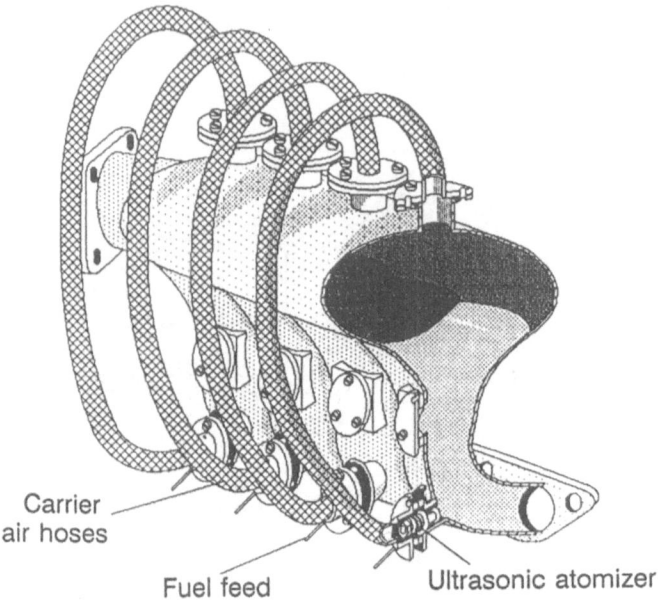

Carrier
air hoses

Fuel feed   Ultrasonic atomizer

Fig. 6.7. Ultrasonic atomizer arrangement in a modified intake manifold of the Ford CVHi engine [6.2]

Improved mixture preparation resulted in better fuel economy during steady-state operation especially at low speeds when compared to traditional multi-point injection systems although, with this system, too, problems are to be expected during transient operation and increased expenditures will have to be taken into account.

The special types of mixture formation described in this chapter are only a few examples of a great number of a widely diversified range of developments in this field. However, in most cases serial production of such systems is impaired by increased design expenditures.

7. Bibliography

Chapter 1

[1.1]   Löhner, K.; Müller, H.: Gemischbildung und Verbrennung im Ottomotor. Springer-Verlag Vienna (1967).

[1.2]   Pischinger, A.; Pischinger, R.; Krassnig, G.; Taucar, G.; Sams, Th.: Die Thermodynamik der Verbrennungskraftmaschine. Springer-Verlag Vienna (1989).

[1.3]   Schmidt, F.A.F.: Verbrennungskraftmaschinen, Springer Verlag Berlin (1967).

[1.4]   Lenz, H.P.: Vorlesung Verbrennungskraftmaschinen-Grundzüge.- Textbook - Institute for Internal Combustion Engines and Automotive Enginieering of the TU Vienna. 5th Edition (1989).

[1.5]   Vibe, I.I.: Brennverlauf und Kreisprozeß von Verbrennungsmotoren , VEB Technik Berlin (1970).

[1.6]   Lange, K.: Berechnung von Druckverlauf und Wirkungsgrad im Verbrennungsmotor, Motortechnische Zeitschrift 30 (1969), Nr.5, p. 173-176

[1.7]   Woschni, Gerhard; Anisits, Ferenc: Eine Methode zur Vorausberechnung der Änderung des Brennverlaufes mittelschnellaufender Dieselmotoren bei geänderten Betriebsbedingungen, Motortechnische Zeitschrift 34 (1973), Nr.4, p. 106-115.

[1.8]   Nakai, Meroji; et al: Stabilized Combustion in a Spark Ignited Engine through a Long Spark Duration, SAE Paper 850075 (1985).

[1.9]   Moser, Winfried: Vergleichende Untersuchung von Brennraumsignalen bei Ottomotoren im Hinblick auf Motorregelung, Doctoral Thesis, Technical University Vienna (1988).

[1.10] Akhlaghi, Mehdi: Laufunruhe, Abgasemissionen und Verbrauch eines Ottomotors bei Magerbetrieb und Möglichkeiten zur Ausweitung des mageren Betriebsbereiches. Doctoral Thesis, Technical University Vienna (1978).

[1.11] Lawton, B.; Elmore, K.G.; Render, M.E.J.: Performance of a Small-Scale Turbulence Generator in a Rotary Piston Engine. Conference on fuel economy and emission of lean burn engines. London (1978).

[1.12] Lucas, G.G.; Brunt, M.; Petrovic, S.: Lean Mixture Running of the Spark Ignition Engine by the Generation of a Vortex System within the Intake. SAE Paper 780964 (1978).

[1.13] Göschel, B.: Maßnahmen zur Verbrennung extrem magerer Gemische in Ottomotoren mit äußerer Gemischbildung. Doctoral Thesis, Technical University Stuttgart (1977).

[1.14] Prescher, K.: Brennraumform Ottomotor. FVV Heft R 390 (1980).

[1.15] Quissek, Friedrich: Über den Einfluß außen erzeugter Ladungsbewegung auf das Betriebsverhalten eines hochverdichteten Ottomotors, Fortschrittsbericht VDI Nr.165 (1985).

[1.16] Hofbauer, P.; Hoffmann, E.: Einfluß der Ladungsbewegung und der Brennraumform auf das ottomotorische Arbeitsverfahren. ÖIZ, 17. Jahrgang (1974), Heft 4, p. 128-137.

[1.17] Nagayama, I.; Yasushi, A.; Yasuo, I.: Effects of Swirl and Squish on S.I. Engine Combustion and Emission. SAE Paper 770217 (1977).

[1.18] Thring, R.H.; Overington, M.T.: Gasoline Engine Combustion - The High Ratio Compact Chamber. SAE Paper 820166 (1982).

[1.19] Overington, M.T.; Thring, R.H.: Gasoline Engine Combustion -Turbulence and the Combustion Chamber. SAE Paper 810017 (1981).

[1.20] Bloss, W.H.; et al: Experimentelle und theoretische Analyse der Verdichtungserhöhung bei Ottomotoren mit dem Audi - Brennverfahren. Automobil-Industrie 27. Jahrgang (1982), Heft 1 bis 3.

[1.21] Carstensen, Hartmut: Über den Einfluß von Brennraumform und Einlaßdrall in einem hochverdichteten Ottomotor, Diplomarbeit Technische Universität Wien (1984).

[1.22] Lee, Wenpo; Schäfer, Hans-Jürgen: Verbrauchsreduzierung am Ottomotor durch Optimierung von Brennraumform und Verdichtungsverhältnis, Motortechnische Zeitschrift 82 (1982), Nr.6, p. 279-284.

[1.23] Gruden, Dusan; Höchsmann, Günther: Betriebsverhalten des thermodynamisch optimierten Porsche-(TOP)-Motors 924 bei Betrieb mit M 15-Kraftsoff, Motortechnische Zeitschrift 42 (1981), Nr. 4, p. 133-137.

[1.24] Bender, Karl-Heinz; et al: Der neue BMW-Vierzylinder-Motor. Motortechnische Zeitschrift 48 (1987), Nr. 12, p. 493-500.

[1.25] Abthoff, Jörg; et al: Daimler-Benz 2.3 Litre, 16-Valve High-Performance Engine. SAE Paper 841226 (1984).

[1.26] Brandstetter, W.R.; Decker, G.; Reichel, K.: The Water-Cooled Volkswagen PCI-Stratified Charge Engine, SAE Paper 750869 (1975).

[1.27] Gruden, Dusan; Lange, Karlheinz: Betriebsverhalten und Abgasemissionen eines Verbrennungsmotors mit dem Porsche-Schichtlade-Kammer-System, Motortechnische Zeitschrift 35 (1974), Nr.10, p. 307-313.

[1.28] May, Michael G.; Spinnler, Fritz: Betriebserfahrungen mit hochverdichteten Ottomotoren nach dem May Fireball-Verfahren, Motortechnische Zeitschrift 39 (1978), Nr. 6, p. 243-246.

[1.29] Date, T.; Yagi, S.; Ishizuya, A.; Fujii, I.: Research and Development of the Honda CVCC engine. SAE Paper 740605 (1974).

[1.30] Beitz W.; Küttner, K.-H.: Taschenbuch für den Maschinenbau/Dubbel. 14th new and fully revised Edition. Springer-Verlag Berlin Heidelberg New York (1981).

[1.31] Bertling, H.: Untersuchung der Streuungen im Energieumsetzungsverlauf eines Ottomotors. Doctoral Thesis, TU Braunschweig (1974).

[1.32] Näser, Karl-Heinz: Physikalisch-chemische Meßmethoden. VEB Deutscher Verlag für Grundstoffindustrie Leipzig (1981), 3rd Edition.

[1.33] N.N.: Deutsche Normen, DIN 51900 Teil 1.

[1.34] Ißler, J.: Zündung und Abgas, Robert Bosch GmbH. Techn. Zentrum, Autoelektrik Bosch Techn. Berichte 3, Heft 1, Nov. 1969.

Chapter 2

[2.1] Schröter, W.; Lautenschläger, K.-H.; Bibrack, H.; Schnabel, A.: Chemie. 11th Edition, VEB Fachbuchverlag Leibzig (1977).

[2.2] Dubbel: Taschenbuch für den Maschinenbau, 1st Volume. 13th Edition, Springer Verlag Berlin (1974).

[2.3] D' Ans, Lax: Taschenbuch für den Chemiker und Physiker, 1st Volume. 3rd Edition, Springer Verlag Berlin (1967).

[2.4] Rompps: Chemielexikon, Volume 3. 8th Edition, Frankh'sche Verlagsbuchhandlung Stuttgart (1983).

[2.5] Schmidt, A.: Neue Wege bei der Herstellung von Vergasertreibstoffen, Volume 3. Springer Verlag Wien New York (1978).

[2.6] Dabelstein, W.E.A.: Unverbleite Ottokraftstoffe - Möglichkeiten und Grenzen der Herstellung. VDI Berichte 531 (1984) p.289 - 306.

[2.7] N.N.: Kraftfahrtechnisches Taschenbuch Bosch. 19th Edition, VDI-Verlag GmbH Düsseldorf (1984).

[2.8] Hassel, D.; Pauli, E.; Dursbeck, F.: Das Abgas-Emissionsverhalten in der Bundesrepublik Deutschland im Bezugsjahr 1980. Berichte 9/80 Umweltbundesamt, Erich Schmidt Verlag Berlin (1980).

[2.9] Rau, B.: Einfluß wechselnder Gasqualitäten auf den Betrieb von Gasmotoren. MTZ Motortechnische Zeitschrift 44 (1983) 6.

[2.10] Pischinger, F.; Adams, W.: Abgasemissionen bei Verwendung alternativer Kraftstoffe für Kraftfahrzeug-Ottomotoren. VDI-Berichte 531 (1984). p. 325 - 340.

[2.11] Mollenhauer, K.; Pucher, H.: Der Stadtgasmotor - ein umweltfreundlicher Stationärantrieb.

[2.12] Fessler, H.; Pischinger, R.: Alternative Ottokraftstoffe. 5. Zwischenbericht zur Studie U 601 A "Schadstoffemissionen von Kraftfahrzeugen sowie zu erwartende Entwicklungen". Institut für Verbrennungskraftmaschinen und Thermodynamik an der TU Graz (1985).

[2.13] Kampelmühler, F.Th.: Untersuchung und systematische Verbesserung des Kaltstart- und Warmlaufverhaltens eines 4-Takt-Ottomotors mit Benzineinspritzung. Fortschritt-Berichte der VDI-Z., Reihe 6 (1982) 100.

[2.14] Gruden, D.; Brachert, T.-F.; Höchsmann, G.: Motorinterne Maßnahmen zur Minderung der Abgas-Emissionen. VDI-Berichte 531 (1984) p. 169-187.

[2.15] Moser, F.: Über das Verhalten von Fahrzeug-Ottomotoren, insbesondere Vergasermotoren, im nicht betriebswarmen Zustand, und Wege zur Verbesserung der Abgasemission und Kraftstoffverbrauch in der Warmlaufphase. Doctoral Thesis, TU Vienna (1977).

[2.16] Schmillen, K.; Dübel, F.M.; Schiffgens, H.J.; Pischinger F.: Nutzung von Biogas in Gaszündstrahlmotoren. MTZ 50 (1989) 7/8.

[2.17] Müller, H.; Rohde, S.; Klink, G.: Gemischbildung, Verbrennung und Abgas im Ottomotor. Fachbibliographie mit Referaten bis 1965, Braunschweig (1972).

[2.18] Brauer, H.: Grundlagen der Einphasen- und Mehrphasenströmungen. Verlag Sauerländer, Aarau und Frankfurt am Main (1971).

[2.19] Kaskas, A.: Berechnung der stationären und instationären Bewegung von Kugeln in ruhenden und strömenden Medien. Diplomarbeit am Lehrstuhl für Thermodynamik und Verfahrenstechnik der TU Berlin (1964).

[2.20] Pachta-Reyhofen, G.: Wandfilmbildung und Gemischverteilung bei Vierzylinder-Reihenmotoren in Abhängigkeit von Vergaser- und Saugrohrkonstruktion. Fortschritt-Berichte der VDI-Zeitschriften, Reihe 12 (1986) 62.

[2.21] Nati, V.: Untersuchung der Tropfenflugbahnen in Ansaugsystemen von Vierzylinder-Vergasermotoren verschiedener Bauart. Diplomarbeit am Institut für Verbrennungskraftmaschinen und Kraftfahrzeugbau der TU Wien, (1985).

[2.22] Demel, H.: Mögichkeiten zur Verbesserung der Gemischverteilung an Vergasermotoren unter besonderer Berücksichtigung des Einflusses des Kraftstoffilmes. Fortschritt-Berichte der VDI-Zeitschriften, Reihe 6 (1982) 107.

[2.23] Diem, E.: Dreidimensionale Berechnung der Tropfenbahnen mit Randwertelementen. Internal Report of the Institute for Internal Combustion Engines and Automotive Engineering of the TU Vienna, B 1359 (1986).

[2.24] Lo, R.S.; Lalas, D.P.: Parametric study of fuel-droplet flow in an idealized engine induction system. SAE-Paper 770645 (1977).

[2.25] Eisfeld, F.: Der Einfluß der Brennstoffeinbringung auf die Aufbereitung und Verbrennung. Vortrag bei der Tagung: Verdampfung von Brennstoffen und Brennstoffilmen in Motoren, Turbinen und Feuerungen. Haus der Technik, Essen, 1980.

[2.26] Chapman, M.: Two dimensional numerical simulation of inlet manifold flow in a four cylinder internal combustion engine. SAE Paper Series 790244 (1979).

[2.27] Binder, A.: Berechnung der Tropfenbahnen und des Tropfenwachstums in einer Keilströmung. Diplomarbeit an der TU Wien (1980).

[2.28]   Achleitner, E.; Czerwinski, J.; Ullrich, W.: Vergleich der Serienzündkerze und der Wirbelkammerzündkerze bezüglich Druckverlauf, Lichtverlauf und Wärmefreisetzung am hoch-verdichteten Vierzylinder-Reihenmotor. Internal Report of the Institute for Internal Combustion Engines and Automotive Engineering of the TU Vienna, B 1022 (1981).

[2.29]   Kauba, H.: Untersuchung von Einflußgrößen auf die Wandniederschlagsbildung in Saugrohren von Vergasermotoren. Diplomarbeit am Institut für Verbrennungskraftmaschinen und Kraftfahrzeugbau der TU Wien, (1983).

[2.30]   Elsayed-Ahmed, M.Y.: Verdunstung, Tropfen- und Filmbildung in Vergaseranlagen. Doctoral Thesis, TU Braunschweig (1971).

[2.31]   Löhner, K.; Elsayed-Ahmed, M.: Verdunstung, Tropfen- und Filmbildung in Vergaseranlagen. MTZ 35 (1974) 6, p. 186-192.

[2.32]   Rajkow, I.J.; Jerschow, W.W.: Zur Frage der Filmbildung in Vergasermotoren. Translation from the Soviet Magazine "Avtomobilnaja promyzlennost" 11 (1964) p. 6-10.

[2.33]   Fisher, C.H.: Carburation. Vol. 1, 3rd edition, Chapman and Hall, London (1951).

[2.34]   Scherenberg, D.: Einfluß der Gemischaufbereitung auf die Emissionen beim Start und Warmlauf sowie beim Übergang von Last in Schubbetrieb eines Ottomotors bei verschiedenen Motortemperaturen. Doctoral Thesis, TU Karlsruhe (1979).

[2.35]   Klink, G.: Transport und Verteilung von Kraftstoff durch die Saugleitung eines Vierzylinder-Reihenmotors mit einem Vergaser. Doctoral Thesis, TU Braunschweig (1973) and ATZ 78 (1976) 4, p. 155-161.

[2.36]   Boam, D.J.; Finlay, I.C.: A Computer model of fuel evaporation in the intake system of a carburetted petrol engine. Confer. London 1979, C 89/79.

[2.37]   Stebar, R.F.; Everett, R.L.: New emphasis on fuel volatility-effects on vehicle warm up with quick-release chokes. SAE-Paper 720934 (1972).

[2.38]   Dörges, E.A.: Gemischverteilung an einem Vierzylinder-Vergasermotor. ATZ 59 (1957) 7, p. 194-201.

[2.39]   Sawa, N.: Das Verhalten des Kraftstoffes in der Ansaugleitung während der Verbrennung. Translation from the Japanese Magazine "Iidosha Gijutsu" Volume 27 (1973) 4, p. 342-349.

[2.40]   Mramor, F.: Gemischbildung in Mehrzylinder-Ottomotoren mit Vergaserbetrieb. Deutsche Kraftfahrtforschung 82 (1944).

[2.41]   Jante, A.: Über die Gemischverteilung an Vergasermotoren. ÖIZ 7 (1964) 1, p. 7-17.

[2.42]   Liimata, D.R.; Hurt, R.F.; Deller, R.W.; Hull, W.L.: Effects of mixture distribution on exhaust emissions as indicated by engine data and the hydraulic analogy. SAE-Int. Mid-Year Meeting, Montreal/Canada, 1971.

[2.43]   Asire, H.W.: Das Brennstoffgemisch im Ansaugkrümmer. "Der Motorwagen", p. 294-301.

[2.44]   Dika, R.J.: A discussion of V-8 engine intake manifold design. General Motors enginee-
         ring journal Jan-Feb-March (1959).

[2.45]   Gruden, D.; Richter, H.; Korte, V.: Möglichkeiten zur Verbesserung der Wirtschaftlich-
         keit von Ottomotoren. Automobil Industrie 2 (1984) p. 179-185.

[2.46]   N.N.: DIN 70020, November 1976, Deutsches Institut für Normung e.V., Berlin.

[2.47]   N.N.: ÖNORM V 5003, 1. Juni 1978, Österreichisches Normungsinstitut, Vienna.

[2.48]   N.N.: DIN 6270, Mai 1970, Beiblatt 1, Deutsches Institut für Normung e.V., Berlin.

[2.49]   N.N.: DIN 1940, Dezember 1976, Deutsches Institut für Normung e.V., Berlin.

[2.50]   N.N.: DIN 51600, Juli 1984, Deutsches Institut für Normung e.V., Berlin.

[2.51]   N.N.: DIN 51607, Deutsches Institut für Normung e.V., Berlin: 1984.

[2.52]   N.N.: ÖNORM C 1103, Österreichisches Normungsinstitut, Vienna, 1984.

[2.53]   N.N.: ÖNORM C 1102, Österreichisches Normungsinstitut, Vienna, 1985.

[2.54]   N.N.: DIN 51751, Deutsches Institut für Normung e.V., Berlin.

[2.55]   N.N.: ÖNORM C 1160, Österreichisches Normunsinstitut, Vienna.

[2.56]   Mikulic, L.; Lenz, H.P.; Bamer, F.: Verbrauchsminderung bei Einspritz-Ottomotoren
         durch spezielle Saugrohrgestaltung. ATZ Automobiltechnische Zeitschrift 86 (1984) 10,
         p. 437-442.

[2.57]   Seifert, H.: Die charakteristischen Merkmale der Schwingrohr- und Resonanzaufladung
         bei Verbrennungsmotoren XIX. Int. FISITA-Kongreß Melbourne, 8.-12. Nov. 1982, SAE-
         Paper 82032.

[2.58]   Cser, G.: Ein neuartiges Verfahren zur Verbesserung der Abgasturboaufladung. MTZ 32
         (1971) 10.

[2.59]   Engelmann, K.W.: Design of a tuned intake manifold. ASME-Publication (1973) 73-
         WA/DGP-2.

[2.60]   Anisitz, F.; Spinnler, F.: Entwicklung der kombinierten Aufladung am neuen Sauer-Fahr-
         zeugdieselmotor D4KT. MTZ 39 (1978) 10.

[2.61]   Indra, F.: Kombinierte Aufladung an einem Personenwagen-Ottomotor hoher Literlei-
         stung. MTZ 40 (1979) 12.

[2.62]   Zürner, H.: Entwicklung von aufgeladenen MAN-Fahrzeug-Dieselmotoren in Sechszy-
         linder-Reihenbauart. MTZ 41 (1980) 2.

[2.63]   Fraidl, G.: 3. Zwischenbericht zum Forschungsauftrag: "Einfluß der Gemischaufbe-
         reitung auf das motorische Betriebsverhalten bei intermittierender Einspritzung".
         Unpublished Report of the Institute for Internal Combustion Engines and Automotive
         Engineering of the TU Vienna, B 1098 (1982).

[2.64] Fraidl, G.: 7. Zwischenbreicht zum Forschungsauftrag: "Einfluß der Gemisch-aufbereitung auf das motorische Betriebsverhalten bei intermittierender Saug-rohreinspritzung". Internal Report of the Institute for Internal Combustion Engines and Automotive Engineering of the TU Vienna, B 1136 (1983).

[2.65] Robinson, J.A.; Brehob, W.M.: The Influence of Improved Mixture Quality on Engine Exhaust Emissions and Performance. Journal of the Air Pollution Control Association 17 (1967) 7, p.446-453.

[2.66] Harrington, J.A.: A Study of Carburettion Effects on Power, Emissions, Lean Misfire Limit and EGR Tolerance of a Single-Cylinder Engine. SAE Paper 760754 (1976).

[2.67] Lindsay, R.; Thomas, A.; Woodworth, J.A.; Zeschmann, E.G.: Influence of Homo-geneous Charge on the Exhaust Emissions of Hydrocarbons, Carbon Monoxide and Nitric Oxide from a Multicylinder Engine. SAE Paper 710588 (1971).

[2.68] Hardenberg, H.O.: Thermodynamische Betrachtungen zum Mercedes-Benz Methanol-Gasmotor-Konzept. Automobilindustrie (1983) 3, p. 297-310.

[2.69] Polach, W.: Verbesserung der Verbrennung bei hoher Verdichtung und großer Turbu-lenz durch Benzineinspritzung. MTZ 44 (1983) 7/8, p. 269-273.

[2.70] Almstadt, K.: Einfluß des Gemischzustandes an der Zündkerze auf die Entflammungs-phase im Ottomotor. Doctoral Thesis, TU-Braunschweig 1985.

[2.71] Matthes, W.R.; Mc. Gill; R.N.: Effects of the Degree of Fuel Atomization on Single-Cylin-der Engine Performance. SAE Paper 760117 (1976).

[2.72] Trösch, H.A.: Die Zerstäubung von Flüssigkeiten. Chemie -Ing. -Techn. 26 (1954) 6, p. 311-320.

[2.73] Schmidt, P.; Walzel, P.: Zerstäubung von Flüssigkeiten Chem. Ing. Tech. 52 (1980) 4, p. 304-311.

[2.74] Ohnesorge, W.v.: Die Bildung von Tropfen an Düsen und die Auflösung flüssiger Strah-len. Zeitschrift f. angewandte Mathematik und Mechanik 16 (1936) 6, p. 355-358.

[2.75] Walzel, P.; Michalski, H.: Strömungszustände an Düsen bei kleinen Flüssigkeitsdurch-sätzen. Verfahrenstechnik 14 (1980) 3.

[2.76] Rayleigh, Lord: On the Instability of Jets. Proc. London Math. Soc. Vol.10 (1878/79), p. 4-13.

[2.77] Weber, C.: Zum Zerfall eines Flüssigkeitsstrahles. Zeitschrift für angewandte Mathema-tik und Mechanik 11 (1931) 2, p. 136-154.

[2.78] Haenlein, A.: Über den Zerfall eines Flüssigkeitsstrahles. Forschung 2 (1931) 4, p. 139-149.

[2.79] Chawla, T.C.: Droplet Size Resulting from Breakup of Liquid at Gas-Liquid Interfaces of Liquid Submerged Subsonic and Sonic Gas Jets. International Journal of Multiphase Flow 2 (1975) p. 471-475.

[2.80]   Ostrach, S.; Koestel, A.: Film Instabilities in Two-Phase Flows. A.I.Ch.E. Journal 11 (1965) 2, p. 294-303.

[2.81]   Tatterson, D.F.; Dallman, J.L.; Hanratty, T.J.: Drop Sizes in Annular Gas-Liquid Flows. A.I.Ch.E. Journal 23 (1977) 1, p. 68-76.

[2.82]   Holfelder, O.: Zur Strahlzerstäubung bei Dieselmotoren, Forschung 3 (1932) 5, p. 229-240.

[2.83]   Lenz, H.P.: Zur Berechnung des Zerfalls von Flüssigkeitsstrahlen niedriger Austrittsgeschwindigkeiten. Fortschritt-Berichte VDI-Z. Reihe 3, Nr.20, VDI-Verlag Düsseldorf 1967.

[2.84]   Sterling, A.M.; Sleicher, C.A.: The Instability of Capillary Jets. Journal of Fluid Mechanics 68 (1975) 3, p. 477-495.

[2.85]   Huang, J.C.P.: The Break-Up of Axisymmetric Liquid Sheets. Journal of Fluid Mechanics 43 (1970), part 2, p. 305-319.

[2.86]   Fraser, R.P.; Eisenklam, P.; Dombrowski, N.; Hasson, D.: Drop Formation from Rapidly Moving Liquid Sheets. A.I.Ch.E. Journal 8 (1962) 5, p. 672-680.

[2.87]   Dombrowski, N.; Johns, W.R.: The Aerodynamic Instability and Disintegration of Viscous Liquid Sheets. Chemical Engineering Science (1963) 18, p. 203-214.

[2.88]   Dombrowski, N.; Hooper, P.C.: The Effect of Ambient Density on Drop Formation in Sprays. Chemical Engineering Science 77 (1962), p. 291-305.

[2.89]   Dombrowski, N.; Hasson, D.; Ward, D.E.: Some Aspects of Liquid Flow through Fan Spray Nozzles. Chemical Engineering Science (1960) 12, p. 35-50.

[2.90]   Dombrowski, N.; Hooper, P.C.: A Study of the Sprays Formed by Impinging Jets in Laminar and Turbulent Flow. Journal of Fluid Mechanics 18 (1963) part 3, p. 392-400.

[2.91]   Fraser, R.P.; Dombrowski, N.; Routley, J.H.: The Atomization of a Liquid Sheet by an Impinging Air-Stream. Chemical Engineering Science 18 (1963), p. 339-353.

[2.92]   Clark, C.J.; Dombrowski, N.: An Experimental Study of the Flow of Thin Liquid Sheets in Hot Atmospheres. I. Fluid Mechanics 64 (1974) 1, p. 167-175.

[2.93]   Parlange, J.Y.: A Theory of Water-Bells. Journal of Fluid Mechanics 29 (1967) 2, p. 361-372.

[2.94]   Taylor, G.: The Dynamics of Thin Sheets of Fluids. Proc. Roy. Soc. Volume A1253(1959) p. 289-321.

[2.95]   Lee, S.Y.; Tankin, R.S.: Study of a Liquid Spray in a Non Condensable Environment. Int. J. Heat Mass Transfer 27 (1984) 3, p. 351-361.

[2.96]   Lee, S.Y.; Tankin, R.S.: Study of a Liquid Spray in a Condensable Environment. Int. J. Heat Transfer 27 (1984) 3, p. 363-374.

[2.97] Fraidl, G.K.: Gemischaufbereitung und motorisches Betriebsverhalten bei intermittierender Saugrohreinspritzung, Doctoral Thesis, TU Vienna 1987. VDI Fortschritt-Berichte, Reihe 12, Nr. 90, Düsseldorf 1987.

[2.98] Lenz, H.P.: Über Zerstäubung von Flüssigkeiten und Anwendung instationärer Hydrodynamik zur Brennstoffeinspritzung bei Verbrennungsmotoren. Doctoral Thesis, ETH Zürich 1966.

[2.99] Lenz, H.P.: Über Grundlagen der Zerstäubung und ihre Anwendung bei Vergasern von Ottomotoren, Fortschritt-Bericht, VDI-Zeitschrift, Reihe 6, Nr. 14 (1967).

[2.100] Hinze, J.O.: Forced Deformations of Viscous Liquid Globules. Applied Scientific Research (1948) A1, p. 263-272.

[2.101] Hinze, J.O.: Critical Speeds and Sizes of Liquid Globules. Applied Scientific Research (1948) A1, p. 273-288.

[2.102] Hinze, J.O.: Fundamentals of the Hydrodynamic Mechanism of Splitting in Dispersion Processes. A.I.Ch.E. Journal 1 (1955) 3, p. 289-295.

[2.103] Lane, W.R.: Shatter of Drops in Streams of Air. Industrial and Engineering Chemistry 43 (1951) 6, p. 1312-1317.

[2.104] Simmons, H.C.: The Correlation of Drop Size Distribution in Fuel Nozzle Sprays, Part I and II. ASME Journal of Engineering for Power (1977) 3, p. 309-319.

[2.105] Simmons, H.C.: The Prediction of Sauter Mean Diameter for Gas Turbine Fuel Nozzles of Different Types. ASME Journal of Engineering for Power 102 (1980) 3, p. 646-652.

[2.106] van Lier, J.J.L.; van Paassen, C.A.A.: Überblick über die Forschungsarbeit "Einspritzkühlung", VGB Kraftwerkstechnik 60 (1980) 12.

[2.107] Sangeorzan, B.P.; Uyehara, D.A.; Myers, P.S.: Time-Resolved Drop Size Measurements in an Intermittend High-Pressure Fuel Spray. SAE Paper 841361 (1984).

[2.108] Rizk, N.K.; Lefebvre, A.H.: Spray Characteristics of Plain-Jet Airblast Atomizers. ASME Journal of Engineering for Gas Turbines and Power 106 (1984) 3, p. 634-638.

[2.109] Rizk, N.K.; Lefebvre, A.H.: The Influence of Liquid Film Thickness on Airblast Atomization. ASME Journal of Engineering for Power 102 (1980) 3, p. 706-710.

[2.110] Jasuja, A.K.: Airblast Atomization of Alternative Liquid Petroleum Fuels under High Pressure Conditions. ASME Journal of Engineering for Power 103 (1981) 3, p. 514-518.

[2.111] Jasuja, A.K.: Atomization of Crude and Residual Fuel Oils, ASME Journal of Engineering for Power 101 (1979) 2, p. 250-258.

[2.112] Rizkalla, A.A.; Lefebvre, A.H.: Influence of Liquid Properties on Airblast Atomizer Spray Characteristics. ASME Journal of Engineering for Power (1975) 2, p. 173-179.

[2.113] Rizkalla, A.A.; Lefebvre, A.H.: The Influence of Air and Liquid Properties on Airblast Atomization. Journal of Fluids Engineering, Sept. 1975. p. 316-320.

[2.114] Lorenzotto, G.E.; Lefebvre, A.H.: Measurements of Drop Size on a Plain-Jet Airblast Atomizer. AIAA Journal 15 (1977) 7, p. 1006-1010.

[2.115] Rizk, N.K.; Lefebvre, A.H.: Influence of Atomizer Design Features on Mean Drop Size. AIAA Journal 21 (1983) 8, p. 1139-1142.

[2.116] Kim, K.Y.; Marshall, J.R.: Drop-Size Distribution from Pneumatic Atomizers. A.I.Ch.E. Journal 17 (1961) 2, p. 312-318.

[2.117] Gretzinger, J.; Marshall, J.R.: Characteristics of Pneumatic Atomization. A.I.Ch.E. Journal 7 (1961) 2, p. 312-318.

[2.118] Löhner, Müller: Gemischbildung und Verbrennung im Ottomotor, Springer Verlag Vienna - New York 1967.

[2.119] Huber, E.W.: Untersuchung der Neigung von Flugmotorkraftstoffen zur Dampfblasenbildung unter den möglichen Bedingungen, 1. Zwischenarbeit TH Graz, Lehrstuhl f. Verbrennungskraftmaschinen und Wärmelehre (1944), p. 1-24.

[2.120] Orlicek, A.F.; Pöll, H.: Hilfsbuch für Mineralöltechniker. Springer-Verlag, Vienna 1951.

[2.121] Henning, H.: Zur Berechnung des Zustandes im Vergaser, ATZ 63 (1961), p. 323-330.

[2.122] Widmaier, O.: Kraftstoffe für Ottomotoren. Zeitschrift für Chemie und Technik von Kohle, Öl und Gas 50 (1969) 4, p. 97-132.

[2.123] Jin, J.D.; Borman, G.L.: A Model for Multicomponent Droplet Vaporization at High Ambient Pressures. SAE Paper 850264 (1985).

[2.124] Law, C.K.: Multicomponent Droplet Combustion with Rapid Internal Mixing. Combustion and Flame 26 (1976) p. 219-233.

[2.125] Lui, X.Q.; Wang, C.H.; Law, C.K.: Simulation of Fuel Droplet Gasification in SI Engines. Journal of Engineering for Gas Turbines and Power 106 (1984) p. 849-853.

[2.126] Scherenberg, D.: Einfluß der Gemischaufbereitung auf die Emissionen beim Start und Warmlauf sowie beim Übergang von Last in Schubbetrieb eines Ottomotors bei verschiedenen Motortemperaturen. Doctoral Thesis, TH Karlsruhe 1979.

[2.127] Maisch, Wolfgang: KE-Jetronic - Ein elektronisch gesteuertes Einspritzsystem mit Notlauffähigkeit. ATZ Automobiltechnische Zeitschrift 84 (1982) 11.

[2.128] Willumeit, Hans-Peter, Bauer, Martin: Emissionen und Leistung eines Ottomotors bei sauerstoffangereicherter Verbrennungsluft. MTZ Motortechnische Zeitschrift 49 (1988) 4.

[2.129] Grossmann: Zum Flüssigkeitsbetrieb von PKW. Vortrag beim ÖIAV in Wien (1982).

[2.130] N.N.: Österreichische Normung EN 288.

[2.131] Leiter, E.: Strömungsmechanik: Nach Vorlesungen, K. Oswatlisch. Bd. 1, Vieweg Braunschweig 1978.

[2.132] Prandtl, L.: Strömungslehre, 6th Edition, Friedr. Vieweg & Sohn, Braunschweig, 1965.

[2.133] Bruno Eck: Technische Strömungslehre, Springer Verlag, Berlin 1944.

[2.134] Marhold, A.: Vergaserkraftstoffzusammensetzung und Emissionen. 4. Informationsse-
minar praxisbezogene Qualitätsanforderungen an Kraftstoffe, TU Wien. Forschungsin-
stitut für Chemie und Technologie von Erdölproduktion.

[2.135] Schmidt, F.A.F.: Verbrennungskraftmaschinen, Springer Verlag Berlin (1967).

Chapter 3

[3.1] Moser, F.: Über das Verhalten von Fahrzeug-Ottomotoren, insbesondere Vergasermotoren, im nichtbetriebswarmen Zustand und Wege zur Verbesserung der Abgasemissionen und Kraftstoffverbrauch in der Warmlaufphase. Doctoral Thesis, TU Vienna
(1977).

[3.2] D'Alleva, B.A.; Lovell, W.G.: Relation of Exhaust Gas Composition to Air-Fuel Ratios.
SAE-Journal 38 90 (1936).

[3.3] Weatherford, W.D.: Find Fuel-Air Ratio Graphically. Petroleum Refiner, 1959, p. 181.

[3.4] Spindt, R.S.: Air-Fuel Ratios from Exhaust Analysis. SAE-Paper 650507 (1965).

[3.5] Eltinge, Lamont: Fuel-Air Ratio and Distribution from Exhaust Gas Composition. SAE
680114 (1968).

[3.6] Lange, Karlheinz: Verfahren zur Berechnung der Luftzahl aus der Abgaszusammenset-
zung. MTZ 37 (1976).

[3.7] Brettschneider, Johannes: Berechnung des Luftverhältnisses von Luft- Kraftstoff-Gemi-
schen und des Einflusses von Meßfehlern auf Lambda. Bosch Technische Berichte 6
(1979), Nr.4, p. 177-186.

[3.8] Pucher, Ernst: Meßprinzip und Aufbau des neu entwickelten Lambda-Meßgerätes. Fortschritt-Bericht VDI Reihe 6 Nr. 173. Düsseldorf: VDI-Verlag 1985; p. 4-15.

[3.9] Vassell, W.C.; Logothetis, E.M.; Hetrick, R.E.: Extended Range Air-to-Fuel Ratio Sensor.
SAE 841250 (1984).

[3.10] Pucher, E.; Lenz, H.P.: Entwicklung eines neuartigen Lambdameters mit Hilfe eines Ein-
Komponenten-Meßverfahrens. VDI-Berichte; 553. Düsseldorf: VDI-Verlag 1985; p. 127-
140.

[3.11] Takeuchi, K.; Murayama, H.; Senda, J.; Yanada, K.: Droplet Size Distribution in Diesel
Fuel Spray. Bulletin of JSAE 26 (1983) 215, p. 797-804.

[3.12] Ueno, S.; u.a.: Wide-Range Air-Fuel Ratio Sensor. SAE 860409 (1986).

[3.13] Wiedemann, H.M.; u.a.: Beheizte Zirkondioxid-Sonde für stöchiometrische und magere
Luft-Kraftstoff-Gemische. Bosch Technische Berichte 7 (1984) 5.

[3.14] Pucher, E.; Lenz, H.P.: Theoretischen Grundlagen und Entwicklung eines instationären
Luftzahlmeßgerätes. VDI Berichte Nr. 681, 1988.

[3.15]   Kohoutek, Peter: Luftverhältnisbestimmung an 2-Taktmotoren auf Basis einer Restsau-
         erstoffmessung. Diplomarbeit an der TU Wien. 1989.

[3.16]   Bürkholz, A.: Meßmethoden zur Tropfengrößenbestimmung, Chemie - Ing.-Techn. 45/1
         (1973), p. 1-7.

[3.17]   Michel, B.: Tropfengrößenbestimmung bei Zerstäubung durch einen Ölbrenner. Öl- und
         Gasfeuerung. 14 (1969) 2, p. 137-144.

[3.18]   Trösch, H.A.: Zerfall von Flüssigkeiten und Tropfengrößenbestimmung. Chemie - Ing.-
         Technik 31 (1959) 10, p. 667-673.

[3.19]   Jandl, G.; Bell, O.: Fotographische Methoden zur Bestimmung von Tropfenspektren.
         Brennstoff-Wärme-Kraft 26 (1974) 7, p. 313-319.

[3.20]   Atkinson, D.S.F.; Strauss, W.: Droplet Size and Surface Tension in Venturi Scrubbers.
         Journal of the Air Pollution Control Association 28 (1978) 11, p. 1114-1118.

[3.21]   Peters, B.D.: Laser-Video Imaging and Measurement of Fuel Droplets in a SI Engine. I.
         Mechanical Engineers (1983), 81/83, p. 23-30.

[3.22]   Pishkoff, J.M.; Hammand, O.C.; Chraplyvy, A.R.: Diagnostic Measurements of Fuel
         Spray Dispresion ASME Paper 80-WA/HT-35 (1980).

[3.23]   Seger, G.; Sinsel, F.: Untersuchung einer Zerstäubungsvorrichtung mit Hilfe der Kurz-
         zeitmikroholographie. Staub-Reinhaltung der Luft. 30 (1970) 11, p. 471-475.

[3.24]   Michel, B.; Seger, G.: Die Holographie als Hilfsmittel zur Untersuchung der Gemisch-
         aufbereitung in Verbrennungsmotoren und Gasturbinen. MTZ 32 (1971) 7, p. 252-255.

[3.25]   Rottenkolber, H.: Holographie als Meßmethode. Werkstatt und Betrieb 103 (1970) 3, p.
         189-193.

[3.26]   Fraidl, G.K.: Gemischaufbereitung und motorisches Betriebsverhalten bei intermittieren-
         der Saugrohreinspritzung, Fortschritt Berichte VDI, Reihe 12, Nr. 90, 1987.

[3.27]   Hamidi, A.A.; Swithenbank, J.: Treatment of Multiple Scattering of Light in Laser Diffrac-
         tion Measurement Techniques in Dense Sprays and Particle Fields. Journal of the Insti-
         tute of Energy 101 (1986) June, p. 101-105.

[3.28]   Swithenbank, H.; Beer, J.M.; Taylor, D.S.; Abbat, D.; Mc. Creath, G.C.: A Laser Diagno-
         stic Technique for the Measurement of Droplet and Particle Size Distribution. AIAA paper
         Nr. 76/79, AIAA, 14. Aerospace Sciences Meeting, Washington,DC., 1976.

[3.29]   Dodge, L.G.: Calibration of the Malvern Particle Sizer. Applied Optics 23 (1984) 14, p.
         2415-2419.

[3.30]   Demel, H.; Diem, E.; Lenz, H.P.: Quantitative Messung des Kraftstoffilmes in Saugroh-
         ren von Ottomotoren. Special Publication MTZ 44 (1983) 7/8, p. 275-278.

[3.31]   Mramor, F.: Gemischbildung in Mehrzylinder-Ottomotoren mit Vergaserbetrieb. Deut-
         sche Kraftfahrforschung. H. 82 (1944).

[3.32]   Jante, A.: Über die Gemischbildung an Vergasermotoren. ÖIZ 7 (1964) 1, p. 7-17.

[3.33] Elsayed-Ahmed, M.Y.: Verdunstung, Tropfen- und Filmbildung in Vergaseranlagen. Doctoral Thesis, TU Braunschweig 1971.

[3.34] Jakobs, R.; Rosemann, G.: Modellversuche zur Gemischaufbereitung bei Ottomotoren. Fa. Pierburg, Neuss (1977).

[3.35] Kurabayashi, T.; Karasawa, T.: Measurement of quantity of liquid attached on a wall of an air flowing pipe by means of a rotary ring separator Jari Technical Memorandum Nr. 1 (1971), p. 23-30.

[3.36] Kurabayashi, T.; Karasawa, T.: On the method of measure of liquid quantity attached on air flowing pipe wall. Jari Technical Memorandum Nr. 1 (1971) p. 41-50.

[3.37] Gruber, F.: Aufbau und Erprobung eines instationären Luftverhältnismeßgerätes, Diplomarbeit, Institut für Verbrennungskraftmaschinen und Kraftfahrzeugbau der TU Wien, 1989.

[3.38] Hayashi, S.; Sawa, N.: Measurement of fuel liquid-film flow in intake pipe of two-stroke motorcycle engine using conduction-probe. SAE Paper Series 840554 (1984).

[3.39] Böhme, W.: Abschlußbericht zum Forschungsbericht: "Entwicklung eines Strahlbild-Vermessungsprüfstandes" TU Vienna, Internal Report of the Institute for Internal Combustion Engines, B 1605, 1988.

[3.40] Richter, H.: Meßeinrichtung zur kontinuierlichen Ermittlung der Schadstoff-Massenemissionen bei stationären und instationären Motor-Betriebszuständen. Doctoral Thesis, TU Vienna, 1978.

[3.41] N.N.: Kraftfahrtechnisches Taschenbuch/Bosch. 20th Edition. Düsseldorf, VDI Verlag, 1987.

[3.42] Brand, F.L.: Akustische Verfahren zur Durchflußmessung. Messen Prüfen Automatisieren. April 1987.

[3.43] N.N.: Massendurchflußmesser für Gase. Degussa AG, Datenblatt Nr. 8498.

[3.44] Leiter, E.: Strömungsmechanik. Volume 1, Grundlagen und technische Anwendungen. Braunschweig: Vieweg, 1st Edition 1978.

[3.45] Bandel, W.: Ermittlung des stationären und dynamischen Verhaltens einer geometrischen Kraftstoffverbrauchseinrichtung. Dipomarbeit an der TU Wien 1984

[3.46] Wichart, K.: Möglichkeiten und Maßnahmen zur Verminderung der Ladungswechselverluste beim Ottomotor. Fortschritt-Berichte VDI, Reihe 12, Nr. 91. Düsseldorf, VDI Verlag 1987.

[3.47] Matsumura, T.; Nonyoshi, Y.: New Fuel Metaing Technique for Compensating Wall Flow in a Transient Condition Using the Model-Matching Methode. JSAE Review Vol. 10, No. 3, p. 5-9. 1989.

[3.48] N.N.: Präzisions-Durchfluß-Meßgerät PLU 103 A. Betriebsanleitung. Pierburg Luftfahrtgeräte Union GmbH. Neuss.

Chapter 4

[4.1]   Kretzschmer, F.: Strömungsform und Durchflußzahl der Meßdrossel. VDI-Forsch.-Heft Nr. 381 (1936).

[4.2]   Hansen, M.: Düsen und Blenden bei kleinen Reynolds-Zahlen. Forschg. Ing.-Wesen, Bd. 4 (1933).

[4.3]   Schultz-Grunow, F.: Durchflußverfahren für pulsierende Strömungen. Forschgs.-Ing.-Wesen, Bd. 12 (1941) H.3, p. 117-126.

[4.4]   Eck, B.: Technische Strömungslehre. 7th Edition, Springer-Verlag Berlin (1966).

[4.5]   Ackeret, J.: Grenzschichten in geraden und gekrümmten Diffusoren. JUTAM Symposium Freiburg/Br. 1957, Grenzschichtforschung, hrsgg. v. H. Görtler, Springer (1958).

[4.6]   Sprenger, H.: Experimentelle Untersuchungen an geraden und gekrümmten Diffusoren. Mitt. Inst. f. Aerodynamik, Zürich Nr. 27 (1959).

[4.7]   N.N.: Interner Versuchsbericht der Firma Zenith, Lyon Nr. 152 vom 22. 3. 61 mit Anlage.

[4.8]   Pierburg, Alfred; Lenz Hans-Peter: Vergaser für Kraftfahrzeug-Motoren, 4th Edition, VDI-Verlag GmbH. Düsseldorf (1970).

[4.9]   Schmid, E.: Thermodynamik, Springer-Verlag (1962).

[4.10] Dutta, Rathindra: Optimierungsaufgabe - System Lufttrichter-Vorzerstäuber und Hauptkanal des Vergasers, Deutsche Vergaser Gesellschaft, Neuß (1977).

[4.11] N.N.: Bosch-Mitteilungsblätter K3/EEV-ZEE-Europa-312-7 K3/EEV-EV10-303-1.

[4.12] Schauer, W.: Internal Report of the Deutsche Vergaser Gesellschaft, Neuß, of 20. 12. 1967.

[4.13] Linzer, V.: Ein Beitrag zur Theorie des Vergasers. Doctoral Thesis, TU Vienna, 1963.

[4.14] Internal test report of Bendix-Stromberg Carburetor Co. South Bend, Indiana, USA; BX, Sect. II, Part II.

[4.15] Bockelmann, Wilfried; Steinbrink, Rainer: Der neue Pierburg-Registervergaser 2E, Motortechnische Zeitschrift 43 (1982), Nr. 10, p. 493-498.

[4.16] Bolt, J.A.; Boerma, M.: Der Einfluß des Einlaßluftverhältnisses auf die Vergaserdosierung. (Englisch) SAE-Abhandlung Nr. 660 119 (1966).

[4.17] Lenz, H.P.: Zur Entwicklung von Höhenkorrektoren für Vergaser von Ottomotoren. Fortschritt-Berichte VDI-Zeitschrift, Reihe 3, Nr. 26.

[4.18] Furuyama, Mikio; Ohgane, Hiroaki: A Comparison of Pulsating and Steady Flows in Terms of Carburetor Characteristics, JSAE Review Vol. 8 (1987), No. 3, p. 18-23.

[4.19] Lenz, H. P.: Carburation and Mixture Pretreatment in the Mercedes-Benz Twin Camshaft In-Line Engine, Instn Mech Engrs, Conference Publication 19 (1973), p. 119-125.

[4.20] Härtel, Günter R.:Neues Gemischbildungssystem für Ottomotoren, Automobiltechnische Zeitschrift 83 (1981), Nr. 5, p. 219-222.

[4.21] Robert Bosch GmbH: Kraftfahrtechnisches Taschenbuch. 19th Edition, VDI-Verlag (1984).

[4.22] Robert Bosch GmbH: Autoelektrik, Autoelektronik am Ottomotor. 1. Ausgabe, VDI-Verlag (1987).

[4.23] Jwamoto, A.; Sato, K.: Recent Single-Point Injection System. JSAE Review Vol. 9 (1988), No. 1, p. 22-28.

[4.24] Olaf von Fersen: Das DPI-Einspritzsystem, MOT 18 (1988), p. 106-109.

[4.25] Großmann, D.: Zum Flüssigkeitsbetrieb von PKW. Lecture at ÖIAV, Vienna (1982).

[4.26] Borrmeister, J.: Untersuchungen am Audi-Motor M118. Wiss.-Z. TU Dresden 18 (1969) 2, p. 539-546.

[4.27] Zander, W.: Die Entwicklung von Saugrohren als Beitrag zur Abgasentgiftung der Vergasermotoren. MTZ 31 (1970) 2, p. 52/53.

[4.28] Lange, K.: Einfluß verschiedener Vergaser- und Saugrohrformen auf die Gemischverteilung an einem Vierzylinder-Reihenmotor. Fortschr.-Berichte VDI-Zeitschriften, Reihe 6 (1969) 24.

[4.29] Pachta-Reyhofen, G.; Lenz, H.P.: Möglichkeiten und Grenzen der nachträglichen Optimierung der Gemischverteilung in Ansaugsystemen mit zentraler Gemischbildung. XXI. FISITA-Congress, Belgrade (1986), Yugoslav SAE Paper 865029.

[4.30] Smetana, G.: 3. Zwischenbericht zum Forschungsauftrg "Vergleich elektronischer Vergaser zur Zentraleinspritzung". Unpublished Report B 1375, Institute for Internal Combustion Engines and Automotive Engineering of the TU Vienna (1986).

[4.31] Freche, F.: Gemischverteilung am BMW 310 mit Quetschkopf. Kraftfahrzeugtechnik 10 (1960) p. 42-45.

[4.32] Löhner, K.; Müller, H.; König, H.C.; Klink, G.: Untersuchungen über die Auswirkungen verschiedener Arten der Gemischbildung auf die Zusammensetzung der Abgase der einzelnen Zylinder sowie der Sammelprobe eines Viertakt-Ottomotors. Fortschritt-Berichte VDI-Z., Reihe 6 (1967) 15.

[4.33] Takahama, Kawamura; Mase, Oda: Der Einfluß der Drosselklappe des Vergasers auf die Gemischverteilung in Benzinmotoren. Translation from the Congress Paper of the 43rd Conference of the Japanese Society of Mechanical Engineers, No. 143 (1965) p. 133-135.

[4.34] Borrmeister, J.; Birkit, A.: Probleme der Kraftstoff-Zuteilung und Gemischaufbereitung bei der Verwirklichung eines wirtschaftlichen und schadstoffarmen Motorbetriebes von Viertakt-Ottomotoren. Kraftfahrzeugtechnik 4/79 (1979) p. 108-112.

[4.35] Müller, F.: Entgiftung der Abgase von Ottomotoren durch konstruktive Maßnahmen an Vergaser und Zündlage. Wiss.-Z. der Technischen Universität Dresden 17 (1968) 5, p. 1377-1388 und "die Technik" (1970) 1.

[4.36] Smetana, G.: 2. Zwischenbericht zum Forschungsauftrag "Vergleich elektronischer Vergaser zur Zentraleinspritzung". Unpublished Report B 1338, Institute for Internal Combustion Engines and Automotive Engineering of the TU Vienna (1986).

[4.37] Lenz, H.P.: Über Grundlagen der Zerstäubung und ihre Anwendung bei Vergasern von Ottomotoren. Fortschritt-Bericht VDI-Z., Reihe 6 (1967) 14.

[4.38] Fa. Opel: Opel-Multec-Zentraleinspritzung, Technical Paper (1988).

[4.39] Lembke, M.: Ausführungen heutiger und zukünftiger Einspritzanlagen. Vortrag anläßlich des Seminars "Gemischbildung bei Ottomotoren", Institute for Internal Combustion Engines and Automotive Engineering of the TU Vienna, Nov. 1988.

[4.40] N.N.: Bosch und Pierburg System OHG., Technische Information, Ecotronik-Lambda = 1-Regelung.

[4.41] Schürz, W.: Ursachen für "hängende Gemischverteilung" bei Verwendung von Luftführungseinsätzen und Zerstäubungsvorrichtungen. Unpublished Report B 1562, Institute for Internal Combustion Engines and Automotive Engineering of the TU Vienna (1988).

[4.42] Duelli, H.: Luftführungseinsätze und Zerstäubungsvorrichtungen. Unpublished Report B 1514, Institute for Internal Combustion Engines and Automotive Engineering of the TU Vienna (1987).

[4.43] N.N.: Bosch Technische Unterrichtung. Diesel-Einspritzausrüstung. Robert Bosch GmbH. Abt. techn. Druckschriften KH/VDT (1970).

[4.44] Duelli, H.: Einspritzzeitpunktvariation. Unpublished Report B 1463, Institute for Internal Combustion Engines and Automotive Engineering of the TU Vienna (1987).

[4.45] N.N.: Bosch Technische Unterrichtung. Benzineinspritzung D- und L-Jetronic. Robert Bosch GmbH. Abt. techn. Druckschriften KH/VDT (1975).

[4.46] Sauer: Luftmassenmessung. Vortrag anläßlich des Seminars "Gemischbildung bei Ottomotoren". Institute for Internal Combustion Engines and Automotive Engineering of the TU Vienna, 1988.

[4.47] Greiner, M.; Romann, P.; Steinbrenner, U.: Bosch Fuel Injectors-New Developments, Robert Bosch GmbH., SAE Techn. Paper 870124 (1987).

[4.48] Fraidl, G.K.: Gemischaufbereitung und motorisches Betriebsverhalten bei intermittierender Saugrohreinspritzung. VDI-Fortschrittsberichte, Reihe 12, Nr. 90, Düsseldorf 1987.

[4.49] Nomura, T.; Irino, H.: Non-Linearity at Low Flow Rates from Electromagnetic Fuel Injector. JSAE Review Vol. 8, No. 4, 1987.

[4.50] Krappel, A.; Guggenmos J.; Goldbrunner W.: Elektronikkonzept beim BMW-12-Zylindermotor. Automobil-Industrie Nr. 1/88.

[4.51]   Vogt, R.: Anforderungen der Mehrventilmotoren an die Gemischbildung. Seminar "Ge-
         mischbildung bei Ottomotoren". Institute for Internal Combustion Engines and Automo-
         tive Engineering of the TU Vienna, Nov. 1988.

[4.52]   Nakamura, N.; Namura, K.; Suzuki, M.: Key Factors of Fuel Injection System to Draw
         Out Good Response in 4-Valve Engine. SAE Paper 870126.

[4.53]   Mikulic. L.; Quissek, F.; Fraidl, G.K.; Carstensen, H.: Sequentielle Einspritzung. Mehrfa-
         cheinspritzung Gemischbildung an schnellbrennenden Ottokonzepten. Seminar: "Ge-
         mischbildung bei Ottomotoren", Institute for Internal Combustion Engines and Automo-
         tive Engineering of the TU Vienna, Nov. 1988.

[4.54]   Andrighetti, J.P.; Gallup, D.R.: Design-Development of the Lucas CAV multipoint gaso-
         line injector. SAE 870127 (1987).

[4.55]   Matsuhara, M.; Ando, T.; Takada, S.; Takeuchi, H.: As an fuel injecor for multipoint in-
         jection system. SAE 860486 (1986).

[4.56]   N.N.: New fuel injector design lowers cost society of Automotive Engineers. Inc. March
         1985.

[4.57]   N.N.: Mitteilungsblatt Bing-Schiebervergaser, Fa. Fritz Hintermayer GmbH.-Bing. Nürn-
         berg, 1981.

[4.58]   N.N.: Mitteilungsblatt. Bing-Membranvergaser 48, Fa. Fritz Hintermayer GmbH.-Bing.
         Nürnberg, 1989.

[4.59]   Pachta-Reyhofen, G.: Wandfilmbildung und Gemischverteilung bei Vierzylinder-Reihen-
         motoren in Abhängigkeit von Vergaser- und Saugrohrkonstruktion. VDI-Fortschritt-Be-
         richte, Reihe 12, Nr. 62, 1986.

[4.60]   Kling, G.: Transport und Verteilung von Kraftstoff durch die Saugleitung eines Vier-
         zylinder-Reihenmotors mit einem Vergaser. TU Braunschweig (1973) und ATZ 78 (1976)
         4, p. 155-161.

[4.61]   Mramor, F.: Gemischbildung in Mehrzylinder-Ottomotoren mit Vergaserbetrieb. Deut-
         sche Kraftfahrtforschung 82 (1944).

[4.62]   Lenz, H.P.; Dutta, R.: Durchfluß an Nadeldüsen. VDI-Z-Fortschritt-Berichte, Reihe 6, Nr.
         31, April 1971.

[4.63]   Duelli, H.: Luftführungseinsätze und Zerstäubungsvorschriften. 7. Zwischenbericht zum
         Forschungsvorhaben "Abstimmung einer Sauganlage mit Mono-Jetronic", B 1514, In-
         stitute for Internal Combustion Engines and Automotive Engineering of the TU Vienna,
         Nov. 1987.

[4.64] Schürz, W.: Ursachen für "hängende Gemischverteilung" bei Verwendung von Luftfüh-
        rungseinsätzen und Zerstäubungsvorrichtungen. 10. Zwischenbericht zum For-
        schungsvorhaben "Abstimmung einer Sauganlage mit Mono-Jetronic", B 1562, Institu-
        te for Internal Combustion Engines and Automotive Engineering of the TU Vienna, May
        1988.

[4.65] Demel, H.: Mögichkeiten zur Verbesserung der Gemischverteilung an Vergasermotoren
        unter besonderer Berücksichtigung des Einflusses des Kraftstoffilmes. Fortschritt-Be-
        richte der VDI-Zeitschriften, Reihe 6 (1982) 107.

[4.66] Jante, A.: Über die Gemischverteilung an Vergasermotoren. ÖIZ 7 (1964) 1, p. 7-17.

[4.67] Liimata, D.R.; Hurt, R.F.; Deller, R.W.; Hull, W.L.: Effects of mixture distribution on
        exhaust emissions as indicated by engine data and the hydraulic analogy. SAE-Int. Mid-
        Year Meeting, Montreal/Canada, June 1971.

[4.68] Smetana, G.: Vergleich von elektronisch geregeltem Vergaser und intermittierender Zen-
        traleinspritzung. VDI-Fortschritt-Berichte, Reihe 12, Nr. 134, 1989.

[4.69] Toyoda, T.; Inoue, T.; Aoki, K.: Single point electronic injection system Higashifuji Tech-
        nical Center, Toyota Motor Corp. SAE 820902.

[4.70] Knapp, H.; Lembke, M.: A new low pressure single point gasoline injection system Robert
        Bosch GmbH. SAE 850293.

[4.71] Takoda, K.; Shiozawa, K.; Oishi, K.; Inoue, T.: Toyota central injection (Ci) system for
        lean combustion and high transient response. Toyota Motor Corp. SAE 851675.

[4.72] Lohr F.W.: Marktorientierte Kraftfahrzeugentwicklung. 2. Aachener-Kolloquium, Aachen,
        Oct. 1989.

[4.73] Bamer, F.: Einfluß der Gemischeinbringung bei Saugrohr-Einspritzung. XX. Fisita-Con-
        gress, Vienna, 1984. SAE-845034.

Chapter 5

[5.1]   Pachta-Reyhofen, G.: Wandfilmbildung und Gemischverteilung bei Vierzylinder-Reihen-
        motoren in Abhängigkeit von Vergaser- und Saugrohrkonstruktion. VDI-Fortschritt-Be-
        richte, Reihe 12, Nr. 62, 1986.

[5.2]   Mramor, F.: Gemischbildung in Mehrzylinder-Ottomotoren mit Vergaserbetrieb. Deut-
        sche Kraftfahrtforschung 82 (1944).

[5.3]   Jante, A.: Über die Gemischverteilung an Vergasermotoren, ÖIZ 7 (1964) 1, p. 7-17.

[5.4]   Klink, G.: Transport und Verteilung von Kraftstoff durch die Saugleitung eines Vierzylin-
        der-Reihenmotors mit einem Vergaser. Doctoral Thesis, TU Braunschweig (1973) and
        ATZ 78 (1976) 4, p. 155-161.

[5.5]   Dörges, E.A.: Gemischverteilung an einem Vierzylinder-Vergasermotor. ATZ 59 (1957)
        7, p. 194-201.

[5.6]   Borrmeister, J.: Untersuchungen am Audi-Motor, M118. Wiss. Z. TU Dresden 18 (1969)
        2, p. 539-546.

[5.7]   Löhner, K.: Gemischbildung im Ottomotor. VDI-Berichte 42 (1960) p. 37-44.

[5.8]   Müller, H.: Gemischbildung und Gemischverteilung bei Ottomotoren mit Vergaserbe-
        trieb. MTZ 28 (1967) 8, p. 313-319.

[5.9]   Müller, H.: Ursachen und Umfang der ungleichen Gemischverteilung an Mehrzylinder-
        Ottomotoren. MTZ 28 (1967) 9, p. 335-339.

[5.10] Dörges, E.A.: Gemischverteilung an Mehrzylinder-Vergasermotoren. ATZ 61 (1959) 6,
        p. 167-175.

[5.11] Bartsch, F.: Motorenentwicklung auf neuen Wegen. Motoren-Rundschau H. 9 und 10
        (1967).

[5.12] Lenz, H.P.: Forschung und Entwicklung in der Deutschen Vergaser-Gesellschaft. Auto-
        mobil-Industrie 14 (1969) 2.

[5.13] Inoue, M.: Einige Betrachtungen über die Ansaugleitungen der Benzinmotoren für Kraft-
        fahrzeuge. Translation from the Japanese Magazine "Nainen Kikan" (Die Brennkraftma-
        schine) Volume 5 (1965) 51, p. 11-19.

[5.14] Sawa, N.; Hori, S.: Distribution of fuel Liquid-film flow in intake pipe. Jari Technical Me-
        morandum Nr. 10 (1972).

[5.15] Moser. F.: Über das Verhalten von Fahrzeug-Ottomotoren, insbesondere Vergasermo-
        toren, im nicht betriebswarmen Zustand, und Wege zur Verbesserung der Abgasemis-
        sionen und Kraftstoffverbrauch in der Warmlaufphase. Doctoral Thesis, TU Vienna,
        (1977).

[5.16] Bulaty, Thomas: Ein Programmsystem zur Berechnung des Zusammenwirkens von Ver-
        brennungsmotoren und Abgasturboladern. MTZ 43 (1982) 11, p. 535-543.

[5.17] Ryti, Matti: Ein Rechenprogramm für den Ladungswechsel aufgeladener Dieselmoto-
        ren. Brown Boveri Mitteilungen Volume 55 (1968), Nr. 8, p. 429-439.

[5.18] Woschni, Gerhard: Elektronische Berechnung von Verbrennungsmotor Kreisprozessen.
        MTZ 26 (1965) Nr. 11, p. 439-446.

[5.19] Geringer, Bernhard; Duelli, Heinz: Programm zur Ladungswechseloptimierung unter Be-
        rücksichtigung von Saugrohrschwingungen und Restgasgehalt. VDI Bericht 537 (1984)
        p. 53-68.

[5.20] Woschni, Gerhard; Fleger, Johann: Experimentelle Bestimmung des örtlich gemittelten
        Wärmeübergangskoeffizienten im Ottomotor. MTZ 42 (1981), Nr. 6, p. 229-234.

[5.21] Collatz, L.: Numerische Behandlung von Differentialgleichungen. Springer-Verlag,
        Berlin/Göttingen/Heidelberg 1955.

[5.22] Ohatat, Akira; Ishida, Yasuhiko: Dynamic Inlet Pressute and Volumetric Efficiency of
        Four Cycle Cylinder Engine. SAE Paper (1982), Nr. 820407.

[5.23]   Fiala, W.; Willumeit, H.P.: Schwingungen in Gaswechselleitungen von Kolbebenmaschi-
         nen. MTZ 28 (1967), Nr. 4, p. 144-151.

[5.24]   Engelmann, H.W.: Design of a Tuned Intake Maniford. ASME (1974) Nr. 73-WA/DGP-2.

[5.25]   Ohata, Akira: Development of Toyota variable induction System. VDI-Fortschrittbericht
         (1985) Nr. 173, p. 32-44.

[5.26]   Ryti, Matti: Schwingungen in den Luftleitungen aufgeladener Dieselmotoren. Brown
         Boveri Mitteilungen 52 (1965) Nr. 3, p. 190-203.

[5.27]   Seifert, H.: Instationäre Strömungsvorgänge in Rohrleitungen an Verbrennungskraftma-
         schinen. Springer Verlag (1962) Berlin, Göttingen, Heidelberg.

[5.28]   Seifert, H. und Mitarbeiter: Die Berechnung instationärer Strömungsvorgänge in den
         Rohrleitungs-Systemen von Mehrzylindermotoren. MTZ 74 (1972), Nr. 11, p.421-428.

[5.29]   Lax, P.D.; Wendroff, B.: Systems of conservation laws. Comm. Pure Appl. Math. Vol 13
         (1960).

[5.30]   Seifert, Hans und Mitarbeiter: Beschreibung des Programmsystems PROMO, FVV For-
         schungsbericht, Hefte 160-1 bis 160-9 (1974) und 238-1 bis 238-6 (1977).

[5.31]   Seifert, Hans: Erfahrungen mit einem mathematischen Modell zur Simulation von Ar-
         beitsverfahren in Verbrennungsmotoren. MTZ 39 (1978), Nr. 7/8, p. 321-325.

[5.32]   Görg, K.A.: Verfahren zur Berechnung der Gasströmung durch Drosselstellen in einem
         instationär durchströmten Rohrsystem. Bericht des Lehrstuhls Konstruktionstechnik
         (1978), Ruhr Universität Bochum.

[5.33]   Demel, H.: Möglichkeiten zur Verbesserung der Gemischverteilung an Vergasermoto-
         ren unter besonderer Berücksichtigung des Einflusses des Kraftstoffilmes. Fortschritt-
         Berichte der VDI-Zeitschriften, Reihe 6 (1982) 107.

[5.34]   Lo, R.S.; Lolas, D.P.: Parametric Study of fuel-droplet flow in an idealized engine induc-
         tion system. SAE Paper 770645.

[5.35]   Elsayed-Ahmed, M.Y.: Verdunstung, Tropfen- und Filmbildung in Vergaseranlagen. Doc-
         toral Thesis, TU Braunschweig (1971).

[5.36]   Rajkow, I.J.; Jerschow, W.W.: Zur Frage der Filmbildung in Vergasermotoren. Transla-
         tion from the Soviet Magazine "Avtomobilnaja Promyzlennost" 11 (1964) p. 6-10.

[5.37]   Scherenberg, D.: Einfluß der Gemischaufbereitung auf die Emissionen beim Start und
         Warmlauf sowie beim Übergang von Last in Schubbetrieb eines Ottomotors bei ver-
         schiedenen Motortemperaturen. Doctoral Thesis, TU Karlsruhe (1979).

[5.38]   Boam, D.J.; Finlay, I.C.: A computer model of fuel evaporation in the intake system of a
         carburetted petrol engine. Confer. London June 1979, C89/79.

[5.39]   Stebar, R.F.; Everett, R.L.: New emphasis on fuel volatilityeffects on vehicle warm up
         with quick-release chokes. SAE Paper 720934.

[5.40] Zander, W.: Die Entwicklung von Saugrohren als Beitrag zur Abgasentgiftung der Vergasermotoren. MTZ 31 (1970) 2, p. 52/53.

[5.41] Nefischer, A.: Möglichkeiten zur Verbesserung der filmförmigen Kraftstoffverdampfung in Saugrohren von Vergasermotoren. Diplomarbeit an der TU Wien (1980).

[5.42] Eberan-Eberhorst, R.: Wege zu sauberen Autoabgasen in naher Zukunft. ATZ 74 (1972) 11.

[5.43] Kennmann, H.: Anforderungen heutiger Ottomotoren an die Gemischbildungseinrichtung. Adam Opel AG. Technical Paper (1978).

[5.44] Bedale, P.W.: Mixture formation. Translation of a special edition of Amt. Div. Inst. Mech. Engs. 8/59 (1959) p. 3-14.

[5.45] Jakobs, R.; Rosemann, G.: Modellversuche zur Gemischaufbereitung bei Ottomotoren. Fa. Pierburg, Neuss (1977).

[5.46] Bockelmann, W.: Saugrohrentwicklung, Saugrohrbeheizung. Internal Investigations, Fa. Pierburg, Neuss (1979).

[5.47] Brandstetter, W.: Spezielle Eigenschaften des Luftfilters und ihre Wirkung auf das Fahrverhalten von Ottomotoren. Internal Report. Volkswagenwerk Wolfsburg (1982).

[5.48] Kay, I.W.: Manifold fuel film effects in an SI-engine. SAE Paper Series 780944 (1978).

[5.49] Härtel, G.; Wobky, P.; Rosemann, G.: Zusammenhänge zwischen erreichbarer Gemischaufbereitung, Abgasrückführung, Abgasemission und Kraftstoffverbrauch am Ottomotor. Internal Publication, Deutsche Vergaser GmbH., Neuss (1975).

[5.50] Bond, D.W.: Quick heat intake manifold for reducing cold engine emission. SAE Paper 720935 (1972).

[5.51] Kraus, L.: Maßnahmen zur Treibstoffeinsparung bei Personenwagen - Neuentwicklung. Automobil-Revue 39 (1974).

[5.52] Jones, J.H.; Gaghardi, J.C.: Vehicle exhaust emission experiments using a pre-mixed and pre-heated air fuel charge. SAE Paper 670485 (1967).

[5.53] N.N.: Der Einfluß des Verfahrens zum Anwärmen der Ansaugleitung auf die Leistung des Motors ZIL 130 bei nicht stationärem Betrieb. Sowj. Zeitschrift "Konstruieren, Forschen, Prüfen" (1965).

[5.54] Croft, B.H.: Automatic control of petrol engine fuelling during cold start and warming-up. ISATA 1979.

[5.55] Pao, H.C.: The measurement of fuel evaporation in the induction system during warm-up. SAE Paper Series 820409 (1982).

[5.56] N.N.: Industrie-Information der Fa. Pierbaurg GmbH. Neuss, 4 (1978) p. 1-5.

[5.57] Barthalomew, E.: Potentialities of emission reduction by design of induction systems. SAE Paper 660109 (1966).

[5.58]   Berg, W.: Aufwand und Probleme für Gesetzgeber und Automobilindustrie bei der Kontrolle der Schadstoffemissionen von Personenkraftwagen mit Otto- und Diesel-Motoren. Doctoral Thesis, TU Braunschweig (1982).

[5.59]   YU, T.C.: Fuel distribution studies - a new look to an old problem. SAE Transdictions, Vol 71 (1963).

[5.60]   Grossmann: Exhaust gas improvement and fuel saving by improved methods of preparing and distributing the fuel/air mixture. Automotive Fuels Symposium, Fa. Shell, 23/44 Sept. 1976 in Eddelsen (BRD).

[5.61]   Curtis, T.: Meet the vapipe. Motor week ending June 16 (1973).

[5.62]   Harnded, J.L.: Heat pipe early fuel eyaporation. SAE Paper 760565 1976).

[5.63]   Harrow, G.A.; Mills, W.D.; Thomas, A.; Finlay, I.C.: The vapipe - a practical system for producing homogeneous gasoline-air mixtures. SAE Paper 760564 (1976).

[5.64]   Zusammenfassung der Vorträge der Internationalen Heat Pipe Konferenz, 15.-17.Oct. 1973. Stuttgart (BRD).

[5.65]   Beier, R.; Fränkle, G.; Haahtela, O.; Hiereth, H.; Kordon, H.; Pundt, D.; Riesenberg, K.O.: Verdrängermaschinen, Teil II Hubkolbenmotoren, Handbuchreihe Energie; Volume 2, München: Technischer Verlag Resch; Köln Verlag TÜV Rheinland (1983).

[5.66]   Berg, P.: PTC honeycomb heater for improved fuel vaporization SAE Paper Series 810156 (1981) p. 69-74.

[5.67]   Paganelli, J.: PTP-ceramic heaters in automative control. SAE Paper Series 840143 (1984).

[5.68]   Aquino, Ch.F.: The effect of local heating on A/F ratio control. SAE Paper Series 820411 (1982)

[5.69]   Hofbauer, P.: Gemischbildung und Verbrennung des Fahrzeug-Ottomotors im stationären Prüfstandsbetrieb und im instationären Fahrbetrieb. Diplom-Arbeit TU Wien (1967).

[5.70]   Hires, S.D.; Overington, M.T.: Transient mixture strength excursions - an investigation of their causes and the development of a constant mixture strength fuelling strategy. SAE Paper Series 810495 (1981).

[5.71]   Pachta-Reyhofen, G.; Lenz, H.P.: Möglichkeiten und Grenzen der nachträglichen Optimierung der Gemischverteilung in Ansaugsystemen mit zentraler Gemischbidlung. XXI. FISITA-Congress, Belgrade (1986), SAE Paper Series 865029.

[5.72]   Nishimura, Y.; Ohyama, Y.; Sasayama, T.: Transient response of fuel supply system for carburetor engine. SAE Paper Series 810788 (1981).

[5.73]   Huhn, J.; Wolf, J.: Zweiphasenströmung. 1st Edition, VEB Fachbuchverlag Leipzig (1975).

[5.74]   Jacubzig, J.: Ladungswechsel, Vorhaben Nr. 160. Teilbericht 9: Durchflußzahlen, Rohrverzweigung. FVV-Forschungsberichte Verbrennungskraftmaschinen, H. 160-9 (1978).

[5.75]   Lange, K.: Einfluß verschiedener Vergaser- und Saugrohrformen auf die Gemischvertei-
          lung an einem Vierzylinder-Reihenmotor. Fortschritt-Berichte VDI-Zeitschriften, Reihe 6
          (1969) 24.

[5.76]   Nati, V.: Untersuchung der Tropfenflugbahnen in Ansaugsystemen von Vierzylinder-Ver-
          gasermotoren verschiedenster Bauart. Diplomarbeit am Institut für Verbrennungskraft-
          maschinen und Kraftfahrzeugbau der TU Wien, B 1272 (1985).

[5.77]   Brauer, H.: Grundlagen der Einphasen- und Mehrphasenströmungen. Verlag Sauerlän-
          der, Aarau und Frankfurt am Main (1971).

[5.78]   Dubbel: Taschenbuch für den Maschinenbau, 1. Volume. 13th Edition, Springer Verlag
          Berlin (1974).

[5.79]   Smetana, G.: 3. Zwischenbericht zum Forschungsauftrag "Vergleich elektronischer Ver-
          gaser zur Zentraleinspritzung". Unpublished Report B 1375. Institute for Internal Com-
          bustion Engines and Automotive Engineering of the TU Vienna (1986).

[5.80]   Dika, R.J.: A discussion of V-8 engine intake manifold design. General Motors enginee-
          ring journal Jan-Feb-March (1959).

[5.81]   Müller, H.: Die Auswirkung einer Beschichtung der Innenoberfläche an einer Sauglei-
          tung auf die Gemischverteilung. Internal Report (1972).

[5.82]   Müller, H.: Analyse eines Gemischverteilungskennfeldes. Internal Report (1972).

[5.83]   French Patent No. 7442953 of 12.27.1974. Invernter: M.Y. Serruys.

[5.84]   Marsee, F.J.; Olree, R.M.; Hill, J.L.; Temby, A.L.: Lean burn - a rational approach to emis-
          sion control. SAE Paper 82031, XIX. FISITA Congress Melbourne (1982).

[5.85]   Marsee, F.J.; Olree, R.M.: Distribution factors that influence emissions and operation of
          lean burn engines. Paper C99/79, Inst. of Mech. Eng., London (1979).

[5.86]   Seifert, Hans: Die charakteristischen Merkmale der Schwingrohr- und Resonanzaufla-
          dung bei Verbrennungsmotoren. XIX. Fisita Congress, Melbourne, Australien, SAE
          Paper 82032 (1982).

[5.87]   Duelli, H.; Geringer, B.; Bauer, F.: Möglichkeiten der Saugrohrentwicklung zur Verbes-
          serung des motorischen Betriebsverhaltens durch Berechnung und Versuch. 6. Wiener
          Motoren Symposium. Institute for Internal Combustion Engines and Automotive Engi-
          neering of the TU Vienna. VDI-Fortschritt-Berichte, Reihe 6, Nr. 173.

[5.88]   Cser, Gyula: Ein neuartiges Verfahren zur Verbesserung der Abgasturboaufladung. MTZ
          32 (1971) Nr. 10, p. 368-373.

[5.89]   Anisits, Ferenc; Spinnler, Fritz: Entwicklung der kombinierten Auflladung am neuen
          Saurer-Fahrzeugdieselmotor D4KT. MTZ 39 (1978) Nr. 10, p. 447-451.

[5.90]   Löhr, Joachim: Neue Aufladeverfahren zur Erhöhung von Wirtschaftlichkeit und Leistung
          bei Nutzfahrzeug-Dieselmotoren im Vergleich zu Saugmotoren. Verkehrsunfall und
          Fahrzeugtechnik 21 (1983) Nr. 11, p. 323-326.

[5.91]   Indra, Fritz: Kombinierte Aufladung an einem Personenwagen-Ottomotor hoher Literlei-
         stung. MTZ 40 (1979) Nr. 12, p. 581-584.

[5.92]   Indra, Fritz: Entwicklung eines aufgeladenen Ottomotors für Personenwagen mit 73,5
         kW Literleistung. ATZ 80 (1978) Nr. 4, p. 141-146.

[5.93]   Dorsch, H.; Rutschmann, E.; Ulrich. J.G.; Zickwolf, P.: 20 Jahre Porsche 911-Auslegung
         und Daten der neuen 3.2-Liter-Motoren. MTZ 44 (1983) Nr. 9, p. 317-322.

[5.94]   Duelli, H.: Berechnungen und Versuche zur Optimierung von Ansaugsystemen für Mehr-
         zylindermotoren und Einzylinder-Einspritzung. VDI-Fortschrittberichte, Reihe 12, Nr. 85,
         1987.

[5.95]   Yamaguchi, Jack: Variable intake passage improves torque characteristics. Automoti-
         ve Engineering 92 (1984) No. 7, p. 78/79.

[5.96]   Engels, Hans Rainer; Main, Jeremy John: Neue Entwicklungen auf dem Gebiet der An-
         triebsregelung für Personenwagen. MTZ 46 (1985) Nr. 12, p. 481-487.

[5.97]   Gassmann, Hans: Ansaugleitung von Brennkraftmaschinen. Deutsche Patentschrift
         (1957) 957 802.

[5.98]   Ueda, Kazuhiko; Hitomi, Mitsuo; Sasski, Junzo: Ansaugvorrichtung für Kolben-Verbren-
         nungskraftmaschinen. Letters of Patent (1984) DE 34 46377 A1.

[5.99]   Gartner, Jurij; Kaindl, Wolfgang: Register-Luftansauganlage für Brennkraftmaschinen,
         insbesondere Mehrzylinder-Einspritz-Brennkraftmaschinen. European Patent Applica-
         tion (1983) 0098 543 A1.

[5.100] Takeda, Keiso: Luftansaugvorrichtung für eine Brennkraftmaschine. Letters of Patent
         (1984) DE 34 16950 A1.

[5.101] Yamaguchi, Jack: Honda chooses 90-degree V-8 to lower hood height: Automotive En-
         gineering 94 (1986) Nr. 2, p. 125-131.

[5.102] Lenz, H.P.; Duelli, H.: Neue stufenlos längenvariable Sauganlage für optimalen Drehmo-
         mentverlauf eines Einspritzmotors. ATZ 89 (1987) 6.

[5.103] Mikulic, L.; Lenz, H.P.; Bamer, F.: Verbrauchsminderung bei Einspritz-Ottomotoren
         durch spezielle Saugrohrgestaltung. ATZ 86 (1984) 10.

[5.104] Geringer, B.: Untersuchung des Einflusses von Schwingungsvorgängen im Saugrohr
         eines Vierzylinder-Ottomotors auf den Ladungswechsel. Diplomarbeit am Institut für Ver-
         brennungskraftmaschinen und Kraftfahrzeugbau der TU Wien, B 1159, Sept. 1983.

[5.105] Duelli, H.: Gestaltung von Saugrohren für Einspritzmotoren. Seminar: "Gemischbildung
         bei Ottomotoren" Institute for Internal Combustion Engines and Automotive Engineering
         of the TU Vienna, Nov. 1986.

[5.106] Burghardt, H.M.; Arnold, G.: Rechnerische Auslegung des gestalteten Ansaugsystemes
         Dual Ram Automobil-Industrie, 5/89.

[5.107] Smetana, G.: Stand der Saugrohrtechnik. Seminar: "Gemischbildung bei Ottomotoren" Institute for Internal Combustion Engines and Automotive Engineering of the TU Vienna, Nov. 1988.

[5.108] Bamer, F.: Saugrohre zur Erhöhung von Drehmoment und Leistung. SDP-Technik Technisch - Wissenschaftliche Berichte, Steyr-Daimler-Puch AG. 1985.

Chapter 6

[6.1] Emmenthal, D.: Sonderformen der Gemischaufbereitung. Workshop: "Gemischbildung bei Ottomotoren", Institute for Internal Combustion Engines and Automotive Engineering of the TU Vienna, November 1988.

[6.2] Eigenbord, C.; Poeschl, G.: Ultraschallvernebler als Gemischbildner für Ottomotoren im Vergleich zu herkömmlichen Einspritzdüsen. MTZ 49 (1988) 7/8.

[6.3] Binder Alan, M.: Throttle is only moving part. SAE-Journal, August 1967.

[6.4] N.N.: US. Patent 3.389.894 of 06-25-1968.

[6.5] N.N.: Entwicklungsring KRM, Leverkusen.

Subject Index